PATTERNS OF BEHAVIOR

D1483753

PATTERNS OF BEHAVIOR

KONRAD LORENZ, NIKO TINBERGEN,

AND THE FOUNDING OF ETHOLOGY

Richard W. Burkhardt, Jr.

The University of Chicago Press
Chicago and London

RICHARD W. BURKHARDT, JR., is professor of history at the University of Illinois at Urbana-Champaign. He is the author of *The Spirit of System: Lamarck and Evolutionary Biology.*

The University of Chicago Press, Chicago 60637
The University of Chicago Press, Ltd., London
© 2005 by The University of Chicago
All rights reserved. Published 2005
Printed in the United States of America

14 13 12 11 10 09 08 07 06 05   1 2 3 4 5
ISBN: 0-226-08089-7 (cloth)
ISBN: 0-226-08090-0 (paper)

The University of Chicago Press and the author gratefully acknowledge a publication subsidy awarded by the Campus Research Board of the University of Illinois, Urbana-Champaign.

LIBRARY OF CONGRESS CATALOGING-IN-PUBLICATION DATA
Burkhardt, Richard W. (Richard Wellington), 1944–
    Patterns of behavior : Konrad Lorenz, Niko Tinbergen, and the founding of ethology / Richard W. Burkhardt, Jr.
        p. cm.
    Includes bibliographical references and index.
        ISBN 0-226-08089-7 (cloth : alk. paper)—ISBN 0-226-08090-0 (pbk. : alk. paper)
        1. Animal behavior—History.   2. Lorenz, Konrad, 1903–   3. Tinbergen, Niko, 1907–   I. Title.
    QL750.5.B87 2005
    591.5'09—dc22

                                                            2004004083

⊗ The paper used in this publication meets the minimum requirements of the American National Standard for Information Sciences—Permanence of Paper for Printed Library Materials, ANSI Z39.48-1992.

To Jayne

{CONTENTS}

{ACKNOWLEDGMENTS}

I began research on this book in 1978. Since then I have benefited greatly from the assistance of many kind and thoughtful people and from the research support provided by several scholarly foundations and my own university. I am very pleased to acknowledge this help.

My research was supported by fellowships from the National Science Foundation (grant numbers SOC78-05922 and SBE9122970) and from the John Simon Guggenheim Foundation (1992–1993), plus a "Travel to Collections" award from the National Endowment for the Humanities (1985). The Research Board of the University of Illinois at Urbana-Champaign provided me with funds for a research trip to Britain. I also wish to thank the many archivists and librarians who have helped me along the way. Their expertise and assistance contributed greatly to the pleasures of the research process. In the bibliography I have listed the archives I used in writing the book. I am particularly indebted to Agnes von Cranach, daughter of Konrad Lorenz, and Janet Tinbergen, daughter of Niko Tinbergen, for their openness to my project and willingness to allow me to quote from letters written by their fathers.

To maintain as much of the authenticity of the original writings as possible, my practice has been to transcribe archival materials carefully so as to preserve original idiosyncratic spellings and constructions. The reader should keep in mind that the writers, for some of whom English was a second (even if familiar) language, had no proofreaders for their letters. I have made one standard alteration, however. In those instances where Konrad Lorenz emphasized a word or phrase by spacing out the letters in it, I have represented the emphasis instead by italics. I apologize if beyond this I have accidentally introduced mistakes without noticing them.

For his long term encouragement of this project—and for quite literally making this project possible in a variety of ways—I am deeply indebted to Ernst Mayr. Ernst's knowledge, insights, and energy have been an inspiration to me since he codirected my doctoral dissertation in the late 1960s and early 1970s on the French evolutionary biologist J.-B. Lamarck. Thirty years before that, he was providing support, encouragement, and constructive criticism to the founders of ethology. Happily for the historian of biology of today, Ernst preserved his correspondence with Niko Tinbergen, Konrad Lorenz, and others. Given that Tinbergen, in contrast, saved little of his own professional correspondence throughout the greater part of his career, the letters that he sent to Ernst (and that Ernst preserved) now constitute a better source for reconstructing Tinbergen's activities in certain periods than the Tinbergen Papers at Oxford. In addition, the fact that Lorenz and especially Tinbergen were both good friends with Ernst was a factor, I know, when both men cordially agreed in the spring of 1979 to talk with me about the history of their field. Later Ernst sent me a collection of diverse materials on ethology he had collected when he was writing his big book on the history of biology, *The Growth of Biological Thought*. He had once thought of writing a book of his own on the history of biological views of animal behavior. Had he proceeded, he would no doubt have written it in the strong, clear, history-of-ideas mode that has distinguished his historical writings for the last four decades. I am sure he did not imagine he would have to live to be a hundred to see the present book.

In addition to Ernst Mayr, Konrad Lorenz, and Niko Tinbergen, other biologists and ethologists have helpfully responded to my questions about the history of ethology. I would like to mention especially G. P. Baerends, Edwin Banks, Patrick Bateson, Colin Beer, Gordon Burghardt, Marian Stamp Dawkins, Robert Hinde, Hans Lissmann, Aubrey Manning, Desmond Morris, W. M. S. Russell, Claire Russell, Wolfgang Schleidt, William H. Thorpe, and Wolfgang Wickler.

The Office of the Vice Chancellor as well as the Research Board of the University of Illinois provided me with a number of excellent research assistants for part of the dozen years or so when I was chairing the Department of History and then directing the Campus Honors Program at the university. I cite with gratitude the help of Joachim Wernersbach, Pamela Sutherland, Barry Mehler, Ki Soon Kim, Fernando Elichirigoity, Valeri Cholakov, and Anja Doil. Especially in the periods when administrative duties made it difficult for me to disappear for any length of time into the library, these excellent collaborators helped me keep thinking on a regular basis about the work I wanted to do most.

Among the colleagues at home and abroad who have provided additional advice, insights, and assistance of different kinds I am pleased to name John Beatty, Keith Benson, Kerstin Berminge, Ingo Brigandt, Philippe Chavot, Robert Croker, Donald Dewsbury, Paul Farber, Jonathan Harwood, Lillian Hoddeson, Veronika Hofer, Theodora J. Kalikow, Robert Kohler, Jane Maienschein, Evan Melhado, Gregg Mitman, Diane Paul,

Philip Pauly, Will Provine, Ronald Rainger, Robert Richards, and Marga Vicedo. Colin Beer and Gordon Burghardt provided me with thoughtful comments on the entire manuscript. Greg Radick offered incisive criticisms on chapters 1 through 5 and Klaus Taschwer provided me with a much welcome reading of chapter 5. I am grateful as well to Wolfgang Schleidt, who offered comments on chapters 6 and 7. To my colleague Harry Liebersohn at the University of Illinois I want to express special thanks for his thoughtful reading of chapter 5 and for an ongoing conversation about German history and culture, naturalist voyagers, and more.

Eva Karner, Astrid Juette, Werner Callebaut, and Gerd Müller of the Konrad Lorenz Institute for Evolution and Cognition Research gave very generously of their time and energy in responding to my inquiries with regard to manuscripts and photos in the KLI Archives, and to related matters. In searching for appropriate illustrations for the book, other kind colleagues aided me as well. Colin Beer, Irenaeus Eibl-Eibesfeldt, Wolfgang Schleidt, Rae Silver, and Lary Shaffer generously shared wonderful pictures they had taken, and Kenneth Boyer, Agnes von Cranach, Hans Kruuk, Desmond Morris, Klaus Taschwer, and Jaap Tinbergen were likewise very considerate in providing me with pictures from their own collections. For their help in tracking down particular items I thank too Jochen Dietrich, Colin Harris, Kurt Kotrschal, Debra Levine, Benjamin Proud, David Simmons, and Rudolf Reinhard. To Barbara Meyer of the University of Illinois I owe special thanks for her artistic and computer wizardry with respect to assembling the whole collection in a manageable and attractive form.

Of the many helpful people at the University of Chicago Press who have made this a better book than it might have been otherwise, I pay special tribute to the late Susan Abrams. For her friendship, wise counsel, and long-standing encouragement of this project I am deeply indebted. I am likewise grateful to Christie Henry, Jennifer Howard, and Erik Carlson for their invaluable help in bringing the book to completion.

Finally, it gives me great pleasure to acknowledge the support of my family and especially my wife, Jayne, to whom this book is dedicated. We have different opinions on whether the completion of the book was accelerated or delayed by the periodic rearranging of notes and books in my study so that various surfaces could be rediscovered and properly dusted. We are at least in agreement that certain stacks of photocopies and files will now be able to leave the house for good. I thank my sons, Rick and Fritz, for their continued interest in the book's themes and progress and in particular for the inspiration they have provided me as they have become adults and embarked on their own careers. I can only hope that the present book has been improved by the time I have had to reflect upon its various themes, as well as to consult archival sources that were not available when my research began. The enthusiasm that colleagues, friends, and family are expressing as I send this manuscript to press leads me to think that finishing my next book only a few years from now would be a wonderful idea.

THEORY, PRACTICE, AND PLACE IN THE

STUDY OF ANIMAL BEHAVIOR

In October 1973, the Royal Karolinska Institute in Stockholm announced its decision to award the 1973 Nobel Prize in Physiology or Medicine to three specialists in the study of animal behavior. One was Karl von Frisch of the University of Munich, famous for his remarkable discovery of the "dance language" of the honeybee. The second was Konrad Lorenz of the Max Planck Institute for Behavioral Physiology in Seewiesen, Germany, best known for his studies of geese and for having directed scientific attention to the phenomenon of "imprinting." The third prizewinner was Nikolaas Tinbergen of the University of Oxford. His field studies of gulls had illuminated the intricate ways in which the behavior of animals is adapted to the ecological settings in which they live.

The ethological community responded jubilantly to the news. The colleagues of the new laureates lost no time interpreting for the rest of the world the significance of the selection. R. A. Hinde and W. H. Thorpe of Cambridge University told readers of the British journal *Nature:* "The award of the Nobel Prize for medicine and physiology to Karl von Frisch, Konrad Lorenz and Niko Tinbergen marks the full emergence of the study of animal behaviour from one of the less respectable corners of natural history to the forefront of the biological sciences." The award, they further claimed, had "not been made for the solution of a particular problem in an established field, but for the creation of a new science—ethology, the biological study of behaviour." [1]

Had the prize committee of the Royal Karolinska Institute itself seen things in just this way? The published record proves ambiguous in this regard. Yet we need not endorse the ethologists' gloss on why the prize was awarded to ask the question that constitutes the central concern of this book: How did the science of ethology come into being? Or, to put it in other terms: How did the biological study of behavior come to take its place as a major new component of the biological sciences?

The study of animal behavior would seem to be an obvious component of biology. Ironically enough, however, throughout much of the nineteenth century, the grand study of animal life was based primarily on an examination of specimens that were dead. In 1802, when the French zoologist

1 Niko Tinbergen receiving the Nobel Prize from King Carl XVI Gustaf of Sweden, with fellow prizewinner Konrad Lorenz in the background. (Photo courtesy of the Konrad Lorenz Institute for Evolution and Cognition Research.)

Jean-Baptiste Lamarck coined the term "biologie," his scientific practice and that of his illustrious colleagues Georges Cuvier and Étienne Geoffroy Saint-Hilaire depended almost entirely on the extraordinary collection of zoological specimens housed at the Muséum d'Histoire Naturelle in Paris. Not meadows, forests, or tidal pools, but instead cabinet drawers, display cases, and dissecting tables were the immediate habitats of the animals of the golden age of French zoology.

The same by and large held true for zoology across the Channel in England, later in the century. Notwithstanding the examples of Charles Darwin, Alfred Russel Wallace, and others who first made their names as naturalist travelers, much of the work of zoologists or biologists continued to be done indoors, through the examination of specimens that were no longer living. In 1890 the Oxford zoologist E. B. Poulton complained: "It is a very remarkable fact that the great impetus given to biological inquiry by the teachings of Darwin has chiefly manifested itself in the domain of Comparative Anatomy, and especially in that of Embryology, rather than in questions which concern the living animal as a whole and its relations to the organic world." [2]

Yet the situation described by Poulton was itself in flux. The landscape of biology was continuing to evolve. Biologists were becoming increasingly convinced of the importance of studying *living* animals. But the most visible and vocal of these scientists were not inclined to view things just as Poulton did. They were disposed instead to believe that the optimal site for biological research was the laboratory. As they saw it, the attractions of studying living animals in a controlled setting, by means of experiment, outweighed those of studying living animals out of doors and in their relations to the rest of "the organic world"—*or* of studying dead specimens in the laboratory with the traditional approaches of the comparative anatomist and embryologist.

The experimentalists had considerable political success, particularly within the university setting, identifying themselves as the representatives of biology's future. The extent to which they did so is reflected in numerous accounts of biology's early twentieth-century history. This view has been challenged, however, and as the historian of science Lynn Nyhart has recently argued, life-history studies, with their roots in traditional natural history, continued to flourish in this period, albeit sometimes in new institutional settings like fisheries, marine stations, or zoos. [3] In any case, issues of practice and place were central to the discussions of which directions zoology and biology should take as the twentieth century began. [4] It was highly plausible to claim a role for the study of animal lives and behavior within the broader whole of the life sciences, but exactly where such work would fit—be it epistemologically, methodologically, or institutionally—was by no means a foregone conclusion.

The American zoologist William Morton Wheeler was among those who believed that the study of animal instincts, intelligence, "habits," and "habitus" would soon occupy a central position in the life sciences. In 1902, he proposed that the term "ethology" be used for this domain of investigation, and he declared that scientists were "on the eve of a renascence in zoology." [5] The British amateur field naturalist Edmund Selous was not nearly so sanguine about zoology's immediate future, but he too was eager to see the study of animal behavior take its proper place among the biological sciences. As he acidly remarked in 1905—leaving no doubts about his own unhappiness with the zoological community of his day—"The habits of animals are really as scientific as their anatomies, and professors of them, when once made, would be as good as their brothers." [6]

It took nearly seven decades before Selous's prophesy about professors of animal behavior being "as good as their brothers" received science's ultimate confirmation. How did one get, so to speak, from the chronically alienated Englishman Edmund Selous, grumbling alone in his birdwatcher's hide, to the discipline-defining "International Ethological Congresses" of the 1950s and 1960s, burgeoning to the point that by the tenth of these meetings, held at Stockholm in 1967, there were more than 250 scientists in attendance? What scientific theories, practices, subjects, and settings were constitutive of ethology's construction as a scientific discipline? What personal, institutional, social, political, ideological, and other factors figured in the scientific worlds of the different individuals whose efforts created this enterprise? How, in detail, did specific people in specific settings create the concepts, practices, and institutions of ethology? And what can we learn from a close attention to "ethology's ecologies"? The goal of this book is to analyze historically the construction of ethology as a scientific discipline, paying particular attention to the ways in which in local and broader settings the founders of ethology generated, developed, contested, and refashioned the concepts and research practices of their newly emerging field.[7]

PEOPLE

The key actors in the founding of ethology as a discipline were Konrad Lorenz and Niko Tinbergen. It was Lorenz who was primarily responsible for laying the field's early conceptual foundations in the 1930s. Focusing in particular on instinctive behavior in birds, he showed how instinctive behavior patterns could be used like structures to reconstruct phylogenies. In addition he introduced a whole series of concepts to account for the physiological causation of instinctive behavior: releasers, innate releasing mechanisms, action-specific energy, threshold lowering, and the like. Although he was not the first person to observe the phenomenon of "imprinting," he was the first to highlight its scientific significance. His best-known work involved jackdaws and then greylag geese, raised from eggs and then allowed to fly freely. This practice, as he represented it, put him in a position to observe the full range of the birds' naturally occurring instinctive behavior.

Tinbergen, for his part, contributed experimental and analytical talents that beautifully complemented Lorenz's early theory building. Tinbergen was first and foremost a field naturalist, not an animal raiser, though he also did laboratory work. His best-known studies were on stickleback fish and herring gulls. Of the two men, Tinbergen was arguably the more important in furthering ethology's conceptual and practical development after the Second World War, and the one who showed himself to be more open to new directions. In the years immediately after the war, it was he who set the important example of putting aside deeply felt wartime grievances for the good of the field's postwar recovery. He was also the ethologist who worked hardest for ethology's coordinated, balanced growth in the 1950s and 1960s. Defining ethology as *the biological study of behavior,* he strove to make this definition a guiding vision for the field.

Early in my researches on the history of ethology, in the spring of 1979, I wrote to Tinbergen to ask if I could visit him in Oxford to talk about a variety of issues concerning the development of his field. He wrote back, expressing his willingness to discuss what he called "that curious ragbag that is now called 'ethology.'"[8] This characterization of ethology took me by surprise. I thought it very strange for a man who had recently won the Nobel Prize to refer to his field as a "ragbag." I assumed his words were simply a reflection of the "pathological modesty" I had heard was such a conspicuous feature of his personality. Only later did I come to appreciate how concerned Tinbergen had been in the 1950s and 1960s to develop and maintain ethology as a coherent whole—and how problematical were the prospects of keeping that whole intact.

Both Lorenz and Tinbergen were exceptionally able figures when it came to attracting disciples and allies to their field. Lorenz was the more charismatic of the two, at least in the more familiar senses of that term. Ebullient and egocentric, he enjoyed being a showman and the center of attention. He liked to charm an audience and to preach to it at one and the same time. Tinbergen had an entirely different style, but he too attracted devoted followers. His students came to refer to him as "the maestro." He was a superb popularizer of animal behavior studies, producing stunning nature films as well as books and articles to reach a broad audience.

But Tinbergen and Lorenz were by no means the only actors involved in constructing the biological study of behavior in the twentieth century. Other academic scientists, students at various levels of their education, zookeepers, animal raisers, amateur naturalists, officers of granting agencies, and various friends, family, critics, competitors, and patrons played roles in the history of ethology. Some of these were contemporaries of Lorenz and Tinbergen. Others preceded them. Here we will pay careful attention to a number of late nineteenth- and early twentieth-century investigators of animal behavior. Their aims, ideas, and methods proved of genuine importance to their successors, albeit at times only in piecemeal or indirect ways. Significantly, they did not imagine themselves primarily in the role of *precursors*. They were actors functioning in their own right, and their activities need to be considered in their own contexts. We will look closely at the American scientists Charles Otis Whitman and Wallace Craig and the English field zoologists Edmund Selous, H. Eliot Howard, and Julian Huxley (among others). The cases of Whitman and Craig illustrate the problems of pursuing animal behavior studies of a naturalistic, experimental, and evolutionary nature at a time when the institutional support for such studies was yet to be put in place. In their day, other modes of research were more successful in laying claim to the resources that the discipline of biology, the American scientific community, and the broader American society had to offer. Selous and Howard for their part were classic English amateurs who contributed pathbreaking researches but remained outside academic biology. Huxley was an academic insider. He made important contributions to field studies of behavior early in his career, but he ultimately chose other directions for his scientific energies.

The founders of ethology, Lorenz and Tinbergen, did not simply proceed along paths established by the students of animal behavior who came before them. With respect to the ideas, observations, and practices of their predecessors, they selectively appropriated what was handy and helpful to them and rejected or ignored what was not. This is particularly striking in Lorenz's case in terms of what he took and what he did not take from Whitman and Craig. It was also true for Tinbergen in the immediate postwar period, when he promoted what he considered to be the essential core of Lorenz's theorizing and put aside other features of Lorenz's thinking he deemed scientifically unfounded, philosophically beyond him, or politically abhorrent.

Readers may be surprised to find that Karl von Frisch, the corecipient with Lorenz and Tinbergen of the 1973 Nobel Prize for Physiology or Medicine, plays only a marginal role in the present book. Certainly there is no gainsaying Frisch's importance for the study of animal behavior in the twentieth century. His brilliant work on the dance language of the honeybee is a classic of modern science. However, he essentially took no direct part in the efforts that constituted ethology as a new discipline in the middle decades of the century. His work deserves a major study in its own right, but that is not something that will be attempted here.[9]

PLACES

Both individually and collectively, the scientists who came to identify themselves as ethologists were constantly engaged in situating their work in a variety of contexts. They sought to create a special space for themselves within the broader landscape of existing and emerging scientific disciplines and specialties. Central to their claims of distinctiveness was their special focus on animal behavior, but it was not just their subject matter that made them unique. The material settings that they chose for their researches served as a basis for their claims to authority. They insisted that these material settings, together with the practices they pursued there, afforded them special insights into the *natural* behavioral repertoires of their animal subjects.

The material settings of their work ranged from laboratories to field sites, with intermediate settings such as zoos or field stations in between. Among these were the Zoological Laboratory of Leiden University (where Tinbergen and his students experimented on the stimuli eliciting the courtship behavior of stickleback fish), the Dutch dunes where Tinbergen undertook his studies of the orientation behavior of digger wasps (as well as the social behavior of herring gulls), and the special research installation Lorenz established at his family home in Austria for studying tame, free-flying jackdaws, night herons, greylag geese, and other birds.

Each of the broad sorts of settings just mentioned had its particular virtues and liabilities, and each had its champions and critics. Not satisfied with any of the basic types of research setting named above, Charles Otis Whitman promoted at the beginning of the twentieth century the establishment of what he called a "biological farm," a special research institution that would have advantages over the laboratory, the field, and the zoo simultaneously. "For the continuous study of living organisms, under conditions that can be definitely known and controlled," he explained, the laboratory was "too narrow," the world was "too wide," and zoos were simply much too subject to the intrusions of the public.[10]

Whitman's biological farm was never built, though in the 1950s the creation of the Madingley Ornithological Field Station at Cambridge and Lorenz's Max Planck Institute at Seewiesen, Germany, bore at least some resemblance to what Whitman had had in mind. The field, the laboratory, and zoos and aviaries proceeded to play their diverse roles in the constitution of modern ethology. The ways these different material settings were conducive to different kinds of practices—and the consequences of these practices for the course of ethology's development—will be a major theme of the present study.

It will also be important here to consider where animal behavior studies fit within the broader landscape of twentieth-century science. When the ethologists Robert Hinde and W. H. Thorpe remarked in 1973 on how ethology had moved within the life sciences from a position of insignificance to a position of prominence, they knew full well that the study of animal behavior had not been the province of life scientists alone. Psychologists too had laid their claims to the subject. Konrad Lorenz and Niko Tinbergen were continually at pains to distinguish themselves from the "subjectivist" animal psychologists, whom they regarded as their major competitors in Europe, and the "behaviorist" psychologists, who dominated animal behavior studies in the United States. Nevertheless, lines between ethologists and animal psychologists were by no means as sharp as the rhetoric surrounding these various traditions might suggest.

Indeed, the contingent circumstances of an individual career could determine whether—or when—an investigator chose to define himself or herself as a biologist or a psychologist. The American Wallace Craig received his Ph.D. in zoology but found employment teaching psychology in a philosophy department. Similarly the Austrian Lorenz, having earned doctoral degrees in medicine and zoology, was willing in the 1930s to think of refashioning himself as a psychologist when this seemed the most promising way of furthering his research interests and career. The first professorial appointment he received was in the Department of Psychology at the University of Königsberg in 1940. Craig and Lorenz's cases should not be seen, however, simply as instances of biologists masquerading in psychologists' clothing. Not only did each of these men genuinely believe he had a contribution to make to psychology; each significantly refashioned his thinking and identity as he adapted to and in some measure reshaped the circumstances in which he found himself. In his new identity as a psychologist, Craig promoted psychological over biological explanations of various behavioral phenomena. As for Lorenz, the ideas that later earned him a reputation as a founder of evolutionary epistemology were ideas that he originally developed in the process of securing and then justifying his appointment as a psychologist at Königsberg.

An additional respect in which place has been important for ethology has been in terms of the local, national, or international settings in which animal behavior studies have been pursued. The "geography" of animal behavior studies evolved in interesting ways over the first half of the twentieth century. We can gain a sense of this by comparing a French animal psychologist's perception of the "new animal psychology" early in the century with an American animal psychologist's comments on Lorenzian animal behavior studies at midcentury. The Frenchman was Georges Bohn, director of the laboratory of biology and comparative psychology at the École des Hautes Études in Paris. In his book of 1911, *La nouvelle psycholo-*

gie animale, Bohn cited four examples of research initiatives that were, in his opinion, revolutionizing the field of animal psychology. These consisted of (1) several hundred studies done in the United States on invertebrate and vertebrate behavior, (2) Bohn's own work in France (which he characterized as the application of "the ethological method to the study of the activities of the lower animals"), (3) researches conducted by Ivan Pavlov and Pavlov's disciples in St. Petersburg, and (4) Edouard Claparède's work in Geneva. Bohn went on to note that not all countries were participating equally in the reformation of the field. In his words, "Germany and especially England have remained a bit outside of the great renovation of animal psychology." [11]

Bohn was not a neutral observer. His comments cannot be taken as an infallible guide to the whole range of animal behavior studies that existed in his day. What is clear, nonetheless, is that by the 1950s the geography of animal behavior studies bore little resemblance to Bohn's map of four decades earlier. The scientists who were doing the most exciting new work in the 1950s were using the word "ethology" to denote their approach to behavior, but they were not located in the centers of activity identified by Bohn. They were working primarily in Germany and Great Britain—the countries Bohn had seen as lagging behind in behavioral studies—and also in the Netherlands, a country that Bohn in 1911 had not mentioned at all.

One index of the growing significance of this new "ethology" was its ability to stimulate others to attack it. The most consequential of the early attacks was launched by the American comparative psychologist Daniel Lehrman in 1953. In his now-famous assessment of Lorenzian ethology, Lehrman noted both the geographical and the disciplinary distinctiveness of the new work. He characterized ethology as being distinctively European in its origins. He observed further that a key factor in ethology's success in Europe was that most of the students of animal behavior in Europe were "zoologists, physiologists, zoo curators or naturalists." In America, in contrast, the majority of students of animal behavior were psychologists. Where ethology was beginning to make inroads in the United States, Lehrman noted, was primarily among the ranks of zoologists, ecologists, and especially ornithologists.[12]

The importance of place in ethology's development would manifest itself not only in differences among countries and among disciplines but also in differences among institutions. An instructive example of this is the way that the ethologists at Oxford and Cambridge established spheres of research that complemented more than they competed with each other. Generally speaking, the ethologists at Oxford focused more on questions of behavioral evolution and function, while the ethologists at Cambridge focused more on questions of behavioral causation and development. Such differences were a function in part of long-standing, diverse research strengths at the two institutions, but they were also, at least in part, the result of an implicit if not explicit understanding of how best to divide the labor. Other research centers likewise had their own particular characteristics, and, to varying degrees, their own distinctive research programs.

PRACTICES

As already suggested, places intersected with practices in interesting and important ways in the history of animal behavior studies. Different investigators or schools promoted the special virtues of their approaches. "Objectivist" versus "subjectivist" approaches; the relative importance of observation, experimentation, and quantification; and the merits of fieldwork and laboratory studies all came to be contested. An example from the literature of American comparative psychology can help us get closer to this issue. In 1917 an aspiring Japanese ethnologist (*not* ethologist) named N. Utsurikawa, working in the Harvard laboratory of the American comparative psychologist Robert Yerkes, reported in the *Journal of Animal Behavior* the results of a study he had been conducting on temperamental differences in white rats. In his paper Utsurikawa invoked a contrast between "naturalistic data" and "experimental data." He identified naturalistic data as the observations made "under the *natural* cage conditions" (italics added).

"*Natural* cage conditions"? The juxtaposition of the words "natural" and "cage" takes us aback, even if we can identify a certain logic to the distinction Utsurikawa was making. In the history of animal behavior studies in the twentieth century, the ethologists were the investigators who laid the strongest claims to studying what was natural in animal behavior. Central to their studies, and indeed privileging their researches over those of others (at least as the ethologists saw it), was their focus on naturally occurring behavior patterns in more or less natural situations. They prided themselves on watching animals over long periods of times under natural conditions as a means of developing inventories or "ethograms" of all the behavior patterns characteristic of a species. Nonetheless, there was a heterogeneity of practices among the ethologists themselves. This was tied to the diverse settings in which they conducted their research. Lorenz and Tinbergen were significantly different from each other in their styles and places of work. As Lorenz liked to put it, he was essentially a farmer, while

Tinbergen was essentially a hunter. To be more specific, Lorenz liked raising and breeding animals, nurturing them when they were ill, and having them as companions. Tinbergen preferred stalking animals in the field, matching wits with them, and discovering how the details of their behavior contributed to their survival. If one asks where Lorenz was most "at home" as a researcher, the answer, literally, was at his own home, that is, at the research station he built at his father's home in Altenberg, Austria. There, with jackdaws nesting in the attic and geese and other birds moving relatively freely in and about the house and surroundings aviaries, he made the vast majority of the observations on which his theories were based. Tinbergen, in contrast, was most at home as a scientific researcher when he was out in the field. Although he also conducted important laboratory studies, it was as a field naturalist that he felt most complete as a researcher and as a person. Here practices and places were inextricably intertwined. The differences in the practices of Lorenz and Tinbergen ultimately led the two men to make appreciably different contributions to the biology of behavior.

POLITICS

We have already seen how the ethological community greeted the 1973 Nobel Prize for Physiology or Medicine as testimony that their field had come of age. The popular press found two other aspects of the awards more intriguing. The first was that Tinbergen was actually the second Tinbergen son to receive a Nobel Prize: his brother Jan had already received the Nobel Prize in economics. The second was that Konrad Lorenz had a Nazi past, the precise dimensions of which were uncertain.

Niko Tinbergen remarked on both of these in a letter that he wrote to his friends Peter and Jean Medawar some three months after the award ceremony. When the press asked Tinbergen how to account for two Nobel Prizes in the same family, "the ethologist," in Tinbergen's words, "found himself pooh-poohing the idea of exceptional genetic endowment, and pointing to fortunate conditions, as did my brother, independently." As for the rumors that Lorenz had a Nazi past, these, Tinbergen acknowledged, had some basis in fact: "Konrad Lorenz had, as a (protestant) Austrian been in favour of the Anschluss and had also fallen for Hitler. Unfortunately he published some paragraphs in scientific papers in which he professed to believe in the racially pure society and although I know that he turned round as soon as he saw what really happened, he has never withdrawn what he wrote (he never does anyway), so it was not astonishing that he was singled out for a (very virulous) attack. And because I was known to have been on the other side of the fence, and also knew in detail how he came to derail, I had to urge him to make a clean breast of it ('I was politically very stupid and gullible') and then, in a series of interviews (the last one in Stockholm) to defend him." [13]

It will be instructive to keep both of the above points in mind as we proceed. That is, it will be helpful to remember the ethologist, Tinbergen, explaining his accomplishments in terms not of heredity but instead of environment. It will also be necessary to look closely at the much-discussed issue of Lorenz's politics under the Third Reich, and to ask what difference this and Tinbergen's having been "on the other side of the fence" made for ethology's subsequent development. We will find further that in the wake of Daniel Lehrman's critique of Lorenzian ethology, the two, opposing sides that formed in the debate over the "innateness" of behavior patterns were each highly suspicious of the ideological commitments of the other. And we will want to look at other kinds of "politics" as well, including that involved in competing for resources and authority and in dividing up scientific turf.

APPROACH

One could write the history of ethology as a story of progressive successes, starting early in the century and culminating with the Nobel Prize for Physiology or Medicine of 1973. One could simultaneously treat the development of ethology as if it had a certain inevitability to it, rather along the lines suggested by Konrad Lorenz in 1978, in the sweeping fashion that had become characteristic of his writing: "Ethology, the comparative study of behavior, is easy to define: it is the discipline which applies to the behavior of animals and humans all those questions asked and those methodologies used as a matter of course in all the other branches of biology since Charles Darwin's time." [14]

However attractive this definition might at first appear, it confuses issues of scientific *domain* and historical *development*. The history of ethology is not adequately represented as a story of progressive successes or as the natural or logical extension of Darwinian biology to the study of behavior. What is more, simply *defining* ethology was never the easy matter that Lorenz suggested. Indeed it was Tinbergen more than Lorenz in the 1950s and 1960s who insisted on identifying and pursuing ethology as "the biology of behavior." Tinbergen's classic formulation of this in 1963 served to firm up the field's identity, and it has provided a general guide for students of behavior ever since. Yet neither before nor after 1963 can it be said that the biological study of behavior evolved according to a set program or inner logic. The course of ethology's development has been more responsive to contingencies, more "ecological" in its relations to the specific and diverse settings of its ongoing construction, and thus more interesting historically.[15]

A generation ago, the history of science tended to be written primarily in terms of scientific theories and their conceptual development. More recently, historians, sociologists, and anthropologists of science have paid increasing attention to matters of scientific practice and to science's broader social and cultural history. This work has led to a much richer understanding of "science in action."[16] In some of this writing, however, the conceptual content of science has almost disappeared from sight. Here we will strive for an integration of "content" and "context," recognizing that any firm distinction between the two is ultimately misleading. Concepts, practices, people, and places intertwined dynamically in the course of ethology's emergence as a scientific enterprise. The major actors in this story did not simply install themselves in existing niches in the cognitive, material, disciplinary, and institutional landscapes of their day. They created new niches and relationships, and they modified their own ideas and practices in the process. The present study is in effect a work of comparative scientific biography, aimed at illuminating how the efforts of different investigators in diverse times and places came to constitute the modern scientific discipline known as ethology. Examining these investigators and the settings in which they operated—and the way their activities to greater or less extents reshaped the conditions of their scientific existence—is what this study is about. We will thus look not only at the scientific achievements of our ethological actors but also at their visions and understandings of their roles as scientists in a complex, modern world.

In the course of my research and writing on the history of ethology (a project I began twenty-five years ago), I have consulted a vast array of books and journals, interviewed many of the field's early practitioners, and examined a wide range of manuscript correspondence, field notes, and other items. Each of these sources has its advantages and its limitations. Published scientific papers and books stand as the public record of scientific accomplishment, but as Peter Medawar has explained in an often-cited observation, the scientific paper serves to disguise how scientific knowledge is produced.[17] Interviews in turn can provide invaluable insights into aspects of the scientific enterprise that are usually not recorded in scientific papers, but memory is a tricky resource in itself, inevitably in-

volving considerable selection and reconstruction. Manuscript correspondence and other documents might seem a surer means of reconstructing
what a person thought or did at a specific time, but these too require analysis and interpretation, as do the circumstances through which some
things are preserved for posterity and some are not. I have used a wealth of
manuscript sources in writing this book, but I am conscious that their testimony needs to be put in context and interpreted in much the same way
as any other source of information.

One particular archival "find" testifies to the point just made. This
study has taken me sufficiently long to complete that I have on more than
one occasion, in studying the manuscript correspondence of a deceased
scientist, found myself reading a letter that I myself wrote to the person in
question. I recognize this is an experience that is bound to happen from
time to time to historians or biographers working on modern topics. This
did not prepare me, however, for a strange twist on the just-mentioned
scenario. In March 1992, working through the Tinbergen Papers in the
reading room of the Department of Western Manuscripts at the Bodleian
Library at Oxford, I came upon a copy of a letter that Tinbergen wrote to
me but never sent. The letter was dated 16 June 1982. It was three pages
long, single-spaced. Nearly ten years after Tinbergen wrote the letter, I in
effect received it for the first time.[18]

The special interest of this unsent letter was that Tinbergen laid out in
writing there, more explicitly than in any other place, his view of the differences between his and Konrad Lorenz's contributions to ethology. He
did so in response to a draft of a paper I had sent him. I had written something in that paper that struck a nerve with him, and he was writing back
to set me straight. The usually self-effacing Tinbergen spelled out in detail,
and with some passion, how his ideas and his contributions to ethology
differed from those of his longtime friend and colleague Lorenz.

I had sent Tinbergen a draft of a paper entitled "Towards an Evolutionary Ethology." I was scheduled to give the paper at the Darwin centenary conference in Cambridge late in June 1982. What upset Tinbergen
about my draft was my use of the quote by Konrad Lorenz that I have just
cited above—the quote in which Lorenz defined ethology as "the discipline which applies to the behavior of animals and humans all those questions asked and those methodologies used as a matter of course in all the
other branches of biology since Charles Darwin's time." My purpose in using the quote then was the same that it is now—namely, to call it into question as a guide to the history of the field. I wanted to insist that ethology's
actual development did not proceed "as a matter of course" along a well-

groomed path, but instead had its own distinctive history which needed to be investigated. To Tinbergen, however, it appeared that I was crediting Lorenz with defining ethology as the biological study of behavior. Ready as he was to grant Lorenz the primary role in founding ethology as a new field in the first place, he felt (correctly) that it was he, and not Lorenz, who deserved the real credit for guiding ethology's later development by insisting on all the ramifications of identifying it as the biology of behavior. He did not want to see his efforts and Lorenz's confused in this regard. And he thus spelled out exactly where he felt his ideas and achievements differed from Lorenz's. But he did not send this letter to me. Instead, three days later, he wrote me a two-page, toned-down letter that addressed some of the same issues but was less animated and less explicit about his differences with Lorenz. He gave this second letter to Ernst Mayr, who was visiting him before the Cambridge conference, and Ernst then carried the letter to me in Cambridge. As I have indicated, it was not until a decade later that I first saw the "original" letter, the existence of which I had had no reason to suspect. Among the things I certainly hope to make clear in this book are the key points on which the two major cofounders of ethology did not see eye to eye, and why Tinbergen in 1982 was keen that the historical record be set straight, at least eventually.

Other archival "finds" appear in the following pages. We come upon Wallace Craig lamenting that in order to make ends meet while conducting his comparative studies of pigeons and doves he might have to end up eating some of his bird subjects. We find Julian Huxley watching the courtship of great crested grebes on a lake and suddenly writing at the end of a long flurry of note taking, when he realizes he has mistaken one grebe for another, "Damn—I believe this is wrong ♂." Konrad Lorenz's correspondence includes exuberant descriptions of the joys he experienced in his work. It also reveals some of the temptations he experienced as an ambitious young scientist in the Third Reich striving to secure a professional post for himself. Tinbergen's letters at the end of the Second World War acknowledge that it will take him some time before the wounds of war heal and he is able to work with German scientists again. Later letters find him happily engaged in the researches that earned him recognition as a founder of behavioral ecology. These are but a few examples of the archival traces that will help us make sense of the actions and career paths of individuals whose efforts brought the science of ethology into being.

One striking feature of ethology's history is that the ideas that constituted the conceptual core of "classical ethology" were relatively short lived. The more lasting achievement of Lorenz and Tinbergen and their colleagues was the cultivation of a collection of attitudes, insights, practices, and goals that continued to provide a sense of shape to the field while permitting it at the same time to have considerable flexibility and adaptability. Another striking feature of ethology's history is its heterogeneity over time and across different local sites. The overall story involves a multitude of people, practices, and places. It testifies clearly to ethology's multiple "ecologies." Thinking of "ethology's ecologies" helps us to recognize the highly situated character of science in action. It likewise helps us recognize how much work went into ethology's construction as an important new scientific discipline in the twentieth century.

ONE

CHARLES OTIS WHITMAN, WALLACE CRAIG,

AND THE BIOLOGICAL STUDY OF

ANIMAL BEHAVIOR IN AMERICA

Instinct and structure are to be studied from the common standpoint of phyletic descent.

C. O. WHITMAN, "ANIMAL BEHAVIOR," 1898

I have about decided that comparative psychology is the line for me. Whether there will be bread and butter in it or not, I don't know—that's a subordinate question.

WALLACE CRAIG TO C. C. ADAMS, 1898

In the summer of 1898, at the Marine Biological Laboratory in Woods Hole, Massachusetts, the American zoologist Charles Otis Whitman delivered a lecture containing the insight that Konrad Lorenz would later call the "Archimedean point" of comparative behavior studies.[1] As Whitman put it: "Instinct and structure are to be studied from the common standpoint of phyletic descent."[2] In other words, instinctive behavior patterns could be used just like body parts in studying how animals evolved.

One could write a history of ethology beginning with other investigators in other settings, but Whitman in the United States is as useful a starting point as any other would be. His case opens up a wide range of questions involving theories and practices, actors and objects, allies and competitors, boundaries and bridges, local circumstances and broader horizons. All of these relate in turn to the issues of who could speak authoritatively about animal life and how to study it.

The actors in animal behavior studies in Whitman's day constituted a stunningly diverse fauna, stirring in marvelous ways. Single-celled organisms moved silently toward or away from the experimenter's chosen stimuli. So-called higher animals performed more complicated motions, often accompanied by buzzing, cooing, clucking, or other ways of signaling their presence. Then too there were the human observers of the animals' behavior—watching, wondering, poking, probing, controlling, conversing, writing, and in general constructing systems of relations connecting themselves with their animal subjects and with the additional material, institutional, and social necessities pertaining to their scientific lives. The roster of investigators of animal behavior in the United States included Whitman and his behaviorally oriented students, William Morton Wheeler, Samuel J. Holmes, Oscar Riddle, and Wallace Craig; Jacques Loeb and Herbert Spencer Jennings, paired in history by their famous debate over tropisms and the behavior of lower organisms; Charles H. Turner, the African-American biologist whose experimental investigations of color sense and form sense in bees predated Karl von Frisch's earliest work on these topics; Francis H. Herrick, a pioneer in the study of the domestic life of birds; and C. C. Adams and Victor Shelford, both animal ecologists. Then too there were Margaret F. Washburn, Robert M. Yerkes, and John Broadus Watson, psychologists whose work, at least in the early part of their careers, focused primarily on animal behavior. Beginning in 1911, the United States could also claim the *Journal of Animal Behavior,* the first major scientific journal devoted exclusively to the study of animal behavior.

These various individuals did not constitute a close-knit community of investigators, sharing a common view of the aims and methods of animal behavior study. The biologists and psychologists, not surprisingly, tended to differ from each other in their ideas and their practices. Yet the lines that separated them were by no means hard and fast. Certain biologists and psychologists, at particular moments of their careers, felt more affinity with each other than they did with members of their own disciplines. Instructively enough, though, none of them sought to establish the study of animal behavior as an independent discipline. Each viewed the study of animal behavior as a means by which his or her own parent discipline could be reformed and developed. Each wanted to make animal behavior studies an integral part of the broader, ongoing enterprise that was either biology or psychology.

This was particularly true of Whitman. Whitman was an eloquent proponent of the view that specialized studies should be pursued only as parts of the organized whole, which in this case meant biology. He put this view into practice in his organization of the Division of Biology at the University of Chicago. He likewise articulated this view in his presidential address to the Society of American Naturalists in 1897, urging: "We need to get more deeply saturated with the meaning of the word 'biological,' and to keep renewing our faith in it as a governing conception. Our centrifugal specialties have no justification except in the *ensemble.*" He encapsulated the same vision in his 1902 call for a special experiment station—a "biological farm"—where "the study of life-histories, habits, instincts and in-telligence" would be conducted alongside "the experimental investigation of heredity, variation, and evolution."[3]

To Whitman it was apparent that the success of biology as a governing conception would depend not simply on the idea's intellectual appeal but also on the establishment of the attitudes, practices, and institutions neces-sary to sustain it. He called the approach he championed "experimental nat-ural history." He specifically distinguished this from the narrower, physio-logical, lab-oriented approach of his colleague Jacques Loeb. In Loeb's approach, Whitman complained, "instinct reduces itself in the last analy-sis to heliotropism, stereotropism, and the like" and "the whole course of evolution drops out of sight altogether."[4]

In the new, university zoology laboratories that symbolized modern biology across the country, investigators took up Loeb's tropism studies with enthusiasm. Whitman himself, however, did not believe the labora-tory to be a sufficiently ample setting for promoting biology as properly conceived. Nor did he believe the problem could be solved simply be sup-plementing lab studies with fieldwork. He envisioned his "biological farm" as the special setting where experimental natural history would be pursued and the "governing concept" of biology would be ratified.

This chapter focuses primarily on the work and careers of Charles Otis Whitman and his student Wallace Craig. Their stories illuminate the problems of pursuing naturalistic, experimental, and evolutionary studies of behavior at a time when the institutional support for such studies was yet to be created and when other forms of research were proving more suc-cessful in commandeering the resources and approbation of the American scientific community. They were not the only investigators in the United States in the early twentieth century who pursued animal behavior studies. They were, however, the scientists who contributed most directly to the particular approach to animal behavior studies that came to be known as ethology.

CHARLES OTIS WHITMAN AND AMERICAN BIOLOGY

When Charles Otis Whitman delivered his lecture on "animal behavior" at the Woods Hole Marine Biological Laboratory in 1898, he was arguably the most influential biologist in America. The MBL was the Mecca of Ameri-can biology, the seaside magnet that in the summer months drew Amer-ica's finest biologists to a single, vital center, and Whitman was its first director. He also headed the Division of Biology and the Department of

1.1 Charles Otis Whitman with his beloved pigeons. (Photo by Kenji Toda. University of Chicago Archives, photographic files, series 1, Charles Otis Whitman, informal no. 3. Courtesy of the Department of Special Collections, University of Chicago.)

Zoology at the University of Chicago, where he had assembled as distinguished a biological research faculty as could be found anywhere in the country. Before going to Chicago in 1890 he had served for three years as the first chair of the zoology department at Clark University. In addition, he had founded the *Journal of Morphology* (in 1887) and been the key figure in establishing in 1890 the American Morphological Society (later to be renamed the American Society of Zoologists). The range of his interests and expertise was unmatched among his zoologist or biologist colleagues in America. How he came to his interest in animal behavior and how that interest fitted with the rest of his concerns and career deserve careful attention.[5]

Ethologists have routinely reported that their fascination with animal life began in childhood, long before they had any inkling of what it might mean to be a scientist. Whitman's own case fits the model. Born in 1842, Whitman as a youth was an ardent bird collector. He distinguished himself by his skill in mounting the specimens he shot, but he also kept birds—and amphibians and mammals too—as pets. Among his live birds were pigeons. As he later reported to his student Wallace Craig, his pigeons fascinated him, and he "sat and watched them by the hour, intensely interested in their feeding, their young, and in everything that they did."[6]

The historian Philip J. Pauly has described how Whitman, the son of Adventist parents in rural, western Maine, broke with his parents' extreme religious views and found his way first to Universalism and then to the natural sciences. Pauly argues that the intensity of the religious conflict of Whitman's youth was reflected in the strength of his later attachment to the idea of a law-governed, progressively developing universe, where supernatural events had no place. Whereas Whitman's parents believed that the Second Coming was imminent, Whitman put his own faith in modern science. He embraced a worldview in which gradual, progressive change and the evolution of life itself were the central processes and were to be explained in material rather than miraculous or mystical terms.[7]

After working his way through Bowdoin College, Whitman taught high school in Massachusetts. In 1873 he was attracted to the natural history summer school instituted by the famed Harvard zoologist Louis Agassiz on Penikese Island in Buzzards' Bay. There, at the sea's edge, with living marine creatures directly at hand, Whitman got his first real taste of zoology. This led him to join the Boston Society of Natural History, to spend another summer at Penikese, and then to follow the example of a fellow Penikese student by going to Leipzig and studying zoology under the German parasitologist Rudolf Leuckart. It was under Leuckart at Leipzig that he was molded into a serious researcher.[8]

In Leuckart's laboratory, Whitman mastered the latest techniques of microscopic analysis, particularly as adapted to the study of fertilization and development. For his doctoral thesis, he investigated the early embryonic development of the fish leech, *Clepsine*. His dissertation, published in 1878, reflected his newly acquired technical expertise, his command of contemporary science, and his commitment to a special vision of organic development. "In the fecundated egg," he wrote, "slumbers potentially the future embryo. While we cannot say that the embryo is predelineated, we can say that it is predetermined. The 'Histogenetic sundering' of embryonic elements begins with the cleavage, and every step in the process bears a definite and invariable relation to antecedent and subsequent steps."[9] This recognition of the importance of tracing cells, structures, and even behavior patterns back to their very earliest origins became a hallmark of Whitman's work.

Over the course of the next two decades, Whitman moved often. He worked at Harvard, the Imperial University of Tokyo, the Zoological Station at Naples, the Allis Lake Laboratory (near Milwaukee, Wisconsin),

Clark University, and then finally the University of Chicago and the MBL. Over the same period, his investigations of leeches grew to encompass these organisms in all aspects of their existence from their development to their systematics, phylogeny, ecology, and behavior. This came to be his model of proper biological research: the study of a single species or a group of closely related species exhaustively pursued in every aspect of its existence. In 1899 he defined biology as "the life-histories of animals, from the primordial germ-cell to the end of the life-cycle; their daily, periodical, and seasonal routines; their habits, instincts, intelligence, and peculiarities of behavior under varying conditions; their geographical distribution, genetic relations and oecological interrelations; their physiological activities, individually and collectively; their variations, adaptations, breeding and crossing." Whitman's focus on the living animal in every feature of its existence is epitomized in the claim that his students greeted each other not with the question "What is your special field?" but rather "What is your beast?" [10]

In Whitman's animal behavior paper of 1898, his own special "beasts" featured prominently. *Clepsine,* the fish leech, was there. So too was *Necturus,* the freshwater salamander he began studying when he directed the Allis Lake Laboratory from 1886 to 1889. There also were his pigeons, the birds that had fascinated him as a boy and that he began studying in earnest again around 1895. They would be the almost constant focus of his research for the last decade and a half of his life.

Whitman presented his animal behavior paper in the form of two evening lectures in the Marine Biological Laboratory's summer lecture series of 1898. Here he set forth his views on the proper methods of studying behavior, the nature of instinct, the importance of studying instinct from a phylogenetic standpoint, the means by which behavior has evolved, and the relations—both ontogenetic and phylogenetic—between instinct and intelligence. Of this paper his student Frank R. Lillie later wrote: "No other of [Whitman's] papers illustrates better the qualities of his genius: the selection of a fundamental problem; painstaking study; publication only after years of observation and reflection; skill in laying bare the simple basis of an apparently complex group of phenomena; a grasp of the subject in all its bearings; and the use of the comparative or phyletic method of attack." [11]

Whitman began his lecture with a few remarks on the subject of animal behavior generally. He then turned abruptly to the special topic of "modes of keeping quiet" among animals. *Clepsine* allowed him to make the point that if one wanted to understand the behavior of any animal species, one needed to have a thorough knowledge of the animal's entire behavioral repertoire. The "deceptive quiet" of *Clepsine* was such that an observer unfamiliar with *Clepsine*'s habits would "almost certainly" draw the wrong conclusions about its sensitivity to stimuli. He drew the same lesson from his work with *Necturus*. It had taken him considerable experience with *Necturus* adults, he said—and two whole seasons rearing *Necturus* young—before he appreciated "the extreme timidity of these animals." This timidity, he explained, is "so deep-seated and persistent that one can form only a poor idea of it without considerable actual contact with it." [12] Whitman underscored the importance of studying animal behavior under natural conditions. It was essential, he said, that one "observe and experiment under conditions that ensure *free behavior*." [13]

Whitman paid particular attention in his lecture to *instinctive* behavior and to how instinct and intelligence were related. For him, the "first criterion of instinct" was that it could "be performed by the animal without learning by experience, instruction, or imitation." The fact that an animal's acts were adapted to purposeful ends, he explained, was not in itself proof of intelligence on the animal's part. *Necturus* always sneaked up on its prey, regardless of whether the prey was living or was simply a piece of meat. The creature, Whitman was convinced, had not the slightest appreciation of the importance of stealth. It was "quite blind" to the significance of its actions. Its movements, in Whitman's words, were "those characteristic of the species, not because they are measured and adapted to a definite end by intelligent experience, but because they are organically determined; in other words, depend essentially upon a specific organization." If *Necturus,* as a species, had to depend on its intelligence, he said, it was "difficult to see how it could escape immediate extinction." Its continued existence was assured by its instincts. [14]

Whitman drew the same sort of conclusion with regard to pigeons incubating their eggs. The birds, observation showed, were wholly oblivious to the biological function of their actions. He wrote:

> It is quite certain that pigeons are totally blind to the meanings which we discover in incubation. . . . They sit because they feel like it, begin when they feel impelled to do so, and stop when the feeling is satisfied. Their time is generally correct, but they measure it as blindly as a child measures its hours of sleep. A bird that sits after failing to lay an egg, or after its eggs have been removed, is not acting from "expectation," but

because she finds it agreeable to do so and disagreeable not to do so. The same holds true of the feeding instinct. The young are not fed from any desire to do them any good, but solely for the relief of the parent.[15]

Whitman found nothing mysterious in instincts. As he saw it, instinct and organization were "two aspects of one and the same thing." Like structures, instincts needed to be understood not just in terms of their development in the individual but also in terms of their evolutionary history. It was in this context that he wrote the words cited earlier: "Instinct and structure are to be studied from the common standpoint of phyletic descent."[16]

This had important implications for the zoologist seeking to reconstruct the ancestry of a subgroup of the animal kingdom such as the pigeon family. It meant that the birds' instinctive behavior patterns, as well as their physical characters, could be used to assess evolutionary affinities. If instincts were just like organs in the way they varied or remained unchanged through a whole group of organisms, then instincts and organs could be used as checks upon each other when it came to determining the course of evolution and deciding where different forms should be located on the family tree. Sometimes instincts might prove even more reliable than organs when it came to deciding the relations within a particular group.

Whitman believed that the study of instincts illuminated not only the course of evolution but also the way in which new characters were introduced in the evolutionary process. The majority of his contemporaries, he felt, failed to appreciate just how deeply rooted instincts were, and just how far back the phylogenies of instincts thus needed to be traced. The source of this failure, Whitman believed, was a misunderstanding of how instincts originated in the first place. Numerous biologists still imagined that instincts arose through the inheritance of acquired characters, or more specifically that "an instinct could become gradually stamped into organization by long-continued uniform reactions to environmental influences." But even those biologists who thought in terms of natural selection instead of the inheritance of acquired characters, Whitman maintained, addressed the wrong question. They asked: "How can intelligence and natural selection, or natural selection alone, initiate action and convert it successively into habit, automatism, and congenital instinct? In other words, the genealogical history of the structure basis being completely ignored, how can the instinct be mechanically rubbed into the ready-made organism? Involution instead of evolution; mechanization instead of organization; improvisation rather than organic growth; specific *versus* phyletic origin." For such thinkers, Whitman complained, it did not matter "how long this blunder-miracle had to be repeated before it happened all the time." In such evolutionary scenarios, he observed, "purely imaginary things can happen on demand."[17]

Whitman's own view of the matter, informed by his previous experience tracing organizational differences back to their earliest foundations, was exemplified in his analysis of the way *Necturus* pauses before grasping its prey. This particular behavior pattern, he argued, represented "an instinct the history of which may be coextensive with the evolution of the animal." As he put it, "Very early in the vertebrate phlyum, possibly at its dawn, the chief characters of the instinct, as we now find it, were probably fixed in structural elements differing from those in *Necturus* only in superficial details."

Whitman applied the same reasoning to the phenomenon of "pointing" in dogs. He agreed with Darwin that the "pointing" of dogs was not initially the result of training. He was not satisfied, however, with Darwin's explanation that the original tendency to point in dogs arose as an "accidental" variation, the cause of which was unknown. Such variations, Whitman was convinced, were "manifestations of instinct roots of more or less remote origin."[18]

Whitman's "orthogenetic" interpretation of the origin of instincts thus stood in opposition not only to Lamarckian but also to Darwinian explanations of instincts' being "rubbed into the organism on demand." Still, when he had to choose between Darwinian natural selection and the Lamarckian idea of the inheritance of acquired characters, he opted for natural selection. Darwin, in the *Origin of Species,* had argued that the instincts of neuter castes of insects could not be explained by the inheritance of acquired characters. He believed, nonetheless, that certain other instincts were best explained in terms of "use-inheritance." Whitman for his part took the stronger, "neo-Darwinian" position of August Weismann, claiming: "Repetition may become habit and produce marked effects on the nervous mechanism or other organs; but the individual structure so affected is not continued from generation to generation." There was "no conceivable way," he said, for the characters acquired by use or disuse in one generation could be "stamped upon the germs and so carried on cumulatively.[19]

The importance to Whitman of denying the inheritance of acquired characters, particularly with respect to the understanding of animal behavior, was that it meant that instinct could not be interpreted as "lapsed intelligence." Instinct, in his view, always came before intelligence. Whitman's position was that "instinct precedes intelligence both in ontogeny and phylogeny, and it has furnished all the structural foundations employed by intelligence." Instinct, in other words, was "the actual germ of mind." [20]

To analyze how instinct graded into intelligence, it was necessary, Whitman indicated, to study animals with complex instincts, the automatic character of which was incontestable. Pigeons, he felt, were perfect for this purpose, because they possessed such instincts and they could be studied comparatively. Whitman undertook a series of experiments to test how the wild passenger pigeon (*Ectopistes*), the tamer ringdove (*Turtur risorius*), and the domesticated dovecote pigeon (*Columba livia domestica*) each reacted to having its eggs placed just outside its nest. He found the results instructive: only in the most domesticated of these three species, the dovecote pigeon, would the parent bird reclaim as many as two eggs placed outside the nest. He ascribed this to the way that, under the conditions of domestication, the action of natural selection was relaxed, and this in turn led to a relaxation in the rigor of the "instinctive coordinations" that under normal circumstances would otherwise have prevented alternative actions. As he put it, "Not only is the door to choice thus unlocked, but more varied opportunities and provocations arise, and thus the internal mechanisms and the external conditions and stimuli work both in the same direction to favor greater freedom of action." [21]

Significantly, Whitman did not regard what went on under domestication as the antithesis of what went on in nature. He supposed to the contrary that "domestication merely bunches nature's opportunities and thus concentrates results in forms accessible to observation." His concluding remarks on the subject, supported by statements from Conwy Lloyd Morgan, William James, and Herbert Spencer, bear repeating: "Superiority in instinct endowments and concurring advantages of environment would tend to liberate the possessors from the severities of natural selection; and thus nature, like domestication, would furnish conditions inviting to greater freedom of action, and with the same result, namely, that the instincts would become more plastic and tractable. Plasticity of instinct is not intelligence, but it is the open door through which the great educator, experience, comes in and works every wonder of intelligence." [22]

Such were the major conceptual thrusts and programmatic suggestions of Whitman's animal behavior paper of 1898. Throughout, the paper was rich in observations and insights bearing on fundamental questions regarding instinct, intelligence, and the evolutionary process. Whitman based his comments on careful, long-term observations of naturally occurring behavior, supplemented by experiments. Some of the phenomena he de-

scribed—including the differences between the instinctive behavior patterns of wild and domestic races, the way that instincts "run down," and
certain features of what is now known as "imprinting"—were phenomena that Konrad Lorenz would later invest with special theoretical significance. Whitman, however, had his own theoretical concerns. For him, the
great questions of biology were the questions of heredity, development,
and evolution.

On the face of it, Whitman's paper would appear to have had the potential of serving as a powerful model for further biologically oriented researches on animal behavior. Whitman's intellectual framework and research methods, however, were not ideally suited for rallying disciples. The
enterprise of reconstructing phylogenies had passed its prime, at least in
the United States. Furthermore, Whitman's organismic and orthogenetic
perspectives lacked the attractive simplicity of the new views of heredity
and evolution that were firing the attention of the new generation of laboratory experimentalists. Doubts about whether the animal mind was a subject for scientific investigation may also have cast a shadow on Whitman's
interest in the evolution of animal intelligence. And had other investigators
found Whitman's *ideas* about animal behavior to be inspiring, they might
still have found his research *practices* too daunting. Detailed, comparative
studies of the behavior of higher animals required considerable time and
money, just what young biologists seeking to make names and academic
careers for themselves typically felt they lacked.

Whitman recognized full well the time-consuming nature of his kind
of animal behavior research. In 1899 he cast a critical eye on recent efforts
by psychologists to establish comparative psychology as a science in its own
right. As he saw it, "any attempt to soar to 'the nature and development of
animal intelligence,' except through the aid of long schooling in the study
of animal life, is doomed to be an Icarian flight." He explained: "The qualification absolutely indispensable to reliable diagnosis of an animal's conduct is an intimate acquaintance with the creature's normal life, its habits
and instincts. Little can be expected in this most important field of comparative psychology until investigators realise that such qualification is not
furnished by parlor psychology. It means nothing less than years of close
study,—the long-continued, patient observation, experiment, and reflex-
ion, best exemplified in Darwin's work." [23]

Whitman allowed that while Darwin's *ideas* had come to be generally
appreciated, the "real secret of [Darwin's] success" was yet to be generally
recognized:

He was no hustler on the jump for notoriety, no rapid-fire writer; but a cool, patient, indefatigable investigator, counting not the years devoted to preliminary work, but weighing rather the facts collected by his tireless industry, and testing his thoughts and inferences over and over again, until well-assured that they would stand. Such a method was altogether too laborious and searching to be imitated by students ambitious to reach the heights of comparative psychology through a few hours of parlor diversion with caged animals, or by a few experiments on domestic animals. We are too apt to measure the road and count the steps beforehand. Darwin allowed the subject itself to settle all such matters, while he forgot time in complete absorption with his theme.[24]

In point of fact, Whitman's portrait of Darwin fit Whitman better than it fit Darwin. It was Whitman who "forgot time" as he became wholly absorbed in his pigeon work. Darwin produced a prodigious amount of important scientific writings over the last decade and a half of his life. Whitman, in contrast, published very little. As his study of the evolutionary history of the pigeons led him from one fundamental biological question to another, his publications became fewer and fewer.[25]

Whitman began his systematic study of pigeons in the 1890s. Frank Lillie described Whitman's enterprise as follows:

He gradually collected a large number of species of pigeons from all parts of the world, and in the latter part of his life the collection comprised some 550 individuals representing about thirty species. His house was surrounded by pigeon cotes, and he always had some birds under observation indoors, so that the cooing of doves was for years a dominant sound in his house. He took care of the birds for the most part himself, though he usually had the assistance of one or two maids. He thus actually lived with his birds constantly, and very rarely was absent from them even for a single day. He made observations and kept notes on all aspects of the life and behavior of each species, as well as of such hybrids as he was able to produce. He always had one Japanese artist at work continuously drawing pigeons, and for several years two.[26]

When Whitman's associates expressed amazement at his special "insight" in understanding his birds, his response and explanation were: "Live with the birds day and night year in and year out."[27]

Whitman's pigeon studies testify to the breadth of his conception of what it was to be a biologist. He aimed to reconstruct the evolutionary history of pigeons through a painstaking study of their heredity, variation, and development. He expected this would enable him to reform in turn current understandings of heredity, variation, development, and evolution. The comparative study of the birds' calls and instinctive behavior patterns constituted only part of his research. He also paid attention to the development of the birds' color patterns, the fertility of different hybrid crosses, the results of inbreeding, the phenomenon of dominance, and the production of sex.

It is instructive to consider how the phenomenon now known as "imprinting" figured in Whitman's work. In his study of pigeon phylogeny, Whitman attempted to cross many different bird species. He was remarkably successful in doing so thanks to his exploitation of a particular discovery. He found he could get different species to mate by having the young of one species hatched and reared by adults of the other. His manuscript notes include the following observation:

> If a bird of one species is hatched and reared by a wholly different species, it is very apt when fully grown to prefer to mate with the species under which it has been reared. For example, a male passenger-pigeon that was reared by ring-doves and had remained with that species was ever ready, when fully grown, to mate with any ring-dove, but could never be induced to mate with one of his own species. I kept him away from ring-doves a whole season, in order to see what could be accomplished in the way of getting him mated finally with his own species, but he would never make any advances to the females, and whenever a ring-dove was seen or heard in the yard he was at once attentive.[28]

On the basis of such evidence, one could certainly argue that Whitman knew of "imprinting."[29] What is more interesting is what he actually did with this knowledge. He deployed it not for conceptual purposes but instead for practical ones. He used it as an aid in hybridizing different species. By rearing a pigeon of one species with individuals of another species he could influence the first bird's ultimate mating preferences. One cannot help but note in passing, however, the tragedy reflected in the case Whitman cited. With the passenger pigeon rapidly approaching extinction, a male of the species, imprinted on ringdoves, displayed no interest in mating with his own kind.[30]

The broader stakes of Whitman's pigeon work were manifested in a major address entitled "The Problem of the Origin of Species" that he delivered at the Congress of Arts and Science Universal Exposition at

St. Louis in September, 1904. Sharing the podium with the Dutch biologist Hugo de Vries, Whitman took issue with de Vries's mutation theory and offered instead his own view of orthogenetic evolution, arguing that orthogenesis, properly understood, was neither teleological nor incompatible with Darwinian natural selection. De Vries had based his theory on his studies of variations in the evening primrose, *Oenothera lamarckiana*. Whitman doubted that de Vries's data could be reliably generalized. He offered instead the color patterns of pigeons, and specifically wing bars and their homologues in different pigeon species, as a more promising window on the workings of evolution. In his opinion, studying a specific character in related organisms, over the course of generations, with the help of breeding experiments, was the best way to get at the fundamental issues regarding the origin of species.

Whitman took as his starting point the two distinct color patterns—checkers versus bars—found in the wild rock pigeon (*Columba livia*). Darwin had cited these differences in his work *The Variation of Animals and Plants under Domestication,* but his purpose in doing so had been simply to demonstrate the existence of naturally occurring variations in a species. Whitman found three flaws in Darwin's methodology and thinking: Darwin had not undertaken a comparative study of the color patterns of wild pigeons, he had not attributed any particular importance to the variation in the plumage, and he had gotten the direction of variation wrong. Darwin had supposed that the barred pattern preceded the checkered pattern. Whitman found, to the contrary, that checkers preceded bars. More important, he found support for orthogenesis. "Nature," he said, "has here pursued *one chief direction of color variation.*" It was not necessary to assume, he explained, that checkers had arisen either as chance, slightly useful variations or as mutations. Pointing instead to an ancestral "dark spot" in some of the old-world turtledoves, he proposed that the checkered pattern had been derived "by direct and gradual modification of [this] earlier ancestral mark, which came with the birth of the pigeon phylum, as a heritage from still more distant avian ancestors." [31]

Whitman allowed for the action of natural selection in evolution, but he was confident that of the different processes contributing to organic change, orthogenesis was "the primary and fundamental one." And orthogenesis had additional theoretical advantages. It allowed one to "escape the great difficulty of incipient stages" and to "readily understand why we find so many conditions arising and persisting without any direct help of selection." [32]

Whitman was skeptical about de Vries's mutation theory not only because it was inconsistent with Whitman's own orthogenetic view of the development of characters but also because de Vries had not undertaken a thorough study of his own experimental organism, the evening primrose, in its original home (America). "Persuaded as deeply as I am that we can never draw from a species anything for which no ancestral foundations preëxist," Whitman said, "I anticipate that our wild evening primroses have revelations to make."[33] These words were borne out a decade later by the work of other investigators. Nevertheless, in opposing de Vriesian mutation theory and the increasingly popular Mendelian idea of "unit-characters," Whitman must have looked to many biologists as if he were far from biology's cutting edge.[34]

Whitman was convinced that pigeons were better subjects than primroses or peas for unraveling the secrets of heredity and evolution. His pigeon studies came to dominate his whole existence. For many summers, he took his birds with him from Chicago to Woods Hole and back again. This, however, proved to be worrisome as well as expensive; he always lost some birds in the process. His anxieties on this score multiplied when the railroad companies began refusing to let him take the birds as excess baggage and care for them along the way. According to Frank Lillie, the difficulties transporting the birds to Woods Hole and back were the ultimate reason Whitman in 1908 resigned the directorship of the MBL.[35]

Whitman ended up pouring virtually all of his energies and financial resources into the development of his home pigeon station in Chicago. As his wife later put it, "With the co-operation of his family [he] had let nothing stand in his way." She and he both cashed in their life insurance policies to help pay for the expanded facilities. "Outside friends"—including Mrs. Whitman's wealthy brother, L. L. Nunn (a successful mining and electric power entrepreneur who had already provided financial support for the *Journal of Morphology* and the MBL)—contributed some twenty thousand dollars toward the enterprise. Thanks to this help, Whitman finally had all the facilities he needed for his pigeon work: barns, cages, fountains, and the like. He took to spending very little time at the university. He turned his administrative responsibilities over to Lillie. His graduate students had to seek him out at home.

Whitman planned to write a monograph on the behavior of pigeons after he completed his analysis of the birds' heredity and evolution. Unfortunately, these plans came to naught. On the first of December, 1910, a cold wave struck Chicago. Whitman spent the entire afternoon out of doors putting his beloved birds in their winter quarters. The next day he was found in a coma. He developed pneumonia. On 6 December 1910, he died.[36]

The University of Chicago was not prepared to support either the continuation of Whitman's research or the publication of his manuscript notes. In September 1911 Whitman's former student Oscar Riddle contacted Robert S. Woodward, president of the Carnegie Institution of Washington, to ask whether the Carnegie Institution might be willing to fund the work necessary to keep Whitman's researches from being lost to science. Woodward had little enthusiasm for the idea at first. As he later told the geneticist C. B. Davenport, whose Station for Experimental Evolution at Cold Spring Harbor, New York, had been founded with Carnegie money, he had thought that Whitman "was one of those who potter and who from over-refinement or other reasons do not bring their works to the point of publication." Davenport acknowledged that Whitman had not set the best of examples late in his career: "For 13 years he worked almost without publishing, in a fine disdain of the modern craze for rushing into print, until many of those who knew him best felt some doubt whether he had anything to say; though all had to admit two things, that he had a breadth of view and a thoughtfulness that put him first among biologists and, secondly, that he was everlastingly at his work with the most single-eyed devotion."[37]

Riddle, Davenport, and Albert P. Mathews all lobbied Woodward on the importance of continuing Whitman's work. Mathews, who described Whitman's studies as the most important work done on evolution by an American, explained the existing financial situation to Woodward as follows: "Dr. Whitman left no estate. All his income had gone for years into this work. He left no life insurance. Mrs. Whitman has kept the birds alive and together at great personal sacrifice as she is poor and the University pays her only $1500 a year pension. We cannot permit this splendid work to be lost, but I don't know where to turn for help if you are unable to aid. It is desirable also that the birds be kept alive and together for constant reference during the preparation of the work for publication."[38]

In February 1912 the Carnegie Institution of Washington agreed to support Riddle in preparing Whitman's manuscripts for publication and continuing studies on Whitman's birds. A sum of $2,000 annually was to be provided for the care and maintenance of the pigeons. Riddle was to receive an additional $2,400 as salary. The Carnegie Institution drew up articles of agreement with Mrs. Whitman. These stated that the Carnegie Institution would pay for the maintenance of the pigeons, that Mrs. Whitman would allow Riddle free access to the pigeon facility, and that when the work was done, the pigeons would be turned over to the Carnegie Institution of Washington for transportation to the Laboratory for Experimental Evolution at Cold Spring Harbor.

In the fall of 1913, after experiencing a variety of difficulties with his research and with Mrs. Whitman, Riddle succeeded in having the operation transferred to Cold Spring Harbor. It was not until 1919, however, that Whitman's posthumous works finally appeared in the form of three large volumes published by the Carnegie Institution of Washington. The first volume was entitled *Orthogenesis in Pigeons*. The second was *Inheritance, Fertility and the Dominance of Sex and Color in Hybrids of Wild Species of Pigeons*. The third, which had to be pieced together from very fragmentary notes, was *The Behavior of Pigeons*."[39] Some of the kinds of work embodied in these volumes continued to be carried on by one or two of Whitman's disciples, most notably Oscar Riddle (in his endocrinological studies of pigeon reproduction) and Wallace Craig (in his work on the voices and behavior of pigeons). Whitman's ideas about constraints on variation have received the recent praise of evolutionary biologists.[40] By and large, however, the publication of these posthumous volumes in 1919 was a nonevent for most of the scientific community. The volumes did not inspire other investigators to pursue the problems of heredity, variation, development, evolution, or behavior in the way that Whitman had. By 1919 Whitman's pigeon studies were even farther out of the mainstream of American biology than they had been when Whitman was alive.

WALLACE CRAIG

The case of Wallace Craig is as instructive as that of Whitman. Craig's publications, like Whitman's animal behavior paper, embodied observations and insights that could well have been the building blocks of a new science of animal behavior. Furthermore, unlike Whitman, who took all of biology as his purview, Craig concentrated his attention on behavior. Indeed, Craig later had a direct and significant influence on Konrad Lorenz at a critical stage in the development of the latter's thinking, and Lorenz was happy to acknowledge Craig as "one of my most respected teachers."[41]

Lorenz's borrowing from Craig, however, was more selective than ethologists and historians have generally realized. What Lorenz took from Craig reflected the immediate needs of Lorenz's own intellectual program at the time, not the full range of insights available in Craig's work. It thus does not suffice to consider Craig simply a precursor to Lorenz—any more than it makes sense to think of Lorenz as a successful Craig. In each case, specific local factors served as resources or constraints for the individual scientist seeking to fashion a career for himself.

Craig was born in 1876 in Toronto to a Scottish father and an English mother.[42] At some point in his youth he lived in Scotland, but he went to high school in Hyde Park in Chicago and from there he went on to study at the University of Illinois, where he majored in zoology. One of his teachers at Illinois was Stephen Forbes, the noted ecologist. Another was C. C. Adams, who came to Illinois in 1896 as an instructor when Craig was a junior.

In his senior year at the University of Illinois Craig produced a bachelor's thesis on "the early stages of the development of the urogenital system of the pig." The Chicago meatpacking firm of Armour and Company provided the material resources for his study. At its facilities at the Union Stock Yards in Chicago, Craig was able to collect a sufficient number of embryos from freshly slaughtered sows to allow him to study the earliest stages of development of the pig's nephric system. Broader questions of evolutionary history provided Craig's intellectual justification for the painstaking histological work his study required. He hoped his work would have a bearing "on Haeckel's Law and on the question of the ancestry of the vertebrates."[43]

Craig's results were consistent with the annelid theory of vertebrate descent: the nephric tubules in the earliest stages of mammalian development were comparable to the nephric tubules of amphioxus and the annelids. But this kind of zoological study had no real appeal for him. It represented the only sustained morphological work he ever undertook, and he never referred to it in his later writings. More to his liking were the kinds of research pursued by his teacher Forbes. Like Whitman and numerous other biologists at the turn of the century, Forbes insisted on the importance of studying living nature instead of mere laboratory preparations.

Several years before Craig's completion, in 1898, of his bachelor of science degree, Forbes had established a biological field station at Havana, Illinois. Craig was appointed resident naturalist at the Havana station in the summer of 1898.[44] He continued there in that capacity through the spring of 1899, making systematic collections of fish and plankton at various locations on the Illinois River and adjacent waters. He reported his fish results to Forbes and his plankton results to Charles A. Kofoid. Forbes, in his director's report of the Illinois State Laboratory of Natural History for 1898, wrote enthusiastically about the resources that had been put at Craig's disposal and the aims of the research:

He has been handsomely provided with various kinds of apparatus for the collection of fishes in all the Station situations, including seines of all sorts, fish traps of various size and construction, set nets, and trammel nets. This work is being so conducted as to give us correct ideas not only of the species occurring at the Station, but of their relative abundance and local distribution, their haunts, their habits, their regular migrations and irregular movements, their breeding times and places, their rate of growth, their food, their diseases and their enemies, and, in short, the whole economy of each kind there represented and of the whole assemblage taken together as a community group.[45]

The results of Craig's collecting efforts, unfortunately, fell far short of Forbes's rosy projections. The seines, traps, and nets did not work as well as anticipated, and Craig had difficulty getting the full cooperation of the man who was supposed to help him with mechanical matters. The data that were collected, furthermore, were not always easy to interpret. Craig found that the subject that interested him most, the *lives* of the fish, was not being illuminated by the means at his disposal. In December 1898 he wrote to C. C. Adams stating: "It's a very hard matter to tell much about the lives of fish by just catching them in nets, and not seeing them alive, and yet I'm getting more interested in it all the time." In words that would prove prophetic he also announced: "I have about decided that comparative psychology is the line for me. Whether there will be bread and butter in it or not, I don't know—that's a subordinate question."[46]

In March 1899 Craig wrote to Forbes indicating that he probably would not want to continue his work for a second year. The work involved a great many manual operations, he said, and he was coming to the conclusion that his own manual skills were so deficient that he was not cut out to be a zoologist. He was thinking, he said, of taking up comparative psychology, which interested him at least as much as ecology—"or more so." He elaborated: "The work on fishes, which attracted me here, has, of course been very disappointing in its results. And, even if the new nets prove very efficient, they will surely make a poor substitute for direct observation of the fishes' mode of life. It seems to me it would be better for me to teach school, and use leisure time in observation, than to continue thus working in the dark."[47]

In partial fulfillment of the requirements for his master's degree from the University of Illinois, Craig wrote up the results of his fish studies in the form of a ninety-one-page thesis entitled "On the Fishes of the Illinois River System at Havana, Ill." His introductory comments noted the problems that had beset the collecting operations, the fragmentary nature of the data collected, and "the difficulty of starting an investigation in a new field, with no hypotheses to test, and none to be found in the literature." He expressed the hope that some of his suggestions might prove of use to others who subsequently came to the same field. He himself, however, was ready to move on.[48]

Craig proceeded to teach high school science for two years, first in Harlan, Iowa, and then in Fort Collins, Colorado. In Colorado, in the spring of 1901, he finished writing up his report for Forbes on the fishes of Havana. With the expectation that this work would earn him a master of science degree from the University of Illinois, he began to think about going on for a doctorate. He entertained some thought of going to Harvard on a Chicago Harvard Club scholarship. Nonetheless, he was of the opinion, as he told Adams, "that the best finish I could have to my education, that I know of now, would be working on pigeons under Whitman [at Chicago]." Family matters helped confirm this view. In the fall of 1901 Craig enrolled at the University of Chicago with the intention of studying the voices of the birds in Whitman's collection.[49]

Craig found Whitman to be a welcoming, encouraging, and generous mentor. Soon the younger man was studying pigeons in Whitman's own aviary. He followed Whitman and his pigeons to Woods Hole for the summer of 1903. At the MBL's summer seminar, he reported on his research on vocal expressions in the ringdove (*Turtor risorius*). The ultimate goal of his research, as he described it, was to use the calls and associated movements of pigeons of different species (and of hybrids between the species) to reconstruct the evolutionary history of the pigeon family. "Nearly five hundred species of wild pigeons are known," he explained, "and, so far as observation goes, each species has a perfectly distinct and constant set of notes. These voices have had a common origin, and the problem is to discover this and trace the derivation of homologous elements."

Craig identified three parts to his research program: "(1) A description of the different notes, the attitudes which accompany each, and their whole significance in the life of the bird; (2) [a study of] the development of the voice in the young; (3) a history of the seasonal changes in voice and behavior in the adult bird." Of his species of choice he reported: "The adult ring dove has only three principal calls, but these have a number of modifications, which, together with many expressive movements, afford a considerable variety of expression. All these modes of behavior seem to be strictly inherited. They develop in the young bird very gradually, and in a definite order which is probably also the order of their development in the race." [50]

Craig, quite clearly, had been listening to Whitman's voice as well as the calls of ringdoves. His research program reflected not only Whitman's conviction that phylogenetic studies could be furthered through a consideration of behavioral traits—species-specific call notes, movements, and postures ("attitudes")—but also Whitman's embryological and orthogenetic point of view, with its emphasis on the continuities in both individual and evolutionary development.

After another year's study at Chicago, Craig, without finishing his dissertation, taught for three years, the first at a high school in Coshocton, Ohio, and then two more at the state normal school in Valley City, North Dakota. He spent the summer of 1906 again at Woods Hole. In the fall of 1907 he returned to the University of Chicago to complete his dissertation. He received his Ph.D. in 1908 for a doctoral thesis entitled "The Expression of Emotions in the Pigeon."

In his dissertation he gladly acknowledged the older biologist's guidance. "Professor Whitman," he wrote, "knows the emotions, the voices, and the gestures of the pigeons very much better than I do; he has told me a great many facts about the birds which my more limited experience has not afforded; and he has always given helpful answers to my questions as to what a bird is thinking about when it does a certain act." Craig felt indebted to Whitman not only for factual information and interpretive insights but also "for the influence of his spirit of research." Craig clearly wanted to model his own work on his professor's "enthusiasm and steadiness of labor, sympathetic insight into the animal mind, patience with details, yet a constant reference to general problems." [51]

Craig's dissertation provided a detailed description of the sounds and body movements of a single pigeon species, the blond ringdove. He described—and represented by musical notation—the voice of the ringdove from its time of hatching to its old age, and through its annual cycle and special brood cycle. He planned to follow this study, he said, with a comparison of the sounds and gestures of different species ("showing specific characteristics, homologies, and the possibility of voice and gesture throwing light on problems of phylogeny") and studies of inheritance, variation, selection, sociology, and psychology. [52]

In the same the year that he completed his dissertation, Craig published two substantial papers, one entitled "North Dakota Life: Plant, Animal and Human," the other entitled "The Voices of Pigeons Regarded as a

Means of Social Control." The latter was an important contribution to the study of animal behavior. The former showed Craig's keen eye for the ways life forms are shaped by their environments.

Before he went to teach at the state normal school in Valley City, Craig had lived in a number of settings he considered beautiful. Although he expected North Dakota to suffer in comparison to the hills of Scotland and New England or the forests of Wisconsin, he had no inkling of how monotonous the terrain of North Dakota would seem to him. His first view of his new surroundings gave him "a shock even worse than he was prepared to receive." In time, however, he came to appreciate the North Dakotan landscape and its associated organic forms, aesthetically as well as intellectually.

The primary interest of Craig's paper on North Dakota life resides in his assessment of how the diverse life forms of the area were adapted both behaviorally and structurally to the area's dominant ecological conditions —flatness, aridity, and severe winters. The birds of the plain, the bison, the pocket gophers, the Dakota Indians, the cowboys, and the modern ranchers and farmers had all had to adapt themselves to "the same hard conditions." Craig was familiar with the eighteenth-century English naturalist Gilbert White's observation that "birds that sing as they fly are very few." Craig quickly recognized that White's generalization did not apply in North Dakota. There, found that "every one of the highly volitant songsters of the prairie" did at least some "singing on the wing." He named eight birds that had indeed "developed flight-song to a very high pitch of facility, beauty, and expressiveness." The purpose of singing while soaring, he was confident, was so the male's song would be audible to females from a long way off. What applied to the birds of the plains, he suggested, also applied, broadly speaking, to the mammals and humans. They all displayed special adaptations to long-distance perception and communication. He cited W J McGee's observations on how the Sioux had developed pantomime, gesture, and long-distance signaling codes. In Craig's view, such adaptations to the local environment were displayed not just by lower animals and aborigines. The modern inhabitants of North Dakota, he argued, also showed "a geographic unity, a social solidarity, and a political discipline which they could not possibly show if they were not in a plains country."[53]

Craig never referred to his North Dakota paper in his later writings, but the general ecological insights he derived from his North Dakota experience stayed with him. His paper also made an impression on his ecologist friend Victor Shelford, who applauded Craig's analysis of the way that the flatness, aridity, and severe winters of North Dakota evoked similar behavioral responses in different organisms.[54]

Craig's second publication of 1908, "The Voices of Pigeons Regarded as a Means of Social Control," was based not on field observations but instead on aviary research, specifically on observations he made at Whitman's pigeon facility. The paper appeared in the *American Journal of Sociology*, edited by the Chicago sociologist W. I. Thomas. Its aim was to lay the groundwork for understanding how social influences acting upon the instinctive machinery and limited learning abilities of individual pigeons permit the organization of pigeon society to be remarkably flexible and adaptable.[55]

Craig looked in particular at the function of song in the life of the pigeon. Sexual selection theory, he allowed, could account for some facts of birdsong, but there was much more to the story. The significance of voice in birds was "of a very much wider scope than has ever been suspected. The voice is a means of social control: that is to say, the voice is a means of influencing the behavior of individuals so as to bring them into co-operation, one with another." To understand the social behavior of birds and other lower animals, Craig said, it is not sufficient to regard the individual animal as being endowed with a set of social instincts (any more than it is sufficient to assume that order in human society is a simple function of individuals' having instincts for sociality and good conduct).

Craig did not deny that instincts dominate bird life. His point, rather, was that there are drawbacks to thinking of the individual bird as a distinct entity. It was necessary, he said, "to transcend this individualistic viewpoint, to see that the instincts of the individual can effect their purposes only when they are guided and regulated by influences from other individuals. In a complete explanation of animal society, therefore, the account of the social instincts must be supplemented by an account of the social influences by which the instincts of many individuals are brought into harmonious co-operation."[56]

The "instruments for effecting social control," as Craig explained the situation, include much more than birdsong or even bird voice: "The different utterances of the voice, the varying inflections of each of these utterances, the form and color of the body, the bowing, strutting, bristling of feathers, and all the expressions of emotion, are agencies potent to rouse and direct the activities of other birds." The same is true, he maintained, for "the nest, the eggs, and the young, when they come."[57]

Discussing the social development of young birds, Craig described what he had learned from Whitman and from his own observations of the phenomenon Lorenz would later name "imprinting." Young doves, Craig stated, do not recognize their own kind instinctively. Instead they learn from their parents "the body-form, colors, gestures, and sounds characteristic of the species." As he recounted.

> Professor Whitman has proved this again and again by taking the eggs or young of wild species and giving them to the domestic ring-dove to foster, with the result that the young reared by the ring-doves have ever after associated with ring-doves and tried to mate with them. Passenger pigeons, for example, when reared by ring-doves, refuse to mate with their own species but mate with the species of the foster-parents. Hence we must believe that young doves have no inherited tendency to mate with birds of a particular kind; they learn to associate with a particular kind during the period when they are being fed, when the characteristics of their nursing-parents are vividly impressed upon their young minds.[58]

Throughout his "social control" paper, Craig displayed a keen critical sense informed by an intimate knowledge of the ringdove's behavior. Commenting, for example, on the very strong bond between mated doves, he indicated that this bond was not "blind and inevitable." Doves could form new unions after the death of a partner, or in a subsequent breeding season, or in "other circumstances." To stay united and faithful required not only that the members of the pair remain attracted to each other but also indifferent to the sexual behavior of their neighbors. This was achieved in part, Craig explained, by "a daily and almost hourly communication of affection by means of voice, gesture, and mutual caresses."[59]

Craig took exception to the idea that females have a special instinct of "coyness" which needs to be overcome before copulation can take place. The standard explanation for this coyness, as offered by the German zoologist Karl Groos (among others), was that "the act of copulation should be rendered difficult, in order that it may not be repeated to an injurious extent, and in order that the birds may gradually be made ready for the act." Coyness allegedly served this end. According to this account, the elaborate courtship ceremonies of birds are the means by which the female's coyness is overcome.

Craig dismissed this line of reasoning. He argued against the idea of female coyness and the notion that courtship ceremonies serve to overcome it. In pigeons, he said, where there is always a ceremony before pairing, the females display no special instinct of coyness. Instead, both the female and the male have to be made ready for the mating process. Unlike domestic fowl, said Craig, which copulate at almost any time, pigeons are sexually active only at intervals of four to five weeks during the breeding season. An "elaborate preliminary ceremony" is needed "as a stimulus to bring both birds to the point of sexual activity." Both the female and the male have to be started on "a long, complex cycle of operations" involving copulation, the laying of the eggs by the female, sitting on the eggs by both parents, and feeding the young (again by both parents) once the eggs hatch. The birds have to "go through the whole series of activities in order." The male, for example, will not be ready to sit if he has not gone through the pairing ceremony previously, and he will not be ready to feed the young if he has not sat on the eggs for the preceding fourteen days. What is more, the male and female have to go through these successive changes at the same time. As Craig put it: "Each bird contains in its nervous system, not only a train of explosive material, ready to be touched off, but also an accurate 14-day chronometer; the male chronometer and the female chronometer must be wound up at the same time and set going synchronously, in order that the birds may enter synchronously upon the feeding of the young."[60] Getting the whole process started required the active cooperation of the mate: "Whenever either bird is more ready than the other, the first bird is retarded by the influence of the second, and the second is accelerated by the influence of the first. Thus synchronization is effected by mutual adjustment, not by the adjustment of either bird exclusively."[61]

Craig summed up his analysis of the ringdove's vocal behavior in good, Whitman-like fashion, insisting that birdsong "ought never to be studied (as hitherto it has been studied) without reference to the whole system of vocal and gestural activity." He had found that vocal utterances and/or song play major roles throughout the social behavior of ringdoves. They manifest themselves in the food-begging behavior of young, the recognition by the young of their own kind, the various complex ceremonies involved in the stimulation and coordination of sexual behavior, the selection of a nesting site, the laying of eggs, the maintenance of the pair bond, the defense of territory, and so forth.

Craig did not use the word "ceremonies" casually. He applied it to certain sequences of behavior in the ringdove, but not to all of them. He specifically restricted it to those special "performances" that seemed to be "reserved for more important occasions." "Ceremonies" had special characteristics: "They are highly elaborate, and accompanied by violent gestures or tense attitudes; they occupy a considerable time, often with a certain number of repetitions; and they have a fixed and definite form, which is not sacrificed to meet the petty circumstances of each occasion." He rejected the view of those naturalists who believed that the extravagant songs and ceremonies of birds were simply a means of dissipating surplus energy. Comparing avian ceremonies with social behavior in humans, he observed: "Let [naturalists] remember that similar extravagance appears in the human analog of bird songs—the ceremonies of primitive peoples. Extravagance does not prove that savage ceremonies are useless, no more does it prove that bird songs are useless."[62]

In this paper of 1908 Craig promised to give a more complete account of the forms of social control in pigeons in a later, larger work. He allowed, in fact, that he planned bringing out "within a year or so" an entire book on the subject of the functions of birdsong, as part of a new series of books on animal behavior edited by Robert Yerkes.[63] He soon found himself fully occupied, however, with the duties of a new job. In the fall of 1908 he joined the faculty of the University of Maine as professor of philosophy.

Craig began this new stage of his career with energy and enthusiasm. He needed both, because he was in effect a one-man department of philosophy and psychology, providing the only instruction on these subjects at the university. In his first year he taught four different courses each semester. The next year he taught an additional course each semester. On top of the regular classes he also offered a reading seminar and the opportunity to do individual, directed research. Yet he did not feel particularly overworked—at least at first. Teaching appealed to him, and the university allowed him to develop new courses according to his own interests. He was pleased to feel he was his own master, and he was happy to find that the local people were relatively liberal minded.

In his second year at Maine, Craig introduced a course on evolution, a course on social psychology, and a new "introduction to philosophy." For the university's course catalog he described his evolution course as "An elementary presentation of evolution in all its phases, cosmic, geologic, organic, psychic, and social." His "introduction to philosophy" was a sequel to the evolution course, taking up, among other issues, "the bearing of evolution upon ultimate philosophic problems." As Craig represented the offering in 1913, an introductory course on evolution—a course surveying "theories of biological evolution" plus the evolution of animal behavior, mental evolution, social evolution, and the evolution of man—was the proper introduction and foundation for studies "in psychology, sociology, and allied fields."[64]

At the same time that he was actively developing new courses, Craig was publishing the results of his earlier research. His doctoral study of the expression of the emotions in the ringdove appeared in the *Journal of Com-*

1.2 Wallace Craig as a young professor at the University of Maine. (Photo courtesy of the Special Collections Department, Fogler Library, University of Maine.)

parative *Neurology and Psychology* in 1909. Companion pieces on the expression of the emotions in the mourning dove and the passenger pigeon followed in the *Auk* in 1911.[65] He published an additional paper in 1911 in the *Journal of Morphology* and four more papers in Robert Yerkes' new *Journal of Animal Behavior* between 1912 and 1914.[66]

In these papers Craig, like his mentor Whitman, promoted the comparative study of a few closely related species. Also like Whitman, he demonstrated that to understand any single part of an animal's behavior, one needed to be familiar with the animal's entire life history. Beyond these essentially biological themes, Craig advanced additional themes particularly befitting his new position as a psychologist. As he had already done in his paper on the voices of pigeons as a means of social control, he identified instances where the influence of one bird on another was essentially a psychological rather than a physical or biological one. He discovered with ringdoves, for example, that the female's egg laying did not depend upon actual copulation with the male. The male bird's mere presence, or, alternatively, a stroking of the female by the experimenter, could induce the female to lay. The critical conclusion here, as Craig identified it, was that the process of egg development depended not on the introduction of sperm but rather on a psychological stimulus.[67]

Craig's papers of 1911 on the mourning dove and the passenger pigeon were especially poignant for Craig for two reasons: Whitman had just died, and the passenger pigeon was all but extinct. Craig expressed a faint hope that passenger pigeon specimens might yet be rediscovered in the wild. However, to the best of anyone's knowledge, only a single passenger pigeon remained alive in 1911. That was a female named Martha, sent by Whitman to the Cincinnati zoo in 1902. She would die there in 1914.[68]

Craig based his account of the social behavior of the passenger pigeon on notes he made on the passenger pigeons in Whitman's aviary at Woods Hole in the summer of 1903. Unfortunately, as he put it, that had been "too late to see much of the vanishing birds, yet too early in my own study for me to have a good grasp of the problems."[69] In 1903 there were no successfully breeding pairs among the passenger pigeons in Whitman's collections.

Craig's observations were sufficient, nonetheless, to convince him that the particular poses and other behavioral actions exhibited by the ringdove, the mourning dove, and the passenger pigeon were just as distinctive of these species as were their call notes or their physical appearance. Audubon's representation of the passenger pigeon in *The Birds of America,* Craig allowed, was totally wrong in the attitudes and habits it gave to the birds. The artist had gotten the birds' shapes and colors right, but he had pictured the birds in poses that were wholly uncharacteristic of the species.

Craig detailed the distinctive features of passenger pigeon behavior. The bird had a wing-flapping display in the breeding season that was not to be found in the ringdove or mourning dove. It also had a different kind of nod. The bird was a powerful flyer, but it was awkward on the ground. Among other things, it did not indulge in the strutting or "charging" characteristic of the mourning dove and ringdove. The passenger pigeon's call notes were also louder and harsher than were those of the other two species, and its courtship was likewise distinctive. Craig described the species' courtship as follows:

> There is no bowing as in the Turtles, Ring Doves, etc., no strutting as in the Domestic Pigeon. . . .
>
> When close beside the female, the male Ectopistes had a way all his own of sidling up to her on the perch, pressing hard upon her, sometimes

putting his neck over her neck, 'hugging' her, as professor Whitman expressed it. . . .

When the female becomes amorous, instead of edging away from the male when he sidles up to her, she reciprocates in the "hugging," pressing upon the male in somewhat the same manner that he presses upon her.[70]

These distinctive features of the passenger pigeon's courtship, Craig explained, posed a special problem for pigeon breeders. It was "far more difficult for the breeder to cross the Passenger pigeon with other species than it is to cross many of these other species *inter se.*" He recounted how Whitman had attempted to cross a female passenger pigeon with a male homing pigeon (*Columbia livia*), only to find that when the female reciprocated the male's attentions by sidling up to the male and trying to hug him, "he took this for just so much pugnacity every time, and edged off, with the result that they never mated."[71]

Craig saw the distinctive features of passenger pigeon behavior not simply as arbitrary differences, evolved in the service of species recognition, but instead as actions that were adapted to the particular life of the species. The passenger pigeon's characteristic call notes, for example, he interpreted as "an adaptation to life in a community so populous and hence so noisy that cooing could hardly be heard and the pigeon which could best win a female or warn off an interloper would be the pigeon with the merely loudest voice." As he further explained, "In this ultra-gregarious species, the soft note, the coo, has degenerated; whereas the hard cry, the kah, has been developed and intensified into the loud sounds of kecking, clucking, chattering and scolding."[72]

Craig ultimately concluded that all of the special habits of the passenger pigeon—its voice, its wing-flapping exercises, its awkwardness on the ground, its ceremonial fighting, and its manner of courtship—hung together. They were connected either "directly or indirectly" with the bird's "extreme gregariousness" and its "breeding in vast colonies."[73] Roughly half a century later, Niko Tinbergen explored similar correlations among the species-specific behaviors first of kittiwakes and then of black-headed gulls.

Unfortunately for Craig, the promise of his early publications was not matched by the progress of his career. Though he began his teaching at Maine with enthusiasm, he soon concluded that the students at Maine were "the most unscholarly or antischolarly lot that I ever knew." He was

1.3 Wallace Craig's 1911 rendering, in musical notation, of the "scolding, chattering and clucking of the Passenger Pigeon," which he made "as accurately as [he] could by ear, in Professor Whitman's aviary." By 1914 the passenger pigeon was extinct. By the early 1920s, Craig's hearing had deteriorated to the point that he could no longer study bird vocalizations on his own or interact effectively with students in the classroom.

also unhappy with the limited resources of the library and the lack of any colleagues in psychology. His meager salary compounded his difficulties, especially since the university was unprepared to provide him funds for research.[74]

Like Whitman before him, Craig found that financing a research program based on pigeons required all the resources he could muster. Three years of dealing with an unsympathetic landlord convinced him that if he wanted to work on pigeons without interference, he would have to buy a house and lot of his own. However, once he sank all his savings into a house and lot, borrowed on his life insurance, and secured a mortgage, he could barely make ends meet. He told Adams in May, 1913: "We spend so little on clothes that we look hardly respectable. My research work suffers seriously. It was impossible for me to go to the meeting [of the American Psychological Association] at Cleveland last Christmas. I am too poor to buy the pigeon cages which I ought to have."[75]

Craig's alternatives, as they seemed to him at the time, were to try to secure a better-paying position, to leave comparative psychology to do research in abstract philosophy, or to carry on with comparative psychology by making it pay for itself. He resolved on the last of these. He described his plan for survival in the frankest of terms: "We must keep hens; while I watch their behavior we can eat their eggs, and later we can put the specimens themselves in the pot. I must keep large pigeons as well as doves; we can eat the squabs." Interested in birds that imitated other birds, he concluded it would be best to work with parrots, rather than with the mockingbirds or sparrows he would have preferred, because parrots might later be sold for a profit. "If I thus aim at the financial," he told Adams, "science must suffer somewhat; but I see no other way, for keeping birds and experimenting with them is a mighty expensive business. All scientific work is expensive, but this sort is more expensive than many branches of zoology."[76]

Craig's guarded optimism about making comparative psychology pay for itself did not last long. Seven months later he wrote to Adams indicating that the plan he had entertained no longer appealed to him, because it simply would not allow him to do the research he wanted: "Probably I could maintain a bird farm and make it pay, but it would take every minute of my time, and I should have not time for making observations & experiments, keeping records of them and writing up the results. I could make mass observations just as every pigeon breeder can do; but mass observations are not what are needed in comparative psychology today; what is needed is minute, exhaustiv[e] observation and carefully controlled experiment. I find that the farming business takes all a man's time and brings in just enough money to make it pay, and no more. And I am, after all, more in need of time than I am of money."[77] As for the money, Craig reported: "There is no hope, certainly not for a long time to come, of this university appropriating money for research in animal behavior. It gives no money for research in any field (except the Experiment Station, which is entirely separate)."[78]

Although Craig had decided for financial reasons that he could not make the trip to Cleveland for the American Psychological Association's annual meeting in 1912, he resolved to get to the next APA meeting, scheduled for New Haven in December 1913. His trip there constituted the first time he had been out of the state of Maine for two years and four months. "Hereafter," he later told Adams, "I shall go to the psychological meeting, or some such meeting, every year, for here I am, as a psychologist, absolutely alone. You don't know how paralyzing such isolation is."[79]

At New Haven Craig delivered the single most important paper of his career. He entitled the paper "Appetites and Aversions as Constituents of Instincts." A report of it appeared in 1914 in the *Psychological Bulletin*.[80] The paper was published in final form in 1918 in the *Biological Bulletin* of the Marine Biological Laboratory at Woods Hole. The paper did nothing to advance Craig's own career, but it attracted the favorable attention of two influential psychologists: William McDougall (in his *An Outline of Psychology*—1923 and subsequent editions) and E. C. Tolman (in his 1932 book, *Purposive Behavior in Animals and Men*). Tolman allowed that he borrowed his notion of ultimate drives and how they operated "almost in toto from Craig." A few years after Tolman made use of Craig's paper, Konrad Lorenz drew heavily upon it in developing his own understanding of instincts.[81]

The conceptual highlights of Craig's 1918 paper were his identification of the role of "appetites" in instinctive behavior cycles and his separation of appetitive behavior from the "consummatory actions" by which such cycles are concluded. An "appetite," as he defined it, was "a state of agitation which continues so long as a certain stimulus, which may be called the appeted stimulus, is absent." He further explained, "When the appeted stimulus is at length received it stimulates a consummatory reaction, after which the appetitive behavior ceases and is succeeded by a state of relative rest." An "aversion," in contrast, was "a state of agitation which continues so long as a certain stimulus, referred to as the disturbing stimulus, is present; but which ceases, being replaced by a state of relative rest, when the stimulus has ceased to act on the sense-organs."[82]

At the time he wrote, instincts were commonly viewed as chains of innate reflexes. Craig readily acknowledged that instinctive behavior patterns *included* innate reflex actions, but he insisted there was more to instinctive behavior than that. Most important, appetites, not reflexive reactions, were what set cycles of instinctive behavior into motion. Furthermore, not all the reactions within a cycle of instinctive behavior were innate. Typically, as he saw it, "the reactions of the beginning or middle part of the series are not innate, or not completely innate, but must be learned by trial." It was only "the end action of the series, the consummatory action," that was "always innate."[83]

Craig cited examples of sexual and nest-building behavior in pigeons as evidence that "the bird must *learn* to obtain the adequate stimulus for a complete consummatory reaction, and thus to satisfy its own appetites." This could be demonstrated by experiment: "If a young bird be kept experimentally where it cannot obtain the normal stimulus of a certain consummatory reaction, it may vent that reaction upon an abnormal or inadequate stimulus, and show some satisfaction in doing so; but if the bird be allowed at first, or even later, to obtain the normal stimulus, it will be thereafter very unwilling to accept the abnormal stimulus."

On the basis of his observations on appetitive and aversive behavior in doves, Craig enumerated four phases of the typical instinctive behavior "cycle":

> *Phase I.*—Absence of a certain stimulus. Physiological state of appetite for that stimulus. Restlessness, varied movements, effort, search. Incipient consummatory action.
>
> *Phase II.*—Reception of the appeted stimulus. Consummatory reaction in response to that stimulus. State of satisfaction. No restlessness nor search.
>
> *Phase III.*—Surfeit of the said stimulus, which has now become a disturbing stimulus. State of aversion. Restlessness, trial, effort, directed toward getting rid of the stimulus.
>
> *Phase IV.*—Freedom from the said stimulus. Physiological state of rest. Inactivity of the tendencies which were active in phases I., II., III.

In discussing instinctive behavior cycles, Craig found it natural to talk of "energy" flowing in "channels." For example, he allowed that a bird might show appetitive behavior but never reach phase II of the instinctive cycle "due to fatigue or to drainage of energy into other channels." He similarly spoke of a "discharge" of energy in situations when the consummatory response was elicited by a biologically inappropriate stimulus: "many instinctive appetites are so persistent that if they do not attain the normal appeted stimulus they make connection with some abnormal stimulus . . . ; to this the consummatory reaction takes place, the tension of the appetite is relieved, its energy discharged, and the organism shows satisfaction." Such talk of the "draining" or "discharging" of the energy associated with particular appetites or actions can be found in the writings of other observers of bird behavior—and of Sigmund Freud as well. It would later find expression in Konrad Lorenz's writings of the late 1930s.

Also present in Craig's writing—and likewise destined to be repeated in Lorenz's writings—was the idea that much of human activity has an in-stinctive basis. To Craig it seemed clear that the cycles of instinctive behav-ior found in birds were "fundamentally the same" as those displayed by humans. What is more, he said, what was true of human instinctive behav-ior with respect to such objects as "food, mate, and young" also applied to "the objects of our highest and most sophisticated impulses." Citing the example of a music lover attending a concert, he explained how the music lover experiences first anticipation and then satisfaction, but loss of atten-tion later if the concert is too long, and finally restfulness and relief. The behavior of birds and human beings alike, he said, could be seen as "a vast system of cycles and epicycles, the longest cycle extending through life, the shortest ones being measured in seconds." Craig, the dove expert, closed his paper with the observation that "we, like the birds, are but little able to alter the course of our behavior cycles."[84]

At the University of Maine Craig continued to introduce new courses —anthropology, child psychology (soon renamed by him "genetic psy-chology"), applied psychology, experimental psychology (a lab course), and, among others, a course on the history and philosophy of science. However, he never taught a course on his own specialty, animal psychol-ogy. Nor did he ever teach a course in zoology. Over time, perhaps under pressure to maintain or increase enrollments, his offerings became in-creasingly utilitarian in orientation. As of the 1916–1917 academic year, his introductory psychology course was required of students in home eco-nomics and the professional curriculum for teachers. The following year, he renamed his social psychology course "social and economic psychol-ogy." Among the topics listed for this course were the applications of psy-chology to "daily life, hygiene, education, art, advertising, [and] managing men." It was required of seniors in mechanical engineering. That same year he ceased giving his course on evolution. Only five years earlier he had ad-vertised the course with the claim "The idea of evolution dominates all thought in our era." It no doubt discouraged him to learn that this was not in fact the case at the institution where he was teaching.[85]

In the meantime, Craig's initial plan of publishing a book on birdsong for Robert Yerkes' animal behavior series had come to naught. As early as February of 1909 he had a different book in mind, this one on "the emo-tional expressions of pigeons." That too never saw the light of day, al-though he made some progress on it for a while. In New Haven in 1913 at the meeting of the APA, he told the psychologist Margaret Washburn he had given up the idea of writing a book. She helped him regain confidence in himself, and he was soon telling his friend Adams that he intended to write "a large book on pigeons, with many illustrations and musical notations." He had worried about finding a publisher, he said, and this had made it difficult for him to put words to paper, but he was encouraged by the idea that he might be able pay the publication costs himself in a few years, once he paid off his mortgage.[86]

By 1920 Craig had still not finished his book, and he despaired of ever doing so as long as he remained at Maine. He wrote to C. B. Davenport at Cold Spring Harbor explaining his situation and asking if Davenport could do anything to help him out:

> The results of my study of pigeon behavior, which I have continued in one form or another up to the present time, should be brought together and treated in a monograph. They constitute one whole, the parts all intimately bound together, and they should be treated so. To do this work as it should be done is impossible at the University of Maine. At that institution I have not the time, nor any other conditions suitable for the work. In order to do much in research I must move from the University of Maine to some other university, where research in pure science is encouraged. But before taking up the work of teaching in a new university it would be well if I could devote one year to writing up the research on pigeon behavior. This letter is written to ask you if there is any possibility that within a year or a few years from now I might secure a position on the staff of the Station for Experimental Evolution, the position to be held for one year, and my work to consist solely in finishing and writing up my study of pigeon behavior. The product to be published in monograph form by the Carnegie Institution.[87]

Craig outlined for Davenport the thirteen chapters he planned for the monograph, which was tentatively entitled *Social Behavior and Emotional Expression in the Blond Ring Dove and Other Pigeons*. He indicated he had also drafted six chapters of another book of a "more general and theoretical-psychological and even somewhat philosophical" sort, though he did not imagine that a book as theoretical as this would fall within the Carnegie Institution's purview. Davenport, unfortunately, could encourage neither project. He responded: "On account of depreciated buying power of a dollar the Institution was never so poor and the President's orders are plain; take on nothing new."[88]

Craig remained at Maine for two more years. During this time he published two new papers of theoretical consequence. He devoted the first to the question, Why do animals fight? He focused the second on the limitations of Darwin's analysis of emotional expression in man and animals.[89]

Craig's basic aim in his paper on animal fighting was to contest the way certain theorists had called upon zoological data to justify human fighting. He was not the only American scientist to take on such a project. Vernon Kellogg, David Starr Jordan, William Patten, and Raymond Pearl (who was Craig's colleague at Maine) had already argued that, notwithstanding the opinions of a number of German neo-Darwinist ideologues, Darwinian evolutionary theory did not sanction militarism.[90] Craig wanted to make his argument without pointing a finger at Germany. It was not simply "Nietzscheism or Treitschkeism or Prussianism" he was against—names which in any event, he said, were "invidious and more or less unjust"—but rather militarism itself. "Militarists," he explained, "are at work in every nation, and in every nation they emphasize what they call the 'biological' argument for war."[91]

In addition to leaving out any appeal to nationalist sentiment, Craig's argument was distinctive in the way he focused not on evolutionary theory but instead on the actual facts of animal fighting. He did so, he explained, because the theories of evolution of his day all remained "highly speculative." Allowing that he himself favored "no one proposed solution of the problem of heredity and the method of evolution," and that contemporary theory was in any case inadequate in deciding either for or against the biological argument for war, he proceeded to address the question on empirical grounds.

Craig was willing to identify himself as a pacifist—"a person who longs to see war abolished, and who is willing to labor to the very best of his ability toward that end." He had no patience, however, with the "absurd" argument of some pacifists that "no animal fights its own kind." "The essential truth" of the biological matter at hand, he said, is that "every animal fights its own kind." The question that remained to be answered is, Why do animals fight?[92]

True to his training by Whitman, Craig argued that to understand an animal's fighting behavior, one needed to study this behavior in relation to the animal's "other behavior, to its life history—in short, to its whole economy." It simply would not do to attempt "the old-fashioned method of surveying the entire animal kingdom," collecting bits of incomplete and miscellaneous information along the way. One needed to attack the problem through "an intensive study of one species, or one group of related species." Craig's proposition was clear: "If it can be shown that even one flourishing group of animals has evolved into its present prosperous state without its members engaging in internecine strife, that is enough to prove that warfare is not necessary for evolution."[93]

Craig chose his own specialty, the pigeon family, to make this case. The pigeons, he allowed, were a "properly representative group" insofar as their behavior seemed typical of the majority of birds and the family was spread in a large number of species all over the world. The pigeon family indeed gave every indication of being "in the most flourishing condition and in a state of rapid, progressive evolution." It thus promised to provide a good test for the issue at hand.[94]

With the pigeons as his model, and with the confidence that his knowledge of the behavior of other birds and mammals was sufficient to justify his generalizations from his pigeon studies, Craig set out two main theses:

I. Fundamentally, among animals, fighting is not sought nor valued for its own sake; it is resorted to rather as an unwelcome necessity, a means of defending the agent's interests.
II. Even when an animal does fight, he aims, not to destroy the enemy, but only to get rid of his presence and his interference.[95]

Pigeons, Craig explained, have no special appetite for fighting. A pigeon does not seek the fighting situation, nor does it seek to prolong a fight when engaged in one. Fighting in pigeons was an aversion, "a means of getting rid of an annoying stimulus." He elaborated on this in the clearest possible terms: "The pigeon, unless his temper is aroused, has no appetence for a battle. He has appetence for a great many other objects; as, water, food, mates, nest: if kept without such appeted objects he shows distress, tries to get out of his cage, and in every way makes clear to us that he is seeking the appeted object. But when he is kept without enemies, he never manifests the least appetence for them."[96]

Craig acknowledged fully that, under certain conditions, animals do a great deal of fighting. Pigeons crowded into cages too small for them indeed fight, he said, "to a degree that is cruel and distressing." But this, he insisted, is a function of the artificial situation. It does not happen in nature, where birds manage to space themselves out without fighting. When animals do engage in prolonged fighting, Craig proclaimed, it is because they lack sufficient intelligence to resolve their conflicting interests. "The reason why animals fight," he wrote, "is that they are too stupid to make peace." He recounted how his own intervention in a dovecote had been necessary when two pigeons fought for the possession of the same compartment. After he accustomed one of the pigeons to living in a different compartment, the two were able to live side by side in peace. His observations showed, he said, that pigeons are able to learn from experience how to adjust their differences without fighting.

In an earlier article Craig had described the ceremonial displays that seemed in most cases to settle pigeon quarrels without resort to actual combat. He had also described actual physical clashes between birds. However, as he insisted in his 1921 article, an animal fights not to destroy an opponent but simply to get the opponent out of the way. There was no support in the bird world, he concluded, for "the militarist theory." If that theory were correct, he explained, one would expect to find an animal directing its actions toward "the extinction of the enemy's line of descent," that is, toward the actual destruction of the enemy itself or its eggs or its offspring. This, though, seemed to occur only in the case of certain parasitic species, including some insects and perhaps the European cuckoo. In the latter case, furthermore, the nestmates pushed from the nest were not fellow cuckoos but rather members of the host species.

As Craig saw it, there are basically three things a weaker animal can do when fighting with a stronger one. It can flee, it can submit, or it can continue to fight. The behavior of the stronger animal, in the meantime, is instructive. Typically, when the stronger animal causes its opponent to flee beyond the boundaries of the stronger animal's territory, the stronger animal appears satisfied. In territorial disputes, combat of a ceremonial kind is usually all that is needed.

More surprising, perhaps, is the behavior of the stronger animal when its opponent does not flee but instead submits: "If the enemy submits, the agent ceases fighting. In pigeons this is witnessed again and again. In the heat of battle the agent may rush upon his enemy, jump on his back, peck him with all his might, and pull out his feathers. But if the reagent lies down unresisting, the agent's blows quickly diminish into gentle taps, he jumps off his prostrate foe, walks away, and does not again attack the enemy so long as he is quiet. This behavior is typical, and it proves that the pigeon is devoid of any tendency to destroy his rival." Craig saw here not simply the absence of any impulse to destroy the opponent but instead "a positive impulse to quit fighting a non-resisting bird of his kind." He believed this impulse was related to the mode of sex recognition in birds whereby when a stranger fights, it is treated as a male, and when a stranger refuses to fight, it is treated as a female.[97]

Contrasting starkly with such common and "ceremonial" pigeon fights were the much rarer cases of "brutal physical struggle," where the second bird neither flees nor submits but stays instead to fight. Craig described these as "grim, silent, unrelenting, physical" fights. Yet even these, he said, do not tend to result in the death of a combatant: "At any time when either combatant feels that he has had 'enough,' he needs only to leave the field in possession of the victor; he thereby saves himself from further injury."[98]

Craig concluded his paper by asserting that although "no bird or mammal follows a policy of non-resistance," it remained the case that "aggressive fighting does not pay." Furthermore, there was "no distinctively 'biological' need for fighting." Fighting occurred among humans and animals alike, he insisted, in relation to their interests: "Animals and men fight because they must conserve their interests, and their technique for the adjustment of conflicting interests is too imperfect to adjust all cases of conflict." This, rather than any biological need to fight, was why wars happen.[99] Craig's argument was a cogent one. Interestingly enough, when Konrad Lorenz wrote on the same subject a decade and half later, he ignored Craig's comments completely.

Within a year or so of publishing his analysis of why animals fight, Craig published a critique of Darwin's study of the emotional expression of man and animals. The essence of his critique, which appeared in the *Journal of Abnormal and Social Psychology,* was that Darwin had failed to recognize that the expression of emotion in lower animals serves the same communicative functions for them that it does for humans.[100] Beyond this, Craig proposed that Darwin's theory was too biological and not sufficiently psychological. On the latter score, Craig's comments would appear to have reflected the path of his own academic career. In fact, he had laid the foundations for his position as early as 1908 when he wrote about the voice as a means of social control in pigeons.

In criticizing Darwin's views on emotional expression, Craig had two main theses of his own to offer. The first was "that expressive movements are useful as a means of communication between organisms, a means by which one individual can influence or control another, and they must have evolved as adaptations to this end." The second was "that expressive behavior is of great importance because of its meaning to the agent himself and its effect on his own behavior, and this is one of the most important facts to explain the evolution of expressive behavior."[101]

In underscoring Darwin's resistance to the idea that expressive behavior had evolved in the service of intraspecific communication, Craig was keen to indicate that he was not criticizing Darwin himself so much as the science and assumptions of Darwin's time. Indeed, instead of citing modern scientific evidence to illustrate particular deficiencies in Darwin's understanding of emotional expression, Craig devoted his article to putting Darwin's thought in context. He suggested half a dozen different reasons why Darwin had not connected emotional expression with communication. One was Darwin's need to refute the special creationists (who had insisted, among other things, that God had given man special features to communicate specific emotions). Another was that Darwin, as an Englishman, belonged to a culture where conspicuous gestures tended to be done away with, attention was focused on words and not movements, and the psychology of the time neglected the "margins of consciousness." But Darwin's greatest problem, as Craig saw it, and even more the problem of Darwin's co-workers and disciples (such as August Weismann), was the desire to reach an all-encompassing theory of evolution. The flaw of Darwin's age, Craig believed, was that in working out a theory of evolution and convincing the world of its truth, "it unduly exalted certain biological theories and unduly neglected psychology." The time had come, he indicated, "to reverse that maladjustment and to free psychology from the undue domination of biological speculations."[102]

Craig's own circumstances, however, never allowed him to pursue this goal with any real success. He left Maine in 1922. He had been continually frustrated by his inability to do research there. His immediate reason for leaving, nonetheless, was physical: he was losing his hearing, and this made things increasingly difficult for him in the classroom. His increasing deafness was a tragedy for him not only as a teacher but also as a student of birdsong. He was still only forty-six years old.

Craig taught animal psychology as a visiting lecturer at Harvard in the spring of 1923. At the semester's end, he had little hope of repeating the experience. He wrote Raymond Pearl, "So far as I know now, there will be no opening for me at Harvard for next year. And I am too deaf to secure a position elsewhere. I shall probably have to give up teaching." He wished most of all for some kind of research position as a zoologist, ideally one where he could continue to study animal behavior. He explained to Pearl that he was "willing to begin at the bottom, like a young man just out of college, and learn some specialty. If opportunity offered, I could well accept a half-time position, with half pay, or a position for only a part of each year." He allowed to Robert Yerkes that if he could not get a research job at a university, he would be willing to do something of a more practical nature. He thought he might be able to continue his work on animal behavior by doing something on horses, perhaps for an agricultural experiment station, or for the War Department, or for some commercial concern. But to Pearl he confided he might have to "go into something purely mechanical, e.g., electrical work," leaving animal psychology for his "spare time." [103]

With Yerkes' support, Craig explored the possibility of getting a position with the Animal Husbandry Division of the United States Department of Agriculture. By the end of July 1923, however, he had found a different post entirely, that of "Librarian in the Department of Biophysics of the Cancer Commission of Harvard University," working under Dr. W. T. Bovie of the Harvard Medical School. The job involved "indexing, abstracting and translating literature on biophysics." Craig told Yerkes he enjoyed it very much. The pay, however, was pitifully low. At Maine his annual salary had reached $3,500. At Harvard it was only $1,500.[104]

Craig spent four years (1923–1927) as librarian in Bovie's laboratory at the Harvard Medical School. For two more years he had a job in the Harvard College Library. After that he had no steady employment. He and his wife went to live in Great Britain in August 1935. In June 1937 they moved to Albany, New York, where Craig's friend C. C. Adams, then director of the New York State Museum, helped Craig publish a monograph on the song of the wood pewee. This was a subject Craig had been pursuing for more than a decade, enlisting the cooperation of field observers who could hear what he could not.

For some time Craig had felt that not only his fortunes but also his talents were inadequate for making a major contribution to science. In a letter of 1936 to the psychologist Leonard Carmichael he explained why he had never published a paper he had read to the students in his Harvard seminar in the spring of 1923: "The manuscript which I read to you was written at Maine. When I was at the University of Maine I got an exaggerated idea of the importance of my theories. As soon as I came to Harvard I began to be deflated. I discovered that many of my students, even my undergraduate students, had better minds than my own. Since then I have given my research time to studies that are less ambitious and more exact. I shall keep on studying, and do the best I can." [105]

It was a shame that Craig should feel so little confidence in his own ideas—and particularly at a time when at least a few important thinkers were beginning to appreciate them. As indicated above, his ideas received favorable attention from William McDougall and E. C. Tolman in major psychological texts of the 1920s and 1930s. The linguist and philosopher Grace Andrus de Laguna likewise featured Craig's work in her book *Speech: Its Function and Development* (1927), building in particular on Craig's analysis of the social functions of the calls of pigeons.[106] Most important for the history of animal behavior studies, the young Austrian zoologist Konrad Lorenz also found merit in Craig's ideas.

Thanks to the urgings of the ornithologist Margaret Morse Nice, Lorenz entered into correspondence with Craig early in 1935. Margaret Nice —"Mrs. Nice," as she was commonly known and referred to by a great many ornithologists of the day—played a major role in the history of bird behavior studies through her pioneering field studies of the song sparrow.[107] She was also highly influential in her efforts to make certain that young ornithologists (and older ones too) became familiar with the ornithological literature and the other ornithologists about which and whom they most needed to know. Through her good offices, Lorenz and Craig were corresponding with one another by early 1935. Lorenz at this point was just putting the finishing touches on his major, pathbreaking monograph entitled "Companions as Factors in the Bird's Environment." He had time to insert positive mention of Craig's work in this monograph before it was published, but it was not until his next key paper, "The Establishment of the Instinct Concept," that he made maximum use of Craig's thinking. Of this paper Lorenz later wrote: "It is hardly an exaggeration to say that this paper is about half written by Craig himself because it is really the distillate of the extensive correspondence we had in the years 1935–37."[108] Craig's analysis of appetitive behavior, together with Erich von Holst's physiological studies, helped Lorenz conclude that the active organism is not simply responding to external stimuli through a chain of reflexes but is itself producing centrally coordinated impulses that form the basis of its fixed motor patterns.[109]

Craig's general visibility, however, remained comparatively slight. He moved from Albany to Cambridge, Massachusetts, in 1944. There, with the support of a grant from the American Philosophical Society, he lived an impoverished existence while working on a long manuscript entitled "A Study of the Space System of the Perceiving Self." At the recommendation of C. C. Adams and Leonard Carmichael, the American Philosophical Society renewed Craig's grant twice, temporarily buoying up Craig's spirits. In January 1946 at the colloquium of the Harvard Psychology Department, Craig gave an "introductory statement" of his research results. By the summer of 1947, however, he had not made the progress he had hoped, and the work remained unfinished. The young zoologist Donald Griffin, who would himself become a distinguished figure in the study of animal behavior, lived in the same apartment building as the Craigs from 1943 and 1946. He told Carmichael in 1952 that on his return visits to Cambridge he had found the Craigs "doing the best they could with slender resources." Craig himself, Griffin reported, "was struggling bravely to continue his scietific writing under a considerable handicap of deafness, old age, and poverty."[110]

When the English ethologist W. H. Thorpe lectured on animal behavior at Harvard late in 1951, he expressed appreciation of the important contributions to behavior study made years earlier by the American biologists Whitman, Wheeler, and Craig. Though the audience was a large one, only one or two people in the whole audience seemed to know who Craig was. Thorpe, supposing Craig was dead, was astonished to learn that Craig was not only alive but in the audience. Thorpe paid a visit to the Craigs and was sufficiently distressed about their financial situation to write Griffin about it and to ask Griffin to write to Carmichael.[111]

The Craigs moved from Cambridge to Woods Hole in 1953. Craig died there the following year of pancreatic cancer. He had not been aware of the seriousness of his illness, and his final decline was rapid. Five years earlier he had told his friend Adams: "I believe that if I can get this book finished, it will be an extremely important book. But it has to be written as a whole. The important thing is the way all the different phenomena fit together in one system." At his death the manuscript was still not completed. Whether a copy of it remains somewhere is uncertain.[112]

PROBLEMS OF DISCIPLINARY IDENTIFICATION AND INSTITUTIONAL SUPPORT

Craig's case was unique in a number of respects—his loss of his hearing, for example—but he was by no means the only student of animal behavior to experience problems of disciplinary identification or institutional support. His mentor Whitman, to be sure, had had no problem with disciplinary identification. He clearly identified himself as a biologist. For Craig the problem of disciplinary identification was more complicated. He was trained as a zoologist, and he modeled his research on Whitman's, but Whitman was not the only academic in Craig's environment interested in the evolution of behavior and mind. Craig's first major paper, "The Voices of Pigeons Regarded as a Means of Social Control," was vetted by the sociologist W. I. Thomas and the philosopher G. H. Mead before its publication in the *American Journal of Sociology,* and it is at least plausible that these connections had some influence on his first (and ultimately only) full-time academic post as a philosopher-psychologist at Maine.

If Craig felt any dissonance between his training as a zoologist and the professorial position he obtained, he managed to negotiate this in his researches and the way he interpreted his research results. In his studies of mating and egg-laying behavior in birds, he emphasized the influence of psychological factors on physiological states at least as much as the other way around. Professionally he identified himself a psychologist, joining the American Psychological Association and attending at least one of its national meetings. After leaving Maine the only other academic position he held was as a psychologist—his brief appointment as visiting professor in comparative psychology at Harvard in 1923. However, when he looked for a position after that, his expressed desire was to secure something in zoology.[113]

Other American students of animal behavior also had occasions to ponder where their own professional training and particular interests left them with respect to the existing disciplines of biology and psychology. This was certainly the case with the zoologist Herbert Spencer Jennings, named, ironically enough, after the famous nineteenth-century English philosopher who had not hesitated to write at length on the principles of biology and of psychology alike. Jennings, who distinguished himself in the early years of the century by his work on the behavior of invertebrates, remarked in 1906 to Robert Yerkes: "While I am a sort of a homeless wanderer so far as my subject of investigation is concerned, it is certainly true that the Psychologists come nearer to taking up the matters in which I am interested than do any other of the Societies." Conversely, in 1909 John B. Watson expressed to Yerkes his frustrations with regard to "finding a proper place and scope for psychology." He wrote, "Am I a physiologist? Or am I just a mongrel? I don't know how to get on." Jennings, significantly, did not stay with the subject of behavior. He gave up his early behavioral studies for genetics in 1907. As for Watson, instead of using his animal behavior studies to forge ties between psychology and physiology, he made them his starting point for reforming psychology as a whole. Yerkes himself allowed in 1910 that he had once thought that it did not matter whether psychology was viewed as part of biology or as an independent science, but he had since changed his mind. He had come to believe it was crucial for psychologists to develop their own aims, methods, and abilities. As the historian John O'Donnell has shown, Yerkes' restructuring of his thinking on psychology was intimately related to the insecurity of his position at Harvard. In short, how a student of animal behavior perceived the relations of biology and psychology depended very much upon the different and changing pressures her or she experienced in the course of pursuing a career.[114]

It was no easy matter simultaneously to carry on research; to develop and maintain patrons, disciples, or other allies; to see to the construction of institutional bases to support behavioral studies; and to keep from being viewed as marginal to one's "parent" discipline. Whitman devoted a substantial part of his own career to discipline building and institution building, but in his later years, when he was studying the behavioral (and other) features of pigeons, he retired into the seclusion of his own researches and did little—except for his lecture of 1898—as a spokesman for behavior studies. As for Craig, he was never an institution builder. The psychologist Yerkes, on the other hand, had energy, promotional skills, and administrative abilities commensurate with his own considerable professional aspirations. He recognized early in his career that his prospects of success as an animal psychologist were linked closely to the prospects of providing animal behavior studies with a secure institutional foundation. He made a number of moves to give animal behavior studies increased visibility and support. The most promising of these, early on, was his establishment of a journal devoted specifically to the field. This was the *Journal of Animal Behavior*.[115]

Yerkes appreciated that a journal of animal behavior could survive only if psychologists, biologists, and even "nature lovers" joined together in support of the enterprise. He advertised his new journal as a place to publish "field studies of the habits, instincts, social relations, etc., of animals, as well as laboratory studies of animal behavior or animal psychology." He hoped the journal would "serve to bring into more sympathetic and mutually helpful relations the 'naturalists' and the 'experimentalists' of America." Wanting the journal in addition to be more than just a regional enterprise, he chose the journal's editorial board with an eye to geographical as well as disciplinary breadth.

Over the course of the journal's seven-year existence, it succeeded in attracting biologists and psychologists in almost equal numbers. There was no such balance, however, between lab studies and field studies. In the research of psychologists and biologists alike, laboratory investigations predominated. This was overwhelmingly the case for the psychologists, for whom seventy-four out of seventy-five papers involved laboratory studies. A full two-thirds of these papers emanated from just four university labs (Harvard, Chicago, Johns Hopkins, and Texas), all of which had been established in the previous decade or two. As for the biologists, just under three-quarters of the seventy-two papers they contributed to the journal involved laboratory-type studies, the majority of which were done at universities.

Despite their shared orientation toward laboratory work, however, American biologists and psychologists in fact overlapped very little with respect to the kinds of animals they used and the particular problems they studied. The biologists worked primarily on invertebrates, focusing especially on tropisms and other apparently mechanical responses to stimuli. The psychologists worked almost exclusively on vertebrates, concerning themselves for the most part with problems of sensory discrimination and learning. While 40% of the biologists' papers involved insects, none of the psychologists' papers did. Conversely, while 74% of the psychologists' papers involved mammals, only one (1.4%) of the biologists' papers did.

In studying birds, Craig worked on the class of animals where the interests of the biologists and psychologists intersected most—13% of the biologists' papers and 20% of the psychologists' papers were on birds (Craig's papers have been counted here as biological contributions). Clearly, however, birds did not constitute the preferred class of either discipline. While Craig was studying the interrelations of instinctive and learned elements in the complex behavior of pigeons, most behaviorally oriented biologists were studying the responses of invertebrates to stimuli such as light or galvanic currents, and most comparative psychologists were studying the abilities of mammals to discriminate between stimuli or to learn to run through mazes.

Craig was also out of the mainstream with respect to his institutional setting. At Maine he had no psychological or zoological laboratory. He had to fund his resources with his own meager resources. Unlike Whitman, he did not have a wealthy relative to help support his work. Unlike Yerkes, he was not an aggressive entrepreneur who knew how to further his studies by developing new institutions. Craig's pigeons would presumably have been no more costly to maintain than W. E. Castle's hooded rats, but Castle was studying genetics, not behavior, and the level of support he received was exceptional. His project survived thanks to continuous support from the Carnegie Institution of Washington. Few other investigators enjoyed such assistance.

The Carnegie Institution of Washington, founded in 1902, looked early on as if it might help animal behavior studies as part of its general support for zoology. Its Advisory Committee on Zoology clearly had something on the order of Whitman's biological farm in mind when it called for the establishment of an experimental biological station that would concern itself with "experiments in heredity, in variation, in instinct, in modification, all of which should extend over a series of years and be planned systematically." Indeed, the Carnegie Institution endorsed a proposal by C. B. Davenport that included, as an attachment, a set of paragraphs from Whitman's biological farm paper. These included Whitman's comments on the importance of having a special place for studying "development, growth, life histories, species, habits, instincts, intelligence." Once Davenport's "Station for Experimental Evolution" was established at Cold Spring Harbor, New York, however, it concentrated primarily on Davenport's own interests, which had migrated from morphological and behavioral studies to genetics. Nothing came of the attempts, first by Jennings and then later by Yerkes and Watson, to secure Carnegie funds to establish a behavior station either in conjunction with Davenport's lab or independent of it. Oscar Riddle's work at Cold Spring Harbor in the 1910s used Whitman's pigeons to study the alteration of sex behavior through the injection of sex hormones, but this work was not typical for Davenport's lab.[116]

More important for behavioral studies was the Marine Biological Laboratory established by the Carnegie Institution at Bird Key in the Dry Tortugas, Florida, with Alfred G. Mayer as its director. H. S. Jennings did some early work there, and it was there that John B. Watson conducted the studies on bird behavior that have led some writers to regard him as a pioneer ethologist. What Watson's studies demonstrate most clearly, however, is how contingent animal behavior studies were at a time when institutional support for any kind of research was hard to come by.

Watson's research in the Tortugas on the behavior of noddy and sooty terns was not the result of any long-term commitment on his part to fieldwork. He did not apply to the Tortugas Laboratory. Instead, he was *invited* there. Mayer was eager to justify the new lab's existence by attracting enterprising scientists to it for summer projects. In 1907 he invited Watson to the Tortugas to study the behavior of sea gulls. Watson had tried previously to secure Carnegie funding for research projects of his own devising. In 1903 he applied for support to study the sensory nerve fibers in the spinal roots of man. In 1905 he applied to study "the role which the various sense-organs play in the associations of animals." Both applications were turned down. Failing to receive research funding, and needing to pay back his academic loans, he found himself having to teach summer school.

Watson was thus delighted when Mayer offered him the opportunity to be paid in the summer for doing research. The Carnegie Institution could provide him only with transportation and living expenses for his projected three months in the Tortugas, but Mayer was able to come up with another $55, and the University of Chicago added $150 on top of that.

These funds, plus $40 per month from the Audubon Society for serving as "custodian" of Bird Key, were enough for Watson to be able to make the trip.

Prior to the trip, Watson admitted to Yerkes that he did not know exactly what he was going to do in the Tortugas once he got there. He had read through the *Auk, Bird-Lore,* and other such journals in the hope of finding a model for his research, but he had not found anything he was prepared to call scientific. In a statement that suggests just how low Whitman's profile had become at Chicago by 1907 (though perhaps the narrowness of Watson's view as well), Watson told Yerkes: "Our library is vile on bird literature and the fellows in zoology here don't seem to know a whole lot about the behavior side."[117]

Watson's later work in the Tortugas under Carnegie auspices was supported more generously than his first summer there. He was appointed a research associate of the Department of Marine Biology for three summers (1912–1914) with a stipend of $500 per year. Then, in 1915, he was awarded a $1,400 publication grant for his monograph with Karl S. Lashley entitled "Homing and Related Activities of Birds." This grant and its early support of the work of Jennings were the Carnegie Institution's only outlays for animal behavior research in the first quarter of the twentieth century.[118]

There is no indication that Wallace Craig ever applied directly to the Carnegie Institution for research funding, although, as indicated above, he did in 1920 explore the possibility of getting an appointment at Cold Spring Harbor. As it was, the institution's second president, Woodward, had developed a preference for large enterprises over small, individual operations, and Craig's aspirations were of a very minor scale. In 1914, when he applied to the Elizabeth Thompson Science Fund for a grant of $150, he told the would-be benefactors that "even $50 would be a real aid." Later, in 1923, he applied for a fellowship from the National Research Council but was turned down because the fellowship in question was supposed to be for young scientists just beginning their careers.[119]

CONCLUSION

As prescient as Charles Otis Whitman's animal behavior lecture of 1898 might appear in retrospect, the ideas and practices it represented failed to become an integral part of American biology over the course of the next quarter of a century. The study of animal behavior was not broadly embraced as a vehicle for elucidating phylogenetic relationships, evolutionary mechanisms, or the relations between instinct and mind. Nor did other biological approaches to behavior retain the vigor they displayed at the beginning of the century. The study of animal tropisms lost most of its early enthusiasts as the hopes for a general theory of animal reactions foundered on the actual complexities of animal behavior and on terminological wrangles over how the word "tropism" should be understood.[120]

By the mid-1920s, most of the bright lights of American animal behavior studies from a decade or so earlier had either died, moved off into different fields, or, as in Craig's case, simply faded from view. Whitman, for example, had been dead for more than a decade. So too had the talented amateur naturalist George Peckham (1845–1914), who with his wife, Elizabeth Peckham, had conducted such remarkable studies on the behavior of insects and arthropods. They had demonstrated more convincingly than anyone else had since Darwin's time that the evidence of animal behavior supported Darwin's theories of natural and especially sexual selection. More recently deceased was Charles H. Turner (1867–1923), who had also done fine work on insects. Remaining among the living were William Morton Wheeler, Herbert Spencer Jennings, and Samuel J. Holmes, but none of them was campaigning for the development of behavior studies. Wheeler had continued his studies of ants and his somewhat idiosyncratic interpretation of instincts, but he had given up pushing the word "ethology" as the best designation of what modern biology ought to be about. Jennings and Holmes were devoting their attentions to other subjects. Jennings, as indicated above, had left behavioral studies for genetics. Holmes by the 1910s was devoting most of his energies to eugenics. In contemplating in 1907 a particular experiment on the behavior of bacteria, Jennings had remarked to Yerkes, "amid the great multitude of possible things, one has to choose for doing those that look most promising." What was true of individual experiments applied with equal force to entire programs of research.[121]

Psychologists as well as biologists moved away from animal behavior work to other areas. Jennings allowed to Yerkes in 1916 that Watson "appears to be going into psychiatric work, and more or less into applied psychology in general." He noted further that his colleague S. O. Mast was of the opinion that "there was no demand for work on animals in Psychological departments; that practically every man trained in that line was forced to go into human work . . . and that work on behavior was going to find its home, after all, in the Zoological departments."[122] Mast was wrong about the zoological side of things, at least for the short term, but his perception of what was happening in American psychology departments was accurate enough. By the 1920s, most of Yerkes' students, having found little support for careers in animal psychology, had moved into such fields as education, educational psychology, and vocational psychology. In 1921, of the 424 members of the American Psychological Association, only 26 (6%)—a group including Craig, Holmes, and Wheeler—identified animal behavior studies among their research interests. Three times as many listed education or educational psychology, while four times as many listed "testing." [123]

The *Journal of Animal Behavior* ceased publication in 1917 owing to skyrocketing publication costs and wartime dislocations. The war channeled Yerkes' own interests into more practical activities, most notably devising psychological tests for the army. In 1921 a new journal, the *Journal of Comparative Psychology*, was set up as a "continuation" of the *Journal of Animal Behavior* and the even shorter-lived *Psychobiology* (1917–1920). However, the editor of the new publication, Knight Dunlap, was not a comparative psychologist himself, and he was less interested in theoretical issues than he was in developing better experimental apparatus. Yerkes as late as 1925 was just getting back into academia after having devoted two years to wartime testing and then five more (1919–1924) to the activities of the National Research Council. Watson in this period was forced out of academe for having an extramarital affair and began putting his interests in behavioral control to use as a consultant to the Madison Avenue advertising firm of J. Walter Thompson. The future of animal behavior studies as a part of psychology, it would seem, was no clearer than its future as a part of biology.

Investigators of animal behavior in America in the early twentieth century faced not only the problem of disciplinary identification; they also had the problem of establishing the research settings most appropriate to their work, a problem they often shared with other investigators. The ecologist C. C. Adams, having been essentially unsuccessful in his attempts "to secure facilities for the intensive study of animals in relatively natural conditions," underscored in a letter to Yerkes of 1932 just how important such facilities were for a person's life's work. Congratulating Yerkes on the primate lab Yerkes had finally succeeded in establishing, Adams noted:

It is a strange fate that so many men must spend the best years of their lives before they have a real chance to do what they most desire to do. What would it have meant to you to have had the facilities you now have say, 20 years ago! Ritter had a long struggle to get his marine laboratory, and possibly this came too late for him, as I have heard that he lost interest in the project in later years. Whitman wanted his "biological farm" at Woods Hole, but it has not yet arrived in America, even today. It once looked as if the Desert Laboratory would make such a centre, but it has not developed as I had hoped. Not only do the men migrate but, as well, their interests. It is a severe strain on most men to have a sustained interest that lasts for many years. Of course external pressure has its influence as well.[124]

For a whole complex of reasons—personal, professional, conceptual, methodological, and institutional—the American biologists who took an interest in animal behavior in the first quarter of the twentieth century did not succeed in establishing animal behavior studies as a field that could make special claims on the resources of the biological community. They failed to bequeath to the next generation a conceptually integrated and institutionally grounded program of research. Animal behavior studies became neither an indispensable part of a generalized American biology nor a well-delineated subdiscipline. Nonetheless, in the second quarter of the century a handful of individuals in the States addressed behavioral problems from a variety of angles. The work of Warder Clyde Allee, G. Kingsley Noble, and Karl Lashley, among others, was particularly noteworthy in this regard.[125]

Interestingly enough, Allee, who taught at the University of Chicago, never saw his own work as a continuation of a tradition begun at Chicago by Whitman. Indeed, Allee's physiological approach to the analysis of animal aggregations and animal sociology bore little resemblance to the work of Whitman and Craig, with their focus on the comparative study of instinctive behavior patterns in closely related species. When in 1943 F. R. Lillie read the material Allee had prepared for the historical introduction to Allee's forthcoming, coauthored *Principles of Animal Ecology,* Lillie chided Allee for not having done justice to Whitman's contribution to the study of behavior. Unsure of himself on this score, Allee wrote to Yerkes asking, "How significant was the contribution of Professor Whitman to the work in animal behavior?"[126]

The first American in the 1930s to recognize the importance of the work of the Austrian naturalist Konrad Lorenz was not a professional biologist but the exceptionally capable amateur ornithologist Margaret Morse Nice. She took it upon herself to introduce Lorenz's ethological ideas to English-language readers. Furthermore, as previously indicated, she was the one who put Lorenz and Wallace Craig in touch with each other. It was then Craig who introduced Lorenz to the work of Charles Otis Whitman. This link proved to be of real consequence. On the other hand, its contingent and precarious nature is conspicuous. In the United States in the first third of the twentieth century, the study of animal behavior, despite its promising beginnings, failed to take shape as a robust new discipline. Despite Lorenz's claim that Whitman provided the conceptual "Archimedean point" for comparative behavior studies, the construction of the new discipline of ethology was accomplished not in America but instead in Europe.

TWO

BRITISH FIELD STUDIES OF BEHAVIOR:

SELOUS, HOWARD, KIRKMAN, AND HUXLEY

The real naturalist should be a Boswell, and every creature should be for him, a Dr. Johnson. He should think of nothing but his hero's doings; he should love a beast and hate a gun. That is the naturalist that I believe in, or that I would believe in if ever he appeared on earth. . . . Every man has his ambition. To make a naturalist who shall use neither a gun nor a cabinet, is mine.

EDMUND SELOUS, 1905 [1]

Like yourself, I am very much interested in the question of sexual selection, and it has been one of my objects in the British Bird Book to give all the information available about the sex displays of the various species. But I am sorry to say that so few ornithologists seem to have learnt that these details are best to be found out by getting up early and standing still that the results achieved are exceedingly meagre.

FREDERICK B. KIRKMAN TO H. ELIOT HOWARD, 30 OCTOBER 1910 [2]

The study of animal behavior in the early 1900s in Great Britain followed a different course from the one it took in the United States. There were points of intellectual contact between the two countries, but the settings and practices characteristic of animal behavior study in the two countries were not identical. In the United States, animal behavior studies found their home primarily in the new zoological and psychological laboratories of universities or in such special facilities as Whitman's home pigeon station or Yerkes' primate laboratory. In Britain the most distinctive contributions to animal behavior study at the turn of the century came not from universities or from university scientists (notwithstanding the pioneering efforts of C. Lloyd Morgan), but rather from a handful of amateur field ornithologists. Rejecting the dissector's bench, the morguelike character of natural history museums, and academic zoology in general, these fieldworkers thrived outdoors. Furthermore, unlike most field naturalists before them, they went out into nature not as specimen collectors but rather as animal watchers. Self-consciously intent upon discovering the wonders

of animal life, they promoted the study of animal behavior and they reformed field natural history in ways that were richly innovative and distinctively British.

Three of these individuals—Edmund Selous, Henry Eliot Howard,
and Frederick B. Kirkman—deserve our special attention. These dedicated
observers made bird "habits," "antics," "displays," "scenes," "performances," "ceremonies," "attitudes," and "postures" the focus of their attention. They were not content, however, with simply recording what they
saw. They brought their observations to bear on continuing issues of evolutionary theory—especially Darwinian evolutionary theory and the
much-debated idea of sexual selection. They also sought insights into the
animal mind. None of these three made his living as a scientist, but their
efforts did not go unnoticed by others who would in fact do so. The field
practices of Selous in particular were a special stimulus to the young Julian Huxley, who modeled his own early field studies of bird behavior on
Selous's work.

HISTORICAL ANTECEDENTS

Observations on animal behavior were a featured part of British popular
natural history writing well before the twentieth century. They played an
important role in works of natural theology going back at least as far back
as John Ray's 1691 classic, *The Wisdom of God Manifested in the Works of
Creation.* The "wise Author of Nature," as Reverend Ray explained, had designed not only the structures of animals but also the instincts that went
along with them. It was thanks to God's wise design, Ray wrote, "that poul-
try, partridge and other birds should at the first sight know birds of prey,
and make sign of it by a peculiar note of their voice to their young, who
presently thereupon hide themselves." Ray was confident, however, that
such innate "knowledge" involved no understanding on the birds' part. To
the contrary, as he put it, birds by their instincts are "acted and driven to
bring about Ends which [they] themselves aim not at (so far as we can dis-
cern) but are directed to."[3]

Ray urged his readers to "converse with Nature as well as Books." The
evidence of nature, he argued, testified not simply that God existed but also
that God was powerful, wise, and benevolent. Ray was happy to suppose in
addition that God was pleased with those peoples who had put to good use
the "Wit and Reason" with which He had endowed them. Civilized man,
in Ray's words, had "[adorned] the Earth with beautiful Cities and Castles,

with pleasant Villages and Country Houses, with regular Gardens and Orchards and Plantations of all sorts of Shrubs, and Herbs, and Fruits, for Meat, Medicine or moderate Delight, with shady Woods and Groves, and Walks set with rows of elegant Trees; *with Pastures clothed with Flocks, and Valleys covered over with Corn,* and Meadows burthened with Grass, and whatever else differenceth a civil and well cultivated Region from a barren and desolate Wilderness." In contrast, the "roving" hordes of "Savage and truculent Inhabitants" of Scythia and the "slothful and naked Indians" of America lived under conditions that were scarcely better than that of "brute Beasts." [4]

Ray's confident evaluation of the "planted and adorned" English landscape no doubt struck a chord with his countrymen. Nevertheless, when it came to promoting an intimate familiarity with one's natural setting, Ray did not hold a candle to the Reverend Gilbert White. Indeed White's *Natural History and Antiquities of Selborne,* published in 1789, stands today as the most beloved English field natural history text of all time.

White identified himself as "an *out-door naturalist,*" a man who took his observations "from the subject itself, and not from the writings of others." He held a dim view of armchair natural history—the kind of investigation that could be gotten up "at home in a man's study." Nor did he have much time for classification. He acknowledged that classification had its uses, but he denied that it was the naturalist's principal aim. Far more interesting and important, in his view, was the study of "the life and conversation of animals," the knowledge of which was "not to be attained but by the active and inquisitive, and by those that reside much in the country." Isolated from the great scientific collections of London, Paris, and Uppsala, White proceeded to make a virtue out of his rural circumstances. [5]

Contending that "every kingdom, every province, should have its own monographer," White took upon himself the role of Selborne's natural historian. He learned to recognize the birds of the area by their songs and mannerisms as well as by their physical appearance. As he rode or walked about his business, he carried with him a list of the birds that he expected to see or hear. On the basis of their call notes alone, he identified three distinct species of willow wrens living in his district. Out and about in the field, he was able to observe the habits of animals and at the same time visit with (and keep an eye on) his parishioners. [6]

Later generations of naturalists cherished White's powers of observation, as displayed, for example, in his account of how nightjars catch insects:

On the twelfth of July I had a fair opportunity of contemplating the motions of the caprimulgus, or fern-owl, as it was playing around a large oak that swarmed with *scarabæi solstitiales*, or fern-chafers. The powers of its wing were wonderful, exceeding, if possible, the various evolutions and quick turns of the swallow genus. But the circumstance that pleased me most was, that I saw it distinctly, more than once, put out its short leg while on the wing, and, by a bend of the head, deliver somewhat into its mouth. If it takes any part of its prey with its foot, as I have now the greatest reason to suppose it does these chafers, I no longer wonder at the use of its middle toe, which is curiously furnished with a serrated claw.[7]

The young Charles Darwin was among those inspired by White's *Selborne.* In his autobiography, Darwin recalled how reading White as a youth had led him to take "much pleasure in watching the habits of birds," an activity that left him wondering "why every gentleman did not become an ornithologist." In May of 1843, having moved his family from Great Marlborough Street in smoke-filled London to Down House in the fresh air of the Kent countryside, Darwin began his own "Account of Down." His son, Francis, later identified this as an attempt on his father's part "to write a natural history diary after the manner of Gilbert White." Darwin never completed the project, but his notes for his "Account of Down" carried over four seasons.[8]

Darwin displayed a fascination with animal behavior throughout his career. As a young man aboard HMS *Beagle,* he recognized that his role as a voyager-naturalist involved observing not only the physical aspect but also the *habits* of the different animals and peoples he encountered. He was greatly struck by the unexpected actions of a number of different animals, most notably a ground-feeding woodpecker he observed in Argentina. The appearance and behavior of the natives he witnessed in Tierra del Fuego affected him more profoundly still. Back in England, once the idea of organic mutability had begun to take shape in his mind, behavioral phenomena and queries were prominent in his thinking about "man's place in nature." In the summer of 1838, after visiting the London zoo, he wrote in his "C" notebook: "Let man visit Ourang-outang in domestication, hear expressive whine, see its intelligence when spoken [to], as if it understood every word said—see its affection to those it knows,—see its passion & rage, sulkiness & very extreme of despair; let him look at savage, roasting his parent, naked, artless, not improving, yet improvable and then let him dare to boast of his proud eminence."[9]

Darwin's ideas about behavior enriched his emerging understanding of organic evolution. His evolutionary theorizing in turn affected his understanding of behavior. The special sorts of structures and instincts that John Ray, William Paley, William Kirby, and other natural theologians had seen as evidence of God's wise design Darwin concluded were the result of natural selection. He devoted the entire sixth chapter of his *Origin of Species* to the subject of instinct, concluding the chapter with one of the most powerful statements of the whole book: "To my imagination it is far more satisfactory to look at such instincts as the young cuckoo ejecting its foster-brothers,—ants making slaves,—the larvae of ichneumonidae feeding within the live bodies of caterpillars—not as specially endowed or created instincts, but as small consequences of one general law, leading to the advancement of all organic beings, namely, multiply, vary, let the strongest live and the weakest die." [10]

In the *Origin*, Darwin steered clear of the sensitive subject of human evolution. Twelve years later, however, in his *The Descent of Man, and Selection in Relation to Sex* (1871), he summoned up a wealth of evidence to argue that not only man's physical structure but also his mental and moral faculties had evolved from humbler beginnings. He argued that natural selection had been the primary but not the sole agent in effecting this development. The essential message of his book was that "man with all his noble qualities" displayed his kinship with the rest of the organic world not only in his bodily frame but also in his behavior. He carried this analysis further still in his book *The Expression of the Emotions in Man and Animals* (1872). [11]

Darwin believed that the continuity between the mental faculties of humans and the higher animals, especially the primates, was demonstrated by the instincts that humans and the higher animals had in common. As he explained in the first edition of *The Descent of Man:* "All have the same senses, intuitions and sensations—similar passions, affections, and emotions, even the more complex ones; they feel wonder and curiosity; they possess the same faculties of imitation, attention, memory, imagination, and reason, though in very different degrees. [12] In the second edition of the book he expanded this list, adding "jealousy, suspicion, emulation, gratitude, and magnanimity" to the attributes of the higher animals. Some of these animals, he allowed further, "practice deceit and are revengeful; they are sometimes susceptible to ridicule, and even have a sense of humour." [13]

As the subtitle of *The Descent of Man* indicates, Darwin also elaborated at length in this work on his theory of sexual selection. This process, as Darwin explained it, was "an extremely complex affair, depending as it does on ardour in love, courage and the rivalry of the males, and on the powers of perception, taste, and will of the female." The products of this process were the "secondary sexual characters" of animals and humans. These included "the weapons of offense and the means of defense possessed by the males for fighting with and driving away their rivals." They also included the males' "courage and pugnacity." In addition there were structures designed solely "to allure or excite the female." These consisted of "ornaments of many kinds," "organs for producing vocal or instrumental music," and "glands for emitting odours." [14]

Where *natural* selection, by Darwin's account, made organisms "better fitted to survive in the struggle for existence," *sexual* selection was not so much a matter of survival as it was a matter of securing mates and leaving progeny. It operated through "male combat" and "female choice." It did not manifest itself in the lower classes of animals, Darwin explained, because they lacked the mental power to feel rivalry (which was necessary for male combat) or to appreciate beauty (which was necessary for female choice). Only with the higher invertebrates (the crustaceans, spiders, and insects) did one begin to see sexual selection's results. These results then continued to display themselves, to a greater or lesser degree, all the way up to man. As Darwin saw it, sexual selection was the primary cause of human racial differences.

Most of Darwin's contemporaries had no trouble believing that male combat played a role in the evolutionary process. They tended to be much more skeptical, however, about Darwin's idea of "female choice." Yet on this score Darwin would not budge. He insisted that one had only to consider animal displays—especially those of birds—to recognize that female choice played a key role in developing the secondary sexual characters.

Darwin devoted four chapters (201 pages) of *The Descent of Man* to discussing sexual selection in birds. Employing a rhetorical strategy that Konrad Lorenz and E. O. Wilson would each put to use a century later, Darwin imagined how human behavior would look to a visitor from another planet who could observe only human *actions*. "If an inhabitant of another planet," he wrote, "were to behold a number of young rustics at a fair, courting and quarrelling over a pretty girl, like birds at one of their places of assemblage, he would be able to infer that she had the power of choice only by observing the eagerness of the wooers to please her, and to display their finery." The situation in birds, Darwin maintained, was just the same: "They have acute powers of observation, and they seem to have some taste for the beautiful both in colour and sound. It is certain that the females occasionally exhibit, from unknown causes, the strongest antipathies and preferences for particular males. When the sexes differ in colour or in other ornaments, the males with rare exceptions are the more decorated, either permanently or temporarily during the breeding-season. They sedulously display their various ornaments, exert their voices, and perform strange antics in the presence of the females." [15]

To Darwin it was inconceivable that such displays lacked purpose. It made no sense to him that "all the labour and anxiety exhibited by [the males] in displaying their charms before the females" was essentially to no avail. On the other hand, he did not suppose that the exertion of a "choice" on the part of the female necessarily involved *conscious* deliberation. All he really meant by "choice," he explained, was that the female was "most excited or attracted by the most beautiful, or melodious, or gallant males." The female, apparently, was not quite so active an agent in the evolutionary process as Darwin's expression "female choice" seemed to imply. [16]

Alfred Russel Wallace, the codiscoverer with Darwin of natural selection, balked at Darwin's idea of sexual selection. Wallace maintained that secondary sexual characters were due either to natural selection (which selected for protective coloration in the female) or male vigor (in which case the male's brilliant colors were a manifestation of surplus vital energy). [17] His views, however, did not go unchallenged. Among those who disputed them were the American amateurs George W. and Elizabeth G. Peckham. In two elegant and important papers of 1889–1890 on sexual selection in spiders, the Peckhams showed that the coloration of spiders and the way these colors were displayed in courtship supported Darwin's theory, not Wallace's. [18] This work, however, by no means concluded the matter. At the beginning of the twentieth century, as a new generation of field naturalists ventured out into nature, prominent among their concerns was what one of them tellingly called "the vexed question of sexual selection." [19]

Although the British field naturalists will be the focus of our attention for the rest of this chapter, two historical caveats need to be registered. The first is that they were not Darwin's only intellectual descendants when it came to questions of behavioral evolution. Darwin's ideas were sufficiently rich and fruitful that a variety of thinkers pursuing very different methodologies and problems could claim to be part of a Darwinian heritage. The Continental ethologists Konrad Lorenz and Niko Tinbergen certainly thought of themselves as descendants of Darwin, even if their early work was not as intimately involved with questions of Darwinian theory as was the work of the British field naturalists a generation before them. Likewise, behaviorist psychologists, from whom the ethologists were keen to distinguish themselves, could also claim intellectual lineage from Darwin, as one can gather from the title of Robert Boakes's book *From Darwin to Behaviourism: Psychology and the Minds of Animals.*

The second point is that Darwin was not the only author in the mid-nineteenth century thinking about the evolution of the animal mind. Herbert Spencer also addressed the subject, and it was Spencer rather than Darwin whom Douglas Spalding identified as his intellectual starting point for his own experimental studies on the behavior of young chicks. When Conwy Lloyd Morgan, professor of geology and zoology at University College, Bristol, sought to improve upon the behavioral work of Darwin's disciple George John Romanes, Morgan too decided to do experiments. In or about 1893, he revisited Spalding's work, hatching chicks and ducklings in an incubator and then observing how the young birds' behavior developed in the absence of any contact with adult birds of their species. Through the 1890s Morgan studied the relations between habit and instinct together with problems of animal learning. In the 1930s, when Konrad Lorenz sought to distinguish his own understanding of instinct from the ideas of major instinct theorists before him, he took particular aim at what he called the "Spencer/Lloyd Morgan theory." By this he meant two things: that instinctive behavior patterns could be modified by experience, and that in the course of evolution instinctive behavior graded insensibly into intelligent behavior. In short, not only did Darwin's work inspire a variety of different programs of investigation, but the story of late nineteenth-century studies of instincts, habits, and the evolution of the mind also involved more than the working through of problems set by Darwin himself. This scene has been surveyed elsewhere, and we will not pursue it further here.[20]

Before we turn to the field naturalists of the twentieth century, one additional note is necessary with respect to the broader contexts of nineteenth-century theorizing about animal and human behavior. Although Darwin's *Descent of Man* stressed the continuities between animals and humans, it also had things to say about human differences—differences in the relative reproductive success of the different ranks of society, differences among the human races, and differences between males and females. Like his cousin Francis Galton, Darwin worried that "the reckless, the vicious and otherwise inferior members of society" might increase "at a quicker rate than the better class of man" if various checks did not prevent them from doing so. Darwin also remarked on the prospects of the extinction of "lower" races around the globe, noting: "When civilized nations come into contact with barbarians the struggle is short, except where a deadly climate

gives its aid to the native race." As for the "difference in the mental powers of the two sexes," Darwin allowed that the combined action of natural
selection and sexual selection had resulted in man's being intellectually
superior to woman.[21] In brief, assumptions and anxieties about matters
of class, race, and gender appeared throughout *The Descent of Man,* and
they continued to do so in the writings of a host of other late nineteenth-
century authors who took it upon themselves to reflect upon the relations
between evolution and society. As we shall see, these concerns were not absent from the thinking of the amateur field naturalist Edmund Selous,
while Selous's primary heir among the professional zoologists, Julian Huxley, went much farther and made the question of evolutionary progress
central to his worldview.

EDMUND SELOUS AND BIRD-WATCHING

Edmund Selous described the "shaping and driving forces" of his fieldwork
to be "joy in all wildlife and its surroundings, with another joy in Darwin
and a social-shunning disposition—an intellectual love of truth too—(for
in any other way I mostly hate her)."[22] Sibling rivalry may also have urged
him on in the ways he chose to communicate with nature. He was the
younger brother of Britain's most famous big-game hunter, Frederick C.
Selous.

The older brother killed a great many animals in East Africa, took
President Theodore Roosevelt on safari, and held audiences in rapt attention as he recounted the thrills of the hunt. A caricature of him in 1894
in the *Vanity Fair Album* allowed that his African quarry had included
"elephant, rhinoceros, lion, hippopotamus, giraffe, zebra, quagga, hyaena,
koodoo, hartebeest, duiker, oribi, klipspringer, tsessbe, and antelope of all
kinds; many of which animals are now all but extinct, having been killed
off by railways, by civilisation and by Selous."[23] By one of those ironies of
history, the largest game *preserve* in Africa now bears his name. Edmund
Selous, in contrast, came to be known for his writings against the killing of
wild creatures. In addition, he did more than any other naturalist of his
time to corroborate Darwin's theory of sexual selection, including the controversial idea of female choice. Perhaps sibling rivalry played a role here as
well. A recent biographer of Frederick Selous reports that the Selous brothers both fell in love with the same woman. She, Fanny Maxwell by name,
chose to marry Edmund, but she is alleged to have admitted that Frederick
was the more attractive of the two.[24] Whatever motives may have stirred
Edmund Selous's psyche, he proves to have been just what E. B. Poulton

2.1 The big game hunter Frederick Courteney Selous, portrayed in the *Vanity Fair Album* (1894) as one of the "Men of the Day." The accompanying description allowed that Selous "has killed nearly a thousand head of big game, thirty-three of which are standing witness to his prowess in the Natural History Museum of South Kensington" ("Men of the day: no. 585: Mr. Frederic Courtney [*sic*] Selous," *Vanity Fair Album*, vol. 26 [1894]). (Illustration courtesy of the Rare Book and Special Collections Library, University of Illinois at Urbana-Champaign.)

called for when he discussed the need of more "true naturalists"—that is, "men who would devote much time and the closest study to watching living animals amid their natural surroundings, and who would value a fresh observation more than a beautiful dissection of a rare specimen."[25]

Edmund Selous was born in 1857, two years before the publication of Darwin's *Origin of Species*. He studied law at Pembroke College, Cambridge, and was called to the bar in 1881. Not until he moved with his family to Suffolk in 1898 did he begin to devote himself seriously to watching birds. Thereafter, his primary means of support appears to have been a small, private income and what he earned from writing papers and books on natural history.[26]

His first scientific paper, a study of the nightjar, appeared in 1899 in the *Zoologist*. W. L. Distant, the journal's editor, encouraged contributions on what he called "bionomics"—the study of the activities and mental states of living animals.[27] Distant heralded Selous's paper as a "unique publication." The paper also received favorable attention in the *Saturday Review*. There, in a commentary on a new edition of Gilbert White's *Natural History of Selborne,* the reviewer recalled White's account of the nightjar pursuing fern-chafers around the oak and commented: "Gilbert White's observations of that fascinating bird, the night-jar, are scarcely excelled by anything recorded in the 'Natural History of Selborne.' But we have today, it must be conceded, an observer who has surpassed even White in regard to this bird, Mr. Edmund Selous, who, as his papers on the night-jar in the excellent 'Zoologist' show, is a born field naturalist." [28] Higher praise for an English field naturalist's first paper could not be imagined.

The novelty of Selous's nightjar paper was suggested in the paper's title: "An Observational Diary of the Habits of Night Jars (*Caprimulgus europaeus),* Mostly of a Sitting Pair: Notes Taken at Time and on Spot." The paper began abruptly as follows: "*June 22nd,* 1898.—Crawled up behind a small elder bush some three paces from where a Nightjar had laid her eggs. When nearly there the bird flew down, not on to nest, but close to it. Shortly afterwards the other bird flew down beside it, and immediately I heard a very low and subdued 'churr,' expressive of quiet contentment, I think, and very different from the ordinary loud note of the bird." [29]

Selous went on to report his field observations in uncompromising detail. He described the behavior of the two birds, and he also described the evolution of his own tactics as well. At the end of his first evening of peering from behind an elder bush, he improved his hide by hollowing out a spot behind the bush and adding branches for additional cover. The next day, better concealed this time, and with his pocket watch ready to hand, he recorded what the birds did on a moment to moment basis. For sixteen consecutive days, typically for an hour or two in the evening but occasionally at other times as well, he kept up this surveillance. Eventually the eggs hatched, and he was then able to watch the chicks as well as the adults. His observations ended not by his own choice but rather when the birds left their nesting spot and he could no longer find them.

The emerging taste for this kind of attention to avian life, as distinguished from drier works of classification or identification, was signaled in 1901 in a *Saturday Review* notice of a new edition of J. E. Harting's *A Hand-*

book of British Birds. The anonymous reviewer allowed: "It is no reproach to [Harting's book] . . . to say that many new works of this kind are not likely to be called for by the bird-lovers of the near future." Thanks to the works of Harting and others, the reviewer explained, "the birds have all been admirably arranged and divided up into their proper groups and families: all are neatly ticketed." What remained for the future was to learn something about bird life. It was time to go out into the fields and woods, to discover "exactly how the birds build their nests, exactly why they sing: we want to see them at work and at play, and to know something of their loves and their hates, of their toilette, their travels, their relaxations, their etiquette, their marriage customs." It was time, in other words, to attend to the part of the bird world that was still unknown.[30]

Harting's reviewer did not mention Selous, but Selous became the focus of attention soon thereafter when W. Warde Fowler provided *Saturday Review* readers with an account of Selous's first bird book, *Bird Watching.* Fowler was an Oxford don who was noted not only for his scholarly studies of the political, social, and religious life of ancient Rome but also for a nature-book entitled *A Year with the Birds* (1886).[31] To Fowler, what was especially impressive about Selous's work was what it revealed about the importance of patience when it came to bird-watching. Patience, Fowler said, was the special virtue that, along with other qualifications, had to be brought to bear on "those problems of life and mind which will be the chief work of naturalists when that of collecting and classifying is gradually complete." Selous had devoted hours and days to the patient watching of bird behavior. Wrote Fowler: "Reading Mr. Selous's book I feel that if I were beginning life again, I would give all my spare time to watching as he has watched. He has taken a new departure, and needs to be supplemented and tested. There is a wide field in front of the beginner who will follow in his footsteps."[32]

Selous's *Bird Watching* happened to be published within a few months of his brother's latest book, *Sport and Travel East and West.* The British press at the time was bemoaning the disappearance of big game in Africa but not laying the blame on British hunters. Boers, half-breeds ("Griqua bastards, a race, half Hottentot, half Boer"), and "native gunners" were identified instead as the guilty parties.[33] The *Saturday Review*'s analyst of *Sport and Travel East and West* reassured readers that Frederick Selous was a "true sportsman." Selous had killed a great many elephants and lions in his day, the reviewer acknowledged, but with respect to other game he had "never shot wastefully, never slaying more than sufficient to feed himself and his followers, to secure special trophies and specimens necessary for museums."[34]

Pride in Frederick Selous's accomplishments was likewise something of a family affair, at least if one can take as evidence the fact that one of his sisters, "Miss A. B. Selous," drew ten hunting scenes, under his direction, to embellish his first book, *A Hunter's Wanderings in Africa*.[35] Edmund Selous for his part never criticized directly the hunting exploits of his brother. To the contrary, he confessed that he himself had at one time belonged to "the great, poor army of killers." What reformed him completely, he explained, was the study of bird life: "Now that I have watched birds closely, the killing of them seems to me as something monstrous and horrible; and, for every one that I have shot, or even only shot at and missed, I hate myself with an increasing hatred."[36] Opposing anything that caused needless animal suffering, he launched a scathing attack on the London zoo, castigating the institution as "the bastille of the beasts." The "most unmercifully severe confinement" of the animals there, he maintained, was producing a great deal of "wholly unjustifiable animal misery."[37] In books for children as well as books and articles for adults, he hammered home the message that living creatures should be enjoyed through *watching* them. They should not be tortured or killed.[38]

Selous was not alone in voicing such thoughts. The nature writers William Henry Hudson, Charles Dixon, and Warde Fowler had written eloquently of the satisfactions to be found in watching common animals, and the new Society for the Protection of Birds was committed to halting the wanton destruction of bird life.[39] Hudson in particular found an enthusiastic readership among townspeople who felt themselves increasingly cut off from nature and in need of the spiritual regeneration that a holiday visit to the country could provide. Selous was too testy a writer to be as popular as Hudson was. However, as an animal watcher willing to sit tight in one place and observe all that he could, he had no peer.

Selous dedicated himself to animal watching for more than thirty years. Mostly he watched birds. To be a field naturalist in England, as he once explained, was to be a field ornithologist—there were virtually no wild mammals in England besides rabbits.[40] His watching also encompassed seals in the Shetland Islands, butterflies and hornets in France, and bumblebees in Germany, but birds remained his major interest. He filled the pages of his field diaries with observations on nightjars, great plovers, great crested grebes, peewits, moorhens, arctic skuas, mallards, redshanks, Kentish plovers, oystercatchers, avocets, ruffs, great spotted woodpeckers, blackcocks, sparrow hawks, carrion crows, red-throated divers, shags, herons, swans, rooks, and many more. His periods of serious, day-to-day watching often extended for weeks, and sometimes for months. He watched birds from sites in England, the Shetland Islands, Holland, Sweden, France, Germany, and Iceland.

Selous made it his practice to take his notes on the spot, at least insofar as possible. This was not always easy: "One has . . . often to scribble very fast to keep up with the birds, and so must leave a few things to be added."[41] Back at his lodgings, he would copy out his notes and elaborate upon them. Later he might add something else if it remained fresh in his memory. He prided himself on recording *all* that he saw. As he explained in publishing his diary on the breeding of the ruff: "should anyone think that I had better have left out certain things which I saw, I can only say, frankly, that I am not of that opinion, and that it is not my habit to do so."[42]

Nature writers like Hudson were disposed to emphasize the joys of a weekend's rambling in the field. "Sportsmen" in turn wrote of "the absolute bliss" of training their guns on a lion in Somaliland or, closer to home, of (as an apparent skeptic put it) "the pleasures of lying half frozen in two or three inches of icy water at the bottom of a gunning punt" waiting to blast away at a flock of ducks.[43] Selous too experienced moments of great exhilaration out in the natural world, but he never pulled any punches about the discomforts and difficulties of fieldwork. When he wrote about watching the redshank, for example, he offered a whole litany of the "almost unsurmountable" difficulties that stood in the way of the field naturalist's efforts to make sense of behavior. First there was the bad weather that typically prevailed in the mating season. Then there was the problem of distinguishing one redshank from another. This made it difficult not only to tell different pairs from each other from one day to the next, but also to keep different pairs distinct from each other on the very same day, and even to identify which sex was which in a single pair. Then too there was "the very different spirit shown by the same bird" at different times. There were no markings to aid in this work, Selous complained, and none of the comforts of working at home in one's study: "All, or at least the greater part . . . is wretchedness, cold, and discomfort; such, upon close acquaintanceship, are the charms of early spring in north temperate Europe."[44]

To observe the mating habits of the ruff in Holland in 1906, Selous painstakingly dug for himself a kind of "turf-hut" from which he could view the birds' mating place, or *lek*. He made his observations, he recounted, "under such conditions of cold, wind, water, and sand in my eyes, that I am glad now to think it is over. The first and second of these draw-backs (to which I now add cramp) were almost constant, the third came in after rain, when the excavation in which I sat had to be bailed out, . . . and the last asserted itself whenever the wind was from the east, and blew straight through the opening. . . . For the cold, it was most severe precisely at the times when most was to be seen." His diary for 17 April 1906 indicates that he was up that day at three in the morning and at his observing site be-fore four. By eight he was "almost frozen to death," despite the covering of "a thick motor suit, warm underclothing, woolen face-protector, sheep-skin gloves, two Scotch plaids, and a Shetland-shawl comforter."[45]

Two springs later in England, studying the blackcock, Selous's practice was to leave his lodgings between midnight and 1:00 a.m. in order to cover seven miles of difficult terrain and arrive sometime after 3:00 a.m. at his hide. He had a bicycle, but he still had to walk most of the way, usually through wind and rain. One morning in the hide he could bear the cold no longer and had to get up and run about, regardless of whether this fright-ened more females from coming to the lek. The next morning he was kept home by a violent storm. As he explained: "I had, indeed, often emerged upon the dark moor-top, cycle in one hand and umbrella in the other, but there is a limit to everything."[46]

There were, however, compensations for all of these discomforts. Selous was able to observe the intimate details of bird behavior up close. He paid particular attention to how the birds' displays developed or mani-fested themselves over time and how they functioned in the birds' lives. He described in detail all that he saw the birds do, but he was not content sim-ply with reporting facts. He wanted to understand how specific behavior patterns had evolved over time. He was furthermore keen to relate his ob-servations to contemporary theories, especially Darwin's theories of natu-ral and sexual selection.

Among Selous's earliest attempts at theorizing were his speculations on the origin of birds' nests. He had watched the frenzied motions of pee-wits under "the influence of strong sexual excitement." He supposed that the male bird's whirling motions upon the ground, when localized, consti-tuted a likely place for pairing to take place, and that when such motions actually produced a small depression on the ground, and eggs were laid there, then this was in effect a nest, albeit of the simplest sort. The nest was not the result of any deliberation on the birds' part. It was due instead "to nervous and non-purposive movements springing out of the violence of sexual passion." Even the next stage of nest evolution, in which the de-pression was lined with grass, moss, sticks, and other such material, was not necessarily the result of deliberate action—at least not *initially*. A bird "ecstatically rolling on the ground," Selous explained, could be imagined to peck at small sticks and grasses in its vicinity and thus produce a collection of these things. If such a habit were heritable and furthermore beneficial, it could be increased and modified through evolution.

Selous recognized that his explanation of the evolution of bird nests was highly speculative. He was confident, nonetheless, that he was on the right track when it came to thinking about the evolution of bird behavior in general. He believed that "the key to the unlocking of many of the wonder-chambers of bird doings" was to be found in "the highly nervous and excitable organization which birds, as a class, possess, and, especially, in the extraordinary development of this during the breeding and rearing time." The "nervous sexual or parental excitation" generated in these seasons, he supposed, gave rise to "all sorts of extravagant motions and antics." Although these were "at first quite useless," they constituted the "raw material," on which natural and sexual selection acted, and they were thereby channeled in such useful directions as "nest-building, ruses to decoy enemies from the young, displays of plumage by one sex to the other, and so forth." Insofar as intelligence in such cases would be beneficial, it would be "gradually worked and woven into" these actions, resulting in special degrees of intelligence in particular directions. Still and all, Selous concluded, "many actions of birds which seem now to be altogether intelligent and purposive (and, no doubt, are so to a very large extent) will be found to betray traces of a nervous and non-purposive origin."[47]

Here was a fine example of Selous's confidence in the creative powers of natural and sexual selection. He was, as Julian Huxley later wrote of him, "always . . . alert to see in small divergencies of behaviour one of the means whereby a species may become altered or may split into two."[48] Another example is to be found in Selous's 1905 book, *The Bird Watcher in the Shetlands*, where he wrote:

> I remember once passing unusually close to a cock pheasant, which remained crouching all the while, though nineteen out of twenty birds would, I feel sure, have gone up. It struck me, then, that as all such pheasants as acted in this way would have a greater chance of not being shot than the others that rose more easily, whilst these latter were constantly being killed off, therefore, in course of time, the habit of crouching close ought to become more and more developed, and pheasants, in consequence, more and more difficult to shoot. Some time afterwards I met with some independent evidence that this was the case, for a gentleman

who shot much in Norfolk, remarked, without any previous conversation on the subject, that the pheasants there had taken to refusing to rise,
and that this unsportsmanlike conduct on their part was giving great
trouble and causing general dissatisfaction. That was his statement. He
spoke of it as something that had lately become more noticeable, but
only, as far as his knowledge went, in Norfolk, which, I believe, is an extremely murderous country.[49]

Such passages were typical of Selous's writing. Here he offered simultaneously a shrewd observation on evolution in action and a barbed commentary on the English propensity to murder rather than conserve local
wildlife. But this was not quite Selous at his best. His chance observation
on this one "unsportsmanlike" pheasant did not hold a candle to his long-
term, hard-won observations on sexual selection.

Selous wrote in favor of sexual selection as early as 1901, when he discussed the plumage of the arctic skua and the coloration of the mouth cav-
ities of certain other sea birds.[50] In 1905, in his book *The Bird Watcher in the
Shetlands,* he devoted an entire chapter to sexual selection and more especially to what he termed "intersexual selection." He called upon the latter
to account for instances where the males and females of a species were similar in color and structure and both sexes used these characters the same
way in their nuptial displays.[51] Then, in the spring of 1906, he traveled to
Holland with the specific intent of studying sexual selection in action.
There he watched the display behavior of several different species, in particular the redshank and the ruff. The following spring in Sweden and the
spring after that in England he focused on the blackcock.

Selous's field observations on bird courtship left him in full agreement
with Darwin on the significance of behavioral displays. It seemed wholly
improbable to him, as it had to Darwin, that the elaborate displays performed by male birds in the presence of females were without function.
Watching the male redshank display to the female his white tail, the silver-
grey underside of his wings, and his bright red legs, Selous allowed that he
could think of no way of explaining why the male should go through all
these motions if the female could not be moved by them, and why the female acted as if she were moved by them, if she really was not. He enumerated a host of facts that testified to the reality of "female choice":

(1) conscious and elaborate display of the males of some species before
the females; (2) care taken to show and impress upon the female all that
is best worth seeing; (3) interest—greater or less at different times—

taken by the female in such display, and corresponding effect—sufficient or insufficient—produced upon her by it; (4) repetition of the display when at first insufficient, with results that justify such perseverance and make it intelligible; (5) interest taken by the female—more or less and at different times—in the fighting of males on her account; and (6) her participation, at times, in these encounters, readiness to attack, and competence to drive away one or other of the contending males, or to keep other males from joining in such contention.[52]

The ruffs provided Selous with further, dramatic evidence of male-to-male combat and the females' exercise of choice. From 6 April to 3 May 1906 he watched these birds at their breeding ground. He observed the fighting among the ruffs, describing the vigor of their battles as follows: "The birds literally hurled themselves at each other, biting and kicking with the greatest imaginable fury, and striking showers of blows with their wings, the noise of which was like so many little thunder-claps. They spring very high, and when at an unusual distance from one another, one endeavours to get above the other, and to kick down upon him."[53] He also watched the ruffs' courting of the reeves, and the reeves' selection of particular ruffs.

After days of careful observation, Selous concluded that each ruff had a place of its own at the lek and that the most violent fights occurred whenever a ruff, upon flying in to the lek, lit in the place occupied by another. He found further that the birds that engaged in the most fighting were not necessarily those that were the most successful in mating. One morning, while "furious combats" among three to six birds were occurring at one end of the mating ground, two other ruffs, who fought relatively little, were responsible for the large majority of actual matings. These birds were distinctively colored: one was brown, the other was blue-black. They seemed to owe their success not to fighting or to superior vigor but rather to their appearance. Surveying the activity at the lek, and considering the implications of all that he was observing, Selous reflected disdainfully on the professionals who opposed Darwin's idea of sexual selection: "A wonderful drama, truly, of bird-life, thus unfolding itself before me, in the early, bright, but bitterly cold morning, whilst learned ornithologists all over Europe lie sleeping in their pleasant beds! But they will come down all the fresher to breakfast, and perhaps issue bulls against sexual selection from their studies."[54]

On another day Selous summarized the action at the lek as follows: "There is a considerable amount of activity, nine or ten Reeves being sometimes on the ground together, the Ruffs varying in numbers from about that to some fifteen. Of these only four are paid any attention to by the Reeves, *viz.* the brown, the blue, the black, and the brown-ruffed and black or chocolate-headed bird. . . . The selection on the part of the Reeves is most evident. They take the initiative throughout, and are the true masters of the situation. Quiet and unobtrusive as they are, compared to the Ruffs, their whole manner betrays conscious power."[55]

Selous expressed confidence that Darwin would be "triumphantly and often most strikingly vindicated" when "evidence from the field" took the place of "denial from the chair."[56] Yet he was willing to admit that Darwin could make a mistake. Studying the breeding of the blackcock in 1907 and 1908, Selous found that the famous "war-dance" of the blackcock, which Darwin had cited as an example of a courtship display evolved through sexual selection, was not in fact used in courting. Darwin, Selous explained, had had to depend upon other naturalists' descriptions of the bird's frenzied war dance, and he had naturally supposed that this behavior was addressed to the hen. Selous discovered to the contrary that the actual courtship display was a very sober affair, totally different from the "highly extraordinarily spectacle" of the birds' leaping or "dancing." The war dance, as far as Selous could tell, had "nothing to do with courtship proper" and "no special significance for the female bird." It involved males only. But this in no way invalidated Darwin's idea of "female choice." What that required was that the female bird not be impressed equally by different males but instead showed preferences among them. This, Selous said, was precisely what his own observations confirmed.[57]

Selous's observations on the blackcock's war dance led him to speculate on the evolution of the behavior. He suspected that the dance arose out of two conflicting tendencies in the male bird: bellicosity on the one hand and timidity on the other. He explained: "Animals, in their psychology, are like pictures which resemble us in outline, but want the shading. They have our grossnesses, so to speak, but not our refinements. Thus, a bird might be afraid of another bird, but it would not be ashamed of being so, and so would do nothing on the principle of saving its face, or trying to disguise its own feelings from itself. Still the wishing to fight, and not daring to, would certainly produce mental discomfort, for which some relief must be found." The blackcock's dance was thus not a display for the female. Nor was it simply a challenge to other males, a means of "getting up" courage, or "the mere safety-valve of sexual excitement." Though all these elements might play a part, Selous said, he was inclined to look at the dance as a "substitute for actual battle."[58]

Selous found the true courtship of the blackcock, as opposed to the war dance, to be "a serious, methodical and business-like matter, having for its object the exhibition by the male bird of his plumous and other adornments to the best advantage before the female (which mere 'dancing' or leaping does not effect)." He noted considerable variation in the courtship styles of different males and the responses given to them by different hens. Of three different males he watched one day, one seemed "to go beyond the average of pomp and solemnity in his wooing." In this male, "the wide extent of his circlings" seemed particularly "well adapted for a full hind, as well as front and side, view." The second male was "altogether quicker and brisker, nor were his circumambulations anything like so wide or full—in fact, not a marked feature—his appearance, in consequence, being not nearly so impressive." The third male simply "threw off form altogether, and resorted to force."[59]

As for the behavior of the hen birds, Selous confirmed that they were not indifferent to the males, and that indeed they generally exercised choice:

> My observations . . . show that the female Blackcock is affected by the courting display of the male—sometimes so strongly that one may correctly describe her as fascinated—that she does yield to it, and not to force or martial prowess, and that she exercises choice in regard to the various males. They show also that, whilst being courted, she is extremely jealous of any other hen that may approach, and will pursue and fight with such, fiercely. Yet, at the same time, she is often extremely hard to win, and will resist the charm of the cock's allurements, though exhibiting every sign of being strongly impressed, and indeed fascinated by them. Why this should be so I do not know, but the psychology revealed seems more delicate and less simple, nearer to humanity, or more human-seeming, than, even though accepting the doctrine of sexual selection, one might have anticipated. The hens, also, come to the place of meeting with the evident object of being courted, and for that reason only. When the courtship has been brought to a conclusion, either to their satisfaction or otherwise, or should they tire of it, they fly away.[60]

Having watched the breeding of the blackcock for many weeks, over two successive breeding seasons, Selous was confident he had studied the subject more thoroughly than anyone else. He was also confident in his conclusion. "There is simply no place for the display of the male Blackcock before the female," he insisted, "if it be not for the purpose of winning her, and my notes show it does win her."[61]

Selous's observations and ideas on sexual selection were unparalleled in his day. Their impact on his fellow ornithologists and zoologists, however, was limited. This may have been in part because his contemporaries did not care to hear about females having the upper hand in mate choice. Yet some of the fault was certainly Selous's own. He was a cantankerous character who displayed little interest in endearing himself to contemporary authorities. Instead of seeking allies among them he characterized professional zoologists as murderers—"thanatologists" whose museums were "mausoleums" and whose outmoded theories bore no relation to what actually occurred in nature.

Selous's inability to work for long with other ornithologists manifested itself in his dealings with F. B. Kirkman, editor of the multivolume *British Bird Book*. What distinguished this work from previous ornithological publications was its self-conscious focus on bird *habits*. Selous was originally slated to be a major contributor to the four-volume enterprise, but his participation ended with volume 1.[62] When Selous laced his offerings with damning comments about contemporary ornithology, Kirkman tried to restrain him. This led the two of them to quarrel bitterly. Furious with the thought of an editor "hacking" and "mangling" his writing, Selous told Kirkman in 1910: "I highly resent this constant endeavour to constrain my opinions & make them square with your own, which I do not share & which are nothing to me. It is persecution, & I will resist it as such." Selous insisted that Kirkman had known of his opinions in advance, and had indeed urged him to express them.[63]

Kirkman for his part told Selous how he disapproved of "the whole spirit in which you approach the work of collaboration." In the initial volume of *The British Bird Book,* wherever Selous launched an attack on armchair naturalists, "scientific collectors," or other of his favorite targets, Kirkman appended an editorial footnote stated that these were Selous's personal sentiments and did not represent the opinion of anyone else associated with the publication.[64] Eventually Kirkman decided that Selous's relations with the project had to be terminated. As he told Eliot Howard in 1910: "Selous has been taken off the British Bird Book. He has been a terrible nuisance to us."[65]

Selous paid close attention to the appearance of the subsequent volumes of the work, expressing approval when his observations were cited and dismay when they were not. He was happy to see his theory of the origin of birds' social antics and more elaborate courtship displays tentatively endorsed; he was angry when his theory of the origin of nest building was neglected. What galled him most was when *The British Bird Book*'s treatment of the "grouse subfamily" contained no mention of the facts he had recorded on the blackcock's mating behavior.[66] With respect to this last indignity, he complained in the *Naturalist:* "This silence, from whatever cause proceeding, is not in the interests of truth, and moreover does great wrong to Darwin, whose brilliant and most original theory of sexual selection my observations most strongly confirm." He rehashed all he had done: "I have never had better opportunities for observation. . . . Everything I saw I noted, and also everything I did not see which has been stated or implied for a good many years. . . . I saw everything in full swing, and from the earliest moment, always being on the spot before dawn, which sometimes necessitated starting at midnight. I did everything, within my power, to further scientific truth, and have indeed produced immensely strong evidence in favour of the Darwinian theory of sexual selection. It would seem, however, that since the theory itself is (officially) out of favour, such evidence is not wanted."[67]

Selous once wrote: "I am as much of a hermit as I am mercifully permitted to be."[68] Ultimately, his feelings of alienation simultaneously invigorated and constrained the contributions he was able to make to field natural history. On the positive side, they made it easy for him to take satisfaction in the solitude that his field studies demanded. They also enabled him to make some uncommonly penetrating observations about the way that scientific thought was permeated by prevailing social assumptions regarding male dominance, female passivity, and the superiority of Western civilization over other cultures.

Selous did not simply note that assumptions about the relations of the sexes in human courtship were often transferred to discussions of the relations of the sexes in animal mating as well. He also remarked that the common understanding of human courtship was itself probably incorrect:

> We have the prevailing idea that (even in a civilised state of things) it is man who woos and woman who is won; man who advances and woman who retires; man who seeks and woman who shuns. The reason probably is that the actions of man are of a more downright nature, and easier to observe and follow, than those of woman—who, as a clever writer has remarked, approaches her object obliquely—and, secondly, that it is man mostly, and not woman, who has given his opinion on this and other mat-

ters through the most authoritative channels—for it is man who, by virtue of his intellect and his selfishness, holds the chief places of authority." [69]

As for the assumption that civilized people were happier than uncivilized ones, Selous rejected this idea outright. He had traveled to Africa, and if he had not been especially impressed by the intelligence of the natives, he had been struck by what he took to be their obvious happiness. He rejected Lord Avebury's confident assertion that witchcraft is the "dark cloud that hangs for ever over savage life." And even if that were true, Selous said, "Have we no dark clouds, and have we less or more capacity for feeling them? What is an engagement to dine then, or an enforced call? and consider the dark cloud of having to go every year, *en famille,* to the seaside, that hangs over the civilised married wretch!" [70]

Selous's sense of alienation helped him recognize—to a degree that most others did not—the ways that social assumptions were embedded in scientific thought. This same sense of alienation, however, also prevented him from playing a larger role in the reformation of twentieth-century field biology. His interactions with other naturalists were minimal. He professed to make it a habit not to read their writings on the grounds that he did not want to pick up preconceptions that might creep into his own conclusions. There were some exceptions, to be sure. He praised the Peckhams' work on sexual selection. Occasionally he also mentioned approvingly the work of such British contemporaries as W. H. Hudson, Warde Fowler, and Eliot Howard. And Julian Huxley helped him publish in 1927 a collection of field observations under the title *Realities of Bird Life.* Selous ultimately made Huxley his literary executor, but no additional publications emerged from this relationship. For the most part Selous mentioned contemporary zoologists only to condemn them for their fondness for guns, cages, and taxidermy.

As it was, Selous remained uncompromising in his writing style. He refused to alter his prose to fit contemporary scientific conventions or, truth be told, to present his arguments in a more coherent and economical fashion. He frequently encumbered his observations with long digressions or set them forth in such an undigested form as to work against his analyses of the issues at hand. His last books, *Thought-Transference (or What?) in Birds* (1931) and *Evolution of Habit in Birds* (1933), appear to have had no influence on the biological community, with but one exception. Alister Hardy (of whom we will hear more later) was attracted to Selous's notion that some kind of extrasensory perception or collective thought was responsible for those moments when all the members of a huge bird flock moved suddenly and synchronously together.[71]

Selous belonged to no scientific societies. When he died in 1934, his death went unnoticed in British natural history journals. Two appreciative obituaries appeared elsewhere, one in the United States by Margaret Morse Nice and the other in France by Jacques Delamain.[72] Nice hailed Selous's promotion of field studies, but she also identified weaknesses in his work that could have been corrected had he been willing to profit from the ideas and practices of others: "If he had ringed birds he could have been sure of many things that, as it was, remained in the realm of conjecture (for instance, the matter of mating for life, which he believed to be the general rule). If he had read the best of contemporary life-history work, he might have found enough to study in the present-day meaning of the habits of birds, rather than speculating on their evolutionary history."[73]

Nonetheless, Selous's pioneering researches served as an important model for other field studies of behavior. Both Julian Huxley in England and Jan Verwey in Holland were inspired by Selous's work. Early in his efforts as a field biologist Selous argued that "the habits of animals are really as scientific as their anatomies, and professors of them, when once made, would be as good as their brothers."[74] He himself died before this came to pass, but his work helped prepare the way for the realization of this vision.

HENRY ELIOT HOWARD AND THE BRITISH WARBLERS

A second English amateur whose studies did much to demonstrate the scientific importance of fieldwork was Henry Eliot Howard (1873–1940). Howard earned his living not as a zoologist but as the director of a major Worcester manufacturing concern, the steelworks of Stewart and Lloyd. His duties kept him close to home, but he had no need of going farther to make his most important discoveries concerning bird life. In the wilder reaches of the grounds of his estate, Clareland, was a large pond, surrounded by reeds, rushes, and bushes. Nearby was a broad expanse of moorland and marsh. Here were ample sites for birds to breed and Howard to watch them. He took particular interest in the warblers. From the time the birds began arriving in mid-March until the end of the breeding season in the latter part of June, Howard would rise before dawn for his ornithological vigils. Crouched in a gorse bush or other hiding place, enduring discomforts ranging from cold to thorns to insect bites, he brought new insights to the study of bird behavior and helped reform British field natural history in the process.[75]

Like Selous, Howard published his earliest scientific papers in the *Zoologist*. In 1901 he contributed a brief account of the behavior of the grasshopper warbler. The following year he wrote in support of Selous's theory of the evolution of nest building. In 1903 he addressed the problem of sexual selection. Here he based his comments on his own observations, in the field, of "cases of actual display amongst some of our native species." Darwin, he noted, had relied on others' accounts of bird display, and these accounts had been of either foreign birds or British birds observed in captivity.[76]

Howard was satisfied neither with Darwin's discussion of sexual selection nor with Wallace's counterexplanation of secondary sexual characters. Although he was impressed by Wallace's argument that "the extremely rigid act of natural selection must render any attempt to select mere ornament merely nugatory," Howard did not believe that selection for "health and vigor alone" would necessarily have led to that "wonderful harmony" of color, form, and song that "we call beautiful." Had it been only a matter of health and vigor, he said "our most beautiful songsters might have developed shrieks."[77]

In 1903 Howard had little to offer to the sexual selection debate beyond the opinion that animals have an aesthetic sense and the vague suggestion that this was a function of "creative power." Nonetheless, W. L. Distant, the *Zoologist*'s editor, was happy with the piece. It seemed to show, he said, that the study of the animal mind was "not a forlorn quest." He predicted that "future naturalists" would find much of value in the observations his journal was publishing on the activities and mental states of animals.[78] As for Howard himself, apprehending the bird mind would remain the enduring goal of his efforts as a bird-watcher.

The issues of sexual selection and the bird mind were uppermost in Howard's own mind in 1907 when he published the first installment of what ended up being a nine-part, two-volume, eight-year publication project, *The British Warblers: A History with Problems of Their Lives* (1907–1914). Beginning with a discussion of the grasshopper warbler, Howard acknowledged having been frustrated in his attempts to accommodate his observations to the "picturesque theory" of sexual selection. As he saw it, females did not typically have the chance to compare the displays of two or more males, nor for that matter did they seem to pay close attention to a male's displays. As for the male's display to the female, this appeared to be a reflexive rather than a conscious action. Howard concluded "that sexual selection as a rational explanation of the phenomena is impossible."[79]

The concept for which Howard is today most famous is the concept of "territory." His first mention the importance of territory appeared in 1908 in his discussion on the behavior of the chiff-chaff. There he allowed: "Breeding territory is a matter of the greatest importance to the males, frequently leading to serious and protracted struggles when two of them are desirous of acquiring the same area."[80] By the time Howard finished *The British Warblers* in 1914, he was convinced that the males did not compete with each other for females; they competed with each other for territories.[81] He went on to make territory the central subject of his now-classic study *Territory in Bird Life* (1920). He refined the idea further in *An Introduction to the Study of Bird Behaviour* (1929). By the early 1930s "territory" had captured the imagination of contemporary ornithologists to the point that Margaret Morse Nice complained that "bird students of the world . . . are in danger of going territory-mad."[82]

Beyond his development of the territory concept (of which more will be said below), Howard, like Selous, was a strong advocate of intensive field studies of common birds. Studies of domesticated birds, he believed, simply would not do. In *The British Warblers,* expressing skepticism toward the notion of the "plasticity of instinct," he noted: "Anecdotes of animals in a domesticated or semi-domesticated state have furnished most of the evidence upon which [the theory of the plasticity of instinct] is based. This is much to be regretted, since it tends not only to confuse the issue, but also to transfer attention from the only true source of information available, viz., an impartial investigation of wild Nature." He recommended further that attention be paid to numerous individuals of the same species. The question of individual differences, he believed, was very important, for it bore, at least indirectly, on "the possibility of a progressive development from animal to human intelligence, and consequently the whole subject of the genesis of mind." As he saw it, "the final solution" to the question of the genesis of mind rested with "those naturalists who will devote themselves to the study of one, or at the most a few, species."[83]

For guidance on the subject of the genesis of mind, Howard turned to Britain's leading authority on the subject at the time, C. Lloyd Morgan. Morgan had taken the understanding of the animal mind and its evolution as his primary aim. The two men began corresponding with each other in 1910 and maintained a vigorous exchange for the next quarter of a century, until just before Morgan's death.[84] Early on in their exchange, Morgan told Howard, "what I like about your treatment is the fine basis of accurately observed facts." But the issue of what sense to make of these facts remained. The difficulty was how to draw conclusions about the bird *mind* from these facts. In May 1912, after Howard had commented on the way some bird species incorporate phrases from other birds' songs into their own songs, Morgan suggested:

> The song of the bird is a complex bit of synthetic behaviour carrying a complex bit of synthetic experience. Any intercalated "imitation item" is intercalated within a synthetic whole in which it finds its fitting place, so that the whole is an enriched whole and a continuous whole not a scrappy string of fragments. Probably the songster (and perhaps his mate) feels the synthetic unity of the enriched whole in a manner much more perfect than we do, with our lame attempts at conceptual analyses into these and those constituent factors. I often think that a sort of un-analyzed sympathetic artistic sense sets a man nearer to the secret of the animal mind than scientific thought which is at home in the midst of a more intellectual mode of psychological development. But it's very hard to express oneself comprehensibly in this matter.[85]

Later in the same year, having told Howard once more how much he appreciated Howard's factual observations, Morgan remarked how under-standing the mechanisms of evolution and animal behavior seemed to be getting more and more complicated. The Lamarckian idea of the inheri-tance of acquired characters no longer looked as if it played a role in spe-cies change. Mendelian studies of unit characters, on the other hand, ap-peared to be going somewhere. Meanwhile, new findings about chemical processes were supplementing existing notions of how the nervous system works. "On all sides," as he put it, "new views are swimming into ken. And we need a second Darwin who could embrace all the factors & weave them into an interpretation supplementary of that of the great master. But the field is so vast and extensive that it becomes increasingly difficult to gather into one strand all the many threads.[86]

Concluding *The British Warblers* in 1914, Howard identified himself as someone "firmly convinced of the importance natural selection" but "not one of those who regard it as the exclusive means of organic process." He doubted that all the various details of bird courtship could have been de-veloped by natural selection on the basis of their utility. What was one to make, he asked, of such "trifling but specific forms of behavior" as whether a bird raised its wings a little more or a little less, whether it flapped them slowly or fluttered them quickly, or whether it uttered its song while as-cending or descending? He could not believe that organic differences of this sort depended simply on the selection of fluctuating variations or mu-tations. But neither did he give credence to the idea of the inheritance of acquired characters. He resorted instead to Morgan's hypothesis of "the survival of coincident variations" to explain how individual experience could modify an instinctive behavior pattern and how this modification, though not itself inheritable, could then serve as a kind of "foster-parent" for congenital variations in the same direction.[87]

As Howard struggled to make sense of the animal mind, Morgan offered him steady advice. In 1913 he urged Howard to drop expressions that had "too human a flavour—duty, anxiety, violation of claims etc. These . . . seem to me to savour a little of [W. P.] Pycraft's journalese stuff, and to be better avoided in *your* presentation of the subject."[88] On the other hand, Morgan was satisfied when Howard spoke of a male bird's "disposition to secure a territory." To Morgan, this was better than speaking of "an instinct to secure a territory." As he explained to Howard in a letter of 1914, both McDougall and the psychologist George F. Stout had defined instincts in terms of innate dispositions, but he (Morgan) wanted to distinguish between innate "dispositions" or "tendencies" on the one hand and instinctive behavior on the other:

> I prefer to apply the adjective instinctive to specific modes of behaviour and not to innate tendencies or dispositions so to behave. Pugnacity is dispositional—there is an innate tendency to fight evoked under certain circumstances in certain situations. But the instinctive modes of behaviour are scratching, biting, clawing, kicking & the like—always something which can be described or pictured cinematographically. I sometimes feel disposed to form the distinction somewhat thus—modes of behaviour—whether instinctive or intelligently acquired, are *pictureable,* while the disposition, or innate tendency or proclivity (however one names it) is *unpictureable.*[89]

Howard's most important contribution to ornithology, as indicated above, was his theory of territory. Morgan encouraged Howard to write a book explaining the concept. Howard's theory of territory, briefly stated, is that bird pairs distribute themselves in the environment in such a way that serves to ensure a sufficient food supply for their offspring. The pugnacity of the males is what produces the spacing out of pairs. Upon arriving at their breeding grounds in the spring, males fight with other males primarily over territory, not over females. The occupation of a territory is "the condition under which the pugnacious nature of the male is rendered susceptible to appropriate stimulation." In other words, the male's pugnacity depends upon the location it occupies: when its territory is trespassed upon, it has a strong impulse to fight; but when it crosses the boundary of its territory, its impulse to fight diminishes dramatically. As for the male's bright colors and song, these, Howard contended, serve in the first place as a warning to other males and only after that "to evoke some emotion in the female, which helps to further the biological end of mating."[90]

The idea of territoriality was not new with Howard. The German ornithologist Bernard Altum had set forth the basic notion in the nineteenth century. Altum's work, however, had not secured the widespread attention of ornithologists. For that matter, Howard's own early sketches of the idea of territory had little immediate effect on his contemporaries. *The British Warblers,* an expensive work published in sections just before the First World War began, was not a good vehicle for promoting important new ideas. Even Howard's more accessible *Territory and Bird Life* of 1920 had little influence before 1927, when Max Nicholson espoused Howard's territory thesis in his own widely read bird book, *How Birds Live* (1927).[91]

The idea of territory did not demand of its adherents any particular view of the mental states of animals. Probing the bird mind, however, remained Howard's abiding concern. His last two books, *The Nature of a Bird's World* (1935) and *A Waterhen's Worlds* (1940), constituted his best efforts to comprehend a bird's mental life, or rather the different "worlds" in which it lived. The first book stimulated long, probing letters from Morgan, by then in his mideighties, who wanted a clear understanding of where he and his friend agreed—and where they disagreed. Where Morgan specifically parted with Howard in interpretation was over Howard's idea that even without prior experience a bird could have the inherited "knowledge" of the "meaning" of a stimulus situation.[92] Morgan also told Howard once again that he thought Howard was at his best when Howard was simply describing, in "plain tale," what a bird did. In a letter of February 1935 Morgan wrote: "I note that you seek to make it clear that learning is not a mere motor process but an act of mind . . . —in other words, that you seek to interpret what happens in terms of mind-story since body-story *only* does not suffice to foot the bill." Morgan agreed: "All right. Same here." But he would have preferred it, he said, if Howard had stuck to his "admirable descriptions" of bird behavior and had not used "mind-story" phrases like "as if he meant it," or "lose interest in."[93]

Morgan died in 1936. Howard died four years later, the year his final book, *A Waterhen's Worlds,* appeared. In this last book, Howard explained that when he wrote of a "territory," "sexual," or other "world" of a waterhen, this "world" had no existence of its own. It was something he abstracted "from the bird's life." "A world," he said, "includes a particular feeling, particular images, and particular actions." And further.

> It is a whole within a larger whole, for it is evident that an image is revealed to something, that something feels, something acts. Each world is therefore within the Waterhen; and since he has no power to reflect and so to place himself outside a world, he is within each world. If he had no power to refer a particular feeling to its particular object, no object would be perceived, no feeling felt, no world enjoyed. He is the perceiver with power to refer; which, being interpreted is power to perceive things in relation. But having no power to reflect he perceives no self: memories he has, but perceiving no self he has neither future nor past—and so, no time. His world is always in the present, and mostly full of joy.[94]

Asking such questions as whether birds had "mental images" led Howard, as David Lack put it in 1959, "into exalted metaphysical regions where, as yet at least, he has had no followers." Whether Howard talked seriously with others besides Morgan about the bird's "world" is hard to know. He was in correspondence with the young Austrian ornithologist Konrad Lorenz in or around 1935, just as Lorenz was bursting onto the ornithological scene, but the letters between the two do not seem to have not been preserved. Shy and retiring by nature, and comfortable, one imagines, with his amateur status, Howard felt no need to seek the limelight in the way that Lorenz did. Though he belonged to the British Ornithologists' Union for forty-five years, Howard, as Lack reports, was not well known to the members of the union—so much so that "an important book was actually dedicated to his memory over a dozen years before his death."[95]

After Lloyd Morgan, Howard's main connection with the scientific community was Julian Huxley. Huxley's case requires detailed attention. Before turning to Huxley, however, it may be helpful to consider one additional British author who contributed to field studies of bird behavior early in the twentieth century: F. B. Kirkman.

FREDERICK B. KIRKMAN

Frederick B. Kirkman (1869–1945) lacked the originality of either Selous or Howard, but his efforts in developing field studies of bird behavior in Britain are nonetheless instructive. He studied modern history at Oxford and then served as an assistant schoolmaster until medical reasons led him to give up teaching in 1898. His continuing interests were modern foreign languages (he introduced new methods of foreign language teaching in the schools) and the study of birds.[96] He organized, edited, and contributed articles to *The British Bird Book*. He also devoted years of fieldwork to a single species, the black-headed gull. In the last years of his life he also figured prominently in the early history of the Institute for the Study of Animal Behaviour, Britain's first society devoted to animal behavior studies.

The novelty of Kirkman's *British Bird Book* was its focus. Its aim was to "bring together from every source, foreign and native, all the available information of any importance concerning the habits of British birds." The knowledge of bird habits, as Kirkman explained in the work's preface, remained very sketchy: "It is not possible at present to give, in the case even of many of our commoner birds, a detailed reliable description of the nuptial displays that occur at the beginning of the breeding season." Yet such evidence was crucial, he said, if one were ever to resolve "the vexed question of sexual selection."[97] The reason so little was known about the "sex displays of the various species," he wrote Eliot Howard in 1910, was that "so few ornithologists seem to have learnt that these details are best to be found by getting up early and standing still."[98]

In addition to providing verbal accounts of the habits of different species, *The British Bird Book* contained colored pictures. Each picture was intended to represent "a study of some habit of the bird or of one of its most characteristic and striking attitudes." Earlier works, Kirkman commented, had been inclined to present "the traditional bird perched on the conventional twig." In *The British Bird Book*, in contrast, each picture was designed to show the bird "in its natural surroundings, whether courting, singing, feeding its young, sitting on its nest, angry, pleased, alarmed, or inquisitive, thus combining the realism of the photograph with the added advantage of colour and artistic treatment."[99]

Kirkman's anthropomorphic language should not distract the reader from the important discovery that bird books such as his and Howard's were announcing and indeed displaying. This was the discovery of species-specific "attitudes" or "gestures." Howard's *British Warblers* included carefully drawn pictures of the "attitudes" adopted by male and female warblers during courtship. Kirkman described his own awakening to the distinctive "gestures" of gulls and terns as follows:

I spent a month in 1905 watching Black-headed Gulls and Terns. In the observations noted during that period I find but one reference to the various gestures of these species. Yet the birds must have been posturing

2.2 A species-specific "attitude," part of the display behavior of a male blackcap defending its territory. (From H. Eliot Howard, *The British Warblers: A History with Problems of Their Lives* [London: R. H. Porter, 1907–1914].)

daily before my eyes. I had eyes, but not seeing eyes for the particular set of facts. My photographs testify to the same blindness. They recorded the usual hardy annuals: bird on the nest, going on to the nest, going off the nest, flying to the nest, and so on. On subsequent visits to gulleries and terneries I had in my mind definite questions respecting the expression of animal emotions, and, lo and behold, the birds were posturing before my eyes every few minutes! And I photographed the postures, and let the hardy annuals be. My eyes were, so to speak, unveiled. So full, indeed of the unexplored has the subject of Gulls and Terns become for me that it promises to occupy for many years to come all the spare time I can devote to such matters.[100]

By the time *The British Bird Book* reached its final volume, six years after the project began, Kirkman could claim that the work showed a "considerable increase" over previous works in terms of the knowledge it contained about bird habits (and particularly "sex displays"). He felt keenly, nonetheless, that much still needed to be learned. Previously, he said, only a very few ornithologists had been willing to concentrate on just a few bird species and investigate their life histories in detail. He pointed to Howard's study of the British warblers as a happy demonstration that "the results of this closer study, if rightly pursued, is likely to reveal aspects of bird-life of which our predecessors did not even realise the existence." In addition to pursuing specialized studies of this sort, it would also be necessary to bring the study of bird behavior into "proper relationship" with the study of animal behavior in general. Not only did the ornithologist have much to contribute to this "relatively new study," Kirkman said, but animal behavior studies would in turn would inspire and guide the ornithologist in his own efforts.[101]

To facilitate a rapprochement between ornithological field studies and animal behavior studies in general, Kirkman appended to *The British Bird Book* a supplementary chapter, aimed at the amateur ornithologist, on how to study bird behavior. One needed to distinguish, he said, among three different kinds of behavior—instinctive, intelligent, and rational. Instinctive behavior was behavior that was "inherited or congenital." Intelligent behavior was behavior that was learned or acquired. Rational behavior went further than intelligent behavior in that it involved a capacity "to form a mental image sufficiently free to enter into new combinations, to form part, that is, of a reasoning process." Identifying rational behavior, Kirkman admitted, was fraught with even more difficulty than trying to decide whether a particular action was instinctive or intelligent. It remained uncertain whether any animal "below man" had the power to form a mental image. As Kirkman saw it, behavior in lower animals consists primarily of instinctive and intelligent (but not rational) elements, combined in such a way that it is difficult to tell where the one leaves off and the other begins.

To be a good naturalist, Kirkman allowed, one needed to know the right questions to ask, two of which in particular stood out when one studied animal behavior. The first was whether a particular act was instinctive, intelligent, or rational. The second was how these three kinds of behavior originated. "Outside the scope of our study," as Kirkman put it, was the "question of questions" itself: "what is the relation between behaviour and the associated conscious states, between nervous process and mental process, body and mind?" [102]

In addition to asking good questions, Kirkman said, the field naturalist needed to cultivate the right practices. He had to be able to make observations accurately without being influenced by any preconceived theory. He had to verify that what applied to one individual applied to other members of the species as well. He furthermore should "practice with a religious devotion" the rule of noting and dating everything that he found. And he had to learn to be patient and to keep still. In drawing conclusions, he needed to avoid two dangers in particular: he had to take care not to ascribe the mental powers of man to other animals and not "to balance a pyramid of theory on a pin's point of fact." Otherwise all he required was some minimal equipment: a notebook, binoculars, and a square of waterproof material on which to sit. A camera was useful, but not necessary. Being a good photographer and a good observer, Kirkman remarked, were not the same thing. [103]

When it came to identifying the particular kinds of behavior that deserved study, Kirkman, the former schoolmaster, offered a long list. These were, in order, gestures, song and notes, nest building, incubation, nestlings and fledglings, feeding, protection of young, self-protection, locomotion, social or solitary life, and migration. To elaborate on just one of these, his suggestions under the heading of "Gestures—All Movements Intended to Give Expression to the Bird's Conscious States" began with the subcategory "Expression of Sex Emotions." There he identified the various matters to be noted and questions to be addressed:

> Exact descriptions of the gesture (display, attitude, movements). By which sex. How many individuals take part. Are the gestures before and after mating the same? Gestures performed by a bird when alone. Is there display of special features? (plumes, colours, wattles, etc.). Does the hen select, or is she merely appropriated, or does she simply mate with the cock in possession of the nesting area, whether it be her former mate or not? If the hen selects, what appears to govern her selection; has she chosen the strongest, etc.? Does the species under observation pair for life, and, if so, are there sex displays by one or both of the pair? Relation of sex gestures to the act of coition. Is a love gesture made to serve for the expression of emotions other than sexual? Note differences in the gestures made by individuals of a species, and those of species in the same group. [104]

Here, Kirkman was firmly convinced, were areas where a dedicated bird-watcher could play a continuing role in the scientific study of behavior, serving as an agent for the comparative psychologist and the zoologist alike. The amateur did not need to become an expert in psychology. He was more likely to help the field "by sticking faithfully to his own last, that is, by carefully recording, verifying and classifying the facts of animal behaviour in the wild state." The conclusions made by psychologists in the laboratory needed to be checked against the behavior of animals in the wild. The committed field naturalist was obviously the best person for the job.[105]

Kirkman himself became a dedicated observer of the black-headed gull (*Larus ridibundus*). He studied black-headed gulls for more than twenty breeding seasons at some of the most magnificent colonies in Britain (among them Scoulton Mere in Norfolk, Ravenglass in Cumberland, and Twigmoor in Lincolnshire) before publishing his 1937 book, *Bird Behaviour*. He not only observed and photographed the gulls, paying particular attention to their characteristic displays, he also conducted experiments on them. He studied how they rolled eggs back to their nests after the eggs were displaced and likewise how many eggs—and what other objects, shaped like eggs or not—they would sit on. His experiments in this regard predated similar work Niko Tinbergen would do (in one case in collaboration with Konrad Lorenz) in the 1930s. His attention to the gulls' displays likewise preceded work of the same sort for which Tinbergen would become famous. By the time *Bird Behaviour* appeared in 1937, however, Kirkman was in his late sixties, and his book lacked the youthful energy, programmatic brashness, and conceptual promise of what the much younger naturalists Lorenz and Tinbergen had just begun promoting. It seems that Kirkman's importance for animal behavior studies in this period lay less in his new book than in his service to the Institute for the Study of Animal Behaviour (ISAB), which he helped found in 1936. He served as ISAB treasurer and took care of most of the organization's daily administrative business up until his death in 1945.[106]

Overall, Kirkman seems to have been the British amateur who worked hardest to bridge the gap between amateurs and professionals when it came to the study of bird behavior. From the side of the professionals, the individual who contributed the most in this regard was Julian Huxley.

JULIAN HUXLEY AND THE QUESTION OF BEHAVIOR STUDIES IN A BIOLOGIST'S CAREER

Julian Huxley (1887–1975) bore one of the great names of British biology. His grandfather, Thomas Henry Huxley, was the leading spokesman for Darwinian evolution in the years following the publication of the *Origin of*

Species. T. H. Huxley had also devoted himself to the reform of science education, the scientific enlightenment of the general public, and the promotion of the very concept of "biology." As a statesman of science, his stature in Britain was unsurpassed. His image necessarily loomed large for the grandson who wished to follow in his footsteps.[107]

The challenge of living up to this example was compounded for the younger Huxley by the changes that biology had undergone since his grandfather's day. Not only had science become increasingly specialized over the course of two generations, but the laboratory orientation that T. H. Huxley had encouraged in Britain had also caught on sufficiently to make the earlier naturalist tradition of Darwin and Wallace seem old-fashioned. As much as any biologist of his day, Julian Huxley sought to synthesize areas of research that had grown apart from each other. But exactly where field studies belonged in twentieth-century biology, and whether they might find a place in the career of a highly talented and equally ambitious young biologist, were questions that were not easily resolved.

Julian Huxley's talents were literary as well as scientific, thereby reflecting the maternal, Arnold, side of his family as well as his Huxley heritage (his mother was the niece of the poet Matthew Arnold and the granddaughter of Dr. Thomas Arnold, the distinguished headmaster of Rugby; her sister Mary [Mrs. Humphrey Ward] was a famous novelist). In his second year at Oxford, Huxley won the Newdigate Prize for English Verse. The next year he graduated with first class honors in natural science (zoology). Along with his multiple abilities, however, came the same severe mood swings that had plagued his zoologist grandfather and the likewise temperamental character of the Arnolds. He was alternately buoyed up with great hopes and plans and brought low by crippling anxieties, self-doubt, and depression. As his wife, Juliette, later wrote, Julian and his three siblings carried "the mixed genes of Huxleys and Arnolds, a privileged but heavy inheritance, exacting its crusading trophies, achievements, and punishments."[108]

Modesty was never one of Huxley's defining traits. When he came to write his autobiography in the late 1960s, he looked back with special pride not only on his work on relative growth, his idea of clines, and his book *Evolution, the Modern Synthesis,* but also on his early field studies of bird behavior. His paper of 1914 on the great crested grebe he identified as "a turning point in the study of bird courtship, and indeed of vertebrate ethology in general." He additionally credited himself with "having made field natural history scientifically respectable."[109]

Huxley's contributions to the biological study of behavior were genuinely significant. Nonetheless, his autobiographical claims ought not to be accepted at face value. For one thing, vertebrate ethology was not sufficiently recognizable as an enterprise in 1914 to have even had a turning point then. For another, respectability is a relative quality. It can only be judged in context. Huxley said little in his autobiography regarding how it was that he "made field natural history scientifically respectable." By examining his contributions to field studies of behavior and by exploring how this work fit into his career, we can gain a clearer sense of the scope, and the limits, of his accomplishments. We can also gain a better appreciation of how field biology continued to be a problematic area for professional biologists in Britain in the first three or four decades of the twentieth century.

Huxley's activities as a bird-watcher and observer of animal behavior can be divided into three main periods. The earliest, from 1901 to 1911, was the period in which he first took up bird-watching as a hobby. In the second, from 1911 to 1925, he conducted field studies and published a series of five important papers on the courtship behavior of different bird species. Finally, from 1925 until the end of his career, he wrote a number of synthetic pieces on bird behavior and promoted behavioral studies through a variety of additional activities, even though his main interests by this time were evidently elsewhere.

By Huxley's later accounts, the thrill of seeing a green woodpecker up close, when he was thirteen or fourteen years old, imprinted him on bird-watching for the rest of his life.[110] He began devoting his weekends, holidays, and spare moments to recording all the different species he could spy. Some four or five years later, at about the time he finished at Eton, he came upon Selous's book *Bird Watching*. Selous's work impressed upon him "the enormous amount of interesting facts [about bird behavior] which still remained to be discovered." Without giving up recording the numbers of different kinds of birds he spotted, Huxley began to pay more attention to bird habits. Furthermore, for Huxley as for Selous before him, issues of bird habits inevitably led to the topic of bird mind.[111]

Huxley went up to Oxford in 1906. There he came in contact with Warde Fowler, who was not only a historian, classicist, and subrector of Lincoln College but also, as we have seen, the author of *A Year with the Birds* and an enthusiastic proponent of Selous's field studies. The young Huxley met regularly with Fowler to talk about birds and to join him in field outings.[112]

What we know of Huxley's own thinking on behavior in this period comes from the manuscript of a paper entitled "Habits of Birds," which he presented in 1907 to the Decalogue Club at Balliol. The theme of the paper was how bird habits related to "various evolutionary theories." Interestingly enough, however, the twenty-year-old undergraduate made no mention of any of the major alternatives to Darwinian theory that biologists were then debating. He paid no attention to neo-Lamarckism, mutationism, or orthogenesis. Instead, he showed himself to be a born and bred Darwinian. He used the details of bird behavior to illuminate and endorse Darwin's theories of natural and sexual selection. The "drumming" of the snipe and the lesser spotted woodpecker, for example, he cited as behavior that had developed through sexual selection. "The end or object of such performances," he explained, was twofold. In the first place, it promoted "recognition between individuals of the same species." In the second place, it enabled the male bird to show off special features to the female bird. According to Darwin's theory of sexual selection, Huxley elaborated, "the hen consciously or unconsciously selected as her mate the cock whose performance pleased her most." [113]

In addition to endorsing Darwin's theories of natural and sexual selection, Huxley advanced a number of other ideas of potential importance. These included an explanation of birdsong as an expression of emotion ("a kind of mental safety valve"), the recognition that there may be variations in behavior among individuals of the same species, and the appreciation that some forms of apparently purposive behavior in birds are performed unconsciously. He also discussed how, in birds, there is often a compromise between the need for protective coloration and the development of bright colors for sexual display. In addition, he put forward the claim that reason is "merely an amplification of the instinctive process."

Huxley was keen to discuss the differences and similarities between the behavior of birds and the behavior of humans. He insisted that certain forms of bird behavior that humans are inclined to regard as purposive—such as a mother bird's "feigning" an injury and leading a predator away from her nest—are simply instinctive. On the other hand, he happily attributed certain emotions and motivations to birds and humans alike. He spoke of a snipe "in the ecstasies of love-making." He likewise described a "little love scene" between a pair of gray wagtails. With the smug confidence of his class, he compared the display behavior of a peacock to that of a "greengrocer's boy," stating, "in much the same way as the peacock unfolds his tail before the peahens, so does the greengrocer's boy put on his best clothes, & plaster a curl down on to his forehead preparatory to 'walkin' out.' He is not really conscious why he does it, except that were he to appear disheveled & dirty there would perhaps be an end of his courtship."

Like Darwin, Huxley credited birds with "mental states" and granted them at least a glimmer of reasoning power. Simultaneously he endorsed the view that the parental, conjugal, and fraternal instincts of animals "lie at the base of all our moral nature." Here in 1907, at the very beginning of his scientific career, he already displayed a keen interest in seeing progress in evolution and providing an evolutionary explanation of moral action. He concluded his paper on bird habits with a paean to the biological bases of human nature, the evolution of human sociality, and the morality of actions serving the good of the species:

> We find that our affections are based almost entirely on the parental & conjugal instincts, [while] the fraternal instinct, so unimportant at first, is finally the most important; for it develops into the Social instinct, & on that are founded all our ideas of moral good & evil. In Nature it is the evolution of the race that is the important thing—the individual is quite subservient to the race. Accordingly we find that gradual realization of the fact that individual gratification is immoral—wrong, while self-denial is right, but only when by it the race at large is benefited. Man's reason has showed him that in reason there is strength, & that he will only exhaust his inherent capabilities when every individual is playing his part merely as a unit of a glorious whole.

Five years later he provided a similar conclusion to his first book, *The Individual in the Animal Kingdom,* when he optimistically described the state as "that unwieldiest individual —formless and blind to-day, but huge with possibility." [114]

Huxley was pleased when he detected evidence of evolution leading to moral improvement in the natural world. He hated evidence to the contrary. When he learned from the head keeper at Tring, the Rothschild estate, that seventy or more mallard females were drowned each year at the Tring reservoirs in the course of being mobbed and treaded by numerous drakes, he did not want to believe it. By Huxley's calculations, this meant that some 7%–10% of the mallard females on the reservoirs were killed in this way each year. Personally witnessing a mobbing event at Tring, he tried to stop the drakes' behavior by shouting and by throwing stones at them, but his efforts were to no avail. Fortunately for the female involved, she managed to escape on her own. Nonetheless, Huxley found the whole event "a painful and repulsive sight." The males' behavior, he decided, constituted evidence of a "disharmony in the constitution of the species." [115] This conclusion reflected his deeply held belief—or hope—that evolution normally worked for the "good of the species." The theme of the "good of the species" would continue to appear in his writings on animal behavior and on Darwinian selection throughout his career.

This, however, is jumping ahead in the story. In the summer of 1909, following his graduation from Oxford with his honours degree in zoology, Huxley attended the centenary celebration of the birth of Charles Darwin. He had been invited, he later explained, as both a Huxley and a "budding biologist." Thinking of his grandfather's efforts in the Darwinian cause and hearing more about the impact of natural selection on modern scientific thought led him to a dual resolution: "that all my scientific studies would be undertaken in a Darwinian spirit and that my major work would be concerned with evolution, in nature and in man." [116]

Given Huxley's early ornithological interests and his resolution to follow in his grandfather's and Darwin's footsteps, one might suppose that he viewed the fit between field studies and evolutionary theory to be both scientifically promising and personally congenial. Such was not the case. Field studies were not where he first sought to make his mark. Nor, for that matter, were the early years of his career happy ones. In September 1909, he set off on a year's fellowship to the Marine Biological Station at Naples. Though there is scarcely a hint of it in his autobiography, the time he spent in Naples was for him a period of great psychological anguish.

He confessed this to his student Alister Hardy some twelve years later, when Hardy was in Naples and having doubts about his own future as a scientist. In a long and sympathetic letter, Huxley told Hardy just how unsure of himself he had been when *he* started *his* career. He had gone to Naples, he explained, wanting "to do something big, new, original." Unfortunately, two months of trying to make cultures of various protozoa proved "completely in vain." He worked also on sponge dissociations and then *Clavellina,* but these studies proved unsatisfying because they were "based on other people's ideas." He was worried about his prospects as a researcher because he doubted he was capable of doing good work in physiology. "Distrust of one's own powers combined with a good deal of ambition" left him anxious and miserable. [117]

When Huxley returned from Naples to Oxford in 1910 to become lecturer at Balliol College and demonstrator in the Oxford Department of Zoology and Comparative Anatomy, neither his research at Naples nor his responsibilities at Oxford gave him any particular reason to take up the study of animal behavior. Experimental embryology and genetics were the hot fields in biology. [118] The study of animal behavior, in contrast, was scarcely on the horizon. Huxley found his way into behavioral research by chance.

In April 1911, during the spring vacation of his first year of teaching, he took a "reading-party" with him to the north coast of Wales. There he decided that his original goal, reading Bütschli's big text on the protozoa, was not nearly so engaging as watching the hundreds of birds that haunted the tidal flats of the northern half of Cardigan Bay.[119] He was delighted to see the gray plover and the black-tailed godwit, two species he had never seen before. He was even more pleased to have the opportunity, as he put it, "to study, under the most favourable conditions, the natural behaviour and home life of some of the commoner shore-birds."[120] This appears to have been the first time he experienced the particular benefits of devoting "several spells of watching, day after day" to the study of a single species. He was not the first to make this discovery. Selous, Howard, and Kirkman— among others—had already made it before him. But this was the sort of experience one needs to make oneself. The excitement of making his own observations on bird courtship left a greater impression on him than any vicarious experience could possibly have done.[121]

Of all the birds that Huxley saw on the Welsh coast, the redshank was the species that impressed him most. He was struck not only by the male's exciting display but also by the fact that the female in most cases rejected the male's advances by flying off. Huxley himself at this time was engaged to be married. His practice as a naturalist on this trip was to make field notes on the spot, write them up over the next twenty-four hours, and send them off in the form of a letter to his fiancée. His study of bird courtship and his own courting behavior thus intertwined. Describing the emotional and behavioral parallels between birds and humans became part of his own display to the woman he planned to marry. Here, for example, is how he described the behavior of the female and male redshank in the second, "display" stage of their courtship:

> She stands quite stock still. He, seeing this, also stops, & then raises his wings right up, so as to show their white undersides, & flutters them tremulously in a beautiful way, all the time holding them above his back. At the same time, putting his head rather forward, he begins to call—he opens his bill, & then gives a curious continuous rolling call, vibrating his bill all the time; the call, like the oystercatcher's, rather resembles the Dabchick's, but also has quite a flavour of our own dear Nightjar about it, especially when you hear it, as I did later, on a still moonlight night! At the same time he begins advancing very, very slowly, lifting his feet mincingly from the ground one after the other, often seeming to put them down in the same spot again, or even sometimes retreating a little—I

suppose judging from her expression. She has been quite still all the time; I suppose surveying him critically: after this sometimes she suddenly flew straight off, leaving him alone to fold his wings & pretend nothing unusual had been happening—going on feeding!

Huxley continued:

> This is very interesting to me, as it shows how Darwin's idea of Sexual Selection was undoubtedly right in certain cases, such as this—the hen has the power of choice,—if she doesn't like the cock, away she just goes, & he always has to give up the chase eventually. Not only has she got it, but she exercises it a great deal—all the suitors so far have been rejected! It must be a queer kind of choice, I daresay, scarcely conscious at all, but very decided in its workings.[122]

As Huxley later told it, it was only after he had returned from Wales and had written up most of his observations that he discovered "a fairly complete account" of courtship in the redshank written by Selous.[123] This seems a curious declaration on Huxley's part, given that Selous's account of the redshank's courtship appeared in what was arguably the most important paper on sexual selection in birds published in the previous half dozen years. At a time when sexual selection was a vital concern of British field ornithologists, it is surprising to think that Huxley was not already familiar with Selous's work. In any case, neither the observations Huxley reported nor the conclusions he drew from them departed significantly from Selous's. The main thrust of Huxley's paper on the redshank underscored what Selous had already maintained, "namely that the actions of the birds which lead up to each single act of pairing are explicable only on the Darwinian theory of Sexual Selection, or on some modification of that theory." In the redshank, Huxley indicated, there was obvious display on the part of the male, and "an equally marked *power of choice*" on the part of the female. "Though the male in this particular species has the *initiative*," Huxley concluded, "the *final decision* must rest with the female."[124]

The focus of Huxley's redshank paper was the problem of sexual selection, but several additional features of the paper also deserve mention. One is Huxley's interest in the mental states of animals—combined with his belief that it is possible to surmise what these mental states are. Speaking of the male redshank's display he wrote: "To the human eye the whole action seems the expression of eager excitement tempered by uncertainty, and that, presumably, is what the bird is actually feeling." Also worthy of note is Huxley's observation that in the matter of sexual selection, structural characters cannot be isolated from behavior patterns. "When we speak of secondary sexual characters," he wrote, "we usually think of structure only. In reality the real character is the structure plus the instinct to use the structure, for it is the use of the structure which alone has any significance for the species: it is that which constitutes a unity, it is that which has been really acquired by the species." A third point of interest in Huxley's paper was his agreement with Selous that many of the combats between the males were "merely formal."

The following year, 1912, Huxley again spent his spring vacation watching bird courtship. This time the species on which he trained his binoculars was the great crested grebe. Together with his brother Trevenen (Trev), he watched the grebes on Tring Reservoir for a period of ten days. Once more, upon returning home, he found that Selous had in some measure anticipated him, having written a decade earlier what Huxley described with characteristic egoism as "a welcome paper . . . which exactly dovetailed into my own observations." [125]

What struck Huxley most about the great crested grebe was a phenomenon that, as far as he knew at the time, was unique to that particular bird. The bird had special structures—its erectile neck ruff and ear tufts—which were "not only the common *property* of both sexes," but were "*actually used in display, and used in exactly the same way by both sexes.*" [126] Believing initially that these structures were "only used in courtship," he concluded they had to be the result of sexual and not natural selection. He proposed using a word promoted by E. B. Poulton—"epigamic"—for characters that had arisen through sexual selection and that existed equally in both sexes. [127]

Huxley's first comments on the grebe appeared in a paper of 1912. Two years later he published his longer (and now-famous) paper, "The Courtship Habits of the Great Crested Grebe." By this time, informed again by the work of an amateur before him, he had come to interpret the great crested grebe's behavior in a new light. In this instance it was Eliot Howard, not Selous, whose influence was critical. Having observed that much of the display behavior of warblers occurred only *after* the males and females paired up, Howard concluded that such displays could *not* have developed through Darwinian sexual selection. Huxley reached the same conclusion regarding the great crested grebe. In a footnote on the first page of his paper he acknowledged that the word "courtship" in his title was perhaps misleading, since courtship, strictly speaking, should apply only to antenuptial behavior. "Love-habits," he said, would have been a better term for what he was talking about.[128]

Huxley described at length the birds' display behavior. He gave to the birds' distinctive postures or "ceremonies" a series of evocative names: the "cat-attitude," the "Dundreary attitude," the "ghost-dive," the "shaking attitude," the "discovery ceremony," the "weed-trick," and the "penguin-dance." He interpreted all of these as expressions of emotion. He described the "ghost-dive" or "ghostly penguin attitude" of a male great crested grebe in the following terms: "I could scarcely believe my eyes. He seemed to grow out of the water. First his head, the ruff nearly circular, the beak pointing down along the neck in a stiff and peculiar manner; then the neck, quite straight and vertical; then the body, straight and vertical too; until finally the whole bird, save for a few inches, was standing erect in the water, and reminding me of nothing so much as the hypnotized phantom of a rather slender Penguin."[129]

Impressed as he was by the "ghostly penguin attitude," it was the "penguin-dance" that constituted "the highest development of courtship-actions that I have seen." He portrayed in detail how a pair of birds he was watching engaged in a protracted bout of shaking their heads at each other and preening their wings in a formalized fashion. They raised their crests in such a distinctive fashion that he felt certain "something exciting was going to happen." His description of what happened next deserves quoting at length:

> Sure enough, the hen soon dived. The cock waited in the same attitude, motionless, for perhaps a quarter of a minute. Then he, too, dived. Another quarter of a minute passed. Then the hen appeared again, and a second or two later, some twenty-five yards away, the cock came up as well.
>
> They were in a crouching position, with necks bent forward, ruffs still elliptical, and both were holding in their beaks a bunch of dark ribbony weed, which they must have pulled from the bottom. The hen looked about her eagerly when she first came up; when the cock appeared she put her head down still further and swam straight towards him at a good pace. He caught sight of her almost immediately too, and likewise lowering his head made off to meet her. They did not slacken speed at all, and I wondered what would happen when they met. My wonder was justified: when about a yard apart they both sprang up from the water into an almost erect position, looking somewhat like the 'ghostly Penguin' already described. *Sprang* is perhaps too strong a word; there was no actual

leap, but a very quick rising-up of the birds. The whole process, however, was much quicker and more vigorous than the slow 'growing out of the water' of the ghost-dive. In addition, the head was here not bent down along the neck, but held slightly back, the beak horizontal, still holding the weed. Carrying on with the impetus of their motion, the two birds came actually to touch each other with their breasts. From the common fulcrum thus formed bodies and necks alike sloped slightly back—the birds would have fallen forwards had each not thus supported the other. Only the very tip of the body was in the water, and there I could see a great splashing, showing that the legs were hard at work. The appearance either bird presented to its mate had changed altogether in an instant of time. Before, they had been black and dark mottled brown: they saw each other now all brilliant white, with chestnut and black surrounding the face in a circle.

In this position they stayed for a few seconds rocking gently from side to side upon the point of their breasts; it was an ecstatic motion, as if they were swaying to the music of a dance. Then, still rocking and still in contact, they settled very gradually down on to the surface of the water; so gradually did they sink that I should think their legs must have been continuously working against their weight. All this time, too, they had been shaking their heads violently at frequent intervals, and after coming down from the erect attitude they ended the performance by what was simply an ordinary bout of rather excited shaking; the only unusual thing about it was that the birds at the beginning were still, I think, actually touching each other. The weed by this time had all disappeared: what had happened to it was very hard to make out, but I believe that some of it was thrown away, and some of it eaten by the birds while settling down from the Penguin position.[130]

Apart from the sheer beauty of the penguin dance, two things in particular impressed Huxley about this and other of the great crested grebe's displays. The first was that the displays were *mutual*—that is, both sexes displayed the same behavior patterns. The second was that they were by and large *self-exhausting*—that is, they did not seem to serve as excitants to coition but instead were typically followed by periods of calm, where the birds simply engaged in "swimming, resting, preening, and feeding." [131]

Having concluded that the displays functioned neither in mate choice nor as a stimulus to coition, Huxley was left with the problem of determining what they *were* for, since it seemed clear to him that they had to fulfill *some* function. He decided they probably served "to keep the two

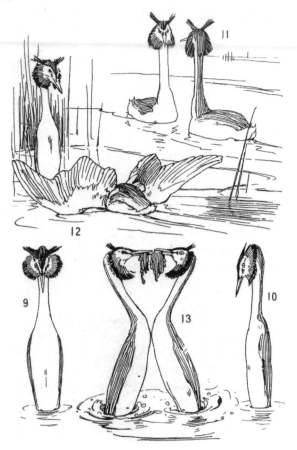

2.3 Display behavior of the great crested grebe. (From Julian Huxley's 1914 paper "The courtship-habits of the great crested grebe [*Podiceps cristatus*]; with an addition to the theory of sexual selection," *Proceedings of the Zoological Society of London* [1914]: 491–562.)

birds of a pair together, and to keep them constant to each other." "From the point of view of the species," he explained, "it is obviously of importance that there should be a form of "marriage"—a constancy, at least for the season—between the members of a pair."[132]

Initially Huxley had supposed that the actions and structures of the male and female great crested grebe had arisen by sexual selection of the males, followed by transference of the secondary sexual characteristics to the females. Now he was of the opinion that these actions and structures were largely a product of what he chose to call "mutual selection." He saw the subjective, emotional side of bird life to be of paramount importance here. The grebes' special rituals, he suggested, represented expressions of emotion, evolved over time: "In the Grebe, [the surplus emotional energy of the bird in springtime] has been diverted into fresh channels through Mutual Selection, and thus pressed more quickly into the service of the species."[133]

Huxley was not apologetic about adopting a psychological point of view in interpreting the courtship behavior of animals. To the contrary, he believed this approach was essential. Mutual selection made sense once one understood that emotions could be developed for a species' good. The "interplay of consciousness or emotion" between a pair of birds was a way of keeping the two birds together. The "selection of accidental variations" in their vocalizations and bodily motions led to different forms of emotional expressions in different species. "Birds," Huxley allowed, "have obviously got to a pitch where their psychological states play an important part in their lives."[134]

Huxley was by no means alone in supposing that a psychological point of view was necessary for understanding the evolution of courtship behavior in higher animals. Selous had posited the same thing, as had W. P. Pycraft in his 1913 book *The Courtship of Animals*.[135] Huxley, in a postscript to his 1914 paper, allowed that "accidental circumstances" had prevented him from reading Pycraft's book until after he completed the text of his own paper. Be that as it may, he agreed with Pycraft that Darwin's main conclusions on sexual selection stood firm, and that the chief modification that needed to be made to them had to do with female choice. Pycraft had shown particularly vividly in the case of newts, Huxley said, that "display and ornament do not act on the aesthetic sense of the female, but on her emotional state; they are—using the words in no narrow or unpleasant sense—excitants, aphrodisiacs, serving to raise the female into that state of exaltation and emotion when alone she will be ready to pair." But Huxley still felt that a kind of choice remained:

> In animals such as Birds, where there is a regular pairing-up season, and where, too, the mental processes are already of considerable complexity, it is impossible to doubt but that mating may be, and in some species is, guided by impulse, unanalysable fancies, individual predilection. There, in a rudimentary state, we find that form of "choice"—intuitive, unreasoned, but none the less imperious, and none the less in its results a true choice—which reaches its highest stage of development in the intensely felt affinities of man and woman—in that condition known as "falling in love," where the whole of the subconscious mental activities become

grafted on to the inherited sexual passions, the whole past of the mental organism is summed up in the present, in the intensely real act of choice which chooses one from among thousands and says, whether in words or no, "that one being, and no others, is the being that I desire for my mate."[136]

Neither the American behaviorist psychologists nor the tough-minded ethologist Niko Tinbergen would have hesitated a minute before dismissing such writing as hopelessly anthropomorphic and subjectivist. Huxley denied his thinking was anthropomorphic on the grounds that he never credited birds with rational faculties comparable to those of humans.[137] On the other hand, with respect to their *emotional* lives, he was quite prepared to believe that birds and humans have much in common. We thus find him suggesting in 1914 that while the physiological foundations of bird behavior were still a long way from being understood, the psychological foundations were more readily accessible: "[By] comparing the actions of the birds with our own in circumstances as similar as possible, we can deduce the bird's emotions with much more probability of accuracy than we can possibly have about their nervous processes: that is to say, we can interpret the facts psychologically better than we can physiologically. I shall therefore (without begging any questions whatever) interpret processes of cause and effect in terms of mind whenever it suits my purpose so to do."[138]

Watching how birds behaved felt to Huxley like watching actors in a play.[139] He had no qualms writing of a "domestic drama" he had witnessed among his grebes, where "some very real jealousy" was aroused. (Significantly, he enclosed the phrase "domestic drama" within quotation marks, but not the word "jealousy" itself.)[140] Parts of his manuscript field notes read like stage directions, for example, "♂ off R. [Male off Right]."[141] He named one of the great crested grebe's distinctive postures the "Dundreary-attitude" after "that famous personage of the drama" (an eccentric English gentleman in the play *Our American Cousin,* made a household name in nineteenth-century England by the acting of Edward Sothern).[142] Some years later, in describing a particular "post-nuptial" ceremony of the avocet, Huxley noted: "The effect was charming, and reminded me forcibly of the little run made by MacHeath and Polly in the 'Beggar's Opera' at the refrain 'over the hills and far away' in their duet."[143]

Huxley's field notes reveal not only his sense of being a spectator at a play but also some of the difficulties faced by the field naturalist attempting to study bird behavior. In his great crested grebe paper of 1914 he claimed to be presenting there "in full" the notes he had made on one pair of grebes over the space of an hour and half. Like Selous before him, he gave a moment-to-moment account of the birds' actions, down to the minute. In the published paper, Huxley's notes, made on 18 April 1912, extend from 10:40 a.m. to 12:18 p.m. What the published paper does not reveal is the vexation Huxley felt at 12:20 p.m. when he realized he was no longer looking at the same two birds. The manuscript notes tell the story: "Damn—I believe this is wrong ♂."[144]

Huxley's descriptions of the great crested grebe's displays inevitably suggest to the modern reader how valuable films could become in the analysis of animal behavior. A performance such as the "penguin dance," captured on film, could be made available for others to see and to analyze frame by frame. The year of Huxley's great crested grebe paper, 1914, was the same year C. Lloyd Morgan wrote to Eliot Howard describing "instinctive modes of behaviour" as the kinds of behavior patterns that could be "described or pictured cinematographically." Within two decades of Morgan's comments and the appearance of Huxley's grebe paper, film sessions had become standard events at ornithological society meetings.[145]

Huxley's great crested grebe paper may well be the best known of all the papers he wrote. In it, in promoting the idea of mutual selection, he took his first step away from the view that characters displayed in courtship were necessarily the result of sexual selection. Here too he first promoted the idea of behavioral *ritualization,* an idea to which he returned late in his career when he organized a symposium on the subject for the Royal Society of London. Nearly four decades after Huxley's grebe study appeared, David Lack told him it had to be "one of the longest-lived of all papers. It's still constantly quoted (by me and others) in lectures." Konrad Lorenz for his part insisted to Huxley how impressed he (Lorenz) had been with Huxley's observation that the grebes' activities were at one and the same time "self stimulating and self exhausting."[146]

In view of this later praise and Huxley's own autobiographical remarks on his paper's historical importance, it is worth noting that Huxley did not introduce this paper with any special fanfare. The tone of his introductory comments was casual and apparently lighthearted. He hoped, he wrote, that the paper would "help to show what wealth of interesting things still lie hidden in and about the breeding places of familiar birds." He continued: "A good glass, a note-book, some patience, and a spare fortnight in the spring—with these I not only managed to discover many unknown facts about the Crested Grebe, but also had one of the pleasantest holidays. 'Go thou and do likewise.'"[147]

Huxley, in short, was not putting his career on the line here. No one, certainly, could fault him for making such industrious and clever use of a holiday. In fact, these breezy introductory remarks masked the crippling anxieties that had plagued him since his time in Naples. He was worried about his worthiness, both personally and professionally In 1913 his fiancée broke off her engagement with him and he suffered his first nervous breakdown.[148] Later that year, not entirely recovered from this breakdown, he traveled to Houston to take up the position of assistant professor and chair of biology at the newly established Rice Institute. At Rice, as he later confided to Alister Hardy, he "led the life of an intellectual semi-invalid— never daring to work hard, only doing my routine of lectures & laboratory, still feeling that I shd. never be able to concentrate enough to do real research."[149] Early in 1914 he wrote up the long version of his great crested grebe paper and sent it off to be read at the Zoological Society of London.

Huxley's continuing self-doubts did not prevent him from watching birds or writing further about them. Visiting Louisiana, he was thrilled by the display behavior of the Louisiana heron. In 1916 he published in the *Auk* a long and valuable paper on the ways in which bird-watchers, naturalists, and biologists could interact profitably with one another to resolve important biological questions.[150] But when he went back to England in 1916 to be part of the war effort, he was still not confident in himself.

After the war, in the spring of 1919, Huxley married Juliette Baillot, a French-Swiss woman ten years his junior. She later recounted in her autobiography how he, on their honeymoon, set up a bird-watcher's hide (with room only for one) and proceeded to be less attentive to her than he was to the grebes on Frensham Pond.[151] Soon after their honeymoon, Huxley was back in Oxford as a fellow of New College and senior demonstrator in the Department of Zoology and Comparative Anatomy. Before the term was over, however, he suffered a second major breakdown. He was taken to Lausanne for treatment. There, in the depths of depression, he "was haunted day & night by the idea of suicide."[152]

Huxley returned to Oxford in the fall of 1919, and he gradually began to gather some self-assurance. By the spring of 1921, when he was encouraging Alister Hardy not to give up the idea of a career in scientific research, Huxley was full of life. Significantly, however, when he described to Hardy how excited he felt about his own work, the studies he described were his new investigations in experimental embryology. Hardy was thinking of taking a job in the fisheries. Huxley tried to lure him back to science with promises of exciting work the two of them could do together in the laboratory. But Huxley said nothing about the interest of fieldwork. And this was despite that fact that Huxley himself was about to set off to Spitsbergen to do field research with an Oxford University expedition.

One outcome of the Spitsbergen expedition of 1921 was Huxley's publication of a new paper on bird behavior, in this case on the courtship of the red-throated diver.[153] Although not as well known today as Huxley's great crested grebe paper, Huxley's red-throated diver paper was comparable to the grebe paper in quality and theoretical significance. Here again Huxley pointed to deficiencies in Darwin's theory of sexual selection. Once more he explained that "most of the sexual ceremonies and adornments" of monogamous birds were "chiefly used in ceremonies which take place after mating-up has taken place for the season" and thus could not be explained as the outcome of male combat or female choice.[154]

Where Huxley's red-throated diver paper advanced most appreciably beyond his earlier work was in its elaboration of another insight Huxley may have gotten from Eliot Howard. This was the recognition that the different modes of courtship in birds are a function of more than just the immediate relations between the sexes. In Huxley's words, "the form of the courtship, not merely in details but in broad lines as well, will depend in the main upon other general biological factors affecting the species."[155] As he explained, while it was biologically advantageous for a pair to occupy a territory early, it was disadvantageous for them to hatch nestlings before a food supply sufficient to sustain the nestlings was available. The dual needs of occupying a territory early and avoiding having offspring too soon had consequences for courtship. "The simplest way" of responding to these needs, Huxley observed, was to have the male and female in different states of endocrine excitement. While the male would be in "a sexually-excited state" impelling him "to seek and occupy territory" and "to pair or attempt to pair with any female who stays in the territory," the female would not be ready to permit copulation.[156]

What this meant, as Huxley saw it, was that male birds would routinely be "stimulated by a powerful emotion" but unable to carry through with the biologically appropriate action for as long as the female remained unreceptive. The upshot was that the males' "psycho-physical energy" had to "discharge into other motor channels."[157] This would be accomplished through various motions, some of which were comparable to copulation movements and others of which were not. These motions thus served as physical ways of releasing emotional tension. Over time they would come to have a stimulative function of their own, often enhanced by the development of associated colors and structures.

In addition to these motions, Huxley observed, there were other motions used in courtship that did not seem explicable in terms of the "originally nonsignificant physical release of emotional tension." He had in mind such actions as preening, head shaking, bill dipping, and picking up nesting material. These, he said, all appeared to have developed initially in connection with other functions and were only later "connected with sex in courtship displays." A certain type of preening in grebes and mute swans, he explained, had come to be employed in courtship in "what I may call a ritual way, and without any of its usual functional significance." He predicted that once attention was drawn "to this 'ritual' use of non-sexual actions during courtship activities," the phenomenon would be found to be "of very wide occurrence." [158]

Huxley concluded his paper by criticizing Darwin's theory of sexual selection but lauding a more general insight on Darwin's part. Much of Darwin's theory of sexual selection had to be rejected, Huxley allowed, because so many of the characters and actions used in courtship could not be explained in terms of competition for females. On the other hand, Darwin was clearly correct in recognizing that the development of characters used in avian courtship displays depended on "the emotional effect" they produced "upon the mind of a bird of opposite sex." [159] As Huxley put it, "once epigamic characters come to be advantageous, the mind of the species (in the females in sexually dimorphic forms, in all individuals in those with mutual courtship) is exerting [an] indirect effect . . . upon the future development of colour, structure, and behavior in the race." [160] Thus, for Huxley, as for many other naturalists and psychologists of his day, Eliot Howard and Lloyd Morgan included, the animal *mind* continued to be a matter of the greatest importance.

Huxley published two more papers on bird courtship, one on the avocet and the other on the oystercatcher. These were based on a trip he made to the bird sanctuary of Texel in Holland in the spring of 1924. [161] He was especially pleased in the case of the oystercatcher to be able to develop an explanation of the birds' piping ceremony. In his view, the piping was probably at first "a mere accidental by-product of the high emotional tone of the breeding-season, the bird having to 'let off steam' by expressing in action its general excitement." Later, the ceremony became stereotyped, serving display and threat functions simultaneously. [162]

In part as the result of interactions with Eliot Howard in the early 1920s, Huxley began thinking about writing a general book on bird courtship. [163] He made a start on the project in 1925. Ultimately, however, the book "never got written." [164] Huxley left Oxford for an appointment as professor of zoology at King's College, London, in 1925. In 1927 he resigned his position at King's to collaborate with H. G. Wells and G. P. Wells on the elder Wells's *The Science of Life,* a large publishing project Wells hoped would be as successful as his *Outline of History.* Huxley never held a full-time academic post again.

Huxley published several additional scientific papers on bird behavior but conducted no further field studies. In 1930 he gave a lecture entitled "Biology of Bird Courtship" at the Seventh International Ornithological Congress, held in Amsterdam. Four years later, at the Eighth International Ornithological Congress, held in Oxford, he delivered another, "Threat and Warning Coloration in Birds." In 1938 he published two major papers on the standing of Darwin's theory of sexual selection in light of contemporary research.[165] As he represented it, that standing was no longer very substantial.

Huxley's primary reason for downplaying sexual selection was the one we have seen already: he believed the great majority of display characters could be explained without having to invoke the mechanism of female choice. But his unhappiness with sexual selection ran deeper than this. At its base was his sense that in those situations where sexual selection did operate, its effects were not necessarily for the good of the species. Particularly in polygamous species, where the competition between males was especially intense, one found "display characters pushed to a limit at which they are clearly useless to the species as such and may sometimes be deleterious to the ordinary struggle for existence."[166] There was, as John Durant has aptly characterized it, "a fundamental tension at the heart of [Huxley's] evolutionary philosophy of nature."[167] Darwinian evolution, when confronted frankly, did not offer unequivocal support of Huxley's ideology of progress. Sexual selection in particular seemed to have the potential to lead to frivolous results or even evolutionary dead ends.

Huxley wrote his long reviews of sexual selection not long after he presented his paper "Natural Selection and Evolutionary Progress" to the zoology section of the British Association for the Advancement of Science in 1936. In this paper, which he delivered as president of the section, he allowed that although evolution had indeed been progressive, progress was by no means inevitable. It was a fallacy, he said, to suppose that natural selection always worked for the good of the species or the good of life in general. He agreed with J. B. S. Haldane "that intraspecific selection is on the whole a biological evil." The most conspicuous examples of this biological evil at work were to be seen in sexual selection ("intra-sexual competition," as he now called it). The tails of the peacock and the argus pheasant, he explained, were "characters of the most bizarre sort which, while advantaging their possessor in the struggle for reproduction, must be a real handicap in the struggle for individual existence." In the course of evolution a balance was typically struck, with the favorable effects marginally outweighing the unfavorable ones, but the situation was nonetheless precarious. Extinction loomed as a possibility, Huxley warned, particularly if a species' conditions of existence changed too rapidly.[168]

Earlier Huxley had been happy to see in sexual and mutual selection clear evidence of the role that mind could play in evolution. He had become less happy with sexual selection, but he continued to view the mind as the real hope for evolutionary progress. The human mind, he believed, had reached the stage where it was capable of planning for the future evolution of the species and for life in general. It was not impossible, he suggested, that the human mind might even eventually be enhanced by the development of the faculties of telepathy and other forms of extrasensory perception, "the bare existence of which is as yet scarcely established." It would not be undesirable, he further mused, if in the course of future evolutionary progress some of man's animal nature might die "so as the more effectually to permit the man to live." Always contrasted with the rosy prospect in which he wanted to believe, however, was the reminder, most visibly manifested in the runaway results of sexual selection, that evolution does not necessarily proceed for the best.[169]

In the 1930s Huxley also published (and in some cases transmitted by radio) a variety of popular offerings on bird-watching and bird behavior. In addition he wrote a long section on animal behavior for *The Science of Life,* and he contributed articles to the *Encyclopaedia Britannica.* In 1934, in collaboration with his naturalist friend Ronald Lockley, he made the film *The Private Life of the Gannets.* This film later won an Oscar for the best short documentary of the year.[170] In 1936 Huxley was instrumental in founding the Institute for the Study of Animal Behaviour, and he became the institute's first president.[171]

Although Huxley's own interests moved into other areas, his early work on bird behavior influenced a number of other naturalists. This was particularly true in Holland, where a concern with nature preservation had recently led to the establishment of nature preserves and bird sanctuaries that provided exceptional opportunities for animal field studies. When Huxley visited the bird sanctuary at Texel in 1924, Dutch naturalists, inspired by the nature study programs of the Dutch schoolmasters Eduard Heimans and Jacques P. Thijsse, were busily observing the behavior of common birds. Among the young Dutch field naturalists getting their start in the 1920s were Jan Verwey, who cited Huxley (and Selous) at length, and Niko Tinbergen, who considered Huxley's papers particularly inspiring.[172]

Huxley's importance in the subsequent development of animal behavior studies was not limited to the influence of his publications. He also played a significant role by encouraging other students of behavior—both amateurs and professionals—and by offering an evolutionary understanding of behavior to the general public. Thus, while Huxley gained much from the amateurs Selous and Howard, he was in turn useful to them. As indicated above, he helped Selous publish some of Selous's bird diaries in the form of a book, *Realities of Bird Life*. He likewise did his best to get Eliot Howard to put more ornithological observations into print, not only by prodding Howard directly himself but also by exhorting Howard's closest adviser, C. Lloyd Morgan, to spur Howard on to publication.[173] Simultaneously, Huxley tried to wean Howard away from issues that seemed scientifically unpromising. Feeling, for example, that Howard was spending too much time worrying whether birds had "memory images," Huxley wrote Howard in 1925: "I am hoping you will content yourself with stating the pros and cons and frankly saying it is not ripe for decision and then getting down to recording the quite invaluable facts which you and you alone possess. Frankly I am looking forward to see you pouring out book after book of straightforward observation reserving a full interpretation . . . for the close." He also urged Howard to provide him with an account of some experiments Howard had done, in order "to show the lab-workers what field naturalists can do."[174] As for his encouragement of professional zoologists, Huxley became a friend of both Konrad Lorenz and Niko Tinbergen, each of whom later looked back with gratitude to Huxley for the support and inspiration he provided them.

Huxley was certainly an important figure in the rise of animal behavior studies in the early twentieth century. His role in this regard was not as large as it might have been, however, nor was it as decisive as he made it out to be in his autobiography. His love for bird-watching and his concern with the problems of Darwinian evolution led him to produce some excellent field studies of bird behavior. On the other hand, he never saw this as the most prestigious area in which to make "big, new, original" discoveries in biology. He also recognized before long that field studies of behavior were not a domain in which novel and major contributions could be made in a hurry. As Selous once complained, what generally befell the field naturalist interested in behavior was "many a weary wandering, many an hour's waiting . . . to see, and seeing nothing."[175] To Huxley, furthermore, it must have seemed that every time he made a fresh field observation, it turned out that Selous had already largely anticipated him, not only in the observation itself but also in its interpretation. Selous was able to devote himself to watching birds over great stretches of time. Huxley, with his professional aspirations and responsibilities, did not have such leisure. He exclaimed to Eliot Howard in 1922, "Alas—I fear that in term-time I get less than no time to think over bird problems—I am full up with pupils, committees, lectures, & other research."[176] The fact is that Huxley's major papers on bird courtship, although written over a period of fourteen years, were the product of a total of at most fifty days of fieldwork—and partial days at that—cobbled together out of vacation time. But Huxley's field studies helped to inspire other field ornithologists, who proceeded to spend many more hours than Huxley had been able to spend studying the lives of birds. Dutch naturalists in particular distinguished themselves in this regard, among them Niko Tinbergen and his students, plus the ardent cormorant watcher Adriaan Kortlandt.

As a teacher at Oxford after the war, Huxley gave a course of lectures on animal behavior. Some of his students, as members of the Oxford Ornithological Society, made field observations which Huxley incorporated in two short papers that he published in *British Birds*.[177] It does not appear, however, that Huxley made fieldwork a significant part of his students' formal instruction. In contrast, when Niko Tinbergen went to Oxford a generation later, training D. phil. students to do fieldwork was one of his primary goals.[178]

Rather than as one of the primary founders of ethology, Huxley is better viewed, as David Lack put it, as the individual who bridged the gap between Selous and Howard in the early years of the century and the scientific bird-watchers of the 1930s. Huxley made the contributions of writers like Selous and Howard more accessible to the scientific community. The cantankerous Selous had thrust his observations before the public in the form of scarcely digested field notes. Huxley, in contrast, was a master at organization, synthesis, and presentation. He worked to provide a coherent biological framework for his observational data and he published his results in the most prestigious of the professional scientific journals, notably the journals of the Zoological and Linnean societies of London. Warde Fowler identified Huxley's role and talents very well in 1916 when he congratulated him on the content and the literary quality of his recent scientific papers, saying: "Darwin never could be dull, & so it is with you. Selous & Howard are both rather heavily conscientious, but now someone has arisen to turn a search-light on their tracks."[179]

Huxley could in truth have played a much larger role in the history of ethology. Although his field observations, methods, and ideas about behavior were largely continuous with those of the amateur field naturalists Selous and Howard, he was still capable of much that they were not. He had both the training and breadth of perspective necessary to place behavioral studies squarely in a broad biological context. He displayed this in his paper of 1916 on bird-watching and biological science, where he distinguished the different points of view the evolutionist, the physiologist, and the psychologist brought to behavior studies, and where he observed how so many disputes in biology were due to a failure to distinguish between ultimate causes, immediate causes, and "mere necessary machinery." [180] He showed this again in 1924 in a special series of lectures he delivered at Rice Institute on "the outlook for biology" and likewise in 1925 in the introductory comments he sketched out for his proposed bird courtship book.[181] The latter, had he completed it, would have been a major landmark in the study of animal behavior. Neither Charles Otis Whitman nor Oskar Heinroth, the two professional zoologists with the greatest interests in bird behavior at the beginning of the century, was a field naturalist, however much each of them insisted on observing animal behavior unfolding under natural circumstances. Furthermore, Whitman died without ever assembling his unrivaled knowledge of the behavior of pigeons, and Heinroth, though he described in detail the instincts of the many different bird species that he and his wife reared by hand, was not a generalist. Huxley, in contrast, was a most able synthesizer. Indeed, in the draft of his bird courtship book, he insisted that the time was now ripe to gather the data from "field observation, animal psychology & behaviour, genetics, & comparative psychology" and consider the problem of animal courtship "from a truly broad & unitary biological standpoint." [182] Unfortunately, he failed to carry through on the project. What is more, in the famous synthetic book that he did write, *Evolution, the Modern Synthesis* (1942), he neither made behavior part of the synthesis nor offered guidelines to suggest how the incorporation of behavior into the synthesis might be accomplished.

Huxley was a leading spokesman for field studies of animal behavior in the early twentieth century, but he did not make animal behavior the central focus of his efforts. His interests and ambitions were too diverse. He acknowledged as much in his autobiography when he wrote: "I have been accused of dissipating my energies in too many directions, yet it was assuredly this diversity of interests which made me what I am." [183] His wife, Juliette, expressed this in another way when she said to him: "So many fingers in so many pies. . . . What a pity you haven't got a few more fingers!" [184]

In recent years, historians of science have developed an increasingly rich sense of science as a social process. Through the influence of the sociology and anthropology of science, they have come to speak of scientists as "actors" and to appreciate that a scientist's success as an actor involves in no small measure his or her ability to induce others to take up the same role. Perhaps the most salient feature of Huxley's example as an actor in the history of animal behavior studies was that, in relation to the whole of his own career, he never treated his behavioral work as more than a bit part. This would not be the case for Konrad Lorenz and Niko Tinbergen, each of whom chose to base the whole of his career on behavioral research.

THREE

KONRAD LORENZ AND THE CONCEPTUAL

FOUNDATIONS OF ETHOLOGY

The study of the ethology of the higher animals—unfortunately a still very untilled field—will bring us ever closer to the realization that in our conduct with family and strangers, in courtship and the like, it is more a matter of purely inborn, more primitive processes than we commonly believe.

OSKAR HEINROTH, 1910 [1]

In light of the new conditions the decision for you cannot be in doubt. You must give up anatomy. Your aptitude in the area of animal psychology is for you such a conspicuous trait that it would be like a self-surgery (and a biologically dangerous one at that!) if you now let yourself be a little cowed and wished to act rationally instead of instinctively. He in whom the instinctive is as strongly imprinted as it is in you, should be heartily glad for this finger of God.

ERWIN STRESEMANN TO KONRAD LORENZ, 7 MARCH 1934,

ENCOURAGING LORENZ TO MOVE OUT OF THE SECOND

ANATOMICAL INSTITUTE AT THE UNIVERSITY OF VIENNA

AND DEVOTE HIMSELF TO THE STUDY OF ANIMAL PSYCHOLOGY [2]

In 1910 the German zoologist Oskar Heinroth described the ethology of the higher animals as a yet untilled field that would one day shed light on the deep animal roots of human social behavior. Some twenty years later, a prodigiously talented and ambitious young Austrian naturalist, Konrad Lorenz, embraced this vision of ethology as his own life's work. He cultivated the field with practices very similar to Heinroth's, but with theoretical and discipline-building aspirations that Heinroth himself never hazarded. In doing so Lorenz laid the conceptual foundations of modern ethology.

Lorenz's was not the only approach to animal behavior in Germany at the time. Germany was the world leader in science at the beginning of the twentieth century. Even though a well-defined science of animal behavior was yet to be constructed, studies of living animals and debates about their behavior abounded. Were insects reflex-machines? Could bees distinguish colors? Could instinctive behavior patterns be used to reconstruct phylogenies? Could horses perform calculations of higher mathematics? The first journal for the study of animal behavior began publication in the United States in 1911, but the first society for the study of animal psychology was founded in Germany in 1912. Its aim was to cultivate "experimental animal psychology." Members of the society directed their attention in the first place to studies of the mental abilities of horses, but they anticipated extending their investigations to dogs, apes, and elephants. In brief, a wide variety of questions about animal behavior were energetically pursued by a similarly wide array of German investigators.[3]

It is simply not the case that Lorenz surveyed the whole of this scene and then built upon the best parts of it. Coming to the study of animal behavior through his experiences as an animal lover and raiser, he effectively developed a research program of his own before he had much knowledge of what many other investigators had been doing. Be that as it may, his ideas and practices soon captured the attention of German-speaking ornithologists, biologists, and animal psychologists. By the late 1930s, he had become the preeminent figure in German and Austrian studies of the behavior of higher animals.

Lorenz's story is not a simple one. He liked to represent his science as having emerged in a strictly positivistic fashion, his theories arising from empirical observations unhindered by any preconceptions. In truth, the development of his research practices and theories reflected more than an uncomplicated engagement with nature. Intent upon making a career for himself as a student of animal behavior, Lorenz worked with an eye to the complex institutional terrain in which he operated as well as to the particular kinds of animals he studied. No less than in other countries, in Germany and Austria in the 1930s the study of animal behavior lacked a clear disciplinary home. Neither the zoologists nor the psychologists saw animal behavior studies as central to their primary concerns. The politics of science in this period were further complicated by politics on a broader scale. Through the 1930s Lorenz moved to situate his work methodologically, conceptually, institutionally, culturally, and politically.

THE PASSIONATE ANIMAL RAISER

Konrad Lorenz was born in 1903, a late arrival to the family of Dr. Adolf Lorenz, professor of orthopedic surgery at the University of Vienna, and his wife, Emma, née Lecher. At the time of Lorenz's birth, his mother was already forty-two, his father was forty-nine, and his only sibling, Albert, was eighteen. The family had the resources and disposition to pamper Konrad, and by all accounts this is what happened. But privilege was also accompanied by responsibilities. Konrad was expected to attain educational and professional distinctions worthy of the family's social and cultural standing.[4]

His father provided the primary role model. Adolf Lorenz had risen from humble beginnings (his own father had been a harness maker) to become a world-famous orthopedic surgeon and professor at the University of Vienna. He was highly educated, well traveled, well published, very wealthy, and greatly pleased with the status he had achieved for himself. He loved hobnobbing with aristocrats and dignitaries. In the course of his distinguished career he was nominated for—but never received—the Nobel Prize for Physiology or Medicine.[5]

Adolf Lorenz gained special attention by inventing a new, noninvasive technique for correcting a congenital hip disorder. The technique earned him a fortune when J. Ogden Armour, president of the world's largest meatpacking company, offered Dr. Lorenz a fabulous sum to come to Chicago to treat Armour's only child, who had been born with the disorder. Dr. Lorenz made the trip to the States, performed the operation successfully, and returned to Austria an immensely wealthy man. The name of Armour thus appears twice on the margins of the history of ethology. The Armour slaughterhouse supplied the fetal pigs Wallace Craig used in his bachelor of science thesis at the University of Illinois, and Adolf Lorenz made a fortune treating J. Ogden Armour's daughter, Lolita. Wallace Craig and Konrad Lorenz surely never discovered they had a mutual tie in the Chicago meatpacking industry.

Adolf Lorenz's newly acquired riches inspired him to conceive of a grandiose building project. In the village of Altenberg, a forty-minute commute by train from his medical practice in Vienna, he put an architect and builders to work transforming a modest peasant's house into an imposing edifice. Sensitive to matters of social standing, he stopped short of building a "palace," fearing the ridicule that would befall "an upstart owner." Nonetheless, he was pleased to think of "Lorenz Hall" not as a large villa but rather as a small castle.[6]

Shortly after Lorenz Hall was completed, Konrad Lorenz was born. His father had worried about the pregnancy. He knew from his medical practice that a woman of his wife's age had a chance of bearing a premature child afflicted with physical defects. He believed, or at least so he reported in his autobiography of 1936, that such situations called for a hard eugenic

3.1 Lorenz Hall, the family home in Altenberg, Austria. (An illustration sent by Konrad Lorenz to Erwin Stresemann in the early 1930s. Courtesy of Staatsbibliothek zu Berlin—Preußischer Kulturbesitz, Nachlass 150 [Erwin Stresemann], Ordner 40: Lorenz, Konrad.)

choice. But as he also allowed, he was not sure whether he would have been able to make such a choice had it been necessary. Whether his wife would have had any say in the decision he did not mention. In any case, Konrad Lorenz was born on 7 November 1903, without any sign of impairment.

As a child, Lorenz had a passion for animals. He credited this in part to his nanny, Resi Führinger, a woman with "a 'green thumb' for raising animals." It was also inspired, he recalled, by Selma Lagerlöf's classic children's book *The Wonderful Adventures of Nils,* which was read to him when he was about six. *Nils* is the story of a boy who is magically changed to the size of an elf and flies off on the back of a barnyard gander with a flock of wild geese. By Lorenz's account, the story led him to want to become a wild goose himself, or, failing that, at least to have one. A look at the book itself allows one to see what Lorenz's autobiographical account does not specify. In Lagerlöf's classic, wild geese are identified as clearly superior to their barnyard cousins. Thus, beginning with the bedtime stories of his childhood, Lorenz was taught that wild animals are stronger and more admirable than their domestic relatives are. This idea would feature significantly in his thinking for the rest of his life.

Lorenz had to settle for his mother's buying him a domestic duckling rather than a wild goose. This alternative nonetheless brought him "intense joy," as he charmingly retold it, for the duckling followed him around as if he were its mother and he in turn became "irreversibly fixated on water fowl." He began collecting different kinds of ducks. His enthusiasm in this regard eventually necessitated the construction of several different ponds on the family property. He also collected amphibians and fish and stocked various terraria and aquaria with them. Margarethe (Gretl) Gebhardt, his childhood playmate who would later be his wife, joined him rearing duck-lings and playing at being iguanodons.[7]

As a mature scientist, Lorenz insisted that his scientific practices were continuous with practices he developed in his youth as an "animal lover." He maintained furthermore that being an animal lover was a *prerequisite* to being a good observer of animal behavior. In his words: "It takes a very long period of watching to become really familiar with an animal and to attain a deeper understanding of its behaviour; and without the love for the animal itself, no observer, however patient, could ever look at it long enough to make valuable observations on its behaviour."[8] This childhood

3.2 Konrad Lorenz credited Selma Lagerlöf's children's book *The Wonderful Adventures of Nils* as one of the special inspirations of what became for him a lifelong passion for raising animals. In this illustration by John Bauer, from the original Swedish version of the book, the young boy Nils is shown after having been transformed to the size of an elf. Domestic geese in the barnyard form a contrast with the wild geese in the sky. (From Selma Lagerlöf, *Nils Holgerssons undebara resa genom Sverige*, vol. 1 [Stockholm: Albert Bonniers Förlag, 1907].)

fascination with animals—seen already in the example of Whitman and virtually universal among the ethologists of the twentieth century—was especially marked in Lorenz's case.

The setting in which Lorenz grew up was one he would later describe as a "naturalist's paradise." Lorenz Hall was set in a beautiful garden and trees on the southern slopes of the Vienna woods, some three-quarters of a kilometer from the Danube. It commanded a magnificent view of forest, river, floodplain, and fields. The hills flanking the river and floodplain were covered with vineyards, meadows, and woods. The floodplain consisted of "dense willow forest, impenetrable scrub, reed-grown marshes and drowsy backwaters." [9] Yet as much as Lorenz drew inspiration from the Austrian countryside, and as much as he insisted on the importance of observing the natural behavior of wild animals, he was not by nature a *field* naturalist. He preferred to collect, keep, and breed animals rather than stalk them in their natural haunts. He found it especially satisfying to raise a fledgling to maturity or to nurse a sick bird back to health. To be sure, he enjoyed being outdoors. He loved walking or swimming or simply lazing about with his geese and ducks along the backwaters of the Danube. But while he took great pleasure in walking over the hills with a tame jackdaw or raven following him, he was not interested in crouching in a hide and spying on a creature that was not in some measure under his care or control. His practices were thus comparable to those of Whitman in the United States and Heinroth in Germany but appreciably different from those of the field naturalists of Britain and the Netherlands.

Lorenz is said to have been an excellent student at the elite, humanistic *Schottengymnasium* in Vienna, which he attended from the age of eleven or twelve to the age of eighteen. His earliest courses were on scientific subjects, but he ultimately received a strong education in the humanities as well. His own autobiographical reflections offer few insights on his gymnasium education beyond his mention of Philip Heberdey, a Benedictine monk who taught his students about Darwinian evolution. Nonetheless, the kind of intellectual formation that the humanistic gymnasiums of Germany and Austria promoted in this period is well known. At the Schottengymnasium Lorenz would have been trained in the classics and in modern science. He would also have imbibed the German cultural ideal of *Bildung,* or "cultivation," that was so central to the self-identification and life goals of highly educated, upper-middle-class German and Austrian intellectuals at this time. The *Bildung* ideal encouraged intellectual or cultural development rather than material acquisition or political advancement. Likewise, it valued comprehensive knowledge over narrow expertise. Over all, it stressed the importance of spiritual *culture* as opposed to the external, worldly, and ultimately superficial aspects of modern *civilization*. The intellectual aspirations, scientific style, and general worldview that Lorenz began expressing in his writings of the 1930s and continued to express for the rest of his career were entirely consistent with the intellectual and cultural ideals he imbibed at home and in school.[10]

Although Lorenz's retrospective comments on his gymnasium days said little about his formal instruction, they included abundant references to his friend and classmate, Bernhard Hellmann. The young Hellmann shared Lorenz's enthusiasm for animals and joined him on collecting expeditions. Together they studied freshwater crustaceans and learned to experiment on the breeding behavior of cichlid fish. Lorenz credits Hellmann with discovering how to manipulate the breeding behavior of a male cichlid fish by placing a mirror in the fish's aquarium. Previously, the male had killed not only other male cichlids but also any female introduced into his tank. With the mirror in the aquarium, however, the male fought his own image to the point of exhaustion, and then immediately courted a new female when Hellmann put it in the tank with him. In their last years of school the two young men also collected Cladocera and speculated on how this group of small crustaceans had evolved.

Upon graduating from the Schottengymnasium Lorenz wanted to study zoology and especially paleontology, but his father insisted that he study medicine instead. In the fall of 1922 Adolf Lorenz sent his son off to the United States to Columbia University to begin premedical studies. Much to Dr. Lorenz's displeasure, the younger Lorenz, with Margarethe Gebhardt on his mind, returned home before the end of the term. He did, however, agree to continue to study medicine. In January of 1923 he enrolled as a medical student at the University of Vienna. It was there that he encountered the comparative anatomist Ferdinand Hochstetter.

Ferdinand Hochstetter was one of the great men of the University of Vienna's medical faculty. Already in his sixties at the time Lorenz entered medical school, Hochstetter was a brilliant comparative anatomist with a vast knowledge of vertebrate zoology, a special interest in comparative embryology, and a deep commitment to conveying to students the value of the comparative method. Lorenz revered Hochstetter. First as a student, then as an instructor, and finally as an assistant in Hochstetter's institute (the second of the two anatomical institutes at the University), he sought to learn from and be like Hochstetter. He imprinted on Hochstetter's model to such an extent that he unconsciously imitated Hochstetter's mannerisms as a lecturer later in his own career.[11]

Lorenz was already familiar with and fascinated by the idea of evolution. He had been introduced to the idea as a child through Wilhelm Bolsche's book *Die Schöpfungstage* (*The Days of Creation*). He had later been taught evolutionary theory in school, and he and Bernhard Hellmann had speculated together on the evolutionary origins of Cladocera. From Hochstetter Lorenz learned much more. In particular, he learned how to study evolution through the rigorous methods of comparative anatomy. Comparative anatomists in this period tended to have less to say about the mechanisms by which evolution operates and more to say about the course that evolution has taken. They sought to identify homologous structures (for example, the horse's foreleg, the bird's wing, and the whale's flipper) and to use these to evaluate evolutionary affinities. With Hochstetter's encouragement, Lorenz decided that the methods of comparative anatomy could be applied to animal behavior patterns just as effectively as they could be applied to body parts. In other words, behavior patterns could be used just as organs to determine common ancestries and reconstruct phylogenies. The idea was not new with him. Charles Otis Whitman in the United Sates and Oskar Heinroth in Germany had already come to it. But Lorenz was yet to learn that he had been preceded in his discovery. He would later call this idea of the "homologizeability" of behavior patterns the "Archimedean point" from which ethology took its origin.[12]

When Lorenz became a full-time medical student, he did not stop raising animals. To the contrary, he now kept them at the family home in Altenberg *and* at the family apartment in Vienna. And not only did he continue to care for the animals he already owned, but he also bought new ones. A young jackdaw (*Coleus monedula spermologus*) was the new acquisition that proved of most consequence for him. He purchased the bird for four schillings from a pet dealer in the summer of 1926. He did so, as he later recounted, not so much for scientific reasons as because he "suddenly felt a longing to cram that great, yellow-framed red throat with good food." Taking the bird back to Altenberg, he placed it at first in an aviary. He found after a few days that the bird had such an attraction for him that he could allow it to fly freely after him, first in the house and then later out of doors. He named the bird Tschock after the sound of its call note.[13]

The bird's attachment to him gave Lorenz special advantages as an observer. Unrestrained either by fear or by the confines of a cage, the jackdaw performed many of its natural behavior patterns in Lorenz's presence. When Lorenz went on walks, the young bird flew along with him. Lorenz watched closely how the bird gradually perfected its flying and landing maneuvers. Combining his talents as a naturalist and his experience as an avid motorcyclist, Lorenz noted that trying to observe how a bird flies is like trying to observe how a cyclist rides: one can more easily recognize and analyze the elementary balancing motions of a novice than the practiced moves of an expert.

Lorenz also paid attention to how his jackdaw responded to the call note of its species, how it behaved in the company of other birds such as hooded crows, and how it found its way back (or, alternatively, failed to find its way back) to him or to the house. He was especially interested to discover that the behavior patterns that a jackdaw would normally direct toward members of its own species were in the case of this hand-reared jackdaw directed toward others. Tschock, who proved to be a female, directed toward Lorenz all the drives that a young jackdaw would normally direct toward its parents, but she directed her courting behavior toward the Lorenzes' housemaid.

UNDER THE WING OF GERMANY'S LEADING ORNITHOLOGISTS

Tschock became the subject of Lorenz's first published scientific paper. Lorenz's fiancée, Margarethe Gebhardt, and his friend Bernhard Hellmann conspired to get hold of Lorenz's jackdaw diary. They copied Lorenz's notes and sent the copy off to Germany's leading authority on bird behavior, Oskar Heinroth. Heinroth was sufficiently impressed to encourage Lorenz to write up his diary for publication. The paper, "Beobachtungen an Dohlen" ("Observations on Jackdaws"), appeared in the October 1927 issue of the *Journal für Ornithologie*.[14] The work earned Lorenz the approval not only of Heinroth but also of Erwin Stresemann, the brightest and most influential figure in German ornithology. Stresemann edited the *Journal für Ornithologie,* served as secretary-general of the German Ornithological Society, and occupied the most important ornithological post in the country as curator of ornithology at the Natural History Museum in Berlin.[15] In the years that immediately followed, both he and Heinroth served as important mentors for Lorenz and helped him in a host of ways in advancing his career.

Lorenz had learned a great deal from having a bird whose social actions and reactions were directed toward him rather than members of its own species, but there were other things that a single bird could not teach him, particularly with regard to how a normal jackdaw colony functioned. He presumed that certain of his bird's seemingly instinctive and "ceremonial" behavior patterns were actions that under normal circumstances must serve to elicit appropriate reactions on the part of fellow jackdaws.

Prodded by Stresemann, in particular, he decided to raise a whole colony of the birds.[16]

In spring 1927, Lorenz set out to establish a jackdaw colony that was sufficiently tame to allow him to observe and manipulate it but sufficiently large that the members of the colony would direct their social drives toward each other rather than toward him. To this end, he converted the attic of Lorenz Hall into an aviary for fourteen young, hand-reared jackdaws. He marked the birds for identification purposes by placing colored bands on their feet. He then allowed the birds to fly freely. The colony was not easy to maintain. More than once, it suffered major losses and he had to replenish it with new, hand-reared individuals.[17]

Lorenz's marriage to Margarethe Gebhardt also took place in 1927. The couple took up residence in the family home in Altenberg. The following year Lorenz received his doctorate in medicine. He was not interested in a career as a physician, however. He wanted to be a zoologist who studied animal behavior. He proceeded to enroll as a doctoral candidate in the university's Zoological Institute, headed by the comparative anatomist Jan Versluys. Zoology at the University of Vienna at this time was dominated by comparative morphology and embryology. Lorenz's graduate training in zoology thus reinforced the intellectual and methodological lessons he had already learned from Hochstetter.[18] Meanwhile, he continued to work for Hochstetter at the university's "Second Anatomical Institute." He also began taking courses taught by the psychologist Karl Bühler, who was keenly interested in the topic of instincts.[19]

While pursuing his graduate studies, he also managed to maintain his jackdaw colony. In the fall of 1930, after three and a half years' experience with the colony, he completed a manuscript detailing his observations on the birds. He added observations on magpies, crows, and ravens. He sent the manuscript to Heinroth, asking for the ornithologist's critical advice. He also invited Heinroth, if Heinroth approved of the manuscript, to send it directly on to Stresemann.

Heinroth liked what he saw. Comparing the manuscript with Mathilde Hertz's fine paper on corvids, he concluded Lorenz's work was better. Hertz had made some beautiful observations, Heinroth told Lorenz, but she had failed to relate the instinctive and learned behavior patterns of her captive birds to the natural behavior patterns of birds in the wild. Heinroth offered Lorenz some minor editorial suggestions and urged him to find a publisher to make his wonderful observations available in book form to a wider audience. Lorenz was greatly encouraged by this response. He had earned the respect of the senior scientist whose interests were most akin to his own.[20]

Heinroth was a full generation older than Lorenz was—thirty-one years older, to be precise. Like Lorenz, he had been an animal lover since his childhood. He is said to have learned to walk in the family hen house, where he watched the birds and mimicked their calls. Later, as an adolescent, especially when his voice was changing, he found he could drive flocks of geese to despair by his rendition of their warning call. The derogatory jibes of his classmates did not dissuade him from pursuing his interests in bird behavior.

Heinroth studied medicine and zoology at the universities of Leipzig, Halle, and Kiel, ultimately earning his doctorate from the Physiological Institute at Kiel in 1895 for a study of urine formation in fish. The following year he went to Berlin for further zoological work. There he divided his time between university and museum studies on the one hand and volunteering at the Berlin zoo on the other. In 1900 he was invited by Bruno Mencke to join Mencke's privately funded scientific expedition to the South Seas as a zoologist and physician. The high hopes for the expedition were dashed in New Guinea when natives attacked the party and mortally wounded Mencke. Heinroth, however, survived the assault and succeeded in bringing back to Berlin a valuable collection of animal skins and living creatures. He wrote up the expedition's ornithological results and went back to studying animals at the zoo.[21]

Heinroth worked in a volunteer capacity at the zoo until 1904, when he was appointed assistant to the zoo's director, Ludwig Heck. That same year, he married Magdalena Wiebe, who became his invaluable co-worker. In 1910 he was put in charge of establishing an aquarium at the zoo. When the aquarium was completed in 1913, he became its first director. Finding any job as a zoologist in Germany in this period was a significant accomplishment, since the supply of trained zoologists far exceeded the number of academic openings. By securing a niche for himself at the Berlin zoo, Heinroth was able at one and the same time to make a living and pursue his interests in animal behavior. Lorenz a generation later would be faced with the same challenge.[22]

Heinroth was especially interested in what he called the "finer details" of bird behavior. He made his concerns clear in a major paper he delivered at the Fifth International Ornithological Congress, held in Berlin in 1910. There he promoted the study of what he called the "ethology" of birds— the instinctive "customs and practices" that were essential to birds' lives as social beings.[23] He argued that "species-specific instinctive actions" (*arteigene Trlebhuudlungen*) could be used just like morphological features to determine the genetic affinities of species, genera, and subfamilies. He seems to have been even more intrigued, nonetheless, by the ways that instinctive behavior patterns functioned in the family life of the ducks, swans, and geese that he studied. He paid special attention to the behavioral displays that serve as social signals among members of a species. He anticipated Huxley in talking about behavior patterns as "rituals." He also described what Lorenz would later call "imprinting." But he was not prone to theorizing, and he did not invest his descriptions of the above-mentioned phenomena with the same kind of theoretical significance that Huxley and Lorenz subsequently did. Still, messages of major import were to be found in his work. He believed that the lives of gregarious birds are largely (though not exclusively) a function of social instincts. He furthermore believed that when bird behavior seems to resemble human behavior, it is not because birds have anything like the intelligence of humans, but because "many of our forms of interactions, such as making others look bad or glorifying our own group members, are nothing more than social instincts."[24]

Heinroth was interested not just in the ducks, geese, and swans that he discussed in 1910. With his wife, he undertook the systematic study of the instinctive behavior patterns of all the bird species of central Europe. Their methodology involved rearing baby birds at home, in isolation from other birds. They believed these deprivation or "Kaspar Hauser" experiments could distinguish innate species-specific actions from species-specific actions acquired by imitating or learning from conspecifics.[25] Basically, they identified as innate all those species-specific behavior patterns that emerge in a developing bird independently of any contact with other individuals of its own species. Emphatically empirical in their approach, they believed that what was innate and what was learned in different bird species could be determined only through experiments conducted on a species-by-species basis.

Magdalena Heinroth gave an early report on this work at the International Ornithological Congress of 1910, but it was not until 1924 that the results of the Heinroths' program began pouring forth in earnest. That was the year they published the first installment of what would eventually be a four-volume classic, *Die Vögel Mitteleuropas in allen Lebens- und Entwicklungsstufen photographisch aufgenommen und in ihrem Seelenleben bei der Aufzucht vom Ei ab beobachtet* (*The Birds of Central Europe, Photographed in All Life and Developmental Stages and Observed in Their Mental Life by Raising Them from Egg Onward*).[26] By 1928, the first three volumes were in

3.3 Oskar Heinroth posing in the 1920s with three juvenile Eurasian eagle owls (*Bubo bubo*) raised from eggs. (Courtesy of Archive Berlin-Zoo.)

3.4 Magdalena Heinroth in the neighborhood of the Berlin zoo, hurrying to keep up with the two tame cranes, Pankraz and Trana, that she and her husband reared by hand. For five years beginning in 1923, the year the birds were hatched, it was the Heinroths' habit every morning to walk (or at times run) with the birds on or near the grounds of the zoo (while the birds walked or flew). (Courtesy of Archive Berlin-Zoo.)

print. These volumes were a glorious testimony to the Heinroths' painstaking labors. The couple's bird-rearing achievements ranged from the tiniest songbirds to the largest raptors and to stately storks and cranes.

Lorenz was thrilled to find such an intellectual mentor as Heinroth. Here was a man with whom he could discuss research practices, the details of bird behavior, and the joys and sorrows of animal raising. Lorenz was soon addressing Heinroth as "my spiritual father" and "leader." [27] As he explained enthusiastically (albeit immodestly) to Heinroth in February 1931, his own ideas were so completely in harmony with Heinroth's that it was impossible for him to estimate just how much he had been influenced by reading *Die Vögel Mitteleuropas*. Other people, he said, had remarked to him on the great similarity between his ideas and Heinroth's own.[28] Four years later Lorenz told Wallace Craig: "It was a remarkable experience for me to read [Heinroth's] book, as even I myself had a queer feeling that I'd written it myself and couldn't remember when I'd done it! In every sentence that Heinroth has written in all his life, there isn't a word that has not been important to me and he never has left anything out, that I would have thought relevant. The consequence of this incredible congruence of viewpoints between Heinroth and me is that I really am in the unique position to 'command another man's experience.'" [29]

Energized by finding an authority whose views were so much like his own, excited by the prospect of establishing animal behavior studies on a new and firm foundation, and considering the implications this could have for human psychology, Lorenz exuberantly asked Heinroth in February of 1931: "Are you aware, Herr Doktor, that you are really the founder of a science, namely, animal psychology as a branch of biology? That *that* is the most profound value of the *Birds of Central Europe*? That this constitutes a way of studying and researching that must actually be extended to 'the animals of the world'? Who knows what will become of today's human psychology if one can only know what is instinctive behavior and what is rational behavior in humans? Who knows how human morals with their drives and inhibitions would look if one could analyze them like the social drives and inhibitions of a jackdaw." [30]

Lorenz's study of the social drives and inhibitions of jackdaws and other corvids appeared in the *Journal für Ornithologie* in 1931.[31] Notwithstanding the difficulties he had experienced in maintaining his jackdaw colony, he had been able to observe a series of phenomena that were of great interest to him.[32] One of these was the rank order in the colony. Another was the way the adults of the colony acted to prevent other members of the colony from straying or flying off with other flocks. A third was how the jackdaws would instinctively attack a "predator" that seized and carried off a fellow jackdaw (or for that matter any shiny black object, as Lorenz found when he walked across the lawn carrying a black bathing suit). Lorenz also observed the pairing of adults, the rearing of young, the "following drive" in the young birds, and much more. In addition, without yet coining the word "imprinting," he noted that a bird reared in isolation from members of its own species could in a brief, early, and critical period of its development become attached to a human being instead of one of its own kind.

One additional feature of Lorenz's 1931 corvid paper merits attention here. This is the way he associated variations in behavior with illness or unnatural conditions. As he saw it, the instinctive behavior patterns of all healthy individuals of a wild, free-living species are essentially identical. Such animals, he claimed, do not differ in their innate responses to specific stimuli. Only in unhealthy or captive animals, he maintained, do the instinctive behavior patterns of the healthy, wild type fail to appear in fully developed form—or sometimes fail to appear at all. Believing the individuals of any wild form to be very homogeneous genetically, he supposed that whenever behavioral differences were observed in captive animals of the species, these were a function of differences in individual experience or state of health. Later he would also associate variation with the degenerative effects of domestication. He was a self-styled Darwinian, but his thinking afforded no real room for the naturally occurring variations that need to be present before natural selection can act. If his training in comparative anatomy gave him a good handle (or so he believed) on studying the *course* of evolution, he was never especially successful when it came to thinking about evolution's *causes*.

Lorenz's success with jackdaws led him to think it would be worth studying the behavior of other social birds as well. He decided the night heron (*Nycticorax nycticorax*) was particularly appealing because of its small size and the ease with which it could be bred in captivity. He wrote to Heinroth in the fall of 1930 asking if it would be possible to get some young night herons from the Berlin zoo. The following spring he got what he wanted—fourteen young birds. He traveled to Berlin to collect the first batch and to meet Heinroth in person for the first time. He was soon able to report to Heinroth that not only was he busy caring for his newly acquired night herons, but he had just received ten young ravens from Bosnia, and even more jackdaws from the neighboring countryside were on their way. Much to his disappointment, however, an order for ten greylag goose chicks from Yugoslavia had not brought results. It was not until the spring of 1932 that he received his first greylag geese, procured for him by Otto Antonius, director of the Schönbrunn zoo. (And it was not until 1935 that he received the egg that provided him with the goose he named Martina, a bird that would make as much of an impression on him as Tschock had done earlier.)[33]

Heinroth meanwhile not only helped Lorenz get new birds. He also in 1931 drew the younger man's attention to two major scientific papers of which Lorenz was unaware. One was Jan Verwey's recent (1930) study of the European gray heron (*Ardea cinerea*). The other was Heinroth's own classic paper of 1910 on the ethology and psychology of the anatids. Lorenz was an unsystematic reader at best, and such suggestions were invaluable to him. He was especially happy to pick up evidence, insights, and references corresponding to his interests (as he indeed did from the papers by Heinroth and Verwey). But he was not especially open to the ideas and approaches of investigators whose studies of behavior differed from his

3.5 The young Konrad Lorenz with his own hand-reared, free-flying ravens, c. 1931. (Courtesy of the Lorenz family.)

own—and whom he did not know personally. As he bluntly put it a few years later, he was willing to countenance only observations that had been reported by people whom he knew and whom he counted among "the limited number of genuine animal observers." [34]

While he continued to acquire new species and made his first moves toward developing a more general understanding of instincts, Lorenz also looked for a better institutional base for his work. Early in 1931 he formulated a plan for an ornithological observation station on an island in the Danube, close to his home. The idea was to create an Austrian bird-watching facility comparable to the stations already in place in Helgoland and at Rossitten. Lorenz appealed to the German Ornithological Society

3.6 A Lorenz self-portrait: the naturalist composing a letter, with ten ravens calling loudly in his ear. (From a letter from Lorenz to Oskar Heinroth in the spring of 1931. Courtesy of Staatsbibliothek zu Berlin—Preußischer Kulturbesitz, Nachlass 137 [Oskar Heinroth], Ordner 27: Lorenz, Konrad.)

and the Austrian Zoological and Botanical Society for joint funding of the operation.

Dr. Moriz Sassi, curator at the Natural History Museum in Vienna, helped construct the official proposal. The plans identified a small island and parts of two other islands in the Danube as the site for the station. A wooden building was to be erected on the first of the islands. Elevated on concrete pylons as protection against flooding, this structure would have four rooms: a laboratory equipped with instruments and a dissecting table; a room for incubating, rearing, and keeping birds; a room for preparing and storing bird feed; and a small sleeping room with a pair of bunk beds. Above these, on the roof, would be an observation room, with windows on all four sides for surveying the surrounding area. Lorenz and Sassi drew up a budget, and Lorenz provided sketches of the facility.

The proposal identified Lorenz as the presumptive director of the Danube station. The station's chief task was to be the investigation of "the finer details of the biology of our native birds," with an emphasis on "the entangled reflex chains that we are accustomed to calling instincts." The reference to "the finer details" of bird biology echoed the work of Heinroth. The attention to "entangled reflex chains" reflected Lorenz's interest in dealing more systematically and more theoretically with the phenomenon of instinct. As the proposal indicated, whereas instincts had "hitherto been studied almost exclusively from a psychological stand-point," the plan here was to study them "physiologically." The station was also promoted as a site for nature conservation. Lorenz aspired to natural-izing a number of waterbirds, including greylag geese and mute swans, which were either rare or no longer native to the area.[35]

Lorenz was sanguine about his chances at first. In April 1931 he wrote Heinroth saying that "it would be 'too good to be true,'" but if the station did materialize, he would find it hard to know what else to wish for.[36] The following month he asked Stresemann to imagine "with what joy I look forward to the development of the station." It would provide him special research opportunities, he explained, that would never have been available to him as a private person. There was so little known about the instincts of most birds, he said, that "really any bird that one can keep in a free flying state promises results!!"[37]

Unfortunately, the plans for the station came to naught. The Austrian Zoological and Botanical Society proved too hard pressed for funds to be able to share with the German Ornithological Society the projected costs of

3.7 Konrad Lorenz in the early 1930s in front of one of the ponds constructed for waterfowl at his family home in Altenberg. Through the 1930s, as other opportunities failed to materialize, Lorenz's private ornithological station remained the primary site of his research efforts. (Courtesy of the Lorenz family. Copy provided by Klaus Taschwer.)

constructing and maintaining the facility. Problems also arose with respect to acquiring the islands. As of the fall of 1931, Lorenz was able to take some small comfort in the fact that Hochstetter had appointed him to an assistantship in anatomy, with a modest salary. But this was not the same as being director of an official ornithological field station operating under the auspices of the Austrian Zoological and Botanical Society and the German Ornithological Society.

IDENTIFYING INSTINCTS

Lorenz had considerably more success making sense of bird behavior than he did establishing a post for himself as director of a field station on the Danube. In 1931 he drafted a paper on bird instincts. Fearing that Heinroth might find the paper insufficiently empirical, Lorenz sent it first to Stresemann. Only after Stresemann responded positively did Lorenz seek out Heinroth's comments as well. The paper appeared in the *Journal für Ornithologie* in 1932 under the title "Methods of Identification of Species-Specific Drive Activities in Birds."[38]

The phrase "species-specific drive activities" (*arteigene Triebhandlungen*) had been coined by Heinroth. Lorenz used it rather than the word "instincts," he explained, because the latter had been used in too many different and conflicting ways, and because Heinroth's phrase seemed better to convey the nature of the process in question. Lorenz also drew upon the authority of the German zoologist and instinct theorist Heinrich Ziegler. Ziegler had distinguished between instinctive and intelligent behavior by claiming that instinctive activities "depend upon inherited pathways of the nervous system" whereas intelligent activities "depend on individually acquired pathways."[39] Lorenz liked the way Ziegler had replaced a psychological definition of instinct with a histological one. He followed Ziegler's lead, stating: "My concept of an instinctive behavior pattern is that of a behavioral sequence based upon inherited pathways laid down in the central nervous system, such that the pattern is just as invariable as its histological foundations or any named morphological character."[40]

Lorenz was not offering a hypothesis to be tested; he was declaring a fundamental assumption. He had no intention of cutting into his animals in an attempt to correlate particular behavior patterns with particular neural structures. What counted for him were the ideas that instinctive behavior patterns are innate and invariable and that such patterns can be just as helpful as bodily structures when it comes to reconstructing phylogenies.

Distinguishing instinctive (and invariable) components of behavior from intelligent (and variable) components was a commonplace. Lorenz differed from the majority of his predecessors and contemporaries, however, in how sharply he separated the two. Friedrich Alverdes, professor of zoology at Marburg and author of various articles and books on animal sociology and psychology, had allowed that "instinctive actions" (*Instinkthandlungen*) are behavior sequences that include variable as well as invariable components.[41] Lorenz agreed that entire sequences of animal behavior involve acquired as well as innate components, *intercalated* to form a functional whole, but he insisted that this intercalation in no way altered the more restricted (and "absolutely rigid") "species-specific drive activities" of which he spoke. These, he insisted, are always innate and invariable.

Having spelled out his basic concept of instinctive behavior, Lorenz proceeded to explain the special virtues of his own research practices. Birds, he maintained, are especially suitable subjects for behavioral analysis. One reason for this is that the variable components within birds' behavior chains are few enough in number to be distinctly recognizable. Another reason is that birds are very similar to humans in the senses they employ in dealing with the external world. They are first and foremost visually orienting animals. Beyond that, they rely upon auditory cues, especially for hearing the vocalizations of birds of their own species. Because we humans have similarly well-developed senses of sight and hearing, this makes it relatively easy for us to identify the cues to which birds respond (as opposed, say, to the olfactory cues in which dogs take such interest).

An added attraction of birds was the one that Lorenz (and the Heinroths before him) had come to exploit very effectively. Birds can be raised and maintained in a free-flying state, thereby permitting the observer to study their instinctive behavior patterns at close hand. Lorenz acknowledged that comprehending particular behavior patterns was sometimes difficult, but this was hardly surprising, he said, given that these patterns had evolved in the service of eliciting responses from fellow birds, not from humans. In any case, Lorenz maintained, the special postures birds adopt in performing instinctive behavior patterns are usually conspicuous enough to make the human observer at least *aware* that an instinctive behavior pattern is being performed, and thus deserves attention.[42]

Over the course of his career, Lorenz's colleagues would credit him with having an extraordinary and indeed unrivaled *intuitive* knowledge of animals. He believed this himself, and helped cultivate the image. He placed a high value on intuitive judgments, and he prided himself on his ability to make them, not only in his scientific work but also in assessing the character of other people. His belief in his own instinctive ability to judge the goodness of others served him poorly later in the decade, not only as a part of his theorizing but also with respect to some of the company he chose to keep. In his 1932 paper, however, he kept his focus on instinctive behavior patterns in animals. There he claimed that anyone with extensive experience raising animals would be able to recognize intuitively which behavior patterns were instinctive and which were not.

But Lorenz knew better than to assert that the matter of distinguishing instinctive from noninstinctive behavior patterns could be left solely to intuition. He went on to identify five criteria that could be used in deciding the question. He was inclined to judge that a behavior pattern was instinctive, he said, when it satisfied one or more of the following five conditions:

1. If a young animal reared in isolation from members of its own species displayed the species-specific behavior pattern without having had any models to learn from

2. If all the individuals of a species performed the same behavior pattern in the same, stereotyped way

3. If there was a conspicuous incongruity between the normal intelligent abilities of an animal (as seen in other situations) and the abilities that would be necessary for the animal to perform the behavior in question by insight

4. If the behavior pattern was performed incompletely or elicited in a situation when the appropriate biological goal was not there, thus making it clear that the animal was not conscious of the biological purpose of the action

5. If the rigidity of the behavior pattern and its resistance to environmental influences continued to be displayed under conditions far different from those under which the pattern originally evolved[43]

Lorenz concluded his paper by reaffirming that instinctive behavior patterns could be studied scientifically. He also admonished students of animal psychology to take it upon themselves to learn the *whole* of an animal's behavioral repertoire. "Many otherwise excellent investigations of animal psychology" fell short, he explained, because the researcher was not sufficiently familiar with the instinctive behavior patterns of the species in question.[44]

The publication of Lorenz's *Triebhandlungen* paper in 1932 in the *Journal für Ornithologie* confirmed his arrival as a significant actor in German ornithology. His new visibility was further enhanced by the conspicuous role he played at the fiftieth annual meeting of the German Ornithological

Society (Deutsche Ornithologische Gesellschaft—the DOG), held in Vienna from 1 to 4 October 1932. Oskar Heinroth, president of the DOG since 1926, presided over the meeting, which attracted one hundred or so participants. Also present was Erwin Stresemann, the secretary-general of the society and editor of the *Journal für Ornithologie*. His Majesty King Ferdinand of Bulgaria, a longtime patron of the society (and of European ornithology in general) likewise graced the meetings with his attendance. Thirty-five other members of the society were also there, including the two professors of zoology at the University of Vienna, Paul Krüger and Jan Versluys; the director of the Zoological Garden at Schönbrunn, Otto Antonius; the curator of the Natural History Museum in Vienna, Dr. Moriz Sassi; and the young ornithologists Horst Siewert, Gustav Kramer, and of course Lorenz himself. Among the sixty or so additional guests was Lorenz's friend Wilhelm Marinelli, who later would succeed Versluys as professor of zoology and director of the First Zoological Institute at the university.[45]

Lorenz himself was much in evidence at the meetings. Among the conference highlights was a Sunday afternoon visit to his research station at Altenberg. Eighty or so of the conference attendees made the excursion to Altenberg by bus or train (while King Ferdinand came in his elegant Mercedes). Once at the Lorenz home, the visitors had the opportunity to admire an inspiring array of free-flying birds. These included some twenty-five jackdaws, four night herons, a greylag goose, a magpie, a pair of black-capped parakeets, and an abundance of ducks, both domestic and wild. A honey buzzard was allowed to fly freely at first, but it was later caged, as was a little egret. In the aviaries the guests saw additional night herons, two black storks, two ravens, and more.

Upon completing their tour of the research station, the guests were ushered into Lorenz Hall for tea. There Adolf Lorenz, "as host and master of the house," was pleased to be able to greet the king and the other guests. Writing his autobiography a few years later, he recounted with relish the short speech he addressed to the gathering. He did not mention that his son also spoke to the assembled group.[46]

The younger Lorenz's presentation to the attendees began with a survey of the aims of his free-flight facility. He stressed the value of such a facility for the study of bird instincts. He also stressed the advantages of using free-flying, tame, and healthy birds rather than birds that were caged or frightened. He gladly acknowledged that Oskar and Magdalena Heinroth had pioneered such studies, noting that their great work, *Die Vögel Mitteleuropas,* provided "the foundation and guiding principle" for ethologi-

3.8 Dinner with dignitaries at the Lorenz home, perhaps at the time of the meeting of the German Ornithological Society in Altenberg, October 1932. *Seated from left to right:* Oskar Heinroth; Adolf Lorenz; Ferdinand Hochstetter; Konrad Lorenz's children, Thomas and Agnes; Konrad Lorenz; and Emma Lorenz (Konrad's mother). (Photo courtesy of the Konrad Lorenz Institute for Evolution and Cognition Research.)

cal studies. He then went on to suggest that the student of behavior was not the only investigator who could profit from the study of tame, free-flying birds. So too could physiologists (studying motion or sensory capacities), psychologists, and even functional anatomists. Mentioning that he had a paper on bird flight forthcoming in the *Journal für Ornithologie,* he invited biologists from different specialties to collaborate in pursuing the many research possibilities that his new approach was opening up. He concluded by indicating that he planned to devote himself to ethology, studying various bird groups in the same way that the society's president, Heinroth, had already studied the anatids.

Heinroth responded by wishing Lorenz luck and pledging that the DOG would support his efforts insofar as possible. (At the society's business meeting earlier that morning, Heinroth had allowed that Lorenz had had the society's moral support since 1931, but financial assistance for Lorenz's work could come only from special contributions, not the society's regular funds.) Heinroth then thanked the Lorenz household warmly for its hospitality and signaled to the guests that it was time to return to Vienna. Over the next two days of the conference, Lorenz continued to play an active role. Two weeks after the conference he wrote exuberantly to Heinroth, thanking him for all he owed him, including bringing the DOG to Altenberg.[47]

Later in his career, Lorenz credited the German historian of philosophy Wilhelm Windelband with the view that "the development of any inductive natural science proceeds through three indispensable stages: the purely observational recording and describing of fact, the orderly arrangement of these facts in a system, and finally the quest for the natural laws prevailing in the system."[48] Historians of science today are not satisfied with this as a general model of how sciences develop, and it certainly does not fit well the historical development of ethology. Windelband's formulation does, however, correspond at least in a general way with Lorenz's own record of publication. At first, his writings were primarily descriptive. He then concerned himself more and more with methodology and theory. He also expressed himself with increasing boldness and authority. But this growing self-confidence was not simply a function of his ever-widening knowledge of bird behavior. It also reflected his success in interacting with other scientists and positioning his views and practices with respect to theirs. The encouragement of Hochstetter, Heinroth, and Stresemann (and others) spurred him on. These scientists helped him feel that his particular scientific practices gave him special authority to speak on behavioral questions.

ACADEMIC DEVELOPMENTS

The week after the DOG's meeting in Vienna, Lorenz apprised Stresemann of new developments in his academic situation. These developments, Lorenz was convinced, were the result of Stresemann's having worked some sort of magic on Lorenz's behalf with Professor Krüger. Krüger, in any event, was ready to expedite Lorenz's progress toward his Ph.D. Once he had that degree in hand, Lorenz imagined, he would be able to turn his attention to a more psychologically oriented program of researches on bird behavior. He was now thinking of writing his *Habilitation* thesis (the postdoctoral thesis he needed to qualify to become a university lecturer) in Bühler's Psychology Institute, where issues of Gestalt psychology were a major focus and where Lorenz had been taking courses since 1928. "Perhaps," Lorenz told Stresemann, "I can work toward a psychological *Habilitation,* without taking myself away from my real interests."[49]

For his doctoral *thesis* in zoology, Lorenz was allowed to count the paper on bird flight he had written for the *Journal für Ornithologie.*[50] For his doctoral *exam*, he prepared himself in the fields of zoology, paleontology, and psychology. He felt confident about his major field, zoology, but he worried about the auxiliary fields. He became particularly anxious about paleontology when he learned that his examiner in that field, the paleobiologist (and rector of the university) Othenio Abel, was committed to a position that conflicted with Lorenz's own understanding of the evolution of bird flight. But the exam, held in July 1933, went smoothly for him. He avoided tangling with Abel on the subject of bird flight, the psychologist Karl Bühler was particularly impressed by his performance in psychology, and he passed with distinction.[51]

Earning his doctorate did not, however, change his immediate situation at the University of Vienna. To the contrary—though not as the result of his earning his doctorate—his conditions of existence there actually worsened. The cause of this was a change in the directorship of the Second Anatomical Institute, where Lorenz continued to earn a modest salary. In 1933 Hochstetter retired as professor and director of the institute and was replaced by Eduard Pernkopf. Unlike Hochstetter, Pernkopf was unwilling to allow Lorenz to devote any of his university time to behavior studies. Nor did Pernkopf like the idea of Lorenz writing his *Habilitation* thesis in animal psychology. The relationship between the new professor and the young assistant became increasingly strained.

In the meantime, though, Lorenz's intellectual horizons were continuing to expand. He was learning more and more about psychology through his participation in Karl Bühler's seminar. In addition, in the spring of 1933, he came in contact with Baron Jakob von Uexküll, an eminent theoretical biologist whose ideas helped Lorenz put his wealth of knowledge about bird behavior in the conceptual framework that was to make him famous.

Years later Lorenz reported that it was in studying with Bühler in the early 1930s that he first became acquainted with certain modern works in animal psychology. Bühler, Lorenz said, assigned him to report first on the work of the "purposivist" psychologists William McDougall and Edward Chace Tolman and then on the work of the behaviorist psychologist John B. Watson. By Lorenz's later telling, "Bühler made me read the most important books of both schools, thereby inflicting upon me a shattering disillusionment: none of these people knew animals, none of them was an expert. I felt crushed by the amount of work still undone and obviously devolving on a new branch of science which, I felt, was my responsibility."[52]

This retrospective account owes more than a little to Lorenz's flair for telling a good story. Lorenz may never have had much that was positive to say about the approach of the American behaviorists, but he did appreciate certain features of the writings of the purposivist psychologists Tolman and McDougall. In 1932 he characterized Tolman's book, *Purposive Behavior in Animals and Men,* as "truly excellent." Three years later, in his "Kumpan" monograph, he allowed that "all true animal observers," to the extent that they recognized that instinctive behavior patterns have subjective, emotional correlates, were "followers of McDougall." To be sure, he did not endorse everything Tolman or McDougall wrote. He disagreed with McDougall's identification of thirteen governing instincts and his distinction between first-order and second-order instincts. He likewise sharply criticized McDougall's conflation of the animal's subjective goals with the biological functions of its instinctive behavior patterns. On the other hand (and particularly, it seems, after discussing matters with Wallace Craig), he acknowledged the virtues of McDougall's observation that "the healthy animal is up and doing," instead of simply acting as a reflex machine. He also acknowledged to Craig that his idea of the "innate releasing mechanism" (*das angeborene Schema*) was the equivalent of McDougall's "innate perceptual inlet."[53] On top of this, when he sketched out a physical model to illustrate the physiology of instinctive action, the model Lorenz offered bore a marked resemblance to one already sketched out by McDougall in the latter's classic text, *An Outline of Psychology.*[54]

One may nevertheless ask just how carefully Lorenz read such authors as Tolman and McDougall on his first encounter with them. Tolman in his 1932 book provided an extended and highly approving account of the ideas of Wallace Craig, but Lorenz failed to be inspired by it. It was not until 1935, after Margaret Morse Nice put Lorenz and Craig in contact with each other, that Lorenz came to see the great value in Craig's distinction between the appetitive part of instinctive behavior patterns and the consummatory actions that bring these patterns to a close. It was only then, and with Craig's prodding, that Lorenz looked more closely at McDougall's work and developed a clearer sense of how McDougall's views and his own were both alike and different.

What remains clear from Lorenz's correspondence of the period is that he did feel he was a better observer of animals than almost everyone else who wrote about them was. This applied not only to English-language writers, representing various psychological schools, but also to the animal psychologists writing in German. In letters to Heinroth, Lorenz alternately joked and expressed vexation regarding authors who were not sufficiently acquainted with the behavioral repertoires of the animals about which they wrote (or who made bad choices of which animals to study in the first place). In November 1933 he told Heinroth to read Bastian Schmid if he wanted to become angry. Schmid, he explained, had simply *"no idea"* that the cluck of the laying hen was a warning sound.[55] The following month Lorenz characterized G. H. Brückner's work on chickens as "not at all bad, except for some mistakes from failing to recognize domestication phenomena." Lorenz regretted Brückner's choice of material: "It is terribly sad, that this brilliant research was not undertaken, for example, on golden pheasants. The man would have worked things out much better on them. But these psychologists have simply no heart for biology."[56] In January 1934 Lorenz announced to Heinroth that in his forthcoming monograph on "companions as factors in the bird's environment" he was going to scold the German zoologist Friedrich Hempelmann and the Dutch animal psychologist J. A. Bierens de Haan. Their work, he said, provided "frightening examples" of the fact that "experimentation without knowledge of the instinct system of the species in question" was "worthless." From these and other cases Lorenz concluded: "People simply do not know animals."[57]

MEETING UEXKÜLL

Lorenz's above comments notwithstanding, he was prepared to recognize that at least a few people besides Heinroth and himself "knew animals." One such person was the biologist Baron Jakob von Uexküll. Lorenz met Uexküll in the spring of 1933 when he came to Vienna to deliver two lectures. Twice during the course of his stay he visited Lorenz at Altenberg. He impressed Lorenz as an expert psychologist and philosopher with a genuine appreciation of living animals. He was not particularly knowledgeable about *birds,* however, so Lorenz, as Lorenz explained in a letter to Heinroth, lectured his visitor "from Heinroth" to good effect.[58] When Uexküll published in 1934 his book *Streifzüge durch die Umwelten von Tieren und Menschen (Journeys through the Worlds of Animals and Men),* he included a section drawing upon Lorenz's experience with jackdaws. He entitled this section "Der Kumpan" ("The Companion"). Lorenz in the meantime was making the *Kumpan* the theme and title of the major work on which he was working.[59]

When Uexküll and Lorenz first met, Uexküll was sixty-eight years old and director of the Institut für Umweltforschung at the University of Hamburg, an institute which had been established for him in 1925.[60] Talented as a researcher, and iconoclastic and outspoken in his views, he was a major figure in German biology. Early in his career he had studied muscle physiology and been a pioneer in the study of the comparative physiology of invertebrates. In 1899 he coauthored with the physiologists Albrecht Bethe and Theodor Beer an influential paper on reforming the scientific language used in describing behavior. The three physiologists urged that the subjective language of contemporary animal behavior studies be replaced with a terminology that did not prejudge the mental capabilities of the animals in question. They maintained, for example, that instead of speaking of "sense organs" in lower animals, comparative physiologists should use the more neutral terms "receptions organs" and "receptors." Rather than assuming that lower animals have a sense of sight, they said, one should speak simply of the response of their "photoreceptor organs" to light waves.[61]

Uexküll later stressed that what he and his two colleagues had intended with their 1899 paper was "to ascertain which outside stimuli were important in the life of the animal subject and which were not."[62] This concern for the animal as a perceiving subject, however, did not represent the main direction that animal behavior studies took as the twentieth century began. Jacques Loeb's studies of tropisms, to cite the most prominent example, paid more attention to orienting stimuli than to the animals that responded to these stimuli. By Loeb's account, the animals responded to the stimuli in a wholly machinelike fashion.

Uexküll regarded Loeb's system as too narrowly *physiological.* He adopted in contrast a position that was self-consciously *biological.* In his *Umwelt* theory, the intellectual contribution for which he is best known, he portrayed each living animal not as an object but as a subject, situated at the center of its own world, a world that the animal itself constructed through its own sensory and motor capacities.[63]

Uexküll based his *Umwelt* concept on the recognition that animals react only to select features of their environments. The animal's *Umwelt*—its effective, surrounding world—was the sum total of factors the animal subject perceived and responded to. As a prime example of what he had in mind, Uexküll cited the feeding behavior of the tick. "Out of the vast world that surrounds the tick," Uexküll wrote, "three stimuli shine forth from the dark like beacons, and serve as guides to lead her unerringly to her goal." Each of these stimuli, he explained, releases a specific activity on the part of the tick, and they take place in sequence. First, the chemical stimulus of butyric acid, emanating from a mammal passing beneath the tick's resting place, elicits the tick's letting go and dropping onto the mammal. Next, the mechanical stimulus of the mammal's hair elicits the tick's running about. Finally, the temperature stimulus of the mammal's skin elicits the tick's boring into the mammal's skin and sucking.

Each one of the three steps in the tick's feeding behavior constituted

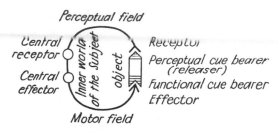

3.9 Jakob von Uexküll's representation of his concept of the "functional cycle." (From Jakob von Uexküll and Georg Kriszat, *Streifzüge durch die Umwelten von Tieren und Menschen* [Berlin: J. Springer, 1934], translated in *Instinctive Behavior: The Development of a Modern Concept*, trans. and ed. Claire H. Schiller [New York: International Universities Press, 1957], p. 10.)

what Uexküll called a *Funktionskreis* (functional cycle), which represented for him the basic way in which an organism interacted with an object in its *Umwelt*. He represented the *Funktionskreis* with the diagram shown in figure 3.9.

In this system of mutual relations between animal and *Umwelt*, the animal's receptor and effector organs determine what kinds of cues the animal can receive from its environment (its *Merkwelt*, or perceptual field) and what kinds of responses it can carry out (its *Wirkwelt*, or motor field). In the example of the tick, where the tick is the subject and the mammal becomes the source of diverse cues, the mammal is in the first place the producer of the sign stimulus of butyric acid. This stimulus impinges upon the tick's receptor organ, which then leads to impulses in an effector organ, which cause the tick to drop from its resting place. The tactile cue of encountering the hairs of the mammal then initiates the next functional cycle. This cue simultaneously inhibits the olfactory stimulus of the butyric acid and calls forth the motor action of running about. When the tick encounters the receptor cue of heat, this inhibits the previous receptor cue and initiates the third functional cycle, leading to the effector action of boring into the skin.[64]

Although Uexküll asserted that the perceptual cues of the tick "prescribe the course of her actions so rigidly that she is only able to produce corresponding specific effector cues," he refused to view the tick as an *object* to be analyzed like a machine. The message of his *Umwelt* theory was that each individual animal is a perceiving and acting *subject*, living in an *Umwelt* determined by the animal's own nature.[65]

When he pursued the topic of the perceptual worlds of animals in his book *Journeys through the Worlds of Animals and Men*, Uexküll happily drew upon Lorenz's remarkable, firsthand knowledge of bird behavior. He reported Lorenz's discovery that jackdaws cannot recognize as prey grass-hoppers that are not moving. He also cited Lorenz's observation that jack-daws will adopt an attack posture when confronted with an animal or per-son carrying a black object. He was especially impressed, though, by the insights on jackdaw social relations that had emerged from Lorenz's obser-vations on his pet jackdaw Tschock. Describing these relations, Uexküll introduced the word "companion" (*Kumpan*) to designate the individual or individuals with which Tschock carried out a given set of activities. Tschock treated Lorenz as her "mother companion" (*Mutterkumpan*), the Lorenz's housemaid as her "love companion" (*Liebeskumpan*), a young jackdaw as her "adoptive companion" (*Adoptivkumpan*), and flying crows as her "flight companions" (*Flugkumpane*). Uexküll illustrated this in his book with a set of four pictures that Lorenz drew for him.[66]

Once again, Lorenz had encountered a senior scientist whose appreci-

3.10 Sketches by Konrad Lorenz of four types of "companions" in the life of his jackdaw Tschock. *Clockwise from upper left:* "mother companion," "love companion," "flight companions," and "adoptive companion." (From Jakob von Uexküll and Georg Kriszat, *Streifzüge durch die Umwelten von Tieren und Menschen* [Berlin: J. Springer, 1934].)

ation of Lorenz's work was greatly encouraging to him. In this case, furthermore, Uexküll, unlike Heinroth, was in no way reluctant to engage in matters of theory. Just how much Lorenz knew of Uexküll's broader views is uncertain. Certainly he would not have been attracted to the anti-Darwinian and provitalist features of Uexküll's thinking. He might have taken an interest in Uexküll's thoughts on the relation between Kantian philosophy and *Umwelt* theory, but his own interest in Kant was yet to develop, and he never cited Uexküll in this regard. Nor did Lorenz ever mention Uexküll's political ideas, set forth in a book entitled *Staatsbiologie* (*State Biology*) first published in 1920 and then reissued in a second edition in 1933. In a few years Lorenz himself would be drawing upon a medical metaphor that figured prominently in Uexküll's political writings, the metaphor of cancerous elements destroying the body politic. That metaphor, however, was a very common one in the nativist and racist discourses of the day. Lorenz certainly did not need Uexküll to introduce him to it. The ideas that Lorenz did attribute to Uexküll and that were immediately important for him were the ideas of "releasers," "companions," and *Umwelt*.

These ideas were critical for the conceptualization of animal behavior that Lorenz provided in his "Kumpan" monograph of 1935, a work which was arguably the most important publication of Lorenz's whole career. Lorenz corresponded with Uexküll on the topic of the *Kumpan* in 1933. He completed a draft of his monograph by January of 1934. He intended to count the work as his *Habilitation* thesis.

IN SEARCH OF A VIABLE NICHE

In the meantime, his position at the Second Anatomical Institute had become more and more untenable. By March 1934, he felt himself at wit's end. He needed the salary from his assistant's position to be able to pay for food for his birds. However, if he worked all day as an assistant for Pernkopf, this left him no time to observe his birds (except for the night herons, which were indeed, as their name suggests, active at night). Having come to view Erwin Stresemann as his primary mentor in career matters, Lorenz described his predicament at length to the Berlin ornithologist.

> Now I ask, can in today's times a man with two children give up a secure governmental position to follow his daemon? It remains at the same time to consider that we (a.) can make it without the anatomy [position] (300 schillings = 150 reichmarks), if we change somewhat the animal stock (fewer fish eaters, generally cheap-to-feed research animals);

(b.) that I would produce a flood of really accurate animal psychology work; (c.) as docent for animal psychology I would remain a member of the university anyhow; (d.) as long as I remain with anatomy, I would have no chance of any kind of getting a university position in psychology. Which I would achieve sooner or later, freed from the yoke. Last but not least, that I as an anatomist, would anyhow not go far as the decidedly least favorite pupil of my boss, therefore the praised security of the anatomical position is not worth much here. I believe I will take the leap into the starving existence of a *Privatdozent*. Sometime, somewhere, a psychological position will open up. Then I will start at the bottom and work my way still farther down. The present situation is in any case really untenable, and I have a better chance as an attendant in a zoo to work on something and make something out of it than as an assistant of this always easily annoyed heavy-handed neurasthenic. . . . The devil with it, I shall take this step of getting, as long as I am young and the money holds out, a salary-less academic position as a psychologist.[67]

Stresemann was sympathetic with Lorenz's situation. He encouraged Lorenz to make the change. In the words already cited as an epigraph for this chapter, "In light of the new conditions the decision for you cannot be in doubt. You must give up anatomy. Your aptitude in the area of animal psychology is for you such a conspicuous trait that it would be like a self-surgery (and a biologically dangerous one at that!) if you now let yourself be a little cowed and wished to act rationally instead of instinctively. He in whom the instinctive is as strongly imprinted as it is in you, should be heartily glad for this finger of God."

Stresemann was not the only one to encourage Lorenz in this regard. The psychologist Bühler did as well, despite the fact, as Lorenz told Stresemann, that the zoologists had been inclined to view Bühler as the "evil enemy." Knowing he would need Bühler's support if he were to pursue a career in psychology, Lorenz decided to arrange a meeting with the psychologist. Before he even accomplished this, however, he found himself by chance seated next to Bühler at a Sunday evening music concert. Bühler asked Lorenz about Lorenz's *Habilitation* plans and was pleased to learn what Lorenz was thinking. Indeed, as Lorenz reported to Stresemann, Bühler was so excited about the idea of "a psychological biologist" lecturing on behavior theory that he promised to send him "so many students there would be no room in the lecture hall." In the days that followed, Professors Versluys, Krüger, and Bühler all expressed enthusiasm for Lorenz's work and his plans. To top things off, Bühler paid a visit to Altenberg. The whole thing seemed decided. The only drawback was that Bühler had no money to offer Lorenz as a docent lecturing in his institute.[68]

In May 1934, however, Lorenz's *Habilitation* plans received a sharp blow. University regulations, as it turned out, prohibited him from securing his *Habilitation* in zoology by using work from an entirely different discipline. His only viable option was to offer a morphological study based on research he had been conducting on nerves to the brain. He was appalled by the idea of spending another year on a subject that he found completely boring. In contrast, he had been finding his increased contact with the psychologists highly stimulating. And they too, he felt, were benefiting from his interactions with them, with the psychiatrists in particular showing a keen interest in imprinting.[69] Bühler's chief assistant, Egon Brunswick, read the whole of Lorenz's "Kumpan" monograph for him so as to ensure that the work was solid from the psychology side of things.[70] As it was, Lorenz would continue to stretch himself thin between anatomy and psychology until the fall of 1935. Living in Altenberg, he had to be up early every morning in order to make the commute to Vienna and arrive at the Second Anatomical Institute by 7:30 a.m. In addition, twice a week, he participated in Bühler's evening seminars, which lasted until midnight or later.[71]

ANGLO-AMERICAN CONTACTS

Though Lorenz expressed doubts about his chances before it actually occurred, he received permission from Pernkopf to go to Oxford in July 1934 to attend the Eighth International Ornithological Congress. This was a key event for him with respect to establishing contact with British and American students of animal behavior. More than three hundred people attended the congress. German and Austrian ornithology were well represented. Erwin Stresemann was president of the congress. He, Heinroth, and Lorenz all attended, accompanied by their wives. Other participants with whom Lorenz was already acquainted included his friend Horst Siewert (who showed lantern slides of the life history of the osprey) and Moriz Sassi. The German evolutionary biologist Bernhard Rensch came as well. He presented a paper entitled "On the Influence of Climate in Enhancing the Character of Bird Races, with Special Consideration of Wing Form and Egg Number." Also coming from the Continent was the French ornithologist Jean Delacour, who presented a paper on the systematics and behavior of anatids. Among the British participants were Julian Huxley, who presented a paper entitled "Threat and Warning Coloration in Birds" and also showed his remarkable film *The Private Life of the Gannets,* and E. B. Poulton, who showed a film on the behavior of young purple martins. The participants from the United States included Margaret Morse Nice, who presented a paper entitled "Territory and Mating with the Song Sparrow," and the young German ornithologist Ernst Mayr, who had moved in 1931 from Berlin, where he had worked with Stresemann, to a position at the American Museum of Natural History in New York.[72]

Stresemann in his presidential address noted that not only was ornithology attracting more participants and publications, but its practitioners were also addressing increasingly diverse problems. Meanwhile, geneticists, evolutionary theorists, and animal psychologists were coming to see ornithologists as comrades in arms. Consulting the list of papers presented at the conference reveals that ten or so out of the total of sixty-five dealt with matters of behavior. The study of bird behavior was becoming an important part of the ornithological enterprise.

For his own presentation to the congress, Lorenz delivered a thirty-minute paper entitled "A Contribution to the Comparative Sociology of Colonial-Nesting Birds." He described his interest in the problem of instinct and his commitment to the methodology of rearing and observing tame, free-flying birds. He then proceeded to compare his observations on two quite different types of colonial-breeding birds—jackdaws and night herons—paying particular attention to their territorial behavior. Jackdaws, he reported, display territorial behavior only when strange jackdaws come near their nesting colony. Night herons, in contrast, have very strong territorial defense actions, which serve to guarantee the inviolability of the individual nests in the densely settled heronry. Connected to these actions were the "appeasement" ceremonies that allow a bird approaching the nest to join its mate (or chicks) without being attacked by the bird(s) already there.[73]

Lorenz's paper was too brief and sketchy to make any real splash at the conference, but the papers given at a conference are often not the most important events of such gatherings anyway. Meetings of scientific societies are frequently excellent sites for making personal contacts, and Lorenz used the International Ornithological Congress at Oxford in just this way. There he met both Margaret Morse Nice and Julian Huxley, each of whom proved extremely helpful to him.

Margaret Morse Nice had initially made contact with the leaders of German ornithology thanks to the help of Ernst Mayr. The two met at a meeting of the American Ornithologists' Union in 1931, and Mayr put Nice in touch with Erwin Stresemann and Oskar Heinroth. Stresemann in turn

3.11 The American ornithologist Margaret Morse Nice in 1920 at the very beginning of her career as an ornithologist. "Mrs. Nice" went on to distinguish herself through her pioneering field studies of the life history of the song sparrow. She also championed the work of the Continental ethologists, hailing Lorenz's "Kumpan" paper as a revolutionary breakthrough in the study of bird behavior. In the spring of 1938 she spent a month in Altenberg learning Lorenz's techniques of raising young birds. In a letter to Ernst Mayr later that same year she aptly wrote of the Austrian ethologist: "He is a genius, but he evidently undertakes too many things and sometimes goes wrong." (Photo courtesy of Kenneth Boyer.)

encouraged Nice to publish in the *Journal für Ornithologie* her remarkable field studies of the life history of the song sparrow. The first installment of her two-part paper appeared in the same issue of the *Journal für Ornithologie* in which Lorenz's paper on bird flight was published. When Nice arrived in Oxford she already knew of Lorenz's work and he already knew of hers. Indeed he told Stresemann that her work gave him an inferiority complex. He had not attempted the intensive kind of *field* studies she had done, but he was in any case "wholly enthusiastic" about her research practices.[74] The two certainly shared the opinion that, as Nice put it elsewhere, "the trouble with most animal psychologists is that they know only what their white rats do in the laboratory, and they never have studied intensively a wild animal in his natural environment."[75]

By chance, the local arrangements committee assigned Nice and her sister and Lorenz and his wife to the same boarding house. This and other opportunities both during and after the conference gave the two ornithologists ample occasion to become acquainted. The most important consequence of this for Lorenz's own intellectual development was that Nice decided he needed to be put in touch with her old friend Wallace Craig. An additional benefit for Lorenz, particularly after he published his "Kumpan" monograph in 1935, was that Nice became a strong promoter of his work in the United States.[76]

Julian Huxley assisted Lorenz in similar ways. He loaded Lorenz down with a whole trunkful of reprints that he insisted Lorenz read. These included not only a good number of his own writings but works of other English-language writers as well, with which Lorenz was previously entirely unfamiliar. Prior to this time, Lorenz's citations of the scientific literature on bird behavior were limited almost exclusively to sources written in German. When he published his "Kumpan" monograph in 1935, however, he was prepared to cite a significant number of British and American sources.[77] Among these were six papers by Craig, three papers by Huxley, Nice's work on the song sparrow, plus publications by Edmund Selous, A. A. Allen, George and Elizabeth Peckham, and G. K. Noble and H. T. Bradley (among others).

THE "KUMPAN" MONOGRAPH

Lorenz's "Kumpan" monograph represented his attempt to organize his wealth of detailed observations of bird behavior into a coherent framework. The task was not an easy one, for his work had not been focused on a single research problem and the number and variety of birds he had maintained and studied in the free-flying state were extraordinary. He listed the diverse species with which he had worked: "15 Little Egrets, 32 Night Herons, 3 Squacco Herons, 6 White and 3 Black Storks, many mallard, many domestic ducks, many domesticated Muscovy duck, 2 Carolina Woodduck, 2 Greylag geese, 2 Common Buzzards and 1 Honey Buzzard, 1 Imperial Eagle, 7 Cormorants, 9 Kestrels, approximately one dozen Golden Pheasants, 1 Great Black-backed Gull, 2 Common Terns, 2 large Greater Yellow-crested Cockatoos, 1 Amazon Parrot, 7 Black-capped Parakeets,

20 ravens, 4 Hooded and 1 Carrion Crow, 7 magpies, more than 100 jackdaws, 2 jays, 2 Alpine Choughs, 2 Grey Cardinals and 3 Bullfinches."[78]

The monograph totaled 202 pages. Its full title was similarly lengthy. "The Companion in the Bird's World: Fellow Members of the Species as Releasers of Social Behavior" ("Der Kumpan in der Umwelt des Vogels: Der Artegenosse als auslösendes Moment sozialer Verhaltungsweisen"). Lorenz dedicated the work to Jakob von Uexküll in honor of Uexküll's seventieth birthday (which had taken place the previous year). It was Uexküll's personal encouragement, Lorenz allowed, that had given him the courage to set forth the complex materials constituting the subject of the monograph. The work appeared in two successive installments in the *Journal für Ornithologie.*

Lorenz's "Kumpan," was less a reporting of results than an elaboration of concepts on which to build a continuing program of research. But the concepts and programmatic injunctions would not have had the force they did had they not been accompanied by a wealth of supporting examples. This makes it difficult to summarize the work while at the same time retaining a sense of its richness and complexity. When Lorenz himself prepared an abbreviated version of the monograph for English-language readers, a version that appeared in the *Auk* in 1937, he was not entirely satisfied with it.

The key concept of Lorenz's "Kumpan" monograph was the idea of the "releaser." Following Uexküll, Lorenz explained that we humans tend to recognize objects in our environment as things, thanks to a compilation of multiple stimuli emanating from these objects and impinging upon our sense organs. The successful integration of these stimuli allows us to build up an understanding of the causal relationships among the things around us, and this in turn helps us to survive. Lower animals such as birds, in contrast, are adapted to their environments not so much through acquired knowledge as through highly differentiated instinctive behavior patterns. Birds act not according to insight but instead according to instinctively determined responses. These have been built up through evolution as a result of their survival value. To be effective, they need only be elicited, or *released,* by one or at most a very few of the stimuli emanating from the objects in their environment. Lorenz elaborated on this in the following terms:

> It is sufficient that an instinctively determined response, which is developed on the basis of survival value of responses relative to a particular thing, should be elicited by one of the stimuli emanating from this thing

(so long as this stimulus characterizes that thing so clearly that erroneous elicitation of the response by another thing emitting similar stimuli does not reach a frequency sufficient to adversely influence survival value). In order to reduce the probability that the latter effect will occur, it is common that a number of stimuli are combined to provide a pattern of stimulus-data. Such a combination is nevertheless quite simple in nature and evokes a response from an "innate schema" [angeborenes Schema]. The structure of such an innate schema must possess a certain minimum of general improbability for the same general reasons that the shape of a key is made to present a generally improbable pattern.[79]

Such, in brief, was the basic relation Lorenz identified between external, releasing stimuli and an animal's innate schema (later to be called the "innate releasing mechanisms"). Lorenz went on to explain that when the object of the instinctive reaction was a fellow member of the same species, the relation between the releasing stimulus and the innate schema could be fine-tuned by adaptive evolution almost indefinitely. In that case, both the receiving and the issuing of stimuli were subject to evolutionary modification, thereby producing combinations of such overall improbability that instinctive reactions would only rarely be elicited by stimuli from the "wrong" sources. Lorenz called "releasers" (Auslöser) all those characters displayed by an individual that serve to activate the innate schemata of members of the same species and thus to elicit the performance of "chains of instinctive behavior patterns." Such "releasers" could be morphological structures, conspicuous behavior patterns, or, quite frequently, a combination of the two.

Releasers, Lorenz continued, are critical for the functioning of the complex, intraspecific social relations of many animal species. The highly organized social life of jackdaws, for example, is built upon on a remarkably small number of simple, innate responses to releasers provided by fellow jackdaws. Uexküll had formulated the idea of the "companion" on the basis of what Lorenz had told him about jackdaw behavior. Lorenz elaborated further on the notion. Each jackdaw, he said, has a number of different social drives with respect to which other jackdaws typically play the role of "companion." These "companions" provide the particular stimuli necessary to release the instinctive behavior patterns appropriate to its drives. Uexküll, Lorenz pointed out, had not intended the idea of the companion to describe anything beyond the relations in individual, independent, unitary, functional systems. "By 'companion,'" Lorenz explained, "we of course understand a fellow human being to whom we are bound only by the links of a single functional system, which themselves have little to do with higher emotional impulses, as is the case with a drinking or (at the outside) a hunting companion.[80]

When he attempted in 1937 to summarize his ideas for an English-speaking audience, Lorenz noted further: "The German word, *Kumpan*, means a fellow who is our companion so far as concerns but one particular kind of occupation, such as hunting or drinking (*Jagdkumpan, Saufkumpan*). It implies that no deeper and nobler bonds link us to our fellow in this kind of companionship. The word certainly meets the case exceptionally well, although it is hardly translatable into English. The word 'companion' certainly lacks the detracting implication which is so essential for the wonderful way in which Uexküll's term describes this lowest type of companionship."[81] One imagines that the English word "buddy" comes closer than the word "companion" to just what Lorenz and Uexküll had in mind.

Having identified the primary conceptual building blocks of his study, Lorenz set about prescribing how animal psychology should be studied. His message was unequivocal: animal psychology was to be studied as a biological science, grounded on a specific methodology and addressing certain, specific topics before others. "*An extensive period of general observation*" was absolutely necessary, he insisted, before one attempted any experiments. Likewise, one needed a thorough knowledge of an animal's *instinctive* behavior patterns before one tackled questions of *learning*. If an investigator was unwilling to begin by gaining a thorough familiarity of the full behavioral repertoire of the species he was studying, Lorenz said, "should leave animal psychology well alone."

As he had told Heinroth that he would, Lorenz cited studies by Hempelmann and Bierens de Haan as examples of the mistakes that arose when one experimented on animal behavior without knowing the complete behavioral repertoires of the species in question. Hempelmann and Bierens de Haan had both experimented on how tits and goldfinches grasp food with their feet. They concluded that this behavior showed tits and goldfinches to be highly intelligent, as compared with other birds that do not display the same ingenuity in grasping their food. Lorenz maintained that the phenomena described by Hempelmann and Bierens de Haan had nothing to do with intelligence. The behavior in question in each case was simply "an innately, reflexively-determined instinctive pattern."

Lorenz was careful not to portray his own work as exclusively *non*-experimental. Experimentation, after all, was one of the hallmarks of modern science. What he claimed instead was that his work was not biased by any *preconceptions*. In his hours of patient observation of the naturally occurring behavior of birds, he said, he had been able to benefit from "chance" or "involuntary experiments," that is, from the unexpected results of particular situations that he himself did not intentionally create. The virtue of relying on such "experiments," he maintained, was that the observer in such cases is "really impartial," that is, he is "demonstrably completely free of any hypothesis."[82] There was a logic to this argument, admittedly, but it was obviously self-serving, and it could not by itself give Lorenz's approach the prestige of experimentation.

As for focusing first upon instinctive behavior and only later tackling learned behavior, Lorenz explained: "In my opinion, the main emphasis in the analysis of animal psychology should initially be placed on study of innate, instinctive behaviour [*triebmäßig Angeborenen*] rather than on variable behaviour patterns representing acquired products of intelligent processes, largely because . . . one can never judge the extent of learning and intelligent abilities of an animal without prior knowledge of its instinctive behavior [*triebmäßigen Verhaltens*]."[83]

Of all the remarkable findings Lorenz reported in his "Kumpan" monograph, the single most intriguing was a phenomenon that could not be readily classed as either instinct or learning. Following Heinroth's lead, he called it "imprinting" (*Prägung*). Heinroth had remarked that newly hatched greylag goslings do not instinctively recognize adult greylag geese as members of their own species. Instead, if exposed to a human being before they are exposed to a mother goose, they will follow that human as if he or she were their parent and will then direct toward this foster parent the innate responses they would under normal circumstances have directed toward a goose. At the Amsterdam zoo, A. F. J. Portielje had reported the same phenomenon with a South American bittern. Whitman in the United States had relied upon imprinting to facilitate interspecific crosses. It was Lorenz, however, who named the phenomenon, insisted upon its theoretical significance, and became its most conspicuous "discoverer."

Lorenz had encountered imprinting with his jackdaws, his geese, and most of his hand-reared birds. As he represented the process, the young bird, at a very brief, critical period of its early development, becomes "imprinted"—and irreversibly so—upon the object that will then serve to release certain of its instinctive behavior patterns. The phenomenon, he insisted, is quite distinct from "true learning." He likened it instead to inductive determination in embryological development.

In Lorenz's system, the concepts of imprinting, releasers, and companions were closely related. Through imprinting, the young bird, which does not instinctively recognize members of its own species, acquires the information it needs. It is imprinted with the schema of the fellow species member. The conspecific thus comes to serve as a single object providing the various stimuli appropriate to the releasing mechanisms of different instinctive behavior patterns. The releasing mechanisms themselves, however, continue to be mutually independent. A key feature of the *Kumpan* concept was to underscore this mutual independence of releasing mechanisms and thereby deny that the fellow member of the species *appears* as a unitary object in the perceptual world of the individual bird. What unified the reacting agent's instinctive actions, Lorenz explained, was not the reacting agent itself, but the provider of the various releasers.

Lorenz used the example of the social life of the jackdaw to identify five different sorts of "companions" a bird has in the course of its life. A baby jackdaw's innate behavior patterns are triggered by releasers provided by the "parental companion" (*Elternkumpan*). A young bird also responds to innate and acquired releasers provided by its nestmates, a category Lorenz characterized as the "sibling companion" (*Geschwisterkumpan*). As an adult, the bird's innate behavior patterns are triggered by releasers from the "sexual companion" (*Geschlechtskumpan*), the "infant companion" (*Kindkumpan*), and the "social companion" (*soziale Kumpan*), depending respectively on whether the bird is courting, feeding young, or foraging or flying away.

Pursued rigorously, the concept of the releaser undermined Lorenz's particular enumeration of "companions," for as he himself admitted, he had broadened Uexküll's concept of the *Kumpan* beyond that of an object releasing a single function: "Strictly speaking, it is not permissible to refer to 'parental companion' and the like with [certain] birds . . . since the parental bird can represent in the environment of the offspring a 'feeding companion,' a 'warming companion,' a 'guiding companion' and various other autonomous entities."[84]

The same insight was manifested in Lorenz's account of how a mother duck does not respond to a baby duckling as an "infant companion" but rather to specific, independent stimuli emanating from the duckling. In various duck species, he explained, the protective response of the mother will be released by the distress call of ducklings of other species, but her other parental responses are released only by the coloration and marking patterns on the head and back of the ducklings specific to her own species. Thus, a mother mallard will come to the aid of a Muscovy duckling giving a distress call, but not finding mallard markings on the duckling, she will proceed to treat the duckling as a strange animal and attack and even kill it if it tries to join her own ducklings.[85]

In developing the concept of the "releaser," Lorenz made an important contribution to understanding the particular morphological and behavioral features of animals known as "secondary sexual characters." Charles Darwin had sought to account for these with his theory of sexual selection. Lorenz, in truth, was not much interested in the idea of sexual selection. Indeed, contemporary debates about how evolution works seem to have passed him by almost entirely. He was surprisingly silent, for example, on the topic of the inheritance of acquired characters, which remained an issue in his day. He by and large trusted that behavior patterns had evolved according to their selective value, which he represented rather uncritically in terms of "the good of the species." He viewed sexual selection with distaste, it seems, because its more bizarre products—the wings of the Argus pheasant, for example—were arguably instances where organic change had proceeded contrary to the good of the species.[86] Finding the idea of sexual selection repugnant, he was happy to account for the majority of secondary sexual characters in other terms. As he saw it, the majority of conspicuous structures, colors, sounds, and behavior patterns in animals served as *releasers* of social reactions in fellow members of the species and had furthermore evolved for this purpose.[87]

Lorenz maintained that the most common effect of a male bird's conspicuous plumage and display behavior is to elicit not only a positive reaction from females but also a negative reaction from males. Heinroth had used the term "demonstrative behavior" (*Imponiergehaben*) to refer to behavior patterns that served both courtship and threat functions at the same time. Such behavior is typically exaggerated in its character. The courtship actions of a greylag gander provide a fine example of this. Every movement the gander makes, Lorenz said, seems to be invested with an excess of energy and muscular power: he sticks out its chest, he walks upright and with great deliberation, and his normal body rotation and all its wing motions are likewise performed in an exaggerated fashion. Similar behavior, Lorenz went on to note, can be observed in human males who in skiing or ice-skating step up their physical activities to appear "considerably more vigorous and dashing when the number of spectators is increased by one attractive girl." The same thing could be said, he added, for motorcyclists revving up their engines in the presence of individuals of the opposite sex.[88]

Lorenz was enthusiastic about the prospect of being able to reconstruct phylogenetic series of demonstrative behavior in geese. He believed he could start with the primitive behavior of the greylag, which is not enhanced by any special morphological characters, and then proceed through the more elaborate ceremonies of other anatids, where special structures and colors serve to emphasize the basic motor patterns in question. He believed, indeed, that the behavior patterns always preceded the specialized, brightly colored structures. In his words, "the *ceremony is always more archaic than the correlated organ.*"[89]

Lorenz acknowledged the hazards of generalizing from one group to another. The findings of Noble and Bradley on lizards, he insisted, should not be uncritically extended to birds. Likewise, he said, the observations of A. A. Allen on the ruffed grouse, or of Thorleif Schjelderup-Ebbe on the rank order of some bird societies, could not be expected to hold for birds in general. Even within a group of closely related birds such as the ducks and geese, he cautioned, there are significantly different forms of mating. None of this stopped him, however, from going on to offer a highly intuitive schematization of pair formation in birds.[90]

Lorenz identified three basic types of pair formation in birds (while acknowledging there could be more). He called them "the lizard type," "the labyrinth fish type," and "the cichlid fish type."[91] He chose the first of these names after reading Noble and Bradley's study of mating in lizards, where only the male exhibits display behavior, but it does so in the presence of *any* member of the same species, whether male or female. If the other individual flees, the male will pursue it and attempt to copulate with it. If it does not flee but instead exhibits display behavior itself (which happens only if it too is a male), then fighting ensues.

In the "labyrinth fish type" of pair formation, in contrast, males and females both exhibit display behavior. Whether fighting or courting ensues depends upon the response of the second animal to the first animal's display. The first animal will proceed to fight unless the second animal responds with *female* behavior patterns. Female display behavior in this type of pair formation is virtually the opposite of male display behavior. While the male fish spreads his fins and attempts to place himself broadside to the other fish, swimming parallel to it, the female orients herself perpendicularly to the male and draws her fins in. For pair formation of this type to take place, Lorenz stated, the male must be dominant over the female. Among species of this pairing type, he noted further, strong females may behave as males toward weaker females, and sexually motivated males may behave as females toward stronger males. Finally, in the "cichlid fish type" of pair formation, no dominance hierarchy exists in the relations between the individuals of a pair. Both members of the pair continue to display toward each other.[92]

Lorenz's classification of pair formation types proved of minimal consequence for the subsequent development of ethology. Just the opposite was true of the "Kumpan" monograph as a whole. Lorenz's concepts of releasers, innate releasing mechanisms, imprinting, and the roles of each of these in the functioning of animal societies became the core concepts of the new science.

Lorenz went on to urge that everything he had learned about instincts confirmed that instincts could be treated exactly like organs. The practical consequence of this for zoology was that the comparative study of "releasers" could furnish new insights into the genetic affinities of related organisms, a primary interest of zoologists ever since the establishment of the truth of evolution. Discussing the display behavior of ducks and geese, the Austrian ethologist boldly asserted:

> I should like to emphasize that nowhere in the entire field of comparative morphology would there seem to be series which are so utterly indicative of genetic relationships as those discussed here. This is probably to a large extent dependent on the fact that the releaser and the elicited behavior are very little affected by environmental factors in the process of a (so to speak) internal 'arrangement' within a bird species, so that convergence can be excluded with great reliability from the outset. In this case, similarity *always* means homology. In consequence, we are often able to determine genetic relations with a degree of accuracy seldom available to the comparative morphologist.[93]

Although his "Kumpan" was almost exclusively about birds, Lorenz ended it with a glance toward the role of instincts in human behavior. He cited approvingly the words of the psychologist David Katz, who had allowed that "animal psychology may eventually be employed for finding laws which govern the social behavior of human groups." Lorenz rephrased this to say that one must "recognize that instinct, governed by its own laws and fundamentally differing from other types of behaviour, is also to be found in human beings, and then go on to investigate this behaviour."[94] As it turns out, this remark was not just a final, stylistic flourish on Lorenz's part. His private correspondence shows that he fully believed, with Heinroth, that "the highest and best" of all human behavior rested squarely on instinctual foundations.[95]

The "Kumpan" monograph was a tour de force. It established Lorenz's reputation as a world leader in the study of animal behavior. Margaret Morse Nice's response to the work is indicative of the impact it made. The previous year, when Nice met Lorenz in Oxford, she was charmed by the bird stories he told her while they lunched together at the Huxleys and duly impressed by the paper he delivered to the congress. She did not, however, see Lorenz as someone who was establishing a whole new theoretical foundation for the understanding of bird behavior. Indeed, in January 1935 she wrote to G. K. Noble saying she was looking for a theoretical framework to provide structure for her researches. Animal psychologists in general did not impress her, she allowed, but she was finding the views of E. C. Tolman promising, and she was therefore developing "an interpretation of Song sparrow behavior on Tolman's lines." When Lorenz's "Kumpan" appeared later in the year, all this changed. Nice reviewed it immediately for the American journal *Bird-Banding*. Bowled over by Lorenz's concept of "releasers" as well as by his extraordinary knowledge of the behavior of so many different bird species, she promoted the work with enthusiasm. She described the monograph as "great" and as "a most remarkable paper of fundamental importance." It provided, she said a "revolutionary and illuminating viewpoint" and a "solid foundation on which to build."[96]

Uexküll for his part had seen a manuscript version of "Kumpan" before the work's publication. He sent a picture of himself to Lorenz with the inscription: "Many thanks for the fantastically interesting Kumpan! 'Imprinting' and 'inborn schema' are the solution of a riddle."[97] Eliot Howard, who earlier had explained to Lorenz that he himself could not read a word German, wrote Lorenz that Huxley had described "Kumpan" to him as the "most important thing that has come out for a very long time!" Huxley himself also sent Lorenz "a very nice letter" about the "Kumpan" monograph, enclosing with it a copy of Huxley's classic paper on the courtship of the great crested grebe.[98] And Craig in 1937 wrote Lorenz that the "Kumpan" paper was "by far the greatest paper on instinctive behavior of birds that I ever read in my life."[99]

GRANDER AMBITIONS

Although "Kumpan" played a fundamental role in establishing Lorenz's reputation as an observer and a theorist, it did not represent the peak of his creativity or his theoretical ambitions. After sending the final version of his manuscript off to Stresemann for publication early in 1935, he had no intention of resting on his laurels. He returned to a plan he had begun formulating as early as July 1933. Writing to Stresemann at the time, he had asked whether his next project should be a small one, for example, a study of the ethology of the night heron, or whether he should begin his "life's work," which he thought of entitling (in English) "The Behavior of Higher Animals."[100]

Lorenz contemplated giving the book an English title rather than a German one because he thought it would be a good idea actually to *write* the book in English. The International Ornithological Congress in Oxford convinced him further of the importance of doing so. It was clear to him that if he were going to make his ideas known to the English-speaking scientific world—and there seemed indeed to be British and American ornithologists who were eager to know more about his work—he would have to publish it in their language. Eliot Howard knew no German. Huxley could read German, but he was not especially familiar with the German-language literature on bird behavior. He had never read Heinroth. Nor had he, in Lorenz's opinion, read Jan Verwey's important heron work as closely as he should have. Lorenz imagined that if he himself wrote a book in English on animal behavior, it would sell well both in England and America. It would also have "a higher moral value," he told Stresemann, in that "the English chaps would really see that we know just as much as they."[101]

Exchanges of letters with Wallace Craig and Eliot Howard in 1935 encouraged Lorenz in his book plan. Lorenz had received a copy of Craig's paper "Appetites and Aversions as Constituents of Instincts" late in 1934 from Margaret Morse Nice, as part of the big bundle of ornithological reprints she had sent him.[102] He found Craig's ideas extremely stimulating, and he proceeded to send Craig a seven-page commentary on them. In their ensuing correspondence, Lorenz was impressed to find how much of "Kumpan" Craig was able to anticipate before actually seeing it. Lorenz concluded Craig was a "remarkable man."[103]

One particular letter from Craig at the beginning of March 1935 led Lorenz to spend a whole Sunday from 8:00 in the morning until 11:00 at night working furiously to extend his critique of contemporary instinct theorists in the draft of his "English manuscript." Among other things, Craig impressed upon Lorenz the importance of addressing the work of McDougall. Craig's letter led Lorenz to proclaim to Stresemann that Craig, after Heinroth, was the best animal-understanding man Lorenz knew.[104] This was certainly high praise from Lorenz, given that Lorenz viewed Heinroth as his own alter ego in ethological matters (as he had in fact explained in a letter to Craig only a few weeks earlier).

Lorenz still had time to insert some mention of Craig's work in the sections of "Kumpan" that he had not yet sent off to Stresemann. What Lorenz initially found most attractive and significant in Craig's work was Craig's recognition that "the readiness for a particular instinctive behavior pattern increases, if the release of the behavior remains unaccomplished for longer than normal. This results in not only a lowering of the threshold for the releasing stimuli but also in behavior that can be interpreted as a search for these stimuli." [105] Subsequently, and over time, Lorenz came to emphasize that what was especially important about Craig's "discovery" (Lorenz's word) of appetitive behavior was that it made it possible to distinguish clearly between the animal's subjectively sought *purpose* in performing an instinctive action and the "species-preserving *function*" of that action.[106]

Lorenz pushed on with his project of writing a big book on animal behavior. By August 1935, he was wondering whether it would be desirable to have a German edition of the book come out simultaneously with the English version. The English version, as he explained to Stresemann, was the result of his long exchanges with Craig and Howard. Craig, he said, was "one of the sharpest and most precise thinkers and *experts* [*Vielwisser*] in this field." Howard was "a writer who ¾ intuitively has the most accurate insights on instinctive behavior, the *only one* who like me *denies* the influence of experience on the instinctive motor pattern!" All Lorenz needed to do, he told Stresemann, was to copy out his discussions with Craig and Howard, and then the first chapter of his book would be finished.[107]

Six weeks after describing his plan to Stresemann, Lorenz told Margaret Nice about the project:

> I have started writing my book. I am going to write it in German and English simultaneously. I am afraid I must write a theoretical book first, but I really shall write an English book about bird behaviour too, containing more observations than theory. But you must not think that my first, theoretical book will be as hard to digest as the Kumpan is. I must begin from the beginning and put everything in a much less scientific way. All must be easily within the grasp of a student of zoology as well as of one of psychology, which means that I must avoid all technicalities altogether, and write in a way understandable to every educated person. Heinroth urged that I should put a lot of pictures in it, as very few people, least of all psychologists, know what a Mallard or a godwit looks like. He has offered me his whole material of pictures which of course I accepted gladly, (it really is a great boon!) Of course I shall include as much of Heinroth's work as I think feasable." [108]

Much as he was benefiting from the support and encouragement of Heinroth and Craig, and much as he was prepared to express his indebtedness to them, Lorenz wanted to achieve more than they had achieved. Heinroth was not interested in establishing himself as a theorist. Craig was much more theoretically inclined than Heinroth, but his lack of self-confidence, coupled with his inability to find an appropriate academic niche for himself, had left him introverted and isolated.[109] Lorenz had grander plans for himself. He took care to see that that his work was widely known and that he himself was well connected.

Lorenz's increasing visibility and standing within the scientific community in the 1930s were not immediately accompanied, however, by his attainment of a satisfactory scientific post. The first intriguing possibility, the ornithological station on the Danube, had failed to materialize. In the mid-1930s another possibility appeared on Lorenz's horizon. It was the directorship of the Schönbrunn zoo. Lorenz was clearly qualified for the position. He was also greatly attracted to it.

The possibility emerged in the spring of 1934 when Lorenz's friend Otto Antonius was relieved of the post.[110] The grounds for his dismissal were political: Antonius was a member of the Nazi Party. Lorenz was not happy with the events. Antonius had been "a true friend and patron" to him, as Lorenz told Heinroth. "God knows who his successor will be."[111] Other Viennese holders of zoology doctoral degrees wasted no time applying for the post, but Lorenz, as he later explained to Stresemann, refrained from applying for the position for two months, out of consideration for his friend. Even then, he said, he applied only after Antonius strongly encouraged urged him to do so.[112]

The status of the job, however, remained uncertain. As of January 1935 it was still not clear whether Antonius had been dismissed for good or whether he was going to be allowed to return to the post. In the meantime, Lorenz became increasingly excited by the idea of directing the zoo himself. He told Stresemann, "Although I really sincerely wish Antonius everything good (and correspondingly, in order not to harm him, must even leave some patronage unused), I can not help dreaming of Schönbrunn. My God, that would be something! I believe, by the way, that if Antonius still must go, my chances would be good."[113]

Feeling that his chances for Schönbrunn would in fact be greater if he remained a civil servant in Pernkopf's lab than if he moved to Bühler's institute as an unpaid docent, Lorenz for the time being sat tight. This left him, however, with his grueling schedule of commuting early in the morning from Altenberg to Vienna, working there at the Second Anatomical Institute on his brain-nerve research for his *Habilitation* thesis, and regularly staying late to attend the meetings of Bühler's psychology seminar. On top of all of this, he had to carve out time to write a conclusion to his "Kumpan" monograph.[114]

In the fall of 1935, while still nursing hopes for Schönbrunn, Lorenz decided the time had come to move over to Bühler's Psychological Institute. He announced his upcoming move to Margaret Nice, saying, "I am now leaving the anatomical Institute, moving my pennates to the Psychological Institute. I must buckle down to a serious study of Psychology. I am going to start at the bottom and work my way still farther down. As I am even now convinced that modern psychology is standing on feet of clay, lacking any knowledge whatsoever about instinctive reactions of animals and still less of man, little good will come of my studies. But I feel I must give those poor people a hearing, before condemning them alltogether. In justice I must remark, that the director of the Ps. Institute, Prof Bühler *does* know what an instinct is, a glorious exception among man-psychologists!"[115]

During this period, it was not Lorenz, but rather his wife, with a job as director of a department in an obstetrical hospital in Vienna, who earned the money to provide for the Lorenz family and the Lorenz menagerie. Lorenz earned a meager supplement to this by giving popular lectures as part of the adult education programs at the University of Vienna and the Viennese higher educational institution "Urania" and by publishing popular articles in Viennese newspapers and journals.[116] Nonetheless, it was Margarethe Lorenz who was the primary breadwinner for the family for the rest of the decade. Feeling pinched financially, Lorenz worried about the cost of taking trips, and he thought about working with species that would be less rather than more costly to maintain.

In 1936 there appeared to be a chance that Lorenz might succeed Uexküll at Uexküll's Institut für Umweltforschung. Uexküll wrote him asking if he would be interested. However, this opportunity failed to develop. Through this period, nonetheless, Lorenz remained optimistic about his prospects. Confident that his work was genuinely pathbreaking, he found it hard to imagine that support for his work would not be forthcoming sooner or later. He completed his *Habilitation* thesis and passed the associated examination.[117] He was making great strides in his analysis of the instinct problem. His investigations on ducks and geese had come to embrace not only questions of phylogeny but also the whole topic of the behavioral differences between wild and domesticated forms. He was also eagerly exploiting the possibilities that cinematography afforded him. As of December 1935 he had begun to shoot a film provisionally entitled *Social and Family Life of Animals,* in which his object was to show that nine-tenths of the social life of birds was made up of releasing ceremonies of one kind or another. He was also working on a film on the behavior of the grey-lag goose. He anticipated bringing new insights and precision to his work by comparing film sequences of the behavior of closely related species (and of their hybrids).[118] In addition, he was writing a general book that was certain to help spread his ideas. Most exciting of all, he found himself with a chance to present his views in a most prestigious setting—Harnackhaus, the headquarters of the Kaiser Wilhelm Gesellschaft. There, on 16 February 1936, he delivered a major address on the establishment of the instinct concept that served to announce his arrival as *the* primary authority on animal psychology in the German scientific community.

LECTURING AT HARNACKHAUS

Harnackhaus was a new building that spoke to Germany's eminence in the world of science. Completed in 1929, it served the multiple purposes of holding conferences and evening lectures, hosting visiting scientific researchers, encouraging interaction between foreign and German scientists and scholars, and, in general, serving as a major scientific and cultural center. Lorenz's ornithologist friend Gustav Kramer encouraged Max Hartmann, one of the directors of the Kaiser Wilhelm Institut für Biologie in Berlin, to invite Lorenz to give a Harnackhaus lecture. Stresemann may also have prodded Hartmann to do this.[119] In any event, Lorenz was delighted with the chance to hold forth in this special setting. He hoped there might be a meeting of the DOG in Berlin around the same time. If so, he wanted to show two films there: his greylag goose film and his film on releasers. As he told Heinroth, "If I am already in Berlin, I would like after all to make as much ruckus as possible."[120] Heinroth arranged to have Lorenz's Harnackhaus lecture serve also as the lecture for the monthly meeting of the DOG. He also arranged to have Lorenz's films shown in conjunction with the talk. The event was scheduled for the Helmholtzsaal, a room that was large enough to hold three hundred people and equipped with the finest projection equipment available.[121]

This represented a dream come true for Lorenz. He had previously envisioned writing a paper on the modern concept of instinct. He had also imagined publishing the paper in the eminent German science journal *Die Naturwissenschaften.* His Harnackhaus lecture ultimately satisfied both counts. Entitled "The Establishment of the Instinct Concept," it gave Lorenz his chance to offer, in the limelight, an extended critique of prevailing theories of instinctive behavior.

Lorenz began his talk by explaining that there were many different instinct theories in existence and that this had contributed to considerable confusion on the subject. Most of these theories, he explained, were too broad in what they sought to encompass and too insubstantial in the facts on which they were based. He proposed to put in their place a carefully defined concept of instinct built on a solid, factual foundation. The existence of the appropriate facts was evidenced, he claimed, by the way that *"practical students of animals"*—that is to say, "zoo attendants, biologically educated amateurs or field observers"—experienced no difficulty when speaking to each other about instinctive behavior, regardless of the words they used to describe it.

In constructing his lecture, Lorenz drew extensively on his recent correspondence with Craig.[122] He made special use of Craig's distinction between the appetitive behavior that begins an instinctive behavior cycle and the "consummatory act" that brings the cycle to an end. Where Craig's system needed amending, Lorenz believed, was in Craig's identification of instinctive behavior with the entire behavior cycle. Lorenz saw appetitive and instinctive behavior as fundamentally different from each other. He insisted that "functionally unitary behavior patterns" be divided into two different sorts of components: those that were purposive (and modifiable) and those that were not. As in his earlier paper on instinct, he promoted the idea that instinctive and learned elements are "intercalated" in broader behavior patterns but do not affect each other. Failure to separate these distinct components, he said, was what led earlier writers such as Herbert Spencer, Conwy Lloyd Morgan, and William McDougall to think that animals had insight into their instinctive behavior—or that animal instincts could be modified by experience.[123]

Lorenz proceeded to analyze what he considered to be the three most important theories of instinct: the "Spencer–Lloyd Morgan theory," the "McDougall instinct theory," and the "reflex theory" (which he attributed especially to Heinrich Ziegler). He objected to Spencer's and Lloyd Morgan's claims that instincts could be modified by experience and that instinctive behavior graded smoothly into acquired, intelligent behavior. What he disliked about McDougall's work was McDougall's general scheme of overarching, governing instincts directed toward a particular goal, with subordinate "motor mechanisms" employed as a means to that end. When it came to the reflex theory, however, he was less sure of his own stance. On balance he endorsed the reflex theory, but he had some reservations about it as well. One significant difference that he noted between instincts and reflexes was that animals strive to secure elicitation of their instinctive behavior patterns, but they do not strive to secure elicitation of their simple reflexes. For this reason, he suggested, the instinctive behavior pattern should perhaps be regarded as a special category of reflex, namely, "a striven-after reflex action" (*angestrebter Reflexablauf*), though he admitted this definition was attended with "philosophical difficulties." [124]

Later in his career Lorenz liked to tell how one of his auditors in Berlin was the young physiologist Erich von Holst. By Lorenz's account, Holst sat next to Lorenz's wife in the lecture hall. Holst nodded in agreement with everything Lorenz said up until the point when Lorenz gave grudging assent to the reflex theory of instincts. Then Holst put his head in his hands and muttered that Lorenz was an idiot. [125]

In Lorenz's telling of this story, it took Holst only a matter of minutes after the lecture to convince Lorenz that instinctive behavior patterns were better interpreted as the result of internally generated and coordinated impulses than as chains of reflexes set in motion by external stimuli. Memory is a tricky thing, however. Lorenz's correspondence from the 1930s shows that it was more than a year after his Harnackhaus lecture before Holst convinced him that the chain-reflex theory of instincts needed to be abandoned. [126]

If Lorenz in his Harnackhaus lecture at least acknowledged "philosophical difficulties" with the idea of "striven-after reflex actions," he had also begun to pay attention to particular phenomena that posed problems for a chain-reflex view of instincts. Foremost among these were *Leerlaufreaktionen* (later called "vacuum activities") and threshold lowering. To appreciate the problems these phenomena raised, we need to revisit Lorenz's thinking about how releasers elicit the performance of instinctive behavior patterns.

Lorenz defined releasers as instinctive behaviour patterns and their associated supporting structures and colors, the biological function of which is to elicit social instinctive behaviour patterns in members of the same species. In his Harnackhaus lecture, he compared the relation between a releaser and the animal's "innate releasing schema" to the familiar relation between a key (or combination) and a lock: "Specific stimulus combinations often represent specific elicitatory *keys* (signals) for specific responses. These responses cannot be elicited even by stimulus combinations which are only slightly different from the norm. Thus, specific sign ('key') stimuli are associated with a receptor correlate, which only responds to a quite specific combination of stimulatory effects, rather like a combination lock, and then sets the instinctive behaviour pattern in motion. In another publication, I have termed receptor correlates of this kind 'innate releasing schemata' [*angeborene Auslöse-Schematen*]." [127]

The reason that releasers and their corresponding releasing schemata had been able to evolve to the special degree that they had, Lorenz explained, was because they served the biological function of intraspecific communication, and because the organization of many animal societies depended upon them. In his words, "In many animals, particularly birds, complex systems of releasers and innate schemata form the basis of the entire sociological organization and guarantee unitary and biologically-adaptive interaction with the sexual partner, with offspring—in short, with conspecifics in general." [128]

Given his ideas on how closely adapted the innate behavior patterns of animals are to the sign stimuli and releasers that elicit them, it is not surprising the Lorenz took special note of occasions when animals perform their instinctive behavior patterns in the absence of the stimuli that normally release these patterns. The most conspicuous of these cases were what he called *Leerlaufreaktionen*. There the threshold for eliciting a behavior pattern was reduced to such a point that the behavior pattern "went off" without getting anywhere, like an engine running in neutral. As early as 1932, Lorenz had reported how he had hand reared a starling which "although it had never trapped a fly in its whole life, performed the entire fly-catching behavioral sequence without a fly—i.e. *in vacuo* [*auf Leerlauf*]." [129] This showed that the performance of an instinctive behavior pattern could be separated from its biological purpose. At the same time, it suggested that instinctive behavior was not built up simply from conditioned reflexes. "Threshold lowering" was evidently a related phenomenon. In threshold lowering, an instinctive motor pattern became easier to release the longer it had been since the pattern had last run its course.

Leerlaufreaktionen and threshold lowering suggested to Lorenz that inner stimuli increased in intensity in an animal in proportion to the amount of time since an instinctive action had last been released. This led him to envision a kind of "damming up" (*Stauung*) of a "reaction-specific energy." As he put it in his Harnackhaus lecture, "An animal does actually behave as if some reaction-specific energy were *accumulated* during periods when a specific pattern is not employed." He allowed that he was not especially enthusiastic about using physical concepts to model biological processes. Nonetheless, he offered a physical analogue of the biological process in question: "It is as if a gas were continually pumped into a container in which the resultant pressure continually increases until a discharge is effected under quite specific conditions."[130]

From these early notions Lorenz later developed his famous psychohydraulic model of instinctive action. Prior to doing this, he took cognizance of Holst's work on the endogenous generation and central coordination of nervous impulses and gave up his notion of instinctive actions as a special category of reflex. But mention of both of these developments takes us ahead of the story. We need to focus for now on the period just after Lorenz's Harnackhaus lecture, when he was basking in the afterglow of what he felt had been a most successful performance. His self-assurance was reflected in a letter he wrote to Margaret Morse Nice three weeks after the event. She had been promoting the key concepts of his "Kumpan" monograph to a skeptical G. K. Noble. She had then passed Noble's criticisms on to Lorenz. Confident he could respond to all of Noble's complaints, Lorenz also told Nice of the plans for publishing his Harnackhaus lecture:

> I am looking forward to the discussion with Dr. Noble. My long and friendly discussion with Prof Craig was certainly a tremendous help* [in the margin of the letter Lorenz added: "*For which I owe *you* {underlined twice} eternal thanks!"] to formulate my ideas, and it was this discussion that ended in my writing my latest paper 'Zur Kritik des Instinktbegriffes.' It will appear in 'Naturwissenschaften' which is about the most highbrowish and exclusive German paper. It is only owing to my Berlin lecture that they are taking it on! I am criticising what I consider the three most important Theories of instinct, e.g. first Spencer's and Lloyd Morgan's, second McDougall's and third the reflex-theory of instinctive action as put down by H. E. Ziegler and his followers. I think my thesis' [*sic*] are rather unanswerable but as they are tremendously overbearing, a lot of discussion is sure to follow. It seems that now in spring my fighting reactions are unduly strong![131]

Publishing in *Die Naturwissenschaften,* and publishing a paper as long as Lorenz's, was no small achievement. The editor of the journal, Fritz Süffert, told Lorenz the paper was in fact "three times too long," but he was nonetheless prepared to publish it without alteration, because he had seen "nothing so captivating and convincing" for quite some time.[132] Süffert in addition told Julius Springer of Springer-Verlag (Springer Publishing) that Lorenz was writing *the* German book on animal psychology. Springer wanted to publish it. As of October 1936, if not earlier, Lorenz had chosen for the work a provisional title, *Die Deutung tierischen Verhaltens* (*The Meaning of Animal Behavior*), and also a subtitle, *Eine Einführung in die vergleichende Verhaltenslehre und Psychologie der Tiere* (*An Introduction to the Study of the Comparative Behavior and Psychology of Animals*).[133]

REDIRECTING THE AIMS OF ANIMAL PSYCHOLOGY

Lorenz's rise to prominence as the figure most likely to write *the* major German work on animal psychology occurred in precisely the same period that saw the formation of the German Society for Animal Psychology (Deutsche Gesellschaft für Tierpsychologie, or DGT). Lorenz played no role in the creation of the society. Before the society was a year old, however, he had become one of its major actors.

The DGT was established in Berlin on 10 January 1936. Its original orientation toward practical matters reflected the concerns of its founders. The founders were Carl Kronacher, emeritus professor and director of the Institute for Animal Breeding and Genetics of Domestic Animals of the agricultural faculty of the University of Berlin; Professor W. Klein, director of the Institute for Anatomy, Physiology, and Hygiene of Domestic Mammals at the University of Bonn; and Dr. J. Effertz, an agricultural scientist. All of them sought to promote the study of animal psychology along lines that would assist the breeding and training of animals for particular tasks. They were impressed by recent German successes in breeding and training police dogs, and they believed that comparable achievements could be had with larger domestic animals, if one only studied the animals' mental abilities with an eye to improving the animals' *performance*. Pure and applied science, they believed, could here go hand in hand. From its inception, the DGT described its goals as being "the investigation of the animal mind and the practical utilization of animal-psychological findings."[134]

The DGT's founders suggested in addition that the study of animal psychology could be of *ethical* benefit. They maintained that learning more about the animal psyche would help clarify the relations of man and animal in all those circumstances where man either made animals subservient to him or otherwise involved himself with animal life. They predicted that benefits could be expected in the areas of animal raising and animal keeping, animal and nature protection, hunting, national defense (here the founders were still thinking in terms of the cavalry), veterinary medicine, and more.

Although the DGT was founded in January 1936, it was not until 6 November 1936 that it held its first public meeting. That meeting, convened in

Berlin, featured an account by Effertz of the society's founding followed by ten additional papers delivered by various scientific authorities and government officials. Among the latter was a Dr. Thomalla, identified in the reports of the period as a senior civil servant from the Reichs Ministry of National Instruction (*Volksaufklärung*) and Propaganda. He discussed promoting and disseminating knowledge of animal psychology for purposes of avoiding accidents when working with animals. The majority of the papers, however, were devoted to topics of pure rather than applied science. Most of the speakers in this regard were senior scientists who occupied prestigious positions. They included Oscar Heinroth from the Berlin zoo, Baron Jakob von Uexküll from his Institute for *Umwelt* Research in Hamburg, Hans Volkelt from the Psychological Institute in Leipzig, Friedrich Alverdes from the Zoological Institute in Marburg, and Otto Koehler from the Zoological Institute in Königsberg. Heinroth spoke on "communication among birds," Uexküll discussed "*Umwelt* research," Volkelt considered "animal psychology as genetic holistic psychology," Alverdes addressed "the learning abilities of single-celled animals," and Koehler confronted the question, "Can pigeons count?" A highlight of the meeting was the announcement that the publisher Paul Parey had agreed to produce a journal for the society, the *Zeitschrift für Tierpsychologie*, and the first issues were scheduled to appear early in 1937.

Heinroth mentioned the Berlin meeting to Lorenz just before it took place. Lorenz was unable to attend it.[135] He was, however, scheduled to go to Berlin later in the month to lecture to the German Ornithological Society, so he pursued with Heinroth the idea of using that lecture to point the animal psychologists in a direction more congenial to biologists. The basic drawback of the new society, as Lorenz saw it, was that it ignored almost entirely "the *comparative evolutionary* viewpoint," in other words, "the truly biological viewpoint." The foremost task of animal psychology, he told Heinroth, was "the creation of a foundation on which a truly comparative psychology can be built." There were psychologists who considered themselves "comparative," he said, but they did not have "the slightest idea" of the comparative, zoological way of posing questions. What animal psychology needed most, he claimed, was to adopt the comparative and evolutionary kinds of question-framing that Heinroth and Whitman had already brought to modern ornithology.[136]

Heinroth proved to be an able co-conspirator in Lorenz's project. He arranged for his young colleague to speak on 26 November 1936 to a joint meeting of the DOG and the DGT. On the appointed date, in the lecture hall of the Berlin Zoological Museum, Lorenz delivered a lecture entitled

"On Biological Question Posing in Animal Psychology." The audience of 162 persons included members of the DOG, the DGT, and the Society of Friends of Natural History (Gesellschaft naturforschender Freunde). Lorenz assembled the lecture on short notice, but he had no great difficulty doing so. He simply extracted the lecture from the introductory chapter he had composed for his book on animal psychology.[137]

All natural scientists, Lorenz told his audience, share a common concern with causal analysis, but biologists have to consider other kinds of questions as well. When biologists seek to understand an animal behavior pattern, he elaborated, they need to know not only about its physiological causation but also about its purposefulness (in the sense of its survival value), its relation to the whole pattern, or gestalt, of the animal's natural activities, and its evolutionary history. His goal was to impress upon his listeners the importance of giving all of these questions a central place in the science of animal psychology.

Lorenz began by explaining what he meant when he spoke of the purposefulness of animal behavior patterns. This topic, he said, had long been a source of confusion between mechanists and vitalists. Vitalists, it seemed, were apt to embed the phenomenon of behavioral purposefulness in transcendental discussions of final causes, or else mix the phenomenon up with the question of an animal's subjective intentions. Mechanists, on the other hand, tended to deny the existence of the phenomenon altogether. When he talked about the purposefulness of behavior, Lorenz said, he was referring to the way that behavior patterns functioned for "the preservation of the species." He did not deny the value of analyzing behavior patterns in terms of their physiological causation. He agreed, though, with Otto Koehler (who in turn was emerging rapidly as a major champion of Lorenz's work) that it could be "more interesting and important to biologists to know that a certain pigmentation is effective as camouflage, serving to preserve the species, than to know which factors in the metabolism of the animal actually cause it."[138]

Lorenz next stressed the importance of studying an animal and its behavior holistically. One needed to proceed from the whole to the parts, he said, rather than vice versa. Remarking how Pavlov's experimental work with conditioned reflexes had been generalized beyond its proper sphere, he urged: "There are not very many authors who, before they start making experiments, seek to know a species in all its variations of behavior like a zoologist does. Also, there are not many authors who seek to study their animals under natural conditions and in their familiar environment." Lorenz cited approvingly the Gestalt psychologist R. Matthaei, who in his book *Das Gestaltproblem* compared the holistic approach to the approach of a painter. "A casually drawn sketch is worked out more and more, in the process of which the painter improves all parts, as far as possible, at the same time; the picture appears to be complete in every state of its development—until the picture presents itself in its entire, vivid foregone conclusion." Lorenz likened this approach to his own way of studying animals: "The casual sketch [by the artist] corresponds to a general observation of the animal, if possible in its natural environment; then comes the animal keeping under conditions which resemble the natural ones as much as possible, conditions which also enable one to make an 'inventory' of most behavioral characteristics and patterns which the species in question displays. Only then follows a careful, tentative, and only gradually intensifying experimenting, the results of which lead to further and more specific experiments, without ever losing sight of the functional entity of the action system and the organism as a whole." [139]

The final, biologically inspired message Lorenz wanted to put before his audience was the importance of thinking in comparative, evolutionary terms. Central as this idea was to modern zoology, he said, it had not yet penetrated into animal psychology. Most psychologists did not appreciate that specific instinctive behavior patterns could be used for reconstructing phylogenies, or, what is more, that such behavior patterns were often superior to physical characters in this regard. Likewise deserving of appreciation, based on what was already known from comparative morphology, was the extent to which an animal's genetic composition provided "structural restraints" on its behavioral abilities. Lorenz hammered home the importance of studying first an animal's instincts—its *least* changeable behavioral features—prior to examining other forms of behavior. His focus was not on what could then be done with the other forms of behavior, but on how much could be done with instincts. Once one recognized that instinctive behavior patterns are comparable to organs, he said, and that one can apply the concept of homology to them, then one could compare animal behavior and human behavior. Specifically, one could inquire about the similarities of animal experience and human experience in those cases where instinctive behavior patterns in animals had phylogenetic "homologues" in man. One could thereby employ the lessons of comparative phylogeny to go beyond the limited objectives of the behaviorists and create a biologically informed and truly comparative animal psychology.

Such were the main messages of the lecture that Lorenz delivered on 26 November 1936 in Berlin. His rise to a position of power within the DGT followed quickly thereafter. He was appointed one of the three coeditors of the *Zeitschrift für Tierpsychologie,* together with Carl Kronacher and Otto Koehler. Soon he was given the opportunity to deliver another major address, this at the first annual meeting of the DGT, held from 5 to 7 February 1937 at Harnackhaus. This address, entitled "Evolutionary, Comparative Problem Formulation in Animal Psychology," elaborated on his lecture on biological problem-formulation of nine or ten weeks earlier. He was preparing the new lecture, he told Heinroth in January 1937, in an effort "to leave no stone unturned" in his attempt to influence the direction of the new society. A letter from Effertz, the business manager of the society, had led him to hope, as he explained to Heinroth, "that our way of formulating problems *could* seize command of the new society, if we did it skillfully." His spirits were further buoyed at the same time by the possibility that the Kaiser Wilhelm Gesellschaft was thinking of founding an institute for comparative psychology for him in Altenberg. "The inevitability of a truly comparative psychology," he assured Heinroth, *"is in the air."* [140]

Lorenz's maneuvering with respect to the leadership and direction of the DGT could not have been more successful. When Kronacher died, in April 1937, he was replaced on the editorial board of the *Zeitschrift für Tierpsychologie* by Lorenz's friend, Otto Antonius, who in January 1937 had been reinstated in his former position as director of the Schönbrunn zoo.[141] The control of the journal was now in the hands of a triumvirate— Antonius, Koehler, and Lorenz—that was fully in sympathy with Lorenz's ethological program and ready to promote it.[142]

But this again is getting ahead of the story. Lorenz's first lecture to the animal psychologists was on 26 November 1936. The next day he took the train from Berlin to Leiden in order to participate in a conference on instinct where he was to be one of the featured speakers. It was there that he first met Tinbergen, the Dutch naturalist who would become Lorenz's key collaborator in establishing ethology as a scientific discipline.

FOUR

NIKO TINBERGEN AND

THE LORENZIAN PROGRAM

Of what Lorenz and I discussed at that first meeting in Leiden in the winter 1936–1937 I have only the vaguest memories. We certainly were preoccupied with the "innateness" of so much behaviour and with this selective responsiveness, for which Lorenz at that time had mainly suggestive evidence and I could offer the first set of experimental results, obtained in sticklebacks together with Ter Pelkwijk, with the aid of "dummies." For the rest I remember that we were very much concerned with defending our "objectivistic" approach against the then prevailing notion that the aim of animal psychology was to find out what animals experience subjectively (Bierens de Haan, Portielje).[1]

TINBERGEN TO THE AUTHOR, 6 JUNE 1979, REGARDING
THE NOVEMBER 1936 INSTINCT MEETING IN LEIDEN

In Altenberg there is really much new. Tinbergen has been here with a Stipend from the Donders and Van der Hoeven Foundation since March 2 and is making beautiful experiments on inborn schemata. He has a special talent for precise experimental design that is lacking in me (I am learning tons from him!).[2]

LORENZ TO STRESEMANN, 26 MARCH 1937

NIKO TINBERGEN AND DUTCH ETHOLOGY

When Niko Tinbergen and Konrad Lorenz first met in Leiden late in November 1936, it was not by chance. They had begun corresponding with each other the previous year. When Tinbergen learned early in November 1936 that Lorenz was going to make a trip to Antwerp to pick up an automobile for Lorenz's brother-in-law's car business, Tinbergen urged Professor C. J. van der Klaauw, the head of Tinbergen's department, to invite Lorenz to give a talk in Leiden. Drawing upon funds from the recently established Prof. Dr. Jan van der Hoeven Foundation for Theoretical Biology, van der Klaauw quickly arranged a symposium on instinct with Lorenz as a featured speaker. Tinbergen in addition arranged for Lorenz to give a talk to students the night before the symposium.[3]

In the letters they exchanged with each another prior to their first meeting, Lorenz and Tinbergen expressed admiration of one another's work. Lorenz, who had somehow gotten the impression that Tinbergen was an older scientist, wrote to Tinbergen in a tone that was very polite, formal, and indeed deferential. When the two finally met face to face in Leiden, Lorenz was surprised and amused to find that Tinbergen was a young man— younger than Lorenz himself was. Tinbergen was only twenty-nine. Lorenz had just turned thirty-three.[4]

Like Lorenz, Tinbergen had been an ardent naturalist as a youth. Unlike Lorenz, however, he had found his greatest satisfactions as a naturalist not in raising animals so much as in stalking, watching, and photographing them in the countryside around his home in The Hague. More attuned to field studies than to comparative anatomy, Tinbergen brought to ethology an ecological dimension that was lacking in Lorenz's work. Whereas Lorenz's joys were in feeding animals, having them as companions, and, at least to some extent, controlling them, Tinbergen's joys were in being outdoors, watching and photographing animals in their natural settings, and matching wits with them. As Lorenz and Tinbergen later agreed, of the two of them Lorenz was by nature a farmer, while Tinbergen was by nature a hunter. These differences were not simply matters of individual temperament. They reflected different personal and cultural experiences. Tinbergen freely acknowledged that he imbibed his attitudes toward nature from not only the natural but also the social environment in Holland at the time he was growing up. Nature study was a thriving social activity for the Dutch youth of Tinbergen's day, and nature study above all meant field study.[5]

Tinbergen was born in The Hague in 1907, the third of five children of Dirck C. Tinbergen and Jeannette van Eek Tinbergen. Niko's father was a Dutch grammar school teacher and, in his son's words, a "liberal-minded, very hard-working man with many intellectual and social interests." Niko's mother was "the ever-cheerful, understanding, caring center of 'hearth and home.'" The parents, through their own example, set high standards for their children but also gave each child the freedom to pursue his or her own way.

Frequent family outings to the countryside around The Hague contributed to the young Tinbergen's love of the outdoors. The Hague was still small and relatively nonindustrial at the time, and it was thus easy to get to the countryside:

The sandy seashore, the very rich coastal sand dunes behind it, the meadows, hayfields, and the inland waters of the polders—then still largely
unpolluted and extremely rich in wildlife—were all so to speak on our
doorstep; we had an area of some two hundred square kilometers within
easy walking, bicycling, or skating distance. . . .

All this country was our "hunting range," ever-changing with the
seasons, with the time of day, with the weather, with every new encounter
with its natural inhabitants.[6]

Tinbergen much preferred being outdoors to being in school. He was
a hyperactive child (or at least so he later diagnosed himself), and he had
difficulty fitting into the rhythms of the classroom. His primary loves were
sports (especially field hockey and skating), camping, nature photography,
and sketching. He tuned out of classroom instruction, feeling both "ex-
cruciatingly bored during most lessons, [and] apprehensive lest I would be
tested."[7] He had two teachers, however, who encouraged him in his enthusiasm for field studies. One was his biology teacher, Dr. A. Schierbeek,
who oversaw the school nature club. The other was Gerard Tijmstra, a
mathematics teacher and ex–artillery officer with a great passion for field
ornithology. Tijmstra kindled Tinbergen's interest in field studies of herring gulls and, more generally, in the kinds of biological thinking that
could be applied to the study of animal behavior.

The popular books and collectible nature cards generated by two
Dutch schoolmasters, Eduard Heimans and Jacques P. Thijsse, also stimulated Tinbergen's youthful appreciation of nature. Through their engaging
descriptions and illustrations of Holland's native flora and fauna, their promotion of wildlife sanctuaries, and their cultivation of an appreciation of
living organisms in their natural settings, these two men had an enormous
influence on Dutch attitudes toward nature early in the twentieth century.
When the English clergyman, scholar, and bird-lover Charles Raven paid a
visit to the Dutch bird sanctuary at Texel in the 1920s, he was impressed by
the arrival there of not only a "very large gathering of Dutch ornithologists," but also "the older pupils of a school, with their head-master in
charge, [who] turned up at Koog and were shown all round the Muy and
the Waal en Burg." As Raven went on to explain, "My friend Dr. Thijsse,
whose delightful book is one of the most fascinating records of field study,
has done a notable work in arousing enthusiasm for such pursuits, and is
training up a generation of ardent young naturalists."[8]

Heimans' and Thijsse's efforts also included the publication of a non-
technical journal for field natural history, *De Levende Natuur*. Tinbergen

wrote articles for this journal as a budding young naturalist. He would continue to write for it for almost half a century more.

Also crucial for Tinbergen's early development (and for that of his younger brother, Luuk, and for the ethologist-to-be G. P. Baerends), was the Dutch Youth Association for Nature Study (Nederlandse Jeugdbond voor Natuurstudie, or NJN). Founded as a youth organization after the First World War—perhaps, as Robert Hinde has suggested, in "unconscious reaction to the 'mismanagement of the world' by adults"—the NJN's membership was restricted to Dutch youth ranging from eleven to no more than twenty-three years old. The NJN had branches in the high schools. Encouraging nature excursions, camps, exhibits, and other related activities, it played a major role in creating a sense of community among young field naturalists.[9]

Upon finishing secondary school, Tinbergen was not certain that he wanted to go to study at the university level. University biology, at least as it was then taught at the University of Leiden, had a reputation for being methodologically and intellectually stodgy. Tinbergen's teacher, Abraham Schierbeek, however, believed Tinbergen had the potential to become a professional biologist. So did the theoretical physicist Paul Ehrenfest, who was a teacher of Tinbergen's older brother, Jan, and who knew Niko because Niko and his brothers were friends of the Ehrenfest children. Through his brothers, Niko came to know the Ehrenfests. Niko sometimes led the younger Ehrenfest children on nature walks, and the children's subsequent accounts to their parents of these walks were what drew the elder Ehrenfest's attention to Niko's enthusiasm for field natural history. Together with Schierbeek, Ehrenfest persuaded Tinbergen's parents to send Niko for three months in the late summer and early fall of 1925 to the German ornithological field station Vogelwarte Rossitten, near Königsberg, where Niko would have the opportunity to observe modern biological fieldwork firsthand.

The experience at Rossitten proved decisive. It convinced the young Dutch naturalist to pursue a career as a field biologist. He felt he could slog through whatever university instruction was necessary to get him to the point where he could pursue field studies professionally. Back home after his time in Rossitten he enrolled at midyear at the University of Leiden.[10]

The instruction that awaited him was in most respects just as dreadful as he had feared. His introduction to comparative anatomy came from N. van Kampen, an old-fashioned comparative anatomist, whose interests, as it seemed to Tinbergen, were limited to identifying homologies in animal structures. Van Kampen criticized Tinbergen severely for simply men-

4.1 Niko Tinbergen in 1928 constructing a bird hide from an old basket, sacking, and pieces of sod. (Photo by F. P. J. Kooymans, Copy provided by Hans Kruuk and Jaap Tinbergen.)

tioning the topics of biological function and behavior. Tinbergen found the classes taught by the botanist J. M. Janse to be even worse. The teaching and drills in anatomy and taxonomy impressed him as "incredibly dull and intellectually empty." He ended up skipping many of his labs. His botany practical was held on Wednesday afternoons, but he preferred to devote his Wednesdays to bird-watching. His zoology practical was scheduled for Thursday and Saturday afternoons, but on Thursday afternoons he practiced field hockey and on Saturday afternoons he had his field hockey matches. The best he could say about his behavior, when he looked back on it, was that it enabled him to retain his intellectual independence.[11]

Still, his education at Leiden was by no means a total loss. Arriving as he did at midyear, he was placed under the care of a young zoology assistant, Jan Verwey, who helped him catch up in his course work. Verwey, the son of one of Holland's leading poets, was himself a passionate field naturalist. Two years earlier he had begun a long-term study of the social behavior of the gray heron. Tinbergen had the chance to learn about this work firsthand, long before it was published, and the work impressed him deeply.[12] Verwey in turn had the chance to witness Tinbergen hold an audience spellbound with an account of Tinbergen's ornithological experiences at Rossitten. The two young men became close friends. They frequently took field trips together until Verwey left for the Dutch East Indies to take a position at the Dutch marine biological laboratory at Batavia.[13]

Verwey was not the only instructor Tinbergen regarded highly in his student days at Leiden. In the second half of his five-and-a-half-year course of undergraduate study he was taught (and given considerable intellectual freedom) by two very supportive assistant professors, Hildebrand Boschma (an invertebrate zoologist) and C. J. van der Klaauw (a comparative anatomist interested in theoretical biology and ecology). An additional influence on him in his student years was A. F. J. "Frits" Portielje, the charismatic director of "Natura Artis Magistra," the Amsterdam zoo.[14]

Birds were Tinbergen's primary interest when he began his university career. At this time, only a few ornithologists worldwide were interested primarily in bird behavior. However, Dutch observers were well represented within this small population. Verwey's gray heron investigation, cast in the mode of the work of Edmund Selous and Julian Huxley, was not the only such example. The Dutch had developed nature preserves in the early twentieth century, and field ornithology had flourished as a consequence.[15] Important studies had begun to be conducted on birds in captivity as well. In 1925 Portielje began publishing in the Dutch ornithological journal *Ardea* a series of papers on the "ethology" of various bird species at the Amsterdam zoo. His studies showed that a good zoological garden had more to offer to the study of psychobiological phenomena than zoologists and ornithologists usually acknowledged.[16] J. A. Bierens de Haan likewise based his 1926 paper on the dance of the argus pheasant on observations he had made at the Amsterdam zoo.[17]

Tinbergen himself did not much like watching animals in captivity. He found it too dull. The presence of Portielje drew him, nonetheless, to "Artis." Tinbergen was not attracted, it must be admitted, to Portielje's theorizing. Portielje's idea of "teleocausality" and desire to fathom animal subjective experience looked to Tinbergen like steps in the wrong direction. On the other hand, Tinbergen was greatly impressed by Portielje's practical knowledge as an animal handler. Furthermore, Portielje was a warm and generous man. Tinbergen found he could have friendly, spirited arguments with the zoo director. Tinbergen and his friends always appreciated how well Portielje treated them.[18] Portielje also conducted experiments that may well have provided an inspiration and model for some of Tinbergen's later work. To test instinctive behavior in a bittern, Portielje used simple cardboard cutouts to see which configurations released the bird's instinctive attacking response. A decade later, a similar use of "dummies" became a hallmark of the experimental researches of Tinbergen and his students.

Portielje and Verwey were both familiar with the work of Oskar Heinroth and cited it appreciatively. Portielje in addition made use of the ideas of Uexküll, Lloyd Morgan, McDougall, and Herbert Spencer Jennings, among others. It was thus from his own countrymen, and not from Lorenz, that Tinbergen first learned of the work of Heinroth and Uexküll. Indeed, it is possible that he became acquainted with the work of Heinroth and Uexküll a few years before Lorenz did. At least initially, however, Tinbergen was more impressed by British than by German studies of animal behavior. When Verwey and Tinbergen discussed and compared the work of Heinroth and Huxley, Verwey was keen on Heinroth but Tinbergen, to Verwey's surprise, preferred Huxley. Tinbergen regarded Heinroth as a great collector of data, but he saw Huxley as the more penetrating theorist.[19]

As observed in the previous chapter, Lorenz's self-confidence and sense of purpose were reinforced early in his career by his discovery that he knew things about animals that most of the "experts" did not. Tinbergen's psychic makeup was decidedly different from Lorenz's, but he too was stimulated when he found that he knew something that was unknown to a great name in his field. This occurred at the Seventh International Ornithological Congress, held in Amsterdam in June 1930. At the congress's final plenary session, with the young Tinbergen in attendance, Julian Huxley delivered a slide lecture entitled "The Natural History of Bird Courtship." In the course of his presentation, Huxley allowed that in the genus *Sterna* (terns), the male fed the female at the nest. Tinbergen only three weeks earlier had completed a month's study at the Hook of Holland on pair formation in the common tern (*Sterna hirundo hirundo* L.). He recognized that Huxley's statement told only half the story: females fed males at the nest just as males fed females. The discrepancy between Huxley's claim and Tinbergen's own observations moved Tinbergen to publish in *Ardea* in 1931 his first substantial scientific paper, "On Pair Formation in the Common Tern." The paper corrected Huxley's facts while supporting Huxley's broader views. Tinbergen found *Sterna hirundo hirundo* L. to be a species in which Huxley's "mutual courtship" applied fully: all the behavior patterns, ceremonies, and calls were performed by both members of the pair rather than restricted to just one of them.[20]

Through his student years, Tinbergen remained fascinated with birds, especially herring gulls. He regularly observed a herring gull colony, studying how the birds' individual behavior patterns fit them for social life. With his friends F. P. J. Kooymans, G. A. van Beusekom, and M. G. Rutten he published in 1930 a popular book of bird photography entitled *Het Vogel eiland* (*The Bird Island*). He also conducted ecological studies of what certain birds ate.[21] But while birds were almost his sole biological interest when he arrived at Leiden, they did not remain so. Verwey set out to convince him that not just birds but all animals were interesting. Over time, Tinbergen came to agree.[22] He began studying the abundant literature on the behavior of insects, particularly bees and wasps, paying particular attention to the writings of J.-H. Fabre, George and Elizabeth Peckham, Charles Ferton, Karl von Frisch, and Mathilde Hertz. The topic he ultimately chose for his doctoral dissertation was an entomological one: the homing behavior of the digger wasp, *Philanthus triangulum* Fabr., an insect known as the "bee-wolf" because it captures and provisions its nests with honeybees.

Tinbergen's doctoral study of the homing behavior of *Philanthus* was important for a number of reasons in addition to its identification of the cues the digger wasp uses in finding its way back to its nest. Methodologically it was significant because it led Tinbergen at the very beginning of his career to become adept at bringing simple experiments to bear on the behavior of animals in their natural habitats. For models, he turned to the elegant experiments of Karl von Frisch and Mathilde Hertz. By marking individual wasps with different-colored spots of paint (a technique used by Frisch), Tinbergen was able to keep track of the wasps as individuals and see how they responded to his manipulation of items in their environment. By displacing landmarks (such as pinecones) close to a wasp's burrow, and employing suitable control experiments, he established that a *Philanthus* female makes a brief "locality study," or reconnoitering flight, upon leaving her burrow. Later, upon returning from foraging, she uses visual cues to find her burrow again.[23]

His early locality studies on *Philanthus* established for Tinbergen the particular locality in which he himself would conduct summer field studies for more than two decades. The site was Hulshorst, some eighty miles from Leiden near the small town of Harderwijk on the Zuider Zee. Here, sand dunes, pine woods, and eroded heath intermingled and afforded a habitat for a variety of insects, birds, and other animals. Tinbergen's family frequently spent its summer vacations in the area. Tinbergen developed his first real taste for experimental research at Hulshorst, and he subsequently made it a site for summer field camps where young naturalists from Leiden University learned to conduct research on animals in their natural habitats.

Tinbergen's doctoral thesis was remarkably short, just thirty-two pages in length. Had he been able to spend an additional summer conducting observations and experiments, his dissertation would have been longer, but he was presented with a remarkable opportunity he did not want to miss: a Dutch meteorological expedition to Greenland. The expedition was being organized as part of the International Polar Year, 1932–1933. Tinbergen was offered the chance to go along as a zoologist. Under the circumstances, his professor, Hildebrand Boschma, agreed to accept Tinbergen's brief dissertation with the understanding that Tinbergen would do further work on the topic once he returned from the Arctic. Tinbergen was officially awarded his doctorate in 1932. The day after receiving his degree, he married Elisabeth Rutten, a chemistry student, and the two of them set about planning in detail their yearlong stay among Eskimos on the Arctic Circle on the east coast of Greenland.[24]

Tinbergen jumped at the opportunity to go to Greenland not because he had a particular research project that he wanted to pursue there but rather, as he later put it, because this gave him the chance "to see the Arctic and its wild life, the pack ice, icebergs, to live among the Eskimos—it surpassed my wildest dreams."[25] To justify the trip scientifically, he decided to study the territorial behavior of the snow bunting in the spring. Close relatives of the snow bunting had figured in Eliot Howard's original formulation of the concept of territory. Max Nicholson had reported, however, that the snow bunting's behavior differed appreciably from that of Howard's birds. Tinbergen planned to sort the differences out. Once in Greenland, he and his wife also studied territory and pair formation in the red-necked phalarope. In addition, they learned in at least a rudimentary way the language of the Angmagssalingmiut Eskimos, they collected Eskimo artifacts, and they studied the social behavior of the Eskimo sled dogs.

Tinbergen's ornithological studies in Greenland did not overturn any of the prevailing understandings of bird ecology or bird behavior. The snow bunting proved to be just as territorial as Howard's British buntings. Similarly, the red-necked phalarope's atypical color dimorphism—with the male cryptically colored and the female brightly colored—corresponded, as anticipated, to a reversal of the more typical breeding behavior of male and female birds. Nonetheless, Tinbergen gained valuable field experience and a wealth of detailed knowledge about bird courtship. Among other things he saw male snow buntings stop in the midst of territorial fighting and go through the motions of picking up something from the snow (without in fact picking up anything at all). This provided him with his first look at a class of behavioral phenomena that he would later characterize as "substitute" or "displacement" activities and that he would explain in terms of a motivational conflict.[26]

The year before he went to Greenland, Tinbergen was appointed to the position of assistant in the Department of Zoology at Leiden, the same meagerly paid position Verwey had occupied when Tinbergen first came to Leiden as an undergraduate. This post awaited Tinbergen on his return from Greenland in the fall of 1933. His charge was to teach courses in vertebrate comparative anatomy and animal behavior. One conspicuous disadvantage of the post (in addition to its low salary) was that it provided Tinbergen only twelve days of vacation per year. With such a schedule, he could not have continued his summer wasp studies at Hulshorst. The problem was solved when Boschma, then head of the department, agreed to allow Tinbergen to spend July and August in the field at Hulshorst, provided that he take with him two undergraduate assistants whose help to him could count as part of their own required practical work. The first two students to join him under this arrangement were D. J. Kuenen and W. Kruyt, in the summer of 1934. In this way, the tradition of Hulshorst research camps began.[27]

Another teaching-cum-research arrangement of tremendous value to Tinbergen and his students was instituted after C. J. van der Klaauw became professor of zoology and director of the Zoological Laboratory in the fall of 1934. Convinced that the university's traditional emphasis on comparative anatomy and morphology in its zoology courses was much too narrow and stultifying, van der Klaauw set about upgrading zoology education. He urged Tinbergen to develop "block practicals" for third-year undergraduate students that would allow them to devote a concentrated period of time to animal watching. Ultimately, Tinbergen found these block practicals to be the most satisfying form of teaching in which he ever engaged. For six weeks each spring he set aside all his other activities, including lecturing and corresponding with colleagues, to focus on the block practical.[28]

Tinbergen also had a small group of students who, even as beginning undergraduates, accompanied him in his spring gull watching. In the spring of 1934, the first breeding season after his return from Greenland, Tinbergen took up his observations again at a gullery between Leiden and The Hague. He was joined in his early-morning vigils by two remarkable first-year undergraduates whose tastes for field natural history had already been sharpened through their participation in the NJN. One was his younger brother, Lukas (Luuk) Tinbergen. The other was Jan Joost ter Pelkwijk. The following spring this group was joined by two more first-year students:

Josina van Roon and G. P. Baerends. A fifth biology student, D. Kreger, also participated in the gull watching. Tinbergen's earliest papers on the sociology of the herring gull dealt with the observations that he and his young team of collaborators made during the gull breeding seasons of 1934 and 1935. It was also in 1934 that he began his stickleback studies.[29] In addition to his undergraduate students, Tinbergen soon had Ph.D. candidates working in the field and in the laboratory.[30]

Tinbergen thus constructed an educational program that served students at various stages of their careers and provided them with different sorts of research opportunities. Lab studies and field studies were pursued every year according to a rhythm jointly dictated by the activities of the animal subjects and the routine of the university. The intensive, six-week practical course took place during the school year in the spring, when research could be conducted on such topics as the breeding behavior of the three-spined stickleback. Insofar as this overlapped with the breeding season of the herring gull, Tinbergen would get up to watch gulls from before daybreak to about 8:00 a.m. and then make his way to the lab by 9:00. In the summer, when students were free from other course work, Hulshorst became the site of additional researches. The summer was when *Philanthus* and other insects were most active, and when birds could be studied further. In the fall and winter, Tinbergen spent his time teaching, writing up research results, and preparing for the work of the next season.[31]

Tinbergen was not entirely content, however, to settle into this routine. In a letter of December 1934 to his parents he outlined a plan to return to Greenland in the summer of 1937 and spend a year there following up on the studies he had initiated in 1932–1933. His plan included taking with him his students Joost ter Pelkwijk, D. J. Kuenen, and F. P. J. Kooymans, plus his brother Luuk. This particular project did not materialize for him. Neither did a possibility of spending ten months in the East Indies. He was successful, however, in securing leaves to less exotic places. In the spring of 1937 he spent three months working with Lorenz at Lorenz's home in Austria. In the fall of 1938 he traveled to the United States, where he lectured and learned more about American studies of behavior.[32]

Prior to these travels, Tinbergen continued his research in Holland. The summer work at Hulshorst led to further papers on *Philanthus:* one on the senses *Philanthus* uses in hunting its prey and others (published jointly with his students W. Kruyt and R. J. van der Linde) on *Philanthus's* orientation behavior. In addition, with his brother Luuk and his friend George Schuyl, Tinbergen coauthored a study of the ethology of the hobby (*Falco subbuteo subbuteo* L.), a late-breeding bird of prey common to the Hulshorst area. The paper reported on observations accumulated through some 338 hours of watching over seventy-eight days and three different summers, one (1931) before his Greenland experience, and two more (1934 and 1935) after it. When in the late 1930s the Hulshorst *Philanthus* population decreased substantially from the high point it had reached earlier in the decade, Tinbergen and his students set about identifying new topics, including the courtship behavior of the grayling butterfly (*Eumenis semele* L.).[33]

The organism that provided yeoman service for the springtime laboratory practicum was the common three-spined stickleback (*Gasterosteus aculeatus*). This fish's special virtues as an experimental animal were many. It was readily available in Holland's ponds and waterways, it was tame, it was easily cared for, and it displayed distinctive innate behavior patterns. Tinbergen and his students concentrated their researches on the stickleback's courting and reproductive behavior. Joost ter Pelkwijk took the lead in devising models or dummies to test precisely which sign stimuli the fish responds to in fighting, courting, spawning, and so forth. He found that fighting on the part of the stickleback male could be readily released by a crude model of the fish, colored red underneath.

As Uexküll had claimed in the case of the tick, and as Lorenz had found in his work on jackdaws, Tinbergen and his students concluded that much of the stickleback's behavior consists of innate behaviour patterns released by a limited number of special sign stimuli. The full functional cycle of courtship in the stickleback involves a "reaction chain," with the male and female receiving from each other, in sequence, the appropriate stimulus or stimuli at the appropriate time. The appearance of the female in the male's territory elicits the male's "zigzag dance." That "dance" in turn elicits courtship behavior on the part of the female, which releases the male's behavior of "leading" to the nest, which elicits following on the part of the female, and so on. An early account of the behavior of the stickleback, based on lab researches carried out by ter Pelkwijk, Tinbergen, and four biology students during the ethology practicum of spring 1936, was completed in July 1936 and published in the 1 September 1936 issue of *De Levende Natuur*.[34]

Tinbergen thus had an active, varied, and distinctive research program well under way by the time he met Lorenz late in the fall of 1936. His program had grown out of a lively Dutch tradition of field natural history. This tradition had inspired not only his own work but also that of a whole cohort of other young naturalists, among them Jan Verwey, Luuk Tinbergen, Jan Joost ter Pelkwijk, G. F. Makkink, H. N. Kluyver, A. Kortlandt, G. P. Baerends, and Josina van Roon. Most of these young naturalists, Tinbergen

4.2 The courtship "dance" of the three-spined stickleback, from the paper that Joost ter Pelkwijk and Niko Tinbergen published in the 1 September 1936 issue of the Dutch journal *De Levende Natuur* (J. J. ter Pelkwijk and N. Tinbergen, "Roodkaakjes," *De Levende Natuur* 41 [1936]: 129–137.)

among them, had initially been especially interested in birds. The nature-study teachings of the Dutch schoolmasters Heimans and Thijsse had been broader than this, however, and in his own program of zoological teaching at Leiden, Tinbergen turned to insects and fish as well as birds as subjects for observation and experiment. His goal was to understand the behavior of whole animals in their natural environments (or in the case of the sticklebacks, in aquaria that approximated the fish's natural environment).

Tinbergen, in short, did not need Lorenz to introduce him to the study of animal behavior. Nonetheless, Lorenz's influence upon him was profound. Lorenz was providing ethological studies with a powerful theoretical framework and focus. He was simultaneously embarking on a campaign to reform animal psychology by putting the biological study of behavior at its center. In addition, he was showing that behavior studies could enrich the zoologist's long-standing project of identifying the evolutionary affinities of living things. Tinbergen found all this highly inspiring. Years later he modestly maintained that he would "have remained a piece worker without a comprehensive theoretical approach if I had not been taught by [Lorenz]." [35]

THE LEIDEN "INSTINCT" SYMPOSIUM

The Leiden instinct symposium, held on 28 November 1936, was designed to give the two leading instinct theorists of the day—Lorenz and Bierens de Haan—a chance to square off against each other. Professor van der Klaauw invited other leading figures as well, among them the Dutch scientists Verwey, Portielje, H. J. Jordan, and F. J. J. Buytendijk; the Belgian animal psychologist Louis Verlaine; the German animal psychologist Werner Fischel; and the British biologist E. S. Russell. With little advance notice, some of these invitees were unable to attend. Russell, for example, did not make it to Leiden, though he did contribute a paper to the symposium proceedings. But Bierens de Haan came, as did Portielje and Verlaine. They in effect represented an older generation of animal psychologists, all of whom had been born in the 1880s. Bierens de Haan was the eldest of the group at fifty-three; Verlaine was the youngest at forty-seven. Lorenz for his part was fourteen years younger than Verlaine was and twenty years Bierens de Haan's junior. Tinbergen, several of Tinbergen's students, and the young Finnish ecoethologist Pontus Palmgren constituted, with Lorenz, the younger contingent at the meeting.[36]

The symposium was not of equal consequence for all who attended. It did not alter the thinking or change the careers of the older scientists, who were already well set on their individual courses. For Lorenz and Tinbergen, however, it proved pivotal. This is not to say that Lorenz, at least, was any less set in his thinking than were the leaders of the older generation. He had already assembled the greater part of the conceptual framework he would champion for the rest of his career. The symposium's significance was that it provided the occasion for Lorenz and Tinbergen to meet one another. Each of them had the chance to begin to recognize how well the other's strengths complemented his own. Briefly put, Lorenz's bold and largely intuitive theorizing dovetailed beautifully with Tinbergen's strong analytical and experimental talents. And each of them knew an enormous amount about animal behavior. The good start they made in Leiden would be consolidated the following spring, when Tinbergen had the opportunity to work with Lorenz in Altenberg.

Bierens de Haan gave the opening lecture of the Leiden symposium of 28 November 1936. His address, "On the Concept of Instinct in Animal Psychology," reiterated the position he had staked out in his previous writings. The primary concern of students of animal behavior, he insisted, should be the subjective experience of animals. Animal psychology, he said, needed to be independent of physiology. He underscored this by drawing a clear distinction between instinct (*Instinkt*) on the one hand and instinctive action (*Instinkthandlung*) on the other. Instinct, he insisted, was essentially a psychological phenomenon, and it needed to be defined strictly in psychological terms. Neither a physiological definition (in terms of nervous structure, for example) nor a biological definition (in terms of biological need or purpose) would suffice. As he saw it, instinct was an "innate and specific psychical disposition" involving the same general categories of knowing, feeling, and striving that an investigator needed when studying human psychology. He admitted that if one were to choose to focus simply on instinctive *actions* or *behavior patterns,* then one could reasonably restrict oneself to using only biological terms. In short, "instinctive actions" (*Instinkhandlungen*) deserved to be treated as biological phenomena. But he left no doubt about his own opinion: any approach that neglected the subjective side of animal life missed the most important part of animal behavior.

The Dutch animal psychologist was well aware that a challenge to his approach was emerging in the writings of the energetic young Austrian naturalist who was soon to follow him on the podium. He thus devoted most of the second half of his lecture to commenting on what he took to be Lorenz's main discoveries and concepts. He readily acknowledged, for example, the interest and significance of imprinting. He believed, nonetheless, that the facts of the matter undermined Lorenz's sharp distinction between imprinting and learning and also Lorenz's insistence that instincts could not be modified by experience. Imprinting, as Bierens de Haan perceived it, was a special instance that illustrated how an animal's experiences were capable of reshaping its instincts.

It does not appear that a major clash actually took place at this meeting between Bierens de Haan and Lorenz. To the contrary, it seems that they by and large failed to engage each other in debate. At the very least it can be said that when it came to writing up his paper for publication in the symposium proceedings, Lorenz did not structure his comments in response to Bierens de Haan. What he offered instead was a new version of his Berlin instinct paper.[37] Yet if Lorenz did not take Bierens de Haan's position as the target of his remarks, the contrast between his approach and Bierens de Haan's was conspicuous. Whereas Bierens de Haan made his theme "instinct," broadly construed, Lorenz restricted his own focus to the "instinctive actions" which the Dutch instinct theorist regarded as being of only secondary importance. This was not because Lorenz's goals were any less ambitious than Bierens de Haan's. Lorenz too wanted to reform animal psychology. But he had a very different idea of the phenomena and explanatory principles on which a science of animal psychology should be based.

The initial task Lorenz chose for himself in his Leiden paper was to specify the particular features that make instinctive actions a worthy foundation on which to build a science of animal behavior. He named six of these: (1) Instinctive behavior patterns are extremely valuable indicators of taxonomic relationships, in some cases characterizing species, in other cases characterizing higher categories such as genera, orders, or even classes. (2) Instinctive actions manifest themselves in the individual organism without the animal's having the benefit either of previous experience or the example of an older member of the same species. (3) The regulative control to which instinctive behavior patterns are subject may also be independent of learning and experience. (4) The relation of instinctive actions to intensities of excitation is such that simple stimulus-response models of reaction are inadequate to explain their occurrence. (5) The threshold value of the stimuli that release instinctive behavior patterns can be reduced to the point that instinctive behavior patterns are at times performed in the apparent absence of any releasing stimuli at all. (6) An animal will strive through "appetitive" (purpose-directed) behavior to bring about the discharge of its instinctive actions.[38]

Having suggested the importance of instinctive behavior patterns as elements of behavior, Lorenz moved to the issue of how instinctive behavior patterns are released. Releasers, he explained, can be either learned or unlearned, and learning can influence the intensity at which an instinctive behavior pattern is released. He was quick to insist, though, that regardless of the nature of the releaser, the form of the instinctive action released never changes. He went on to review for his audience his concepts of the innate releasing mechanism, instinct-learning intercalation, and the characteristic differences between innate releasing mechanisms on the one hand and acquired, stimulus-releasing situations on the other.

Lorenz was quite prepared to acknowledge that the creation of a science of animal psychology, "in the proper sense of the term," would require parallel investigations in the physiology and the psychology of experience. Nonetheless, he was adamant in restricting his own definition of instinctive action to those units of behavior he considered innate, precisely delineated, centrally controlled, and sought after through appetitive behavior. His duty as a scientist, as he saw it, was to *analyze* behavior, that is, to break it down into its component parts. He acknowledged that instinctive actions, orientation reactions, and learned reactions often functioned together, but he insisted it was necessary to distinguish these from one another. In his view, a broad definition of instinct that jumbled together such demonstrably different types of behavior simply would not further scientific research.

After the conference, Lorenz wrote to Heinroth about the event. He reported that Bierens de Haan ("der Bierhahn"—the "beer-chicken"—as Lorenz facetiously called him) had not understood a word of Lorenz's talk. Portielje, in contrast, Lorenz imagined to be on his way to being converted to the ethologists' point of view, at least insofar as Portielje appreciated that Lorenz's conception of instinctive action was not intended to reduce animals to machines. There was one person, though, about whom Lorenz seemed to have no question but instead only praise. This was Tinbergen: "a terribly nice and wholly excellent clever fellow."[39]

Tinbergen was inspired to find in Lorenz a man with great energy and insight who was daring to challenge the authorities in the field with a bold new conceptual program. Lorenz in turn was ecstatic about what Tinbergen could tell him of the stickleback experiments Tinbergen and his students were conducting. Years later, in an autobiographical account in which he described his first meeting with Lorenz, Tinbergen reported that Lorenz's reaction to hearing of the Leiden stickleback experiments was to repeat again and again: "That is just what we need." In fact, Tinbergen was being diplomatic. Privately what he remembered Lorenz saying was: "those stickleback experiments are exactly what *I* need."[40]

Tinbergen was not put off, however, by this display of Lorenz's ego, and it is easy to imagine why not. Lorenz was a man of exceptional insight, energy, and charm, but more important, he was bent on reforming the study of animal behavior in a direction that seemed to Tinbergen to be entirely promising. To top things off, he was tremendously enthusiastic about what Tinbergen and Tinbergen's students had been doing.

The two men "clicked," both personally and intellectually.[41] They recognized that their talents were complementary. As Tinbergen described this relation to Julian Huxley a decade and a half later: "[Lorenz] is the pioneer who sees principles in a flash. I am his conscience in two ways: I don't immediately understand him, and pester him until a sharp formulation is found, and I have more patience in thinking out observational and experimental methods to check his ideas."[42] Lorenz in turn, when he was sixty-eight (and facing up to retiring in another two years), allowed that his own primary role in the history of ethology had been to realize the broad significance of the inferences that could be drawn from Whitman's and Heinroth's comparative studies of instinctive behavior. "The further progress of ethology," he observed, "was due mainly to Niko Tinbergen whose inspired yet circumspect experimentation did much to make ethology recognized as a 'respectable' branch of biological science."[43]

It is not possible to reconstruct exactly what Lorenz and Tinbergen discussed in Leiden in November of 1936. Late in their careers, when asked about this, neither of them felt particularly confident about sorting out their conceptual contributions at their first meeting. As Tinbergen reported to this writer, in one of the statements that stands as an epigraph to this chapter:

> Of what Lorenz and I discussed at that first meeting in Leiden in the winter 1936–1937 I have only the vaguest memories. We certainly were preoccupied with the "innateness" of so much behaviour and with this selective responsiveness, for which Lorenz at that time had mainly suggestive evidence and I could offer the first set of experimental results, obtained in sticklebacks together with Ter Pelkwijk, with the aid of "dummies." For the rest I remember that we were very much concerned with defending our "objectivistic" approach against the then prevailing notion that the aim of animal psychology was to find out what animals experience subjectively (Bierens de Haan, Portielje).[44]

Lorenz in one of his own autobiographical recollections said of the Leiden meeting: "During discussions that lasted through the nights, Niko Tinbergen and I conceived the concept of the innate releasing mechanism (IRM), although it is no longer possible to determine by which one of us it was born." [45]

This particular statement by Lorenz is clearly inaccurate, for Lorenz had already introduced his concept of *das angeborene auslösende Schema* (later rendered as "the innate releasing mechanism") in his "Kumpan" paper of 1935.[46] In a more detailed account of his first meeting with Tinbergen, Lorenz wrote:

> Our views coincided to an amazing degree but I quickly realized that [Tinbergen] was my superior in regard to analytical thought as well as to the faculty of devising simple and telling experiments. We discussed the relationship between spatially orienting responses (taxes in the sense of Alfred Kühn) and releasing mechanism on one hand, and the spontaneous endogenous motor patterns on the other. In these discussions some conceptualisations took form which later proved fruitful to ethological research. None of us knows who said what first, but it is highly probable that the conceptual separation of taxes, innate releasing mechanisms and fixed motor patterns was Tinbergen's contribution. He certainly was the driving force in a series of experiments which we conducted on the egg-rolling response of the Greylag goose when he stayed with us in Altenberg for several months in the summer of 1937.[47]

Tinbergen for his part remembered Lorenz as the one who made the conceptual distinction between the elicitation and the orientation of behavior patterns. In any event, the two of them studied this problem together when Tinbergen came to Altenberg in the spring (not summer) of 1937 to work with Lorenz.[48] Their three and a half months together in Altenberg cemented a friendship that was a key element in the subsequent history of the field.

TINBERGEN AT ALTENBERG

Tinbergen's visit to Altenberg grew out of his meeting with Lorenz in Leiden, plus an earlier campaign on his part to get a leave to travel to the United States. Tinbergen's department head, Professor van der Klaauw, concluded that since Lorenz was the scientist whose work excited Tinbergen the most, it made much more sense for Tinbergen to have a leave to do research with Lorenz in Austria than to travel to America. Lorenz was pleased to invite Tinbergen to Altenberg. Tinbergen received van der Klaauw's permission for the leave, plus funding from the Jan van der Hoeven Foundation. Tinbergen's goal in coming, as he told Lorenz in a letter of February 1937, was specifically to be able to work with Lorenz. He wanted to discuss things with him and to watch animals together with him. "I come for you," Tinbergen wrote, "and for the animals only secondarily." If it were simply a matter of using dummies to test innate releasing mechanisms, he suggested, he could do that just as well in Leiden. He was coming to Leiden for "scientific communication and exchange of ideas."[49] At the beginning of March, Tinbergen, his wife, and their toddler son, Jaap, traveled from the Netherlands to Austria. They brought with them a consignment of ducks Lorenz had ordered from a Dutch duck dealer.[50]

In Altenberg Tinbergen and Lorenz conducted their joint experiments on the egg-rolling behavior of the greylag goose. They also experimented on how young birds react to simulated predators. In addition, Tinbergen frequently made the short trek from Altenberg to the University of Vienna, where Lorenz was lecturing twice a week, and on at least one occasion Tinbergen lectured on the work of his own research group.[51] At Altenberg Tinbergen also continued his study of the gaping response in young blackbirds and thrushes (work he had begun earlier in Holland with his student D. J. Kuenen). It was likewise during his stay in Altenberg that Tinbergen took the now-famous picture of Lorenz striding down the lane with his gosling charges following behind him.

That spring, Tinbergen and Lorenz each had the chance to witness and

4.3 Konrad Lorenz leading imprinted goslings down an Austrian country lane in the spring of 1937. (Photo by Niko Tinbergen. Courtesy of the Tinbergen family. Copy provided by The Bodleian Library, University of Oxford, MS. Eng. c. 3129, A. 89.)

appreciate the other's scientific style. Always before, Tinbergen had felt pressed to use every minute to the utmost. Perhaps this reflected a certain Dutch Puritanism on his part. Or perhaps it stemmed from the fact that he could conduct his wasp studies only at times when the Dutch weather was good enough for the wasps to be active, which was certainly not all the time. In any event, Lorenz introduced Tinbergen to the art of leisurely watching—even if this typically meant that Lorenz would lie in the grass, watching, while Tinbergen arranged an experiment. Tinbergen never adopted Lorenz's observational style. Nonetheless, he was profoundly impressed by how much Lorenz was able to see. Lorenz in turn was in awe of Tinbergen's talent in designing experiments. As he happily reported to Stresemann late in March 1937, "[Tinbergen] . . . is making beautiful experiments on inborn schemata. He has a special talent for precise experimental design that is lacking in me (I am learning tons from him!)."[52]

Ethologists of a subsequent generation would decide that not all of

Tinbergen's experiments were designed as tightly as Lorenz imagined they were. Tinbergen himself, for that matter, would feel that Lorenz had over-valued his experimental talents, whereas his strongest talent was in thinking biologically from the standpoint of a field naturalist. But in 1937 this was still in the future. Lorenz was thrilled to have someone by his side doing experiments, and these early experiments were of real significance for ethology's early development.

The two naturalists wanted to analyze a situation in which an instinctive behavior pattern and a simultaneous orienting response (a "taxis") were coordinated in a single, adaptive motor activity.[53] The situation they chose was the way a goose returns eggs to its nest after the eggs have rolled out of the nest cup or have otherwise been displaced from it. As the authors explained:

> [This] behaviour pattern incorporates movements which can be imme-diately recognized as balancing actions, and can therefore scarcely be ex-plained as anything other than products of an orienting response, and also movements which quite clearly exhibit all the characteristics of a true instinctive behaviour pattern. Since these movements also occur si multaneously and at a right angle to one another, we were convinced that we had found a particularly suitable object for the study, and attempted analysis, of taxis and instinctive behaviour pattern. When the Altenberg geese began to brood in the spring of 1937, we did not neglect the oppor-tunity to conduct relevant observations and experiments.[54]

Working with geese, eggs, and a variety of egg "dummies" (including a wooden cube and a large, cardboard Easter egg), the men set about analyzing the motor sequence a greylag goose employs in returning a stray egg to its nest. A film of the sequence made by Lorenz's student Alfred Seitz helped greatly in their work. The investigators were able to show that the goose's motor actions consist of two components, an instinctive behavior pattern (which they now called an *Erbkoordination*) and a taxis. The instinctive component of the motor sequence, elicited by the visual stimulus of an egg or egglike object outside the nest cup, consisted of a "ventrally-directed bending of the head and neck, such that the egg lies against the lower side of the beak and is pushed towards the nest." The taxis compo-nent of the motor sequence consisted of lateral-balancing movements that maintained the egg's rolling in the direction of the nest. Together, the two components combined to produce "a functionally unitary and adaptive

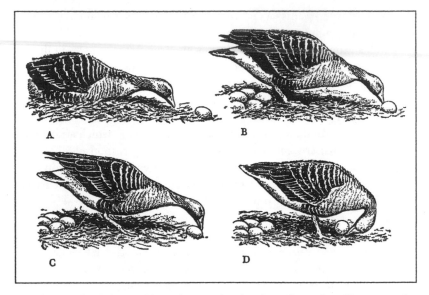

4.4 The egg-rolling behavior of the greylag goose, based on the studies conducted by Lorenz and Tinbergen at Altenberg in 1937. (From Konrad Lorenz, *Studies in Animal and Human Behaviour*, trans. Robert Martin, 2 vols. [Cambridge, MA: Harvard University Press, 1970], 1:329.)

behavioral sequence." The form of the instinctive motor pattern was invariable. It was not influenced either by the shape of the object that was rolled or by the pathway on which the object was rolled.

Before Tinbergen came to Altenberg in March 1937, he had already begun studies distinguishing between instinctive actions and taxes. Furthermore, Lorenz in a letter to Tinbergen had mentioned the egg-rolling reaction as an example of how instinctive actions and taxes were intercalated in one general behavior pattern. What brought added theoretical significance to the greylag experiments in Altenberg in the spring of 1937 was Lorenz's sudden awakening to the importance of the work of Erich von Holst. This occurred in April 1937.

Holst was a student of the distinguished physiologist Albrecht Bethe.[55] Bethe at the turn of the century had endorsed the opinion that there was no compelling reason to consider bees and ants anything other than reflex machines.[56] Later, after studying the regulation of motor coordination in higher animals, he came to see animals as much more than machines reacting passively to external stimuli. His student Holst took this idea further. In the spring of 1937 Holst was still only in his late twenties, but he had already distinguished himself through his brilliant experimental researches on the endogenous production and central coordination of nervous impulses.[57]

As events unfolded, it was only a few weeks after Tinbergen arrived in Altenberg that Lorenz received a letter from Holst, dated 1 April 1937. Holst had just finished reading a copy of Lorenz's paper on the instinct concept, destined to appear in *Die Naturwissenschaften*. It was at this point (and not a year earlier, as Lorenz subsequently mistakenly remembered) that Holst challenged Lorenz's identification of instinctive action as "a striven-after reflex." Holst's critique persuaded Lorenz that instinctive behavior patterns were better interpreted as the result of internally generated and coordinated impulses than as chains of reflexes set in motion by external stimuli. Holst's perspective made sense of appetitive behavior, where an animal, apparently motivated by its own internal state, seeks out particular external stimuli. It also made sense of those two phenomena that the chain-reflex theory had not handled very satisfactorily—"threshold lowering" and *Leerlaufreaktionen*. As Lorenz explained excitedly to Stresemann later the same month, the facts that Lorenz had discovered with regard to "threshold lowering, *Leerlaufreaktion*, intensity decline after being performed one or more times, independence of goal-attainment, etc. etc." could be understood as a function of Holst's "automatic-rhythmic processes."[58]

Despite repeated pleas from Professor van der Klaauw in Leiden, Lorenz had still not submitted for publication a final, revised version of his Leiden symposium paper from the previous November. Now he stalled for more time in order to take into account the way that Holst's work was affecting his theorizing.[59] He no longer felt obliged to endorse a chain-reflex model of how instincts work. Holst, as Lorenz hurried to explain, had demonstrated that "very many movement sequences that are generally understood to be based on reflexes have nothing to do with reflexes." And what Holst had established regarding the central coordination of locomotor rhythms in worms and fish raised the possibility that centrally coordinated processes might also account for "a whole series of peculiarities of the instinctive action."[60] Lorenz interpreted Holst's work on the endogenous production of nervous impulses as highly supportive of the notion of some kind of action-specific energy building up internally in an organism to the point that the corresponding instinctive action would erupt spontaneously. Holst's findings on internally generated and centrally coordinated impulses likewise corresponded to Lorenz's claim that the form of an instinctive action pattern does not actually depend on the receptive processes that set the action in motion.

It was only a short step from this to the idea that the instinctive behavior pattern, once released, maintains its form independently of external stimuli and of the animal's receptors because it still depends upon internal processes in the central nervous system that coordinate impulses to the animal's muscles. Though this was perhaps not the fundamental understanding with which Lorenz and Tinbergen began their study of the egg-rolling behavior of the greylag goose, it was the understanding they conveyed when they finished it. Lorenz in the end was the one who, over the course of the summer, wrote up the theoretical analysis of the experiments. When Tinbergen read Lorenz's account, he wrote his friend, saying, "Your discussion of the eventual identification of the Holstian automatisms and your instinctive actions is wonderful."[61]

When the paper on taxis and instinctive behavior pattern in the egg-rolling behavior of the greylag goose appeared in print in the *Zeitschrift für Tierpsychologie*, it identified itself as the first part of a study that would be continued in 1938. Thus far, the investigators said, they had used hybrid geese in their experiments. They wanted next to use pure, wild-type greylags. The question they wanted to pursue was whether pure-blooded individuals reacted differently from hybrid individuals to the stimuli presented to them.

Experiments had already shown that numerous brooding birds—and some wild forms at that—were quite unselective with respect to the objects that elicited their egg-retrieving responses. Herring gulls, Tinbergen knew, "will brood on polyhedrons and cylinders of any color and of virtually and size, and black-headed gulls appear to be even less selective." Lorenz was the one who was concerned about differences between hybrid and pure-blooded forms. He was becoming increasingly interested in the effects that domestication has on a species' instinctive behavior patterns. Under the conditions of domestication, he believed, the selectivity of innate releasing mechanisms tended to break down, permitting a broader range of stimuli to release a given instinctive reaction. For this reason, he wanted to repeat the egg-rolling experiments with pure-blooded greylags.[62]

In their work together in Altenberg Lorenz and Tinbergen also conducted a set of experiments that involved exposing hand-reared fowl of various species to simulated flying predators made from cardboard dummies.[63] The experimenters strung a rope up between two tall trees and pulled the dummies along the rope to simulate the motion of birds in flight. The dummies were constructed in a variety of bird shapes, and a circle was used as well.

The two naturalists tried their experiment on virtually all the young fowl that Lorenz had at Altenberg in the spring of 1937. Young greylag geese,

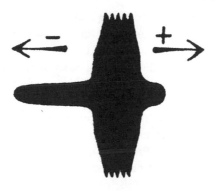

4.5 Tinbergen's illustrations of (1) a "card-board dummy that releases escape reactions [in young turkeys] when sailed to the right ('hawk') but is ineffective when sailed to the left ('goose')," and (2) a selection of the different shapes that he and Lorenz used in testing the instinctive fear reactions of juvenile birds. (From Tinbergen, "Social releasers and the experimental method required for their study," *Wilson Bulletin* 60 [1948]: 6–51.)

turkeys, and numerous species of ducks were all part of the test. The cardboard dummies were pulled along the rope, above the birds, at different speeds and in each direction. The results differed according to the birds being tested. As later reported by Lorenz, it was not until the young greylags were about eight weeks old that they displayed escape reactions to the dummies, and they did so fully only when exposed to a cutout moving very slowly. The shape of the dummy did not seem to be particularly significant. What mattered instead was the "flight pattern." A slow, gliding motion frightened the greylags. Lorenz interpreted this motion as corresponding to the flight pattern of the greylags' chief avian predator, the white-tailed eagle. A swifter motion, on the other hand, disturbed the ducks. This motion Lorenz believed, corresponded to the flight pattern of falcons.

With young turkeys, unlike with the geese and ducks, the shape of the moving dummy seemed to make a difference. Dummies with "short necks" elicited the turkeys' alarm calls much more readily than did dummies with "long necks." Indeed Tinbergen and Lorenz were able to evoke these results with a single, relatively crude dummy constructed with the "wings" located toward one end of the body in such a way as to make one end of the body short and the other long. Which end appeared as the "head" and which appeared as the "tail" thus depended on the direction in which the dummy was pulled. The young turkeys displayed the most alarm when the dummy was moved slowly above them with its short end forward and its long end to the rear. When the dummy was moved with its long end forward, the turkeys were calmer. The two naturalists concluded that this difference in reactions corresponded to the difference in shapes of avian predators and nonpredators. Predators like hawks have short necks relative to the rest of their bodies. Nonpredators like gulls and especially geese have long necks relative to the rest of their bodies. The young turkey's response to the gestalt of the slowly moving, short-end-forward shape, Tinbergen and Lorenz believed, was evidently an innate response forged by natural selection to an environmental cue signaling "predator."[64]

Amusingly enough, the experimenters found that they could not use the same experimental subjects for long in studying instinctive reactions to predators. The young greylags became conditioned to run for cover whenever one of the experimenters climbed into a tree. The birds, it seemed, had learned to associate the experimenter's action with "predators flying over."

Years later, Tinbergen and Lorenz would each refer to this time together in Altenberg as the happiest period of his life. Tinbergen was exhilarated by working jointly with Lorenz on scientific projects, by the sheer boyish fun (Lausbuberei, as Lorenz called it) of interacting with the Austrian naturalist, by the remarkable erudition of Lorenz and Lorenz's family, and by the sophistication of the University of Vienna. It seemed to be a charmed moment—especially so when set against the somber backdrop of the worsening European political situation.

In a letter home to his parents, dated 29 March 1937, Tinbergen indicated that he was finding discussing politics with Lorenz just as interesting as discussing science with him. Lorenz believed that the scientific climate in Germany was looking up, and that it indeed was much better in Germany than in Austria. By Lorenz's account, the Nazis had relaxed the pressures they had initially put on German science. Good scientists whom the Nazis had previously dismissed as untrustworthy were now securely back in their old positions, and the "party swine" who had been inserted in these positions had been thrown out. It seemed that money was available even for research that went against the party line. But in Austria the scene was entirely different. Under the Catholic dictatorship, anyone who was not a Catholic simply did not count. Lorenz's family had converted to Protestantism when Lorenz was a youth, and although they had since reconverted to Catholicism for the sake of Lorenz's career, Lorenz believed his career possibilities in Austria were very limited. His specialty, animal psychology, was not one to which the Catholic educational establishment was sympathetic, and in order to give lectures in this area at the university Lorenz was having to disguise his subject by advertising it under another title.[65] Tinbergen, after his stay in Altenberg, thanked Lorenz enthusiastically for the wonderful time and for how much he had learned there. He encouraged Lorenz not to be desperate about Lorenz's career prospects: "A man of your caliber must somewhere find employment!" If the Kaiser Wilhelm Society or something else did not come through for him, Tinbergen suggested, then maybe something in America would.[66]

However sympathetic Tinbergen may have been with Lorenz about the latter's career difficulties, he did not share Lorenz's optimism about science under the Nazis—or anything else the Nazis were doing. In June, on his way back from Austria, he stopped in Munich to make the personal acquaintance of Karl von Frisch. He recalled years later: "My recollection of that visit is a mixture of delight with the man von Frisch, and an anxiety on his behalf when I saw that he refused to reply to a student's aggressive *Heil Hitler* by anything but a quiet *Grüss Gott.*"[67]

Back in Holland, Tinbergen directed field research as usual at Hulshorst in July and August. In September, he and Joost ter Pelkwijk submitted to the *Zeitschrift für Tierpsychologie* a German version of their article analyzing reproductive behavior patterns in the three-spined stickleback. In contrast to their first published account of this work, which appeared in *De Levende Natuur* in 1936 and where they made no reference to Lorenz's work, they now explicitly adopted Lorenz's theoretical framework. They analyzed the social life of sticklebacks in terms of releasers and innate motor patterns.[68] Tinbergen was fully enlisted in Lorenz's program.

A VISIT TO AMERICA

Tinbergen returned to Leiden from Altenberg convinced that a comprehensive, theoretical approach to the study of behavior was not only possible but in fact at hand. Earlier he had displayed an interest in the instinct theory of William McDougall, without being particularly satisfied by it.[69] Lorenz's instinct theory appeared to him to be immensely more promising. Tinbergen arranged for Lorenz to come back to Holland in February 1938 to give more lectures. Enthusiastic as Tinbergen was, however, about the future of the Lorenzian approach, and eager as he was to contribute to it, he remained interested in learning more about how animal behavior was being studied in the United States. He turned once again to van der Klaauw for a three-month's leave. His idea was to travel to America in the late summer and early fall of 1938. He hoped especially to have the chance to learn more about Robert Yerkes' primate studies, either at Yale or at Yerkes' research center at Orange Park, Florida. He also looked forward to meeting G. K. Noble at the American Museum of Natural History.

The Netherlands-America Foundation agreed to provide funds to get Tinbergen across the ocean and back. Tinbergen imagined he could pay for his expenses in the States through honoraria for public lectures. He wrote to Ernst Mayr and Margaret Morse Nice asking whether this would be possible. He was prepared, he said, to talk about a variety of subjects, including the means of communication among birds, the sociology of herring gulls, territorial and reproductive behavior in snow buntings, the egg-rolling reaction of the greylag goose, and the orientation of bees and wasps. He was also willing to speak to general audiences about his experiences among the Eskimos of east Greenland.[70]

Tinbergen arrived in the States late in July 1938. He remained into October. One of the sites he visited was Yerkes' primate facility in Orange Park, Florida. His hope, as he explained to Yerkes beforehand, was "to get the opportunity to work with several of your men, a week or some weeks with each, in order to get some insight in the way you are putting and investigating those problems that appear in the study of the behaviour of higher mammals."[71] Tinbergen met several of Yerkes' staff and observed their work on chimpanzees, but he did not engage in any chimp work himself. Later, he visited Yerkes' lab at Yale. Before returning to Holland he graciously wrote to Yerkes saying that he had learned a great deal during his visit about the aims and methods of American psychologists.[72] What he did not say was that what he had learned had frankly bewildered him.[73]

Whereas Tinbergen felt out of place among the American psychologists studying rats and chimps, he felt much more at home among ornithologists, particularly a group of them based in and around New York City. He stayed in Tenafly, New Jersey, in the home of Ernst Mayr, "baching" it with Mayr while Mayr's wife and daughters were away in Europe.[74] The Dutch naturalist became good friends with Mayr, William Vogt, Joseph Hickey, and a number of other young ornithologists active in the Linnaean Society of New York. In earlier years the Linnaean Society had been more of a bird-watching club than a scientific organization, but Mayr was energizing and upgrading the society by urging members to do original research and by holding a regular seminar to discuss the latest ornithological literature. In New York Tinbergen also met G. K. Noble and the members of Noble's department at the American Museum of Natural History. He had long talks with Noble about Lorenzian ethology, albeit without ever feeling that he had persuaded Noble of much of anything. In addition he became the friend of a skinny, nineteen-year-old City College undergraduate who was working with Noble and participating in the activities of the Linnaean Society. The young man was Danny Lehrman. A decade and a half later Lehrman would emerge as a major figure in ethology's development.[75]

Tinbergen gave several lectures in the United States. On at least one occasion he spoke about his experiences among the Eskimos of Greenland, but typically his theme was the new, ethological approach to the study of animal behavior. Twice in particular he had the chance to tell special audiences about the latest work in ethology. The first time was at the conference "Plant and Animal Communities," held at Cold Spring Harbor, New York, from 29 August to 3 September 1938. The second time was at the symposium "The Problem of the Individual vs. the Species in Bird Study," held in mid-October in Washington, DC, as part of the annual meeting of the American Ornithologists' Union.[76]

American ornithologists by this time were primed to learn more about ethology, thanks in particular to Margaret Morse Nice's promotion of Lorenz's work. She had reviewed Lorenz's "Kumpan" paper enthusiastically for *Bird-Banding*, put both Wallace Craig and Francis Herrick in touch with the Austrian ornithologist, and challenged G. K. Noble to confront Lorenz's ideas. The contact with Herrick led to the publication in 1937 in the *Auk* of an abbreviated, revised, English-language version of Lorenz's "Kumpan" monograph. Thus, when Tinbergen arrived in the States, recent advances in European ethology were already being noised about in the American ornithological community. Tinbergen, nonetheless, was the first European ethologist to have the chance in America to make the case for ethology in person.

At Cold Spring Harbor Tinbergen presented a paper entitled "On the Analysis of Social Organization among Vertebrates, with Special Reference to Birds." In taking as his subject the way that signals from one individual or group influence another individual or group, he was able to give his auditors a solid overview of Continental ethology's special strengths. He presented Lorenz's latest theoretical views and described observational and experimental work of his own, some of which was yet to be published, on herring gulls, sticklebacks, and European blackbirds and thrushes. He explained how experiments could be used to distinguish between the stimuli that release a behavior pattern and those that direct it, while taking care to stress that experiments were to be conducted only *after* one had learned the full behavioral repertoire of an animal species in its natural environment. He emphasized the attractiveness of the "releaser" concept, allowing that it was an advance over Darwin's theory of sexual selection in its recognition that such structures and movements could serve to release other than sexual or fighting responses.[77]

The most novel part of Tinbergen's paper was his discussion of a new category of releasers, or "signal movements," that he termed at this point "substitute activities."[78] Heinroth had earlier called attention to "intention movements" such as the motions a goose makes before it flies. These movements, which Tinbergen preferred to call "preparatory movements," have communicative value for other members of the species. Portielje had similarly talked of "symbolic movements," for example, the nest-building motions that birds sometime engage in before actual nest building begins. Tinbergen's "substitute activities" were still another kind of "signal movement" that had evidently evolved in conjunction with the communicative function they served. He explained:

> During boundary flights, during the presence of a predator near the nest or young, and during courtship birds often display activities which closely resemble movement belonging to another functional cycle. In the Snow Bunting, for instance, threatening males often show movements highly simulating feeding movements, but which on careful observation appear to be incomplete; food is never taken and the bill does not even reach the ground. . . . This behavior is shown especially during boundary disputes, when neither of the two birds actually attacks, but both indulge in threatening. I propose calling these incomplete feeding movements "substitute feeding," because they replace fighting activities induced by the situation.[79]

Tinbergen offered a quasi-psychological interpretation for substitute activities. In the situations where such activities were displayed, the problem was not that the birds lacked the external and/or internal stimuli necessary to carry through with a reaction (as in the case of "preparatory movements" and "symbolic activities"), but rather that the normal reaction was somehow blocked. This could happen if the animal were simultaneously under the influence of two antagonistic drives, such as fight and flight. It could likewise happen in courtship when one member of a pair was not ready to mate. Finally, it could happen when the normal reaction, having been released several times, was temporarily exhausted. Without claiming that all substitute activities serve a communicative function, Tinbergen allowed that many "signal" movements probably had their origin as "substitute activities."[80]

A month and a half after the Cold Spring Harbor meeting, Tinbergen went to Washington, DC, to attend the fifty-sixth annual meeting of the American Ornithologists' Union. As part of a symposium on the individual versus the species he delivered a slide lecture entitled "On the Sociology of the Herring Gull." The other participants and their paper titles were Francis Herrick ("The Individual vs. the Species in Behavior Studies"), G. K. Noble ("The Role of Dominance in the Social Life of Birds"), Margaret Morse Nice ("The Social Kumpan and the Song Sparrow"), and Frederick C. Lincoln ("The Individual versus the Species in Migration Studies"). Many of Tinbergen's new ornithological friends from New York were also present, including Ernst Mayr, Bill Vogt, Joe Hickey, and Daniel Lehrman.[81]

Nice and Tinbergen made strong showings at the symposium. Each of their papers, in contrast to the papers by Herrick, Lincoln, and Noble, described research that focused on a single species and was informed by a coherent analytical framework. Nice, who herself had spent a month that very spring working at Altenberg with Lorenz, pulled no punches in identifying the framework she was using. Lorenz's "brilliant exposition" in his "Kumpan" monograph of 1935, she said, offered "a sure foundation for the study of bird behavior." Later she received a letter from Lorenz saying: "Tinbergen wrote to me about your talk on the Kumpan and said that it was absolutely excellent. He is a very hard critic, so you may really feel rightly flattered by this! I am very glad you like Tinbergen, he is really one of THE BEST in every respect scientifically and personally![82]

Tinbergen was encouraged to find capable supporters of Lorenzian ethology in the United States. The international political scene, on the other hand, was growing worse and filled him with a sense of foreboding. While still in the States he listened with dismay to the news of the treaty at Munich. He returned to Holland fearful for the future. Writing from Holland back to Mayr in the United States, he expressed his anxieties over how long it would be before Hitler immersed Europe in war. He was appalled by what he had seen of Hitler's treatment of the Jews, who were arriving in Holland so severely beaten that they needed to be taken to hospitals immediately. As he put it, "the unbelievably mean and atrocious deeds which have been committed by the SA and even the Hitler Youth on helpless Jews should be generally known [abroad], and in Germany too."[83]

BACK IN HOLLAND, WITH FEARS OF WAR

Early in 1939, back in Holland, Tinbergen was pleased to learn that the trustees of Leiden University intended to offer him the position of lecturer in experimental zoology, pending the availability of funds. The appointment was confirmed in April, to take effect in the fall. He regarded this as a "good position," and with it in hand he was able to decline an offer from G. K. Noble of a job in the United States, the post of resident naturalist at the newly established Edmund Niles Huyck Preserve in Rensselaerville, New York.[84]

For the spring Tinbergen suspended his work on seagulls and concentrated on his laboratory studies of sticklebacks, working toward a "more or less monographic treatment" of the species' entire "action system." The work, he explained to Ernst Mayr, had its frustrations and its fascinations: "Every day we experience how superficial our knowledge still is. On the one hand this is discouraging, on the other hand it is quite pleasing to see so many different ways ahead. One would like to be able to work faster, though." He was pleased to find sticklebacks displaying "substitute activities" comparable to those he had seen already in birds. Using a dummy to represent a stickleback male in the breeding season, he staged a "fight" with a live male in the male's territory. After the male's fighting and fleeing reactions were elicited several times, the male fought no longer but instead displayed a substitute action: a kind of "substitute eating" where the fish thrust its mouth in the sand at the bottom of the aquarium. Tinbergen interpreted this as a threat display. He found that both of the native species of sticklebacks exhibited it.[85]

At this point in his career Tinbergen was convinced that ethology's main claim to scientific legitimacy rested on its ability to address the question of the physiological causation of behavior. For his ethology practical at Leiden in the spring of 1939 he provided his students with a nineteen-page handout in which he stated: "The scientific approach of this analytic ethology, as it is used in this practicum, is . . . in principle the same as that of physiology. Indeed, to the extent that one could view ethology as the study of animal actions, ethology could be considered 'a branch of physiology.'"[86]

By this time, Tinbergen feared that war was imminent. Although he wrote to Mayr, "Let us not talk too much about international politics," he could still not refrain from expressing anxieties. He did not want to believe that the war was going to start, he said, but he had to admit that the situation at times looked "frightfully black." "One must try," he said, "to not let one's work suffer because of it." [87]

In June of 1939 Tinbergen traveled to Germany for the fifty-seventh annual meeting of the Deutsche Ornithologische Gesellschaft, held in Münster. There he met Erwin Stresemann for the first time and renewed his acquaintance with the Finnish ornithologist Pontus Palmgren. He had hoped to see Lorenz there as well, but Lorenz failed to make it to the meeting. Tinbergen gave what he felt was "a pretty bad lecture" on the topic of ethology as an auxiliary science of ecology. He sought to apply Uexküll's and Lorenz's idea of the innate schema to the issue of what an animal eats. The innate schemata of feeding reactions, he suggested, could be altered through learning. Predators learned to avoid certain forms of prey, especially those displaying warning coloration. He also discussed the function of camouflage. Food choice and camouflage were themes he would take up again in later studies. For the time being, his main message was that "the ethological basis of many ecological questions can be extremely entangled." As he saw it, ecological as well as ethological factors had to be considered in addressing the question "Why does the menu look like this and not like that?" The whole subject, he acknowledged, required much further research.[88]

Tinbergen was signed up for another visit to Germany—a trip to Leipzig in September to participate in the third annual meeting of the German Society for Animal Psychology. He was scheduled along with Frisch, Heinroth, and Koehler to speak on the first morning of the gathering. His topic was "die *Übersprunghandlung*"—the kind of "sparking-over" action one sees when a bird, for example, in the midst of fighting or courting, stops to peck at the ground or preen itself in a stereotyped way.[89] Likewise listed on the program were Lorenz, Lorenz's friend Antonius (the chair of the meeting), and Lorenz's student Alfred Seitz. Lorenz had signed up to give a conference paper on "the so-called inborn schemata." More strikingly, he was also slated to give a special, public, keynote lecture, on the first night of the sessions. Jointly sponsored by the German Society for Animal Psychology and the Leipzig National Socialist people's development program, his lecture was entitled "Rise and Fall in Man and Animal."[90] None of this took place, however. The Leipzig meeting was cancelled when World War II began.

German troops invaded Poland on 1 September 1939. In Holland, many of Tinbergen's colleagues and students were immediately mobilized.

For Tinbergen himself, who remained at the university, there were new demands on his time, such as trying to make the laboratories less susceptible to air raids. He hoped for the downfall of Hitler and the Nazi regime in Germany. He feared and expected, however, that much suffering would take place before this happened. When the time came to send the page proofs of his manuscript on the behavior of the snow bunting back to Mayr in New York, he had to worry about the proofs arriving safely. Boats carrying mail to the United States ran the danger of being torpedoed and ending up at the bottom of the ocean. To increase his chances of conveying to Mayr the corrections he wanted to make, Tinbergen sent a package and a letter, at different times. The package contained the corrected page proofs. The letter listed the same corrections.[91]

Tinbergen was not certain what Mayr would think about the war in Europe, given that Mayr was a German national with relatives in Germany. Tinbergen felt the need, nonetheless, to spell out his own view of the situation. He was convinced that Hitler was bad for Germany and that it would be better for the Dutch to fight alongside the French and British rather than to permit a new Munich to take place. He hoped Mayr would not object to hearing the objective opinion of a "neutral." The reason he was not sure what Mayr might be thinking, he admitted, was the attitude of Lorenz: "I noticed to my surprise that Lorenz, who was relatively sober and moderate before the Anschluß, has now without inducement written an entirely sympathetic defense of German foreign policy. After this happened, I saw that it must be very hard for a German to remain somewhat objective in matters of power politics (which cannot be measured with moral standards as one measures personal behavior)."[92]

Up to and including this letter of October 1939 (which is quoted here in translation), Tinbergen conducted his correspondence with Mayr in German. Beginning in January 1940 he wrote to Mayr in English. He asked Mayr at this time to send him more reprints of his snow bunting paper, which had just appeared in the *Transactions* of the Linnaean Society of New York. He also asked that additional copies be sent to various journals and individuals. He wanted to be sure, for example, that a copy reached Julian Huxley or F. B. Kirkman, because he had heard they had organized "a Society for the Study of Animal Behavior or something like that." He also expressed the hope that Mayr would be able to meet and advise Tinbergen's former student, J. J. ter Pelkwijk—"one of our best young zoologists"— who was on his way to the States and then to Suriname, where he intended to study problems of animal adaptation. Tinbergen concluded by saying:

Here we have a splendid winter and everybody is skating. As you know in our country one can travel along the lakes and the canals for tens of miles continuously. I happen to be able to take some vacation just now and make a long skating tour every day. The greater part of the former Zuiderzee (now "Ijssel meer") is frozen over and we can reach those curious little islands . . . on the skates. At no other season this country's beauty is more impressive: a wide horizont, clear weather, a little hazy, a fine sun. It is rather astonishing one can spend his vacation as carefree as I do in view of the European wars not far from us, but one has to be careless lest one would lose nerves and strength. Yet I still hope that Western and Central Europe will come to some kind of compromis[e] before long, in view of the dangers threatening in the near and the far East.[93]

Tinbergen at this point received another letter from G. K. Noble, offering him once again the post of resident naturalist at the Huyck Preserve. The appointment was for three years at an annual salary of $3,500. Tinbergen once again declined. As "a citizen of a small free country" that stood in grave danger of being overrun and enslaved by another power, he felt obliged to stay where he was. Any day, he said, his country might become the site of battle.[94]

UNDER GERMAN OCCUPATION

The German air attack on Holland began in the early morning hours of 10 May 1940. The Germans rendered Rotterdam defenseless and then bombed the city mercilessly. All over Holland the smoke generated by the bombing could be seen. The Dutch were forced to surrender. The conquest of the country had taken just five days. A month later, Tinbergen wrote Mayr: "So Germany has struck at last. It was a heavy blow and considering the formidable German aircraft and the huge motorized army they sent out it was impossible for us to withstand. Of course it was a bitter experience to see the diving bombers and the parachutists doing their job, especially because the children could see and hear it also, which gave them rather a bad shock." Tinbergen's hurt and anger registered further in the remark: "Now we are living under German rule, which means that organization is perfect."[95]

Tinbergen reported that his students had been demobilized and had returned to their studies. He planned, he said, to hold his summer fieldwork camp "as if there wasn't any war." Then his comments quickly re-

turned to the experience of the invasion. It had been a great disappointment to the Dutch that no allies had come to their aid. They had seen hundreds of German aircraft, but not a single plane from Britain or France.
Tinbergen closed his letter with greetings to Mayr and Mayr's family and
the rueful observation: "Well, I hope we will meet once more in another
phase of world history (or rather geography)."[96]

Tinbergen was able to take his students into the field in the summers
of 1940 and 1941. At Hulshorst he continued to direct studies on the behavior of insects, including work on the dance of the grayling butterfly. Initially, work continued in the lab as well.

The single most impressive study done by Tinbergen's students in this
period was that done by Gerard P. Baerends on the behavior of a species of
the digger wasp *Ammophila*.[97] Solitary wasps had long been a source of naturalists' attentions, from the early work of J.-H. Fabre and George and Elizabeth Peckham to Tinbergen's recent investigations of *Philanthus*. In the
summer of 1936 Baerends and Josina van Roon discovered at Hulshorst an
Ammophila species that had been reported to be capable of maintaining
more than one nest at the same time and provisioning these nests according to the state of development of the larvae in them. This particular capa-
bility, together with long-standing debates on the rigidity or flexibility of
the instinctive behavior of hunting wasps in general, made *Ammophila
campestris* an attractive subject for study.

Baerends undertook with van Roon a systematic study of the wasp's
behavior aimed at distinguishing the typical behavior patterns of the species from individual variations. They studied the wasp's provisioning behavior for five successive summers. Baerends was called up for military
duty in 1939, then demobilized after the Dutch army's defeat in May 1940.
He returned to Hulshorst in 1940 to conclude his data collecting.[98]

Analyzing the data, Baerends concluded that the wasp's behavior is organized in a hierarchical pattern of systems and subsystems in which internal states and external stimuli dovetail with each other. The importance
of the internal state or "mood" of the wasp was made clear by the fact that
a wasp confronted with identical external stimuli might react differently on
different occasions. Baerends used the word "mood" (*Stimmung*—a word
borrowed from Heinroth and Lorenz), simply to designate the state in
which certain behavior patterns can be activated or performed, not to posit
any subjective experience on the organism's part. He identified separate
moods for brood care, food searching, grooming, mating, and sunning. He
observed that moods always exclude each other. A wasp that is feeding on
nectar, for example, is not ready to sting a caterpillar.

Baerends concluded further that a "primary" mood, such as that for brood care, subsumes secondary and tertiary moods. The secondary moods of brood care consisted of three phases or stages. In the first stage, the wasp digs a nest, captures a caterpillar and carries it into the nest, and lays an egg on the caterpillar. In the next stage, the wasp brings a few more caterpillars to the nest with the young larva. In the final stage, the wasp places a larger number of caterpillars with the larva. Later in this stage, after the larva spins a cocoon, the wasp closes the nest and does not return to it.

Each of these stages begins with the wasp's inspection of the nest. The wasp's subsequent behavior depends strictly on what it finds on the inspection visit and the internal state or "mood" primed by these findings. Once entered on the sequence of behaviors appropriate to its internal state, the wasp is unable to alter its behavior until it runs through an entire sequence of subsidiary moods and motor patterns. In phase 2, for example, this involves the moods and motor patterns associated with closing the nest, hunting and capturing a caterpillar, and provisioning the nest. "Provisioning" in turn subsumes the moods and motor patterns of "advancing the caterpillar to the nest," "putting it down," "scraping," "digging," "turning around," and "dragging in."

Baerends covered many things in his masterly, 208-page monograph on *Ammophila*, ranging from the variation in the wasps' stinging of caterpillars to the means by which the wasps orients themselves to their nest sites, hunting grounds, and so forth. The intellectual highlight of the work, however, was his analysis of the hierarchical arrangement of the moods and motor patterns of brood care behavior. He represented this with the diagram shown in figure 4.6.[99]

Baerends summarized his findings by saying: "Most activities are not only caused by external factors (sensory stimuli), but are also ruled by internal factors. . . . Many of the internal factors rule only single movements, others have a multitude of different activities at their disposal, and additional, external or internal factors determine which movement will actually be performed. These facts prove that internal factors, and therefore the moods, form a hierarchical system in which moods of different order are effective."[100]

Baerends believed that his observations helped explain why previous observers of the hunting wasps had disagreed about whether instinctive actions were rigid or modifiable. His findings showed that a primary or secondary mood, encompassing many different potential activities, can find expression in a more plastic, adaptive sort of behavior than can a subordinated mood that has at its disposal a more limited number of options. At

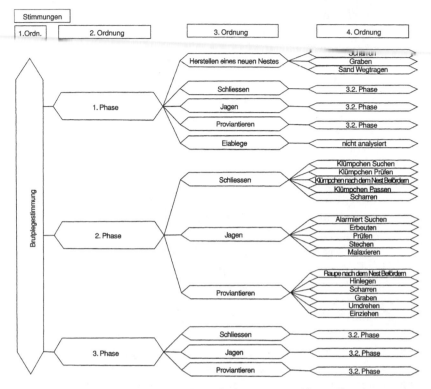

4.6 G. P. Baerends's representation of the hierarchical arrangement of the moods and motor patterns of brood care behavior in the digger wasp *Ammophila campestris* (1941).

the level of the instinctive motor pattern (*Erbkoordination*) itself, behavior proves to be extremely rigid, but the wasp's behavior as a whole is a very plastic sort of affair, combining numerous *Erbkoordinationen,* taxis components, and appetitive behaviors.

Baerends noted that these results pointed toward a revision of Lorenz's ideas about appetitive behavior. Lorenz, following Craig, had identified appetitive behavior as the modifiable and variable search behavior in which an animal engages in order to bring it to the proper situation for releasing a particular instinctive motor pattern. Baerends's observations of *Ammophila* showed that "appetitive behavior does not always lead immediately to the outflow of the *Erbkoordination.*" Sometimes it leads instead to the beginning of a new, subordinate mood and new appetitive behavior appropriate to that mood.[101]

Baerends worked out his diagram and his account of the hierarchical relations among instincts in discussions with Tinbergen, who, independently of Baerends, had come in his stickleback studies to the same general conclusion about the hierarchical ordering of instincts.[102] It is possible that their ideas had been influenced in some measure by discussions with the young Adriaan Kortlandt, who had been introduced to the study of animal psychology by Frits Portielje. Kortlandt, a student of geology and psychology at the University of Utrecht, was carrying the practice of bird watching to new heights, literally as well as figuratively. In 1938–1939 he constructed a twelve-meter-high observation tower from which he could watch at close range the behavior of some thirty pairs of cormorants and their broods without disturbing them. From this special vantage point he was able to study a whole sequence of behaviors beginning with pair formation in the adults and proceeding onward through the development of the chicks, their fledging, and more.[103]

Tinbergen seems to have been very impressed by Kortlandt's ingenuity and diligence as a bird watcher, but not by Kortlandt's psychological and subjectivistic frameworks for interpreting what he saw. Kortlandt in the late 1930s had begun to think and talk in terms of hierarchies of instincts and appetitive behaviors, but Tinbergen at this time was not well disposed to imagine that his own emerging thoughts on instinctive hierarchies could benefit much from someone in the psychologists' camp. He viewed his own objectivistic approach to animal behavior studies as the proper course for the future—and as a clear break from the older, psychological and subjectivistic approaches of Bierens de Haan and Portielje. Kortlandt subsequently felt that the ethologists had unfairly neglected the contributions he had made in this area.[104]

Tinbergen published in 1942 his own diagram of the hierarchical organization of instinctive behavior, in this case with respect to reproductive behavior in sticklebacks. Stickleback males, he explained, are brought into the reproductive "mood" through internal hormonal changes and the perception of a suitable territory. In this state, they are capable of a variety of different activities, including fighting other males, searching for nest materials, and courting females. Whether a fish comes into the fighting, building, or courting "submood" depends upon the additional stimuli it receives. The appearance of an opponent, for example, will bring a fish into the fighting submood. Its precise reactions depend on its opponent's actions. The reactions at this level—fighting, chasing, threatening, biting, and so forth—constituted the specific motor responses Lorenz called the *Erbkoordinationen*.

This was a more structured view of instinctive action than Lorenz had

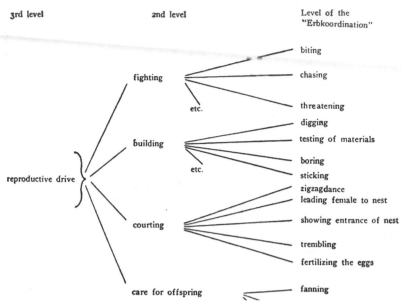

4.7 Niko Tinbergen's 1942 representation of the hierarchical arrangement of drives.

been promoting. In Lorenz's scheme of things, each instinctive behavior pattern had its own action-specific substance, its own appropriate sign stimuli or releasers, and its own internal releasing mechanism. Lorenz had explicitly argued against William McDougall's endorsement of a hierarchical view of instincts, where minor instincts were subordinated to major ones. Tinbergen and Baerends were in effect returning to a view that at least in this particular respect was more like McDougall's. The behavior of Baerends's digger wasps and Tinbergen's sticklebacks suggested that many of the reactions elicited by releasers are not simple "motor responses" but instead internal reactions which make the animal ready for a variety of more specialized activities. Although Tinbergen acknowledged the heuristic value of Lorenz's notion of an energizing factor building up in a reservoir, he disagreed with the assumption that such a factor would be *specific* to a single reaction pattern. He also presumed that nervous "centers" and nervous "impulses" rather than energy reservoirs would ultimately be found responsible for the behavioral phenomena that he and Lorenz were addressing.

As early as the second month of the German occupation of Holland, Tinbergen began nourishing the hope of a postwar, international symposium on animal behavior. He imagined that German scientists would have a role to play in it. In February 1940, he wrote to David Lack saying: "There are so few really serious students of animal behaviour, and yet there is so much to do. When the war is over, it will be highly necessary to reconstruct international cooperation in our science as soon as possible and the first thing to do will be to organize a kind of symposium with Lorenz and some other Germans to discuss a broad long-range program." [105]

Tinbergen and his students managed to do field research at Hulshorst in the summers of 1940 and 1941. For a while they were also able to continue their laboratory studies. When the war began, Tinbergen had thirteen species of cichlid fish for comparative studies. Lack of electricity, however, caused him to give the work up. [106] Politically, too, matters at the university worsened. In November 1940, the German occupiers began dismissing Jewish officials in Holland, professors included. The dean of the Law School, Professor R. P. Cleveringa, denounced the dismissals and spoke up for the university's intellectual freedom. Inspired by his speech, the students at Leiden went out on strike. Student strikes also took place at the Technical Institute of Delft. Classes were cancelled at both universities, though Leiden later reopened for examinations. The faculty remained officially on duty.

Tinbergen continued in correspondence with Lorenz in the first years of the war. [107] He also sent to Otto Koehler, in the latter's capacity as co-editor of the German *Zeitschrift für Tierpsychologie,* a paper reporting on the results of the fieldwork he and his students had conducted at Hulshorst on the courtship behavior of the grayling butterfly. He had promised the paper to Koehler early on, and then, despite his increasingly hostile feelings toward the Germans, reluctantly decided he should not go back on his word. [108]

Tinbergen's most important publication in the early years of the war was a strong, programmatic overview of ethology that he wrote in English and published in March 1942 in the Dutch journal *Bibliotheca Biotheoretica.* Entitled "An Objectivistic Study of the Innate Behaviour of Animals," the monograph constituted Tinbergen's most forceful statement up to that time concerning what ethology was all about. This was the monograph where he set forth his diagram of the hierarchical organization of instincts. It was also the place where he hammered home the notion that what ethology was doing above all else was studying the *causes* of behavior. [109]

Tinbergen's monograph identified a school of animal behavior—"Lorenz's school"—distinguished from other schools of animal behavior by its approach. The principal characteristic of this approach, he maintained, was its "faith in the preeminent value of causal analysis for a better understanding of behaviour." Ethology's foremost concern, in brief, was the study of behavioral causation. Its scientific stance, furthermore, was uncompromisingly "objectivistic." Understanding "the underlying causal relations" of animal behavior required analysis "by the usual methods of the natural sciences: exact and objective description, sharp formulation of questions, when necessary and possible, experimental methods, objective and sharply defined concepts, unmistakable conclusions." What this amounted to was "applying physiological methods to the objects of animal Psychology." [110]

Tinbergen distinguished the ethologists' approach from that of the American behaviorists. He acknowledged that the behaviorists also represented an "objectivistic" approach to animal behavior. He felt, however, that their enterprise was too narrow, insofar as they only dealt with a few animal species, they drew their problems disproportionately from human psychology, and they tended to overemphasize matters of method to the detriment of the study of scientific issues. He also distinguished his own approach as an ethologist from that of the European subjectivist psychologists. He contrasted "the psychologist" (whom he characterized as "the behaviour-student who dislikes consistent causal analysis and objectivism") with "the physiologist" (in other words, ethologist) "who studies the causal structure of the phenomena." He concluded his account of the ethologists' "objectivistic approach" with a reference to its subjectivist critics (here he named Bierens de Haan and Russell) and with an ostensibly modest request: "The only privilege asked by Ethology is, not to be rejected a priori, but to be judged according to its results." [111]

In the spring of 1942 the German authorities set about replacing various incumbent faculty at Leiden University with Nazis. Eighty percent of the faculty resigned in protest, Tinbergen among them. About a third of these were arrested, most of them in July. Tinbergen himself was not apprehended and arrested until September, at Hulshorst. [112] In a long and moving letter written to Margaret Morse Nice at the war's end, he recounted his experiences as follows:

> The Germans began to register the Jews and to outlaw them, then to influence the schools and the teaching. Interference went on progressively and soon a point was reached where we felt we ought to stiffen and resist as much as possible. Our University was, by accident, the first group of

Dutchmen to be tackled by the Germans as a group, and the first to re-fuse to surrender. The Germans wanted to "cleanse" our corps of Jews and of anti Nazi's and proceeded to fire one professor, then another, step by step, on wholly irrelevant grounds. Soon we saw no other way than to resist by refusing to stay in the service of the German-controlled govern-ment, and soon after the University had been closed by the Germans be-cause of anti-German "irregularities" 60 of our professors including my-self laid down their function. This was at the same time our protest and our means to prevent the Germans to nazificate the university by dis-missing only some few of us, for whom they had a nazi-remplaçant, and to keep the rest as "flags" to adorn the planned nazi-university. As a re-prisal, we were captured (that is to say, some 20 of the supposed leaders of the resistance) and put in a camp as hostages, together with about 1300 patriots and internationalists. The Germans threatened to shoot a group of these hostages whenever active resistance or sabotage was threatening some vital part of their war-organization. Twice a group, though a small one, of us were shot in 1942, after that we did not run se-rious risks, though our families and we lived in uncertainty, which had a more or less depressing influence on most of us.[113]

Tinbergen was put in the internment camp at St. Michielsgestel near 's-Hertogenbosch. Among those interned with him were Jan Verwey and C. J. van der Klaauw. Van der Klaauw was released in the spring of 1943 and Verwey in the fall of the same year. Lorenz and his colleagues Otto Koehler and Otto Antonius explored the possibility of using personal connections to free Tinbergen, but Tinbergen refused to take any special favors from the enemy side. It was "just as Lorenz feared," Koehler told Antonius: Tinber-gen had chosen to play "the role of martyr." This led Antonius to observe that "these people too must ultimately find the way in the new era," and Koehler to remark in return about "people who were much too badly in-formed to recognize" that they would get nowhere by making martyrs of themselves.[114] Tinbergen remained a prisoner for two years, until his in-ternment camp was liberated.[115] This occurred in September 1944, when the Allied offensive aimed at capturing the bridge of Arnhem reached the area. Tinbergen rejoined his family, but unfortunately on the "wrong side" —some thirty miles north—of the front. The Allies' attempt to take Arn hem failed, and the war in Holland dragged on for another eight months.

Through the terrible winter of 1944–45, with Holland in the grip of cold and famine, Tinbergen struggled to keep his family from starvation.

At the same time, he assisted war victims, spied for the Dutch resistance, and managed to elude the manhunts of the German "Security Service." A year and a half after the war's end he told Robert Yerkes:

> The German terror was unbelievably cruel and a terrible burden. I never had suspected before that whole generations could be spoiled so badly and so thoroughly in so relatively short a period as the nazi regime had at its disposal. I refrain from writing you details, and can only ascertain you that the worst you have seen in your newspapers is still short of the truth. Many of the refined bodily and spiritual atrocities were beyond imagination, and defy description. When the allied troops rescued us in April 1945 we wept for joy and still, nearly 18 months after our liberation, we are still easily moved to tears by remembrances of those days.[116]

FIVE

LORENZ AND NATIONAL SOCIALISM

When the Nazis came to power in Austria in 1938, Konrad Lorenz welcomed their arrival. He did so because he believed the new political order would transform the local ecologies most relevant to his own career. He was glad to have the chance to make himself visible to a regime that identified its destiny in biological terms. The kind of professional post that had previously eluded him in Austria now seemed within reach, and he immediately set about adapting to the new circumstances. He did not have to revise the conceptual foundations he had laid for the science of ethology or his practice of raising and observing animals in captivity. These could thrive, he felt certain, in his new ideological and political surroundings. He believed that doing good scientific research, serving the National Socialist state, and advancing his own career could readily proceed hand in hand.

The present chapter explores the development of Lorenz's scientific thought and practice from just before the *Anschluss* to the end of the Second World War. The subject is a complicated one. The burgeoning scholarly literature on science under the Nazis demonstrates that the simple categories of Nazi or anti-Nazi, party member or non–party member, collaborator or resister, rarely provide an adequate framework for evaluating the ideas and actions of the individuals in question.[1] Ambiguities attend the cases of most of the scientists who worked in Germany or Austria during the Third Reich, and Lorenz's case is no exception. Lorenz was an aspiring young scientist who sought to advance his research and his career within the complex terrain of National Socialist biology.

Scholars have handled Lorenz's career as a scientist in the Third Reich in a variety of different ways. Some have lauded his science without mentioning at all that his wartime behavior and writings raise issues about ways in which his science may have intersected with politics.[2] Others have portrayed him as one of the more politically active biologists in the Third Reich.[3] The American ethologist Peter Klopfer has suggested that Lorenz's ideas about releasers, innate releasing mechanisms, action-specific energy, and the like may have been inspired as much by Lorenz's National Socialist sympathies as by his work with animals.[4] But Lorenz's former student Paul Leyhausen has insisted to the contrary that Lorenz was not "an active or an occasional promoter of Nazi ideology." Says Leyhausen of Lorenz's work in this period: "There really never was any question of bending ethology to suit Nazi ideology, but only a vague and—as things stood—ill-founded hope of applying the former to correct the latter." Where some have seen in Lorenz's wartime writings clear evidence that Lorenz endorsed Nazi thinking, or at least no evidence that he tried to distance himself from such thinking, Leyhausen claims Lorenz's writings actually represented "very cautious" attempts "to swim against the current." The problem in interpreting these essays today, Leyhausen maintains, is that "people in those days were keener at hearing such undertones than the subsequent critics of Lorenz were and are."[5] The general historical principle implied here is a worthy one—that is, that one should seek to make sense of an individual's behavior within the context of his or her own time. The difficulty with Leyhausen's gloss on the situation is that it acknowledges neither that Nazi biology accommodated a diversity of views nor that Lorenz's writings might have conveyed different meanings to different audiences. Ambiguities and cautious undertones can indeed be found in his writings. However, one such ambiguity involves precisely what Lorenz meant when he wrote about the "weeding out" (Ausmerzung) of social "parasites" and "carcinomas." The historian looking back on the period must indeed inquire whether "people in those days" recognized that they should dismiss such talk as mere rhetoric.

Seeking to understand this part of Lorenz's career is a complicated project, and interpretations of the question are bound to differ. The goal in this chapter is to chart the interrelations and intersections of Lorenz's scientific thought, practice, and career aspirations as he set about negotiating the complicated political terrain of Nazi biology. There can be no denying that he claimed on numerous occasions that his research and ideas on animal behavior had a contribution to make to the race-political aims of the Third Reich. The logic and the context of such claims need to be explored in some detail.

FROM SPRING 1937 TO THE ANSCHLUSS OF MARCH 1938

In chapters 3 and 4 we saw how an increasingly confident Lorenz was further emboldened in 1936 and 1937 by his interactions with Niko Tinbergen and Erich von Holst. The former excited Lorenz with evidence that Lorenz's ideas were amenable to experimental verification; the latter provided a critical new element for Lorenz's conceptualization of instinct. Lorenz's reach was extended further early in 1937 when he assumed a leadership role in the new German Society for Animal Psychology. As of April the editorial board of that society's journal consisted of two of his supporters (Otto

Koehler and Otto Antonius) and Lorenz himself. In the *Zeitschrift für Tier-psychologie*, in the review journal *Berichte über die wissenschaftliche Biologie* (the German counterpart of the American review journal *Biological Abstracts*), and elsewhere, he and his comrades spread the word that his new approach was revolutionizing the study of animal behavior.[6]

Nonetheless, Lorenz's growing authority among biologists and animal psychologists was not sufficient to secure him gainful academic employment. He remained an unpaid lecturer in Karl Bühler's Institute of Psychology at the University of Vienna. He began seeking research support from Germany. He applied for funding from the Deutsche Forschungsgemeinschaft (the German Research Organization [henceforth the DFG]), which had a division for aiding Austrian scientists. He also nurtured a much grander dream. He hoped the Kaiser Wilhelm Society would establish for him a special institute of comparative behavior studies in Altenberg. He worked feverishly in February 1937 to freshen up the facility prior to a visit there by Friedrich Glum, the Kaiser Wilhelm Society's secretary general. Glum's inspection of the site in March 1937 corresponded almost exactly to Tinbergen's arrival there. No doubt Lorenz's optimistic portrayal to Tinbergen of the improved scientific climate in Germany (as opposed to that in Austria) was a reflection, at least in part, of Lorenz's high hopes after Glum's visit.[7]

Lorenz began exploring the possibilities of research funding from Germany as early as January 1937. He wrote to Walter Greite, head of the DFG's section for biology and medicine, indicating that he was coming to Berlin for the spring meeting of the German Society for Animal Psychology and that he hoped soon thereafter to submit a grant application to the DFG. Three weeks later, having made his trip to Berlin, he submitted his proposal. He proposed to study the courting behavior of different species of ducks. The instinctive motor patterns involved in courting, he explained, were of considerable value when it came to reconstructing the evolutionary history of the group. Furthermore, the ease with which different duck species could be hybridized made it possible to analyze their behavior patterns through the breeding techniques of the geneticist. Filming (including the new technology of slow-motion cinematography) would also be critical to the project. He was already working with the Reichsfilmstelle (the government's film board) on the production of two films, one on animal behavior patterns and the other on the greylag goose. He asked the DFG for funds to purchase film and an extra camera lens (he already had a Zeiss-Kinamo movie camera, thanks to the Reichsfilmstelle). He also requested funds for an incubator, materials for the construction of bird-rearing boxes, some twenty-two males and females of four different duck species, and duck feed.[8]

Lorenz's relationship with Greite quickly developed beyond that of a mere applicant for research funds. In March 1937, following up on an earlier conversation, he sent Greite a lengthy, confidential appraisal of the work of the leading students of animal behavior working in Germany. He evaluated the merits of Werner Fischel, Jakob von Uexküll, two of Uexküll's students (Friedrich Brock and Heinz Brüll), and Oskar Heinroth. Lorenz, it would appear, had gained the ear of a strategically placed scientific power broker.[9]

These early interactions with Greite, however, did not bring Lorenz any immediate, tangible benefits. When Greite left the DFG in the spring of 1937, Lorenz's application for funding got lost in the shuffle. Lorenz learned the bad news from his friend Holst. He wrote to the DFG in great consternation. The study of bird behavior, he fumed, could not be undertaken whenever one wished to do so. He needed to observe phenomena that took place only in the spring, during mating season. Precious and indeed irretrievable research time was slipping away from him while he awaited the DFG's decision. It would be extremely distressful to him, he said, to have his researches cut short just when they were close to reaching their goal.[10]

Lorenz submitted a new application to the DFG, explaining at greater length than he had done previously the importance for zoology of his comparative behavior work. His main purpose, he indicated, was not one of phylogenetic reconstruction. Rather, it was to prove decisively to animal psychologists that hereditarily established behavior patterns in animals could be viewed as organs, and indeed could only be understood "from the standpoint of comparative and genetic research."[11] Lorenz asked Erwin Stresemann to write a letter to the DFG on his behalf. The dean of German ornithologists obliged, touting the younger man as "a pathbreaking scientist of quite exceptional ability."[12] But the new application and accompanying recommendation brought him no windfall. In June the DFG's Austrian research support division informed him that his proposal had been turned down, for financial reasons.[13]

The DFG's official excuse, Lorenz soon learned, was not the whole story. Evidence in the organization's archives suggests that the DFG turned down Lorenz's application for research support because someone called both his politics and his ancestry into question. In short, the purported, reestablished immunity of German science from ideological influences was not all that Lorenz himself had hoped it would be. Hearing that he had been maligned, Lorenz quickly convinced himself who the culprit was. In letters to Stresemann and Heinroth he identified the slanderer as "O. A." and "a personal enemy emigrated from Austria." The man he had in mind was Othenio Abel, the paleontologist from the University of Vienna who had been the source of so much anxiety for Lorenz at the time of his doctoral exam. Lorenz believed that Abel bore a grudge against him because Abel was later left off Lorenz's docent exam committee. Lorenz bemoaned the misfortune of having his application end up in Abel's hands.[14]

But Lorenz had no intention of giving up. He took his case to Fritz von Wettstein, director of the Kaiser Wilhelm Institute for Biology in Berlin-Dahlem. Wettstein pursued the matter, soliciting the opinions of five scientists who knew Lorenz from Vienna and who were in a position to comment on his scientific, political, and hereditary worthiness.[15] Ferdinand Hochstetter, Otto Antonius, Alexander Pichler, Eduard Pernkopf, and Fritz Knoll all provided testimony on Lorenz's behalf. They praised his scientific work and personal character. They also identified him as a scientist who, without being a member of any political party, was clearly dissatisfied with the clerical regime in Austria. Antonius observed that Lorenz had "never made a secret of his admiration for the new situation in Germany." He added further that Lorenz's and Lorenz's wife's genealogies were fine and that the autobiography recently written by Lorenz's father was "a decidedly Nazi book." Pichler noted that Lorenz in his years as an assistant at the Second Anatomical Institute at the University of Vienna had been a man who lived only for his scientific work, but Lorenz had recently taken more interest in politics. Pichler wrote: "Lately Dr. Lorenz has repeatedly displayed to me his constantly growing interest for National Socialism and has expressed himself positively about its idea. As far as I am acquainted with his biological studies, they are in keeping with the world view prevailing in the German Reich."[16] Knoll contributed the information that Lorenz had shown him his family tree going back as far as his great-grandparents. All these ancestors were Catholic or Protestant. Lorenz swore further to Knoll that he knew of no Jewish blood in his ancestry. Lorenz's wife's family, Knoll reported, had a similarly spotless record.[17]

While the confirmation of his racial and political acceptability was underway, Lorenz was invited by Karl Bühler to fill in at Bühler's institute for Egon Brunswick, Bühler's first assistant, who had been offered a visiting professor's post at the University of California, Berkeley. Much as Lorenz would have liked such a position, he was afraid it would make him look bad to be part of an institute that was both racially and politically suspect. The problem, as he explained to Max Hartmann, director of biology at the Kaiser Wilhelm Gesellschaft, was that if he took this position, he would have to work under the direction of Bühler's wife, Charlotte, who was Jewish, and this was something he was not willing to do. Lorenz went on to say that aside from Bühler there was only one other Aryan in the whole institute, and that man was a communist. Hartmann confirmed that Lorenz's anxieties were not misplaced. If Lorenz were to fill in for Brunswick at Bühler's institute, Hartmann admitted, it might indeed count against him, in some quarters, in his search for research support.[18]

In December 1937, Wettstein wrote to the DFG recommending that it support Lorenz. Wettstein had taken it upon himself, he said, to investigate the questions that had been raised about Lorenz's politics and ancestry because the Kaiser Wilhelm Gesellschaft (KWG) was itself very interested in Lorenz's work. Wettstein remarked that although Lorenz had not been politically active, Lorenz had made no secret of his approval of National Socialism. He added that Lorenz's Aryan ancestry was likewise "in order." He explained that the KWG was not really in a position to help Lorenz because Lorenz's private research station was in Austria, and it was impractical to relocate the facility to Germany. He therefore urged the DFG to assist Lorenz through its special Austrian science division.[19]

In January 1938 the DFG sent Lorenz two forms that he needed to fill out before he could obtain DFG support. One was a personal history form. The other demanded proof of his Aryan ancestry (and that of his wife as well). Lorenz provided the information required. He also elaborated upon the very straitened financial circumstances under which he was trying to do his research. In response to the question "To what career goal do you aspire?" he wrote, "employment at a Kaiser-Wilhelm Institute."[20]

And that indeed was what Lorenz wanted most. Or, to be more precise, what he wanted most was the *directorship* of his own Kaiser Wilhelm Institute, devoted to comparative behavior studies and situated, ideally, at Altenberg. All through the year his hopes were high. In September he told Margaret Morse Nice (who had arranged to come work with him the following spring):

> The leading Biologists of the "Kaiser Wilhelm Gesellschaft zur Förderungs der Wissenschaften," Prof. Hartmann, Prof. Kühn and Prof. v. Wettstein are unanimously trying to create an institute for just the kind of work I am doing. First the plan was to build an institute for me in Rossitten, but now they plan to do the thing which I myself always advocated, which is to finance and develop the "station" which I have already created here in Altenberg. It is all still in the air and it is most questionable if they ever

get enough "Devisen" [currency] for their purpose, but even so I am awfully proud that they try to do it! If, however, they should succeed, I should be the happiest man on earth![21]

At its senate meeting in October the KWG agreed, at least in principle, to establish an institute for Lorenz. As a way of finessing the problem of supporting an institute in Austria, which ran counter to KWG practices, it was proposed that Lorenz be given an appointment at the Kaiser Wilhelm Institute for Biology in Berlin-Dahlem and then sent to Altenberg for three-quarters of the year on "research leave." However, the issue of financing the operation remained unresolved.[22]

LORENZ AND THE *ANSCHLUSS*

Within a matter of a few months, the question of the KWG's supporting an institute in another country was moot. The *Anschluss,* the union of Austria with Germany, took place in March 1938. The German army crossed the border into Austria on 12 March, meeting no resistance. The Austrian leaders, Chancellor Kurt von Schuschnigg and President Wilhelm Miklas, had wanted to preserve Austria's status as an independent nation, but the dual threats of military invasion and civil war were too much for them. They were forced to resign. On Hitler's orders Arthur Seyss-Inquart was installed in Schuschnigg's place. On 13 March Seyss-Inquart signed the documents that made the *Anschluss* official.

A great many Austrians were glad to see the new regime. Huge crowds cheered Hitler as his cavalcade motored to Vienna. When he reached the outskirts of Vienna on 14 March he was greeted by the sound of ringing church bells. The *Times* of London reported that the city displayed "no signs of a people bowing unwillingly to a foreign yoke." To the contrary, Vienna "resembled a town which has just received news of a great victory." On 15 March an estimated 250,000 auditors jammed the Heldenplatz to hear Hitler speak.[23]

Historians have allowed that the enthusiasm with which the majority of Austrians greeted Hitler represented not so much an endorsement of Nazi principles as a rejection of the political, economic, and social woes of the previous two decades. As Radomír Luza explains, "[Nazism's] apparent flexibility, vitality, and dynamism aroused the hopes of various groups for the removal of many old-fashioned bureaucratic, reactionary structures and institutions. The Nazi movement became the most promising non-Marxist force for change for those groups who felt themselves neglected by the corporative regime or socialist doctrine. . . . The formulas of national community and National Socialist revolution were designed to reshape the traditional social forms on the basis of new relations among equal members of the German nation." [24]

However one interprets the enthusiasm of so many Austrians for the *Anschluss,* there is no doubt that Lorenz himself was ecstatic about it. One week after Hitler's triumphant speech in Vienna, Lorenz described and explained his own feelings to Oskar Heinroth: "We all cheer like little children over the 'Anschluss.' For scientists it is a release to belong now to the larger Germany instead of to the damned Jesuit rabble. You cannot have the slightest idea what a festive mood reigned and still reigns in all of Austria, especially on the first day. One apparently has to have thoroughly suffered under the black thing [the Schuschnigg regime], in order to understand *fully* the value of Hitler. The revolutions carried out at the university so far are to that extent clearly to the good, so that one can hope for a golden age for our mutilated departments!!" [25]

Four days later Lorenz wrote Erwin Stresemann virtually the same thing, saying "You cannot have the slightest idea of the enthusiasm that reigned here, and even still reigns, and what an exceptional and festive mood even such apolitical people as we are in." It had taken five years under "the black *Schweinehunde*" to effect this result, Lorenz said, but, as he put it, "I believe we Austrians are the sincerest and most convinced National Socialists after all!" One had to thank Schuschnigg and his cronies for this, Lorenz explained, for without their help the Austrians "would not have so quickly, thoroughly, and enduringly become converted to Hitler." [26]

Prior to the *Anschluss,* travel to Austria from Germany had been restricted. Now Lorenz felt he could look forward to German visitors—especially students. He wanted bright, young researchers to come work with him as Tinbergen had. Not everything was possible, he admitted. The biologist Monika Meyer-Holzapfel had wanted to travel to Altenberg from Switzerland. Lorenz thought highly of her, but she could not come because she was Jewish. Lorenz hoped Stresemann might have some young people to send him. As he put it, "I have *many* interesting questions to attend to!" Already Lorenz's friend and student Alfred Seitz, working under only limited guidance from Lorenz, had produced "such magnificent researches on releasers and inborn schemata in bone fish" that Lorenz was "completely enraptured." Otto Koehler's opinion of Seitz's work, Lorenz said, was the same. Lorenz exulted: "Once the KWG has built me the projected Aquarium room here in Altenberg, I can well occupy 6 men and four horses with really productive researches. With birds again as many." He envisioned putting someone to work studying the pecking order in greylag geese. Tinbergen's example, he said, had shown what a well-qualified researcher could accomplish at Altenberg without much cost.

Lorenz's mood was joyful as he described to Stresemann his immediate surroundings and the new political situation. "Here in Altenberg," he exclaimed, "it is splendid! Geese are brooding, ducks are courting, children are healthy and growing, the jackdaws arrived this year in greater number than ever and again much tamer, since I gave up the experiment of forcibly making them tame by temporarily locking them up. These procedures had the contrary result, exactly as Schuschnigg had to experience with us!" Above all else what Lorenz was now looking forward to hearing was that the KWG would be establishing an institute for him. The KWG senate, as Lorenz understood it, had agreed in principle on establishing an institute for him, and now only currency difficulties stood in the way. He urged Stresemann to pass on any news he might hear.[27]

While awaiting word from the KWG, Lorenz explored the possibility of increased funding from the DFG. In his cover letter he said that he wanted to make sure that the most recent version of his research proposal was in the hands of Professor Hartmann, one of the directors of the Kaiser Wilhelm Institute for Biology. He wanted to ask for items he had not mentioned in his previous proposal. As a new citizen of the Third Reich he signed his letter, "Heil Hitler!"[28]

Events moved very quickly after the *Anschluss*, and Lorenz soon had a striking new possibility to consider. He informed Stresemann of the developments in a letter of 11 April 1938. Karl Bühler, the professor of psychology at the University of Vienna, had been ousted from his position and thrown in jail. Precisely why this happened, Lorenz said, was not clear, but it was "certain that he [Bühler] had lied about the ancestry of his full-Jewish wife," and he might also have engaged in currency trafficking with money from the Rockefeller Foundation. "In addition," Lorenz noted, "he was so intensively red and black [i.e., he was both a Social Democrat and a supporter of the Catholic regime], through and through, even by the standard of the times, that that alone is sufficient explanation. In any case he is not coming back again."[29]

Lorenz expressed no sympathy here for his former mentor, despite the considerable support Bühler had provided him.[30] Bühler had introduced Lorenz to the literature of psychology, served on his doctoral and docent exam committees, encouraged his comparative study of instinctive behavior, and given him a place as docent in the psychological institute. Lorenz mentioned none of this in his letter to Stresemann. Instead he spelled out what this might mean for his own career.

Lorenz was in touch with an old friend, Alfred Prinz Auersperg, a psychiatrist who had just been made provisional director of what up to then had been a Jewish-run neurological institute. Auersperg was a specialist in the study of inborn and acquired automatic motions in humans. He also studied sense perception. Work in the latter area had been a major concern of the now-defunct Bühler Institute. Indeed it had been, in Lorenz's opinion, that institute's most valuable contribution to science. Auersperg's plan was to get hold of the Bühler Institute's equipment and continue the tradition of sense-perception studies but with a stronger, biological foundation. For Auersperg's plan to work, someone whose research complemented Auersperg's would need to join the Bühler Institute.

The person Auersperg had in mind was none other than Lorenz himself. Lorenz professed to Stresemann that he had been at first taken aback by Auersperg's suggestion. Then, upon considering the suggestion objectively, he said, he began to see its merits. What appealed to Lorenz was the prospect of being able to lecture on "comparative psychology in the true sense of the word." He would give human social psychology the attention it merited. "I venture to maintain," he told Stresemann, "that this reading would be ideologically welcome, as much welcome as it earlier was unwelcome."

Lorenz wondered to Stresemann what the KWG was likely to say about such a development. Would it see a professorial chair in psychology in Vienna as being compatible with a directorship of a Kaiser Wilhelm Institute (which remained Lorenz's greatest desire)? A professorship, he said, was not something to look down upon, "because I really feel in me the calling and obligation to found a school and bring order into the confused babble of psychology." He would have to invest time in such a position, of course, but if he had a professorship he would also have assistants, and the research they did for him would offset the time he lost personally. It sounded horrible for him to say it, he acknowledged, but he still felt compelled to say that "a complementary double institute, in which a psychiatrist and neurologist on the one hand and a comparative zoologist on the other squeeze human psychology dead and put something new in its place, would be something *genuine*. Above all, something really 'properly' German, since I must (in the strictest confidence) say, that human psychology in its modern German versions is always still from an expert's point of view noticeably derived from the thought of Jewish-babbling, verbose, Jewish leaders. One of the few cases, where I fully acknowledge the perniciousness [*Schäd-*

lingstum] of the Jews. There are greed-addicted [raffsüchtige] and asocial Aryans enough, but making nonsense of science through multiple discourses, that really [is something that] only Jewish human psychologists bring about."[31]

A week later Lorenz wrote again to Stresemann. He thanked Stresemann for a letter in which Stresemann apparently identified difficulties with the scenario Lorenz had presented him but still offered what Lorenz took to be a "generally affirmative opinion." with respect to Lorenz's idea of simultaneously directing Bühler's former institute and a KWG research institute at Altenberg. Acknowledging again that the demands of a professorship would cut into his time for doing research, Lorenz wrote: "If anything can induce me, really to seek a professor's pulpit, it would be the fact that I viewed it as a social and national life-duty to convert the fully, mentally deranged [in Geist aufgelöste] study of human psychology again to a domain of inductive natural science." He reiterated his belief that having assistants would compensate for whatever time he might lose in administrative work and in teaching. He then returned to his earlier concern: what would the KWG think of such an arrangement? If Professors Hartmann, Wettstein, and Kühn strongly objected to his holding both positions, he knew what he would choose: "*To me the Altenberg research station is substantially more important than the most beautiful professorial pulpit!!!!*"[32]

Lorenz's hopes for the chair of psychology and directorship of Bühler's Psychological Institute at the University of Vienna failed to materialize. He did not mention them again in his letters to Stresemann.[33] He continued to believe, however, that the Kaiser Wilhelm Gesellschaft was about to establish an institute for him.[34]

Had Lorenz in 1938 been living outside the sphere of influence of the Nazi regime, or had he been established in his much-longed-for Kaiser Wilhelm Institute, he might never have floated "in confidence" the image of himself as someone who could replace a predominantly "Jewish" human psychology with a new psychology that was "properly German." Similarly, had he not felt thwarted prior to 1938 by the antievolutionary views of the Austrian Catholics and the Schuschnigg regime, he might not have been so attracted to the new political order that the *Anschluss* brought. But it would be misrepresenting his mood in March and April of 1938 to portray him, at the time of the *Anschluss*, as simply trying to make the best of a difficult situation. To the contrary, he saw the change as extremely promising, and he was ebullient about it. The National Socialists had a clear commitment to viewing human behavior and the body politic in biological terms. His own views on human nature, he believed, were finally going to be "ideologically welcome." He was confident he could coordinate his own interests with those of the new political order in such a way as to ensure the progress of his research and his career. On 28 June 1938 he applied for membership in the Nazi Party. In his application he claimed: "I was as a German thinker and scientist naturally always National Socialist."[35]

That said, it deserves to be noted that most of the Lorenz letters we have from this period were addressed to ornithologists, and that in these letters he talked more about birds than he did about politics. For example, the first letter Lorenz sent to Heinroth after the *Anschluss* contains only a single paragraph (out of five paragraphs total) pertaining to the recent political events. In the rest of the letter he apologized for having damaged a lantern slide he had borrowed from Heinroth for a lecture trip to Holland, he described the progress of his research at Altenberg (he had been successful in filming his ducks and geese, and his greylags and jackdaws were multiplying prolifically), and he announced a lecture he was planning to give at the joint meeting of the Society for Animal Psychology and the German Society for Psychology in Bayreuth in July. One cannot conclude from this, however, that he was keeping his science and his politics separate from one another. The upcoming lecture shows otherwise. What he initially said to Heinroth about the lecture was "I will lecture on domestication phenomena in the behavior of domestic geese and domestic and greylag goose crosses, since that is also of some interest for human social psychology."[36] In fact, the lecture was to mark an important stage in Lorenz's career. It represented his first big chance to show a large audience that his own research interests had a bearing on the race hygiene concerns of the Third Reich.

When Lorenz next wrote to Heinroth he provided a more explicit title for his forthcoming lecture—"Breakdowns in the Instinctive Behavior of Domestic Animals and Their Social-Psychological Meaning." He also offered details on what he planned to cover. The "breakdowns" of which he was speaking, he explained, were not in the animals' instinctive motor patterns themselves but rather in the innate mechanisms that served to release these motor patterns in the first place. His geese showed clearly, he said, that when individual features of an innate releasing mechanism were lost, this made the release of the corresponding instinctive motor pattern easier. Before a "respectable," full-blooded greylag goose would mate, many conditions had to be met. In contrast, a "fat, domestic goose" could be induced to mate with no difficulty. Lorenz extrapolated from this to humans: "I believe man has an inborn abhorrence for humans who have degenerate instincts. This abhorrence has also certainly a species-preserving value, since in humans degenerate mating drives and similar brood-care reactions go along with each other, as, e.g., with my greylag / domestic goose crosses." Noting that greylag geese themselves had an abhorrence for "street-walkers" (*Strassendirnen*), Lorenz concluded: "The social 'morality' of humans is thus most certainly in the greatest part inborn and one must consequently distinguish this most sharply from the traditional taboo- and 'duty-controlled' social behaviors of humans." Lorenz asked Heinroth to critique his manuscript before he sent it off to Professor Erich Jaensch, the President of the German Society for Psychology.[37]

Erich Jaensch was professor of psychology and director of the Institute for Psychological Anthropology at Marburg. He was an ardent Nazi, a racist, and an advocate of developing a distinctly *German* psychology. He also studied animal behavior. He had been scheduled to speak along with Lorenz, Otto Koehler, and J. von Allesch at the first annual meeting of the German Society for Animal Psychology in February 1937. Although he did not end up presenting a paper there, it was he who announced in the pages of the *Zeitschrift für Tierpsychologie* the forthcoming joint meeting of the German societies of psychology and of animal psychology. He noted that the site for the meeting, Bayreuth, coincided nicely with the organizing theme of the meeting, "character and education," since Bayreuth was the site of the central office (*Reichswaltung*) of the National Socialist Teachers Guild.[38]

Among the psychologists with whom Lorenz was to rub shoulders in the Third Reich, Jaensch was certainly one of the most powerful. Lorenz may not yet have met him as of March 1938, but he certainly knew of him. Jaensch, for his part, was greatly impressed by the manuscript Lorenz sent him. He offered Lorenz a third more time for his presentation than he offered to the other speakers.[39]

The address Lorenz delivered to the German societies of psychology and animal psychology in Bayreuth in July 1938, this was the outgrowth of an idea he had been nurturing for several years. As early as his 1932 paper on instinctive behavior patterns in birds, and then again in greater detail in his "Kumpan" paper of 1935, he noted that domesticated or captive animals, like animals in ill health, often exhibit mutations and breakdowns in their behavior chains, inhibitions, or fixed action patterns.[40] Others before Lorenz, including Charles Darwin and Charles Otis Whitman, had remarked on the breakdown of instinctive behavior patterns under the conditions of domestication. Lorenz, however, was drawing on cultural resources appreciably different from those of his English and American predecessors, and he found a correspondingly different significance in the phenomena in question. European and especially German intellectuals around the turn of the century had become increasingly worried about cultural and biological degeneration. Lorenz proposed that the degeneration of instinctive behavior patterns in domesticated ducks and geese corresponded to the cultural and genetic degeneration of civilized man.[41]

Lorenz had not written previously about human cultural or racial degeneration, aside from noting in 1935 in his "Kumpan" monograph that Europeans were much more genetically diverse than Negroes or Chinese (a claim that was wholly unfounded but corresponded to his belief that Europeans represented a more advanced state of civilization).[42] In his Bayreuth address, in contrast, Lorenz focused directly on the "alarming" similarities between domesticated animals and civilized man. He argued that "breakdowns" (Ausfallserscheinungen) in the instinctive behavior of domestic animals were strictly analogous to "signs of decay" (Verfallserscheinungen) in the behavior of civilized man. Both had the same cause: the relaxation of natural selection.

It is easy to understand why Lorenz's paper struck such a sympathetic chord with Erich Jaensch. Lorenz enthusiastically enlisted animal behavior studies in the cause of race hygiene. The "backbone of all racial health and strength," Lorenz insisted, is to be found in the "high value placed on our species-specific and inborn social behavior patterns." The nondegenerate individual, he explained, has an instinctive, intuitive ability to recognize the good or bad ethical (and genetic) character of the social behavior of others, and this instinctive response is truer (and more important for the future of the race) than any reasoned response. The danger to the race, he warned, lay in the undesirable types that proliferated under the conditions of civilization. Summoning up an image with which the Nazis were obsessed, the naturalist who only a few days before had applied for membership in the Nazi Party likened degenerate members of society to cancerous cells in an organism: "Nothing is more important for the health of an entire people [Volk] than the elimination [Ausschaltung] of invirent types, which, with the most dangerous and extreme virulence, threaten to penetrate the body of a people like the cells of a malignant tumor."[43] Jaensch was no doubt pleased to see Lorenz use the word "invirent" because it was a word that Jaensch himself had introduced to designate certain types as "weak" or "disintegrative" while, paradoxically, also highly threatening to the health of a people. He concluded that Lorenz's research on the behavioral differences between wild and domesticated forms was getting at the same sorts of issues Jaensch had been studying himself. He was simultaneously glad to note that Lorenz's observations were consistent with the

Nordic movement. The Nordic movement, Jaensch wrote, constituted, at least in part, "a countermovement against domestication damages in civilized man."[44]

Jaensch expressed his approval of Lorenz in a remarkable paper entitled "The Henhouse as a Means of Research and Explanation in Human Race Questions."[45] The paper bears mentioning here because it illustrates how in the Third Reich the trappings of science were used to support the most ludicrous sort of racial thinking. Jaensch's study appeared in the second volume of the *Zeitschrift für Tierpsychologie*, the same volume that included Lorenz's and Tinbergen's paper on the egg-rolling behavior of the greylag goose. With quantitative tables and photographs making the work look all the more objective, Jaensch set forth the results of his studies comparing the pecking styles and other characteristics of northern versus southern races of chickens. He concluded that the differences between northern and southern races of chickens paralleled the differences between northern European and southern European races of humans. Northern chickens pecked steadily and accurately while southern chickens pecked rapidly but impulsively and inaccurately. This mirrored, he claimed, the calm, measured, and tenacious behavior of northern, Germanic types as compared with the restless, lively, and flexible behavior of Mediterranean types. Jaensch aligned his chicken data with his thesis of two distinctive human personality types, the one "inwardly integrated," the other "outwardly integrated."

Before Jaensch's paper appeared in print, he and Lorenz met again, this time at Harnackhaus in Berlin at the second annual meeting of the German Society for Animal Psychology (22–24 September 1938). The DGT gathering intersected with the fifty-sixth annual meeting of the German Ornithological Society, with the animal psychologists starting a day earlier and likewise ending a day before the ornithologists did. Jaensch and Lorenz both presented papers to the DGT, Jaensch discussing "the psychology of the domestic chicken" and Lorenz lecturing on "taxis and instinct." Lorenz in addition gave two lecture-and-film presentations sponsored jointly by the two societies.

Lorenz's films were the improved versions of the two projects he had been working on since 1935, the first comparing the courtship behavior of ducks, the second detailing the ethology of the greylag goose. The greylag goose film featured footage from the spring of 1937, when he and Tinbergen experimented on the greylag's egg-rolling behavior and Alfred Seitz filmed it all happening. Parts of the greylag film were shown at the German ornithologists' annual meeting in Dresden in July 1937. Lorenz then presented the film in May 1938 at the International Ornithological Congress in Rouen. Now in Berlin it was featured as the postdinner event of Friday evening, 23 September. The DOG's official account of the whole meeting spoke enthusiastically of Lorenz's "outstanding" film and reported that it "was received with great applause."[46]

From all that one can tell from the official reports of these meetings, Lorenz's talks to the DGT and DOG in September 1938 were essentially apolitical. His lecture-and-film presentation on the comparative behavior of ducks stressed the importance of films for phylogenetic and genetic studies, explaining as he had done elsewhere that behavioral studies could be even better than morphological studies for these purposes. Likewise his greylag presentation, according to the DOG official account of it, illustrated "a series of fundamental contributions to the knowledge of greylag psychology." But if Lorenz did not sound the broader themes of domestication and degeneration he was voicing elsewhere, it was not the case that the relations between science and politics went unnoticed at the meetings. Lorenz's friend Otto Antonius, as president of the DGT, observed in his concluding remarks to both societies how happy he was that Austrian scientists no longer faced the hindrances they had encountered in previous times. He simultaneously pointed to the "acute danger" that threatened Germany on the Czech border and expressed his hope that this source of continuing trouble would be dispensed with as soon as possible. His hope was in fact realized in less than a week, when the British and French acceded to Hitler's demands and signed the infamous Munich Pact, sacrificing the Sudetenland of Czechoslovakia to Germany.[47]

IDEOLOGICAL ANTECEDENTS

German biologists had a long tradition of debating the meaning of biological evolution for society. In the 1870s, Rudolph Virchow and Ernst Haeckel had tangled with each other over evolution's social implications.[48] In 1900 the German arms manufacturer Friedrich Alfred Krupp founded a thirty-thousand-mark competition for the best treatise on the subject "What can we learn from the principles of evolution for the development and laws of states?" Sixty entries were submitted for the prize. First prize was awarded to Wilhelm Schallmayer for his work, *Vererbung und Auslese als Faktoren zur Tüchtigkeit und Entartung der Volker* (*Heredity and Selection as Factors in the Ability and Degeneration of the People*).[49] Central to Schallmayer's treatise was the view that civilized peoples were more subject to biological degeneration than were primitive peoples because civilized society spared individuals from natural selection. Between 1903 and 1907, nine volumes of treatises from Krupp's competition were published in a series entitled *Natur und Staat*, supervised by the zoologist and instinct theorist Heinrich Ernst Ziegler. Ziegler himself had a decade earlier written a volume of his own on the subject of "science and social-democratic theory," in which he argued that, contrary to the opinion of many socialists, evolutionary biology offered no support for socialism.[50]

The debates over the sociopolitical implications of evolutionary theory were further complicated by the fact that matters of evolutionary theory themselves remained contested, with political considerations sometimes entering the equation. The German geneticist-anthropologist Fritz Lenz, for example, in the standard text on human heredity that he coauthored with Erwin Baur and Eugen Fischer, claimed that the Lamarckian idea of the inheritance of acquired characters was especially attractive to Jews. The logic of this, he explained, was that "if acquired characters could be inherited, then, by living in a Teutonic environment and by adopting a Teutonic culture, the Jews could become transformed into genuine Teutons." Lenz took care to state that such wishes had nothing to do with scientific reality: the Lamarckian doctrine was "an illusion." He quoted approvingly the comment by F. Kahn: "Jews do not transform themselves into Teutons by writing books about Goethe." Lenz thus managed at one and the same time to attack Jews and discredit neo-Lamarckian thinking.[51]

Lorenz never cited any of the above-mentioned works in his own writings. Whether he knew of them is uncertain. Nor do we know whether he ever read Jakob von Ucxküll's *Staatsbiologie* (*Biology of the State*), where Uexküll treated the "anatomy," "physiology," and "pathology" of the state. Reissued by the Hanseatic League in 1933 after its initial publication in 1920, Uexküll's antidemocratic, antitechnocratic work concluded in 1933 with the hopeful observation that Adolf Hitler's rise to power was bringing an end to the forms of pathological decay Germany had been displaying for years.[52] Lorenz met Uexküll in 1933. We know that they talked about jackdaws, Uexküll's *Umwelt* theory, and the idea of the *Kumpan*. We have no evidence that they talked about politics.

Had Lorenz read the works of Schallmayer, Uexküll, and numerous others, he would have encountered conflicting opinions. Schallmayer, the cofounder of the German eugenics movement and a proponent of Darwinism, was scornful of the idea of Nordic supremacy. Uexküll, in contrast, despised Darwinism but held the views of the Nordic supremacist Houston Stewart Chamberlain in the highest esteem. More locally, Lorenz could have heard Julius Tandler, professor of anatomy at the First Anatomical Institute of the University of Vienna (and a Jew and prominent Social Democrat) promote eugenics within a neo-Lamarckian framework.[53] Amid such differences of opinion, there was still one message Lorenz could have imbibed: German biologists and physicians considered it their prerogative to pronounce authoritatively on biology's implications for society.

To be sure, German biologists and physicians were not unique in this regard. There had been ardent eugenicists in many countries over the first third of the century. After the Nazis came to power in Germany in 1933 and began enacting race purity laws, some non-German eugenicists expressed admiration for these developments. Others, however, complained that science was being perverted for racist purposes. Among the latter was the American geneticist Hermann J. Muller, who wrote in 1933 a review of the recent (1931) English translation of the Baur, Fischer, and Lenz text on human heredity. After first identifying this as "the best work on the subject of human heredity which has yet appeared," Muller went on to observe how Fischer and Lenz in particular, as they ventured "into psychology, anthropology, history and sociology," became "less and less scientific" and ended up "acting as mouthpieces for the crassest kind of popular prejudice." Among other things, Muller noted the authors' warnings about race mixing. He quoted Fischer to the effect that racial hybridization could produce "injury to the constitution." He quoted Lenz as saying "the crossing of Teutons and Jews is likely as a rule, to have an unfavorable effect." Muller observed further: "Hitler is said to have studied the Baur-Fischer-Lenz book very seriously, and to have been won over to it, while Lenz has recently written an article favoring Hitlerism."[54]

Others in Germany, however, stressed the importance of race hygiene without raising the specter of race mixing. Karl von Frisch, never known for any attraction to Nazism, did just this in a popular biology text he published in 1936 entitled *Du und das Leben* (*You and Life*). He concluded the book with a section on race hygiene, voicing there the familiar warning that the relaxation of natural selection in higher cultures was leading to the perpetuation of variations that in the wild would have been "mercilessly weeded out" (*erbarmungslos ausgemerzt*). This amounted in effect, he said, to an "encouragement of the inferior," or, as he put it more bluntly, "A tub of lard [*Fettwanst*] or a blind man finds his table as well set as any other person." Deutsche Verlag and Verlag Ullstein published the book, but it was in addition issued as "volume two" of the "Dr.-Goebbels-Spende für das deutsche Wehrmacht" (the "Dr. Goebbels Contribution to the German Military").[55]

Lorenz owned a copy of Frisch's book, but there is no evidence that he read it. The book remains today on a shelf in his library at Altenberg, without any marginal annotations. But Lorenz did not have to read the book to be exposed, in his own home, to the idea of eugenics. His father, Adolf Lorenz, the distinguished orthopedic surgeon, was, as indicated above, yet another person who maintained that when it came to the biological vitality of the race, the art of medicine was detrimental, insofar as it preserved individuals that natural selection would otherwise have weeded out. This was the basis of Adolf Lorenz's declaration that newborn infants should be fit enough to survive without the aid of incubators or other special medical assistance.[56]

ANIMAL BEHAVIOR, RACE HYGIENE, DARWINISM, AND NAZISM

When Konrad Lorenz promoted the idea of eugenics, he did so in conjunction with his argument that the degeneration of instinctive behavior patterns in domesticated animals paralleled a breakdown in the instinctive behavior patterns of humans in civilized society. The Austrian naturalist trumpeted this claim in a series of papers published from 1938 to 1943. He allowed that the conditions of civilization, including the practices of modern medicine, were the occasion for genetic decay because they allowed the reproduction of types that would otherwise have been weeded out by selection. Moreover, he imagined that the conditions of big-city life might themselves be mutagenic. In conjunction with these warnings, he sounded themes that had been expressed earlier in the century and had since come to figure prominently in Nazi ideology. These included the genetic and the moral superiority of the peasant over the city dweller, the exaltation of instinct over reason, the insignificance of the individual as contrasted with the all-importance of the *Volk,* and the necessity of maintaining racial purity.[57]

But not all of Lorenz's papers in this period were freighted with biopolitical commentary. Although he was confident that his biological view of human psychology and social life fitted well with the worldview of National Socialism, he was not undiscriminating with respect to when and where he set forth his biopolitical claims. His writings on the comparative study of behavior (1939), pair formation in ravens (1940), Kant's view of the a priori in the light of modern biology (1941), comparative studies of motor patterns in Anatinae (1941), and inductive and teleological psychology (1942) said nothing about race or politics.

Lorenz clearly recognized that different moments called for different sorts of performances. We find, for example, that he was slated to speak three different times at the DGT annual meeting that was supposed to take place in Leipzig in September 1939. He was signed up to give a paper of a pure science sort on the topic "the so-called inborn schema." He was also listed as a cospeaker, with the society's president, J. Effertz, on the topic "examples of applied animal psychology and their utilization in animal raising and keeping." But beyond this he had the special, public lecture entitled "Rise and Fall in Man and Animal," cosponsored by a National Socialist organization.[58]

As indicated in the previous chapter, the Leipzig meeting was canceled because of the outbreak of war. But Lorenz had had opportunities previously, and he would have opportunities again, to promote his ideas in "politically visible" settings as well as more strictly scientific ones. In the former category were the German Club in Vienna (where he lectured at the invitation of the rector of the University of Vienna in March 1939) and the Physical-Economic Society in Königsberg (where he lectured both in September 1938 and October 1940).[59]

Lorenz published in 1940 two papers that are particularly noteworthy as examples of his efforts to highlight the ideological value of his research. One of these, entitled "Domestication-Caused Disruptions of Species-Specific Behavior," was an eighty-page, extended version of the address he gave in Bayreuth in 1938. It appeared in the *Journal for Applied Psychology and Character Study (Zeitschrift für angewandte Psychologie und Charakterkunde)*. The second paper, "Systematics and Evolutionary Theory in Teaching," appeared in the *Biologist (Der Biologe)*, the Reich's official journal for biology teachers. Here Lorenz argued that evolutionary theory and Nazi race hygiene concerns were mutually compatible.

The first of these papers combined a detailed account of Lorenz's studies of bird behavior with his argument that domestication-induced behavioral deficiencies in birds had their counterparts in the behavior of humans under the conditions of civilization. His correspondence with Heinroth provides insights on the paper's development. Working "full steam ahead" on the paper late in 1938 and early in 1939, Lorenz asked Heinroth to get him photographs of wild and domestic ducks, chickens, and pigeons. Lorenz wanted to use these to illustrate the bodily signs of degeneration that routinely appear in domesticated animals.[60] Heinroth obligingly provided a collection of photos. Some of these he had taken himself, others he had secured from a friend, and still others came from the archive of the Berlin zoo. Lorenz planned to present pairs of pictures contrasting the wild forms with their domestic counterparts. The wild form would go on the left; the domestic form would go on the right. This, he was confident, would convincingly illustrate that humans perceived wild forms as beautiful and domestic forms as ugly. He imagined he could get from Antonius a photograph of a wolf, which he could then put side to side with a photograph of a pug. "Or should I take the bulldog?" he asked. "Stop! I'll take the Pekinese!"[61]

Heinroth had reservations about Lorenz's claim that humans are instinctively inclined to regard as ugly the changes that animals undergo under domestication. In January 1939 he cautioned his younger colleague that the claim that "'our' aesthetic feelings of beauty regard appearances of domestication as ugly" was not always correct. With the word "our," Heinroth explained, "one must strictly distinguish between Lorenz, Heinroth, etc., the enthusiasts, and the feelings of the people." When he showed the anatid collection at the Berlin zoo to lay people, Heinroth explained, they tended to be very impressed by the domestic form of the Chinese goose. They were not interested in the wild type. "In this case," Heinroth observed, "it is the domestic animal that is felt to be proud and noble."[62]

Lorenz was not dissuaded. What he intended to portray, he replied, were the kinds of mutations that were observable both in domestic animals and in people in big cities: a shortening of the extremities, a loss of muscle tone, a reduction of the base of the skull. When painters or sculptors wanted to portray something ugly or base, Lorenz insisted, they used these characters. When they wanted to portray an ideal type, they used the wild form. Lorenz drew a sketch for Heinroth contrasting wild forms and domestic ones. Geese, ducks, dogs, fish, boars, and humans all degenerated *and became to his mind uglier* as they became increasingly "domesticated." Bursting with self-confidence about the validity of his project, but aware that Heinroth had not committed himself on the subject, Lorenz asked the older ornithologist directly: "What do you think about the attempt to discuss our inborn feelings of beauty in terms of inborn schemata?"[63]

If Heinroth ever answered Lorenz's question, he did not do so in a form that has been preserved. Nor do we have any indication that he responded to another of Lorenz's questions: are mutations more frequent under the conditions of captivity than in the wild, or is it simply the cessation of selection that produces the greater variability one sees in domesticated animals? For his own part, Lorenz thought that something more than the cessation of selection was probably at work. He believed that the conditions of domestication might help induce mutations. He was debating this at the time with Otto Koehler, and he wanted to know where Heinroth stood on the matter. Whether Heinroth declared his opinion on this or

5.1 A sketch by Konrad Lorenz of the degenerative effects of domestication in geese, ducks, dogs, fish, pigs, and humans. (From a letter of 18 January 1939 to Oskar Heinroth. Courtesy of Staatsbibliothek zu Berlin—Preußischer Kulturbesitz, Nachlass 137 [Oskar Heinroth], Ordner 27: Lorenz, Konrad.)

not, Lorenz went ahead in his paper of 1940 to suggest that the conditions of domestication—and the conditions under which humans live in large cities—are in fact mutagenic.

Lorenz associated three different types of disruptions in instinctive behavior with domestication. First were quantitative variations in the endogenous production of action-specific energies. Second was a widening of the range of stimuli (and the simultaneous simplifying of the concatenation of stimuli) capable of triggering an innate releasing mechanism. Third was the falling apart of behavior patterns that in the wild-type organism fit together into a functional, unified whole.

Lorenz discussed all of these in relation to observations he had made on the behavior patterns of pure-blooded wild forms, their domestic counterparts, and hybrid mixes of the two. The domestic and half-breed geese were ready to breed sooner than wild greylags. The former bred in their very first year. The wild form seldom bred even in its second year, and it did not reach maturity until the year after that. The domestic goose and their hybrid crosses were also considerably less discriminating in their mating behavior than were the pure-blooded greylags. Domestic and crossbred females readily allowed themselves to be trod by their brothers, but wild greylags showed an aversion to brother-sister matings. Similarly, in the domestic goose and hybrid forms, copulation and pair-bonding behavior regularly became disassociated from each other. This was not the case in the wild form. To Lorenz the wild greylag goose was in fact a model of virtue, a creature that was by nature monogamous. Indiscriminate, unrestrained, promiscuous, breed-polluting sexual behavior he associated with domestication. He made crosses and backcrosses involving wild and domestic ducks, eventually producing birds that derived only $\frac{1}{64}$ of their blood from a domestic ancestor and all the rest from wild ancestors. He claimed he could safely say that a drake with only $\frac{1}{32}$ of its blood from a domestic ancestor nonetheless engaged in more rape chases of females, and with greater intensity, than did the bird's full-blooded, wild-type counterparts. Thus, at the same time that Lorenz stressed the degenerative features of domestication, he also warned that it took only a small amount of tainted blood to have an influence on a pure-blooded race.[64]

Lorenz happily situated his efforts in a line of thought initiated by Heinroth. Maintaining that the parallels between the social behavior of greylag geese and humans provided an "extremely valuable and inexhaustibly interesting object for social research," he quoted with pleasure Heinroth's observation of 1911 that "the study of the ethology of the higher animals—unfortunately still a very uncultivated field—will bring us more and more to the realization that in our relations with families and friends, in courtship and the like, it is more a matter of purely inborn, more primitive processes than we ordinarily believe."[65] Lorenz proceeded, however, to take Heinroth's general line of reasoning in a specific direction Heinroth had never promoted.

Lorenz's basic argument went as follows. We have inborn feelings, he said, for what is beautiful and what is ugly in members of our own species.

We likewise have inborn feelings for whether someone is ethical or not. Unfortunately, the relaxation of selection in modern times has permitted the breeding of "defective types." Like cancerous growths in the body politic, these need to be recognized and surgically eradicated as soon as possible. How was one to recognize them? Lorenz recommended "our own inborn schemata," that is, "our emotional reactions to degeneration phenomena." Paraphrasing Goethe, Lorenz wrote: "A good man, in his dark striving, knows full well whether or not another person is a scoundrel." In this case, Lorenz allowed, one needed to rely on "the unanalyzed, [deeply] rooted reactions of our best people." Scientists still needed to do causal-analytical research on human inborn schemata, but that could not replace the guidance of those inborn schemata themselves.[66]

Lorenz offered thirty-five photographs or drawings to illustrate his points. Roughly half of these were made up of pairs of photographs he selected to demonstrate the differences between wild and domesticated animals. Among the other pictures were five photographs of human statues or busts from ancient Greece. He included these to illustrate how the Greeks had depicted the ideal of beauty on the one hand and the ugly "signs of domestication" on the other.[67] Significantly, Lorenz believed that certain important inborn schemata in humans were race specific. This, he explained, was why the racial features of an Asian were so difficult for a person from another race, like himself, to decipher.[68]

Lorenz gladly pointed out that his conclusions were fully consistent with the ideals of the Nordic movement. As Jaensch had done the previous year, he observed that the Nordic movement had long been concerned with the deleterious effects of domestication: "The Nordic movement has from time immemorial been emotionally directed against the 'domestication' of the human being; all its ideals are among those that would come to be destroyed by the biological consequences of civilization and domestication described here; it struggles for an evolutionary direction that is directly opposite to that in which civilized, big-city humanity is moving."[69] Lorenz concluded his monograph with a rousing call for "the preservation and care of our people of the highest hereditary goodness [Erbgüter]."[70]

Lorenz likewise signaled the compatibility between his ideas and those of the Third Reich in 1940 in an article he wrote for the Nazi biology teachers' journal, Der Biologe. Walter Greite, Lorenz's former contact at the DFG, had become editor of this journal, and he had urged Lorenz to write something for it. Meanwhile, Oskar Heinroth called Lorenz's attention to an article in the journal by a German pedagogue, Ferdinand Rossner, who had just written a scathing attack on some Nazi critics of evolution. Lorenz did not know Rossner, but he was inspired by what he read, and he decided to follow Rossner's lead. He wrote to Heinroth: The article mentioned in *Der Biologe* is outstanding; Roszner [*sic*] must be a good man. It was entirely new to me, that there are opponents of evolution in the Third Reich! Why then racial care [*Rassenpflege*]? To me it is totally incomprehensible, how one can be a Nazi and an opponent of evolution at the same time. I have from this article launched so to speak a war on two fronts, in which I, possessed by holy fury, have written for Greite (who had just now jarred me about a long-promised contribution) an essay: 'Systematics and Evolutionary Theory in Teaching.' It pleases me that you too find Roszner especially good."[71]

The journal to which Lorenz sent his article was in its tenth year of existence. It had founded in 1931 by Ernst Lehmann, then director of the Botanical Institute at the University of Tübingen. Lehmann was an ardent anti-Semite and a propagandist for race purity. Advertising his monthly journal as an organ "to protect the interests of the German biologist," he called for the foundation of an Association of German Biologists (Deutscher Biologen-Verband, henceforth DBV). As he saw it, the promotion of the interests of German biologists and the biological education of the German *Volk* were part and parcel of the same project. He believed that biologists and National Socialists were natural partners. Once the Nazis came to power in 1933, he applied for party membership and set about having the DBV incorporated into the National Socialist Teachers' League (NSLB), an organization founded in 1929 by Hans Schemm, the ideologue who coined the phrase, "National socialism is politically applied biology."[72]

Lehmann never gained party membership for himself. Nor did he succeed in making himself the spokesman for a distinctively German biology. He was successful, however, in linking *Der Biologe* with the NSLB. By 1934 the journal was advertising itself as not only the "monthly journal to protect the interests of German biology and German biologists" but also the official organ of both the DBV and the biology division of the NSLB. The following year, *Der Biologe* became the outlet for NSLB publications on "biology and school." It began reporting on teacher training camps and other means of developing education about the life sciences as the basis of the Nazi worldview.

The official ties between biology and the Reich were tightened further in 1939 when the DBV was transformed into the Reichsbund für Biologie and made part of the research and teaching division of the SS called Das Ahnenerbe (Ancestral Heritage) under the control of Heinrich Himmler. Membership in the Reichsbund became mandatory for all biologists. SS member Walter Greite was appointed director of the organization. Specialized work groups within this organization included "Biology and Medicine," "Biology and Law," and "Biology and School." Ferdinand Rossner, professor at the college for women teachers in Hannover, was put in charge of "Biology and School."

Rossner wrote numerous articles for *Der Biologe*. The piece that caught Lorenz's attention was entitled "Systematics and Evolutionary Theory in Teaching." In it, Rossner portrayed "National Socialist blood and race teachings" as a "Copernican" revolution. By his account, Copernicus had destroyed the old geocentric worldview, evolutionary theory had destroyed the old homocentric worldview, and National Socialist hereditary and racial theory was destroying the egocentric worldview.[73] Unfortunately, allowed Rossner, there were those who refused to acknowledge the laws of life that modern scientific research had revealed. Not daring to fight directly against modern ideas of race, they launched their attacks instead against evolutionary theory. Among these attackers were such well-known educators as Heinrich Scharrelmann, who claimed that the notion of evolution had spawned socialism and communism, and Ernst Krieck, who railed against the "myths" of evolution. Krieck was the rector of the University of Heidelberg, but Rossner did not hesitate to challenge him. He recited several of Krieck's claims as illustrations of why no one in the present day should allow a man as ignorant and prejudiced as Krieck to claim any leadership in scientific matters. "The belief of educators in Krieck's authority," Rossner wrote, "is meaningless at this point. Biologists can only laugh at the biological revelations of Ernst Krieck!"[74]

Rossner was not the only writer to ridicule Krieck's views in the pages of *Der Biologe* in this period. So too did Gerhard Heberer, associate professor of general biology and human evolution at Jena. Heberer delivered his barbs at Krieck in an article entitled "Present-Day Views of the Family Tree of Animals and E. Haeckel's 'Systematic Phylogeny.'" Krieck had recently attacked Haeckel and Haeckel's evolutionism, dismissively asking for "a clear and precise answer" to the question "What is Haeckel's original creative contribution that still holds up today?" Heberer's response was quick and to the point: "Haeckel is the great classic of phylogenetics, and will be so in the future as well." For anyone knowledgeable about biology, Herberer indicated, Krieck's assault on Haeckel looked "strange if not funny."[75]

Significantly, neither Heberer nor Rossner seems to have imagined that by criticizing Krieck he was challenging Nazi ideology. Heberer himself was a member of the party and an officer in the SS. He was ready to criticize Catholics and National Socialists alike who, by denying human evolution, were undermining, in his view, the very possibility of a sound racial science. His own work demonstrates that a researcher of human evolution and racial origins could enthusiastically endorse ideas of Nordic supremacy and race purity.[76]

The cases of Rossner and Heberer, together with the way *Der Biologe* prided itself as an organ that promoted Nazi ideology and first-rate biology simultaneously, help put into context Lorenz's own critique of Ernst Krieck and other antievolutionists. They show that Lorenz's article was neither as independent nor as non-Nazi as he later sought to represent it.

Lorenz began his article by indicating how surprised he had been to learn, upon reading *Der Biologe*, "that in the system of education of greater, National Socialist Germany, there are men who still continue to reject evolutionary thought and the theory of descent as such."[77] He proceeded to name Krieck and Scharrelmann as two particularly benighted educational authorities. Their scientific beliefs, he said, were essentially worthless. He noted, however, that the "fundamental spiritual stance" (*seelische Grundhaltung*) of the two men was in general identical to his own nature-based, scientific worldview. In clinging firmly to the ideals they had chosen in their youth—even if these ideals were "mere illusions"—they were exhibiting a characteristic found in "the most genotypically respectable people." It is difficult to tell whether Lorenz was being bitingly satirical here or softening his overall critique by endowing his targets with redeeming features. Whatever the case, Lorenz did not treat Krieck and Scharrelmann as sarcastically as Rossner had done. What is more, he quickly shifted his attack in other directions, most notably toward the Catholic Church, the doctrines of which, in his view, constituted an obstruction to the natural biological impulse toward the betterment of "*Volk* and race."[78]

Interestingly enough, Lorenz's response to the antievolutionary stances of Krieck and Scharrelmann was not to rehearse the best evidence for evolution. Instead it was to argue for the essential compatibility of evolutionary theory and National Socialist ideology. The aims of National Socialism, he insisted, were better served by the idea of the continued evolution of the German *Volk* than by the "race-political fatalism" of the belief in a perfect and unchanging race. Countering the charge that Darwinists were value-blind materialists, he claimed that providing young German men with a knowledge of phylogenetic processes would imbue them with a commitment to higher values and a recognition of their "own duty regarding the higher racial development of our *Volk*."

Lorenz insisted that Darwinism provided support for National Socialism, not for socialism or communism. It did so, he said, precisely because it was the *race*, rather than *all of humanity*, which constituted the essential, homogeneous, biological unit. He recounted how in Austria, under the previous regime, opponents of evolutionary theory had treated evolutionary theory and Nazi ideology as if they were equivalent to each other. The Austrian ministry of education cut biological instruction to a minimum and would not permit evolution to be mentioned. "Under these circumstances," Lorenz explained, "it is understandable that even inherently apolitical scientists became stirred up and were led to draw the correct political conclusions."[79] He told how he once had a student who was initially "a complete Marxist" but who became convinced of "the untenability of the dream of 'the equality of all mankind'" when he heard Lorenz deliver "a wholly apolitical lecture on comparative phylogeny." Teaching the facts of natural history, Lorenz insisted, was an excellent way to win students over to the ideals of National Socialism. As he put it, "Certainly socialism and communism spring from a half-digested Darwinism which incorrectly regards the whole of humanity as an equally valuable homogeneous unit. But the self-evident correction, that not the whole of humanity, but only the race is such a biological unit, now makes National Socialism out of socialism! To discover that this correction can succeed in the course of teaching has been, as I have already mentioned, one of the greatest joys of my existence."[80]

As Lorenz represented it, the German people were in a position to choose their own fate. "Whether we share the fate of the dinosaurs," he wrote, "or whether we raise ourselves up to a higher level of development, undreamed of and perhaps inconceivable with the present organization of our brains, is exclusively a question of the biological penetrating power [*Durchschlagskraft*] and life will of our *Volk*." He was pleased to report that "just now, in this race to be or not to be, we Germans are a thousand paces ahead of all other cultured peoples [*Kulturvölkern*]." He warned, however, of imminent dangers, including "degeneration, through the racial and moral decay caused by big-city life, declining birth rate, carcinoma and world capitalism and countless other forces hostile to the *Volk*."

After the war, in 1950, Lorenz's candidacy for the professorship of zoology at the University of Graz was scuttled when opponents brought to the attention of the minister of education Lorenz's 1940 *Der Biologe* article. At the point, Lorenz decided that his immediate chances for any decent academic position in Austria were slim and he ought to look for employment abroad. To this end, he wrote to his fellow biologists in England, W. H. Thorpe and Julian Huxley, offering each of them a gloss on his *Der Biologe* article. As he explained the matter to Thorpe,

In 1940 Ernst Krieck, a very important Nazi-bonze, President of the N. S. Lehrerbund and Rektor of Heidelberg University, and a damn fool into the bargain started a serious compagne to eradicate evolution out of National socialistic teaching, at schools and universities. He wrote a series of articles in a Nazi journal, "Der Biologe." Quite naively, without knowing the importance of this fellow, I wrote a countering article and sent it to the same journal, whose editor, for reasons of his own, published it immediately. This article brought me some very real danger, because, as is my nature, I had ridiculed my antagonist rather pitilessly, a thing which was not usually done to Nazi bonzes. But of course it was written in a generally nazi-like manner and made a point that all eugenic tendencies on which the Nazis laid so much stress, [were] quite senseless, if one assumed the species to be absolutely constant, as originally created by God. So that, if one quotes unconnected passages of this short article, it is actually possible to make me appear as a very dangerous Nazi, although of course, the real trend of the thing, is quite the opposite, (though I cannot pretend it was a plucky thing to write, as I did not know who Krieck was when I wrote it.) . . . If you read that paper, you would actually find that I never had been a Nazi at all, only, fighting for the theory of evolution (which seriously comprises for me much of the creator and creation), I should be ready to talk in Communist, Catholic, or whatever other terminology you want.[81]

Lorenz told essentially the same story to Julian Huxley, saying of his criticism of Krieck:

This really was rather dangerous and for a time it even looked as if it might jeopardize my getting the chair in Königsberg. But of course the whole confounded thing is written in Nazi terminology (I should write in the terms of any old terminology in order to make the value of the theory of evolution understood!) and if one quotes disconnected passages out of this article one can make me a damn Nazi indeed. Also I do not deny that I was, at the time, rather taken with the idea of making Eugenics a sort of state religion—I really did not suspect then that it was only an excuse to kill off Jews and other "racially inferior" peoples.[82]

In writing of the difficulty of analyzing the language of psychology treatises written during the Third Reich, the historian Ulfried Geuter has observed that in some cases "what seem to modern readers to be clearly racist expressions turn out on closer inspection to be artful criticism of

Nazi views."[83] In Lorenz's case, it appears, there was artfulness in his *Der Biologe* article of 1940, but there was also artfulness in his reconstruction of the story for Thorpe and Huxley. As we have seen, his criticism of Krieck and Scharrelmann was not as independent, or daring, or non-Nazi an act as he represented it to his English colleagues. Rossner, Herberer, and Walter Greite, the journal's editor, all promoted Nazi ideology and biology at one and the same time. Rossner respectfully cited Alfred Rosenberg as an authority on obeying "the laws of life." Greite praised the neurologist and psychiatrist Ernst Rüdin on the occasion of Rüdin's sixty-fifth birthday, noting not only Rüdin's special contributions to Germany's sterilization laws but also how the Führer himself had recognized Rüdin's lifework by awarding Rüdin the Goethe Medal.[84] Lorenz, in letters to Stresemann and Heinroth, characterized Rossner and Greite as good men.[85] His own writings never reached the high pitch of Nationalist Socialist zeal displayed in their writings, but he felt enough common interest with them to ally himself with them in the pages of *Der Biologe*.

Was it possible to be a good biologist under the Third Reich? The answer of the editors of *Der Biologe,* of course, was an emphatic yes. They simultaneously insisted on the necessity of a biologized politics and the importance of getting their biology right. "Biology is no playground for Ignoramuses," they thundered in 1934, "and especially not today, when biology is part of the foundation of the National Socialist Weltanschauung."[86] They fully expected German biology to continue to be the best biology in the world. They made it a point to distinguish between good biology and bad biology, setting aside a special section of the journal to call their readers' attention to examples of biological misinformation or downright nonsense that appeared in the German press. As it was, however, the ideology of race purity effectively trumped any evidence from genetics or evolutionary biology that might have served to undermine it.

Like the other authors just mentioned, Lorenz was happy to imagine that promoting his own science and serving the state were congruent enterprises. In 1938 we saw him give Greite his opinion on the strengths and weaknesses of the work of German animal psychologists. In 1939 he was pleased to send Greite his paper for *Der Biologe*. In January of 1940, a new occasion arose for him to make contact with the Nazi bureaucrat. In this case Lorenz wrote Greite asking him to squelch some bad ideas and a new "scientific" society that were masquerading as scientifically legitimate.

Heinroth called Lorenz's attention to the problem in December 1939. It seems that a certain Georg Schwidetzky had undertaken to write about language in animals (especially primates) with an eye to identifying hereditary elements in animal vocalizations that were homologous with hereditary elements in human speech. To this end, he created in Leipzig in the early 1930s a society of sorts that he called the German Society for the Study of Animal and Primeval Languages (Deutsche Gesellschaft für Tier- und Ursprachenforschung). Beginning with a slender volume of 1931 entitled *Do You Speak Chimpanzee? (Sprechen Sie Schimpansisch?)*, and continuing at least as late as 1938, Schwidetzky published a series of pamphlets related to the general theme of "race and language." He associated the vocalizations of different primates with what he assumed were the early vocalizations of different human races.[87]

Lorenz was not against the idea of researching homologies in apes and humans. He had no time, however, for Schwidetzky's particular observations and ideas. He dismissed as fraud Schwidetzky's suggestion that a Neanderthal's speech would have affinities with a chimpanzee's, an Ur-European's with a gibbon's, and a Malay's with an orangutan's.[88]

Lorenz was not especially worried about Schwidetzky's ideas when Heinroth first told him about them. Heinroth, on the other hand, remained concerned. Late in January 1940 he wrote Lorenz, saying he wanted to bother Lorenz once more about Schwidetzky's "unfortunate works." It seems that a journalist from one of the Berlin evening papers had taken an interest in Schwidetzky's claims. Heinroth wanted Lorenz's full opinion on the subject. Heinroth had already discussed the subject with Heinz Heck, the director of the Munich zoo (and son of Ludwig Heck, the director of the Berlin zoo). Heinroth told Lorenz: "In our eyes Schwidetzky is simply an uninformed madman, who understands nothing about animals. . . . I think such people are dangerous and one should not even bring their opinions to the people."[89]

Lorenz wrote back to Heinroth right away, explaining the action he had taken:

> Regarding the poor madman: just as a sex killer is, so is a poor madman, who personally can do nothing about his deficit mutations [*Ausfallsmutationen*]. Since he however is enormously harmful for the people as a whole, one slaughters him justly! I have sent the whole Schwidetzky package to Greite, who is just now very busy with the authoritarian suppression of trash in biological literature. The manner and way Schwidetzky derives *us* from hybrids out of hybrids (what an idea he has of the fertility of crossbreeds! Try once to cross the baboon and the gibbon!) is nothing short of propaganda for crossing human races, racial shame [*Rassenschande*] on the large scale. It will be forbidden immediately, as

well as the whole Society for Primitive Language Research. The *Ahne-nerbe* will make short shrift of it. Fritsche's "Animal Mind and the Mystery of Creation" shall also be forbidden. Finally good sense! Greite is a good man.[90]

When the Viennese biologist Otto Koenig published the Lorenz-Heinroth correspondence in 1988, he left out the above passage. He also omitted a sentence where Lorenz asked Heinroth what Heinroth thought of Lorenz's *Der Biologe* article. When Heinroth replied to Lorenz two months later, he was tactfully vague about this question. He had found it a treat, he said, to read articles by Lorenz and Otto Koehler together in the same issue, but of Lorenz's article itself he simply said "we must discuss it again personally." Later in the same letter he noted: "The Schwidetzky thing will be forbidden on Greite's orders, the "society" will thus be disbanded."[91]

Lorenz in the meantime had written to Stresemann telling him about the Schwidetzky affair. He mentioned he had sent Schwidetzky's work to Greite so that Greite could ban it. "We are now living in an alarming deluge of trashy animal psychology literature," Lorenz observed, "which through National Socialistic camouflage threatens to become really dangerous." He identified Bernhard Hecke's *Animal Mind* (*Tierseele*) and Herbert Fritsche's *Animal Mind and Secret of Creation* (*Tierseele und Schöpfungsgeheimnis*) as two further examples. "Fortunately," Lorenz wrote, "there is the good Greite, who descends upon these fellows like a wild uhlan [lancer], and with the help of good people, e.g., Koehler's sharp words [*Gösche*], wages a successful war of extermination against these knights."[92]

The issue here was who could speak with authority about animal behavior in the Third Reich. Lorenz was consolidating his claims as the leading authority in the field. At the same time, however, his closest friends and scientific advisers were not comfortable with his Nazi sermonizing in *Der Biologe,* and some of them had doubts about his ideas on domestication as well. They were not certain that his pro-Nazi gestures were necessary to advance his career. They worried to the contrary that he was in danger of carrying his politicking to the point of hurting his reputation as a scientist.[93]

APPOINTMENT AT KÖNIGSBERG

Lorenz was called to a chair in psychology at the University of Königsberg in 1940, officially assuming the position in February 1941. An Institute for Comparative Psychology was also established for him. Explanations differ as to why he received the appointment. He himself claimed that he was appointed because of his interest in Kantian philosophy, an interest which was particularly appropriate because the chair in question was one of two chairs descended directly from the professorship originally held by Immanuel Kant himself. He was also aided, as he liked to tell it, by fortuitous circumstances. His friend Erich von Holst was playing the viola da braccio in a string quartet with Eduard Baumgarten, a pragmatist who had just been called to the first chair of philosophy at Königsberg. Baumgarten asked von Holst whether he knew of any psychologist with evolutionary training who might also have an interest in Kant's understanding of the a priori. Von Holst identified Lorenz as just such a person. By Lorenz's account, "Von Holst and Baumgarten approached the zoologist Otto Koehler and the botanist Kurt Mothes, on whose authority the philosophical faculty of Königsberg invited me to the chair of psychology."[94] Historian of psychology Ulfried Geuter offers a different view entirely. He cites the psychologist Hans Thomae's claim that Lorenz's appointment at Königsberg took place to the "astonishment of the psychological world" and occurred only because of "the intervention of Minister Rust against the resistance of the faculty."[95]

Was Lorenz "invited" by the philosophical faculty, or was he appointed against faculty resistance? And if there was resistance to him (not universal resistance, certainly, since Baumgarten and Koehler favored him), was that because of how some of the faculty perceived his politics, or for other reasons, as, for example, an aversion to the idea of appointing a student of animal behavior to a chair of psychology? Conclusive answers to these questions may no longer be recoverable. It is instructive, however, that when the historian Theodora Kalikow asked Baumgarten about Lorenz's appointment at Königsberg, he convinced her that the circumstances of Lorenz's appointment "were so unusual that Lorenz's political leanings were relatively unimportant." But this is not to say that this was an either-or situation. Lorenz had good claims to being both scientifically and politically qualified.[96]

Certainly Lorenz's political leanings were no longer in doubt. Since 1937, when someone had torpedoed his application to the DFG with the suggestion that his politics and ancestry were suspect, Lorenz had worked hard to demonstrate his commitment to the Reich. But politics alone would not have been sufficient to secure him the appointment at Königsberg. Scholars who have looked closely at Third Reich appointment policies have concluded that as of about 1940, if not sooner, the Education Ministry did not award science professorships to individuals with inadequate scientific credentials. While Lorenz had signaled his willingness to put himself at the service of the Reich, he had equally distinguished himself as a pioneering animal psychologist and biologist of the first rank.

Interestingly enough, when Lorenz went to Königsberg and wrote there about Kant, he did not connect this writing to National Socialist ideology. One can only speculate whether this might have been his way of suggesting to local doubters that it was philosophical merit and not governmental influence that had earned him his post. In any case, when he constructed his paper "Kant's Doctrine of the A Priori in the Light of Contemporary Biology," he made no reference to issues of race purity or the degenerative effects of civilized life.

The task Lorenz set for himself in his Kant paper was to compare the different views of the a priori offered by transcendental idealism and modern evolutionary biology, and then to reinterpret Kantian doctrine in the light of the latter. He did not focus on the Kantian categories of space, time, and causality as such. Instead, he concentrated on the concept of the a priori itself. His essential claim was that "human reason with all its categories and forms of intuition" had, just like the human brain itself, evolved through a continuous interaction with nature and nature's laws. The a priori, in other words, was "due to hereditary differentiations of the central nervous system which have become characteristic of the species, producing hereditary dispositions to think in certain forms."[97] Ascribing to Kant the view that the a priori categories of human thought were part of a God-given, immutable system, Lorenz offered instead a strictly naturalistic interpretation of human reason.

Here was an intellectual achievement worthy of the *Bildung* ideal that had been so much a part of the self-identification and life goals of highly educated, upper-middle-class Germans and Austrians during Lorenz's formative years.[98] Building respectfully on Kant fit perfectly with an ideal that valued comprehensive knowledge over narrow expertise, intellectual development over material acquisition or political advancement, and spiritual *culture* over the superficialities of modern *civilization*. Lorenz allowed that Kant himself would not have been averse to the position he was promoting, namely, "that organic nature is not something amoral and godforsaken, but is basically 'sacred' in its creative evolutionary achievements, especially in those highest achievements, human reason, and human morals."[99]

When it came to formulating a research agenda at Königsberg, Lorenz by and large planned to continue to do what he had been doing already. He would study comparatively the behavior of ducks and geese. He would also study cichlid fish—"the little man's anatids" (die Anatiden des kleinen Mannes) as he called them to Stresemann. They constituted "a wonderful group, rich in species, easy to raise, and rich in phylogenetically informative instinctive behavior patterns." They promised to be his "daily bread" in Königsberg.[100] In his Kant paper, he explained the philosophical importance of his animal work. Through "the investigation of pre-human forms of knowledge" he expected "to gain clues to the mode of functioning and historical origin of our own knowledge, and in this manner to push ahead the critique of knowledge further than was possible without such comparisons."[101]

Significantly, Lorenz represented his projects to different audiences in different ways. To the philosophical faculty of the university he described his work as pushing forward the critique of knowledge. To the readers of *Der Biologe,* he predicted that the investigations of animal behavior at his new institute for comparative psychology would be "fruitful not only for theoretical but also for race-political concerns." He made this statement in a short piece praising Oskar Heinroth on the occasion of Heinroth's seventieth birthday. He noted that his mentor's birthday happened to coincide with the founding of his own institute for comparative psychology at Königsberg. He suggested this augured well for animal and human psychology alike.[102]

"THE COMPARATIVE BEHAVIOR OF THE ANATINAE"

Lorenz had a greater tribute to offer Heinroth, however, than a short piece in *Der Biologe.* He contributed a major monograph entitled "Comparative Studies of the Motor Patterns of Anatinae" to a Festschrift for Heinroth published in the *Journal für Ornithologie.* The monograph was a model of what comparative behavioral studies could contribute to traditional phylogenetic questions.[103]

In a work totaling nearly one hundred pages, Lorenz described in detail the characteristic behavior patterns of some twenty different species of ducks he had studied with care. Among these behavior patterns were "chin-raising," "nod-swimming," "sham-preening," "burping," "head-up-tail-up, and "body-shaking," plus such sounds as the "grunt-whistle" and the "'Krick'-whistle." From his analysis of thirty-three such behavior patterns (taken together with fifteen morphological characters), and using the principles of the comparative morphologist, Lorenz arranged the different species according to their apparent affinities. Surveying the table of evolutionary relationships this gave him, he proudly claimed: "Even this provi-

5.2 Display behavior of the Mandarin duck. (From Lorenz, "Vergleichende Bewegungsstudien an Anatiden," in "Festschrift O. Heinroth," Ergänzungsband 3, *Journal für Ornithologie* 89 [1941]: 194–293.)

sional, incomplete table shows clearly the applicability of the phylogenetic homology concept to characters of innate behaviour. This fact, the demonstration of which was a major aim of my investigation, is of the greatest significance for *comparative psychology*." [104]

Lorenz, it bears emphasizing, did not offer his monograph simply as a report of his empirical discoveries or of the phylogenetic relations he identified. He offered it also as a confirmation of the research practices that were as much a part of his science as were his organizing concepts. He formulated the special value of his approach to animal behavior study in the following terms: Those who were inclined to have a special "intuition" for systematics were people like zoo workers who were well acquainted with substantial numbers of live animals of a given group. When such people also knew the anatomy and paleontology of the animals in question, they had an advantage over museum specialists, who tended to be unacquainted with the behavioral characteristics of the animals they studied. Lorenz cited his friends and colleagues Oskar Heinroth and Otto Antonius as zoo biologists who recognized intuitively the natural affinities among the species of a given group. He observed further that zoo workers also had distinct advantages over field biologists. Establishing the necessary catalogues (ethograms) of the behavior of different species was something that could only be done by the investigator who lived with his animals "day after day, year after year." What is more, zoos allowed the biologist to compare, side by side, the behavior patterns of closely related animal species that did not live side by side in nature. Field studies were not suitable for such tasks.

Lorenz went on to explain that it was not only his methods of observation but also the particular organisms he used that made his work especially fruitful. A "special advantage" of the Anatinae was that they could be crossed with each other to produce interspecific hybrids. Many of these hybrids were fertile, and they could thus "be employed for investigation of the inheritance of species-specific behaviour patterns." Here was an arena, apparently, where genetics and phylogenetics could be profitably synthesized. In many of these interspecific hybrids, Lorenz claimed, the behavior patterns and morphological characters of the hybrids were not intermediate between the parental forms but instead were more primitive. This allowed one to draw conclusions about the earlier history of their lines. At the same time, the degree of hybrid fertility could be used as an index of the "degree of relationship between the parental species." [105]

Hybridization had been one of the key themes of Lorenz's proposals to the DFG. It had likewise been one of the major themes of his thinking about racial hygiene. In his monograph on the Anatinae, nonetheless, he did not provide a systematic discussion of his breeding results. [106] He allowed that he had observed many duck hybrids and compiled behavioral inventories for them, but he described the results of only one interspecific pairing, that between a free-flying female mallard and a Chiloë widgeon drake at the Berlin zoo. He promised to assemble his information on hybridity in another paper. He never did so.

LORENZ IN POLAND

Lorenz was drafted for military service in October 1941. He had wondered whether he would be called up as a physician or as a motorcycle messenger. When the time came, it was as a motorcyclist. His talents in this regard initially led him to be appointed as a motorcycle-riding instructor. However, when the military became aware that he had previously been a professor of psychology, he was made a military psychologist. This job, which he performed in Poznan, Poland, consisted by his account "mainly in administering routine psychological tests to aspiring officers." But it was not a job that lasted long. Army psychological testing was abolished in May

1942. Luftwaffe personnel testing had been discontinued the month before that. Some said Reichsmarschall Hermann Göring had stopped the personnel tests after he learned that the famous German fighter pilot Werner Mölders had been judged by the tests to be unfit to be a pilot. Lorenz himself repeated this claim.[107] The most convincing explanation, however, has been offered by the historian Ulfried Geuter. He argues that psychological testing of applicants for officers' positions became superfluous in 1942 when the demand for officers came to exceed the candidate pool (and when successful, prior field experience came to be seen as the primary criterion for identifying suitability).[108] Whatever the case, for a brief period beginning in May 1942, Lorenz was without an official assignment. He was then appointed to the Department of Neurology and Psychology at the reserve hospital in Poznan. He served there as a physician and psychiatrist for almost two years.

The city of Poznan (Posen, in German) was part of the Wartheland, the section of Western Poland incorporated into the Third Reich after the Germans invaded Poland in 1939. The German government destined the Wartheland as an area for German resettlement. Questions of race and race purity were central to the project.[109] The gauleiter of Poznan, Arthur Greiser, ordered that Poles and Germans be strictly separated in the "incorporated" area. The first of six principles he set forth in an official command of September 1940 stated: "Any individuals belonging to the German community who maintain relations with Poles which go beyond the needs arising from service or economic regards will be placed under protective arrest. In serious cases, especially when an individual belonging to the German community has seriously injured the German interests of the Reich by relations with Poles, he will be transferred to a concentration camp."[110] A similar declaration from Dr. Kurt Lück, a high official of the German administration for the area, insisted: "National Socialism demands a ruthless separation of the members of the German nation from those of the Polish nation, a separation which recognizes no false sentiments. For Germanism in the East the order is binding; Germanism from home and abroad must create a homogeneous and solid front. Above all the Germans from home must understand the needs of the East and serve them, if they want to be equal to the task of making this land German in every respect."[111]

A wartime volume entitled *The German New Order in Poland*, published in London by the exiled Polish Ministry of Information, described what happened in Poland (and Poznan) between September 1939 and June 1941. On the most important tram routes of Poznan, the lead cars were reserved exclusively for German use. Poles were not allowed to ride the tramway at all between 7:15 and 8:15 a.m. Numerous restaurants, hotels, and offices posted the order Entrance Forbidden to Poles, Jews and Dogs. Professors at the University of Poznan, like their colleagues in Krakow and Warsaw, were subjected to imprisonment, beatings, and deportation to concentration camps. The Poznan professors "were robbed of everything they possessed: houses, furniture, linen, clothing, money, and also of their private libraries, manuscripts and scientific works."[112]

Lorenz was posted to Poznan in 1942. It was there, in the two months or so between the time his position as a military psychologist was abolished and the time he was reassigned at the reserve hospital, that Lorenz completed his major manuscript, "The Inborn Forms of Possible Experience." One of the themes of the monograph was race purity and the way that, under the conditions of civilization, innate inhibitions against mating with individuals of different racial types break down, leading to a dangerous race mixing.

At the same time that he was completing this monograph, Lorenz also assisted to an unknown degree in studies conducted by the race psychologist Rudolph Hippius on the psychological worthiness and character of the inhabitants of Poznan. With the Wartheland already Germanized by this time, the study was apparently undertaken with an eye to dealing with the sorts of populations the Germans expected to encounter as they moved east into Russia. According to Hippius's report, Lorenz served as an examining psychologist in an honorary capacity (Ehrenamentlicher Mitarbeiter) in Hippius's "völkerpsychologische" studies of local Germans, Poles, and German-Pole "hybrids."[113]

The Germans had by this time prepared a "deutsche Volksliste" (list of German peoples) that sorted the inhabitants of the Wartheland of German origin into four categories according to their prospects of being re-Germanized. Hippius looked in particular at individuals in the more problematic of these groups (groups 3 and 4), subjecting them to a battery of psychological tests. He subjected full-blooded Poles and individuals of mixed German-Polish parentage to the same tests. He published the results of these tests on "national character" in 1943. Hippius's general conclusion was that the partners of German-Polish marriages and especially their "hybrid" descendants tended to become "detached" from the central character values of the two ancestral races. He alleged that the good qualities of both races (but most important of the German race) were lost through race mixing.

The tests in question were administered between 28 May and 25 September 1942. The first two months of the testing period correspond to the time when Lorenz was between jobs and completing his manuscript on "the inborn forms of possible experience." The extent of his involvement in Hippius's project is unclear. Did he, for example, administer tests to anyone who, based on the results of the test, was subsequently deported? Our information is limited to Hippius's statement that Lorenz participated in the project as an examining psychologist. Twice in the course of his report Hippius referred to views of Lorenz's. In the first instance he cited Lorenz's idea of inborn schemata. The second time he referred to Lorenz's ideas on the relation between sender and receiver in the expression of behavior.[114] But Hippius never identified any specific contribution Lorenz made to either the theorization or the conduct of the study. He did not mention Lorenz's pet ideas about domestication-induced genetic degeneration as these related to race hygiene. Nor did he mention that Lorenz had conducted extensive hybridization experiments with ducks, in which he paid close attention to the ways that species-specific instinctive behavior patterns were affected by crossbreeding. However Lorenz may have served the project itself, the appearance of his name in Hippius's book served Hippius as a means of adding scientific credibility to the work. Lorenz, in contrast, made no reference to Hippius's work in the manuscript he completed in July 1942, nor did he ever mention Hippius in any of his later writings.

What did Lorenz think in 1942 of Hippius's project and how it related to the political order in the Wartheland? When in the 1970s the writer Alec Nisbett asked Lorenz when it was that Lorenz first realized the evil the Nazis were up to, Lorenz said "surprisingly late." In 1943 or 1944 near Poznan, he said, he saw transports of concentration camp gypsies and then first "fully realized the complete inhumanity of the Nazis."[115] He did not mention the Nazi's treatment of the Poles or Polish-German "hybrids."

Lorenz completed his manuscript "The Inborn Forms of Possible Experience" in July 1942 and sent it off to the *Zeitschrift für Tierpsychologie*. As indicated above, one of the themes of the monograph was the importance of race purity, for animals and humans alike. Yet that was only one of many themes. Lorenz's primary aim in this 175-page manuscript was to get at the living organism's "inborn understanding" of biologically relevant situations from both a biological and an epistemological perspective. Lorenz in effect set about synthesizing most of the major themes about which he had been writing over the previous eight years. In these pages he insisted that he was not a materialist, that he fully appreciated the wondrous creativity of nature, and that as a student of comparative behavior he actually recognized better than most people the ways in which human beings distinguished themselves from the rest of the animate world. His main argument, nonetheless, was that the way humans experience the world is ultimately a function of their organ systems—organ systems which themselves are the products of a long process of organic evolution. Thus, if one wished to develop a genuine understanding of the higher human faculties, it made good sense to begin by studying the behavior of simpler animals. Elucidating how releasers and innate releasing mechanisms functioned in the social behavior of birds and fish paved the way for recognizing the deep biological roots of aesthetic and ethical judgments in humans.

Lorenz's monograph was a remarkable document. In it he ranged widely over what for him had become an almost unbounded domain. On the one hand he elaborated on specific details about the behavior of animals. On the other hand he challenged Oswald Spengler's interpretation of the decline of civilization. He explained how the social lives of birds are virtually entirely regulated by the interplay of their social releasers and their innate schemata. He likewise allowed that what moviemakers, doll makers, comic book writers and the like were doing when they exaggerated human features was appealing to the inborn schemata possessed by humans. He also maintained that most animals do not kill members of their own species not because they have compassion for their conspecifics but because they are equipped with innate inhibitions that restrain them from doing so.

A good deal of what Lorenz had to say in this monograph was things he had said in previous works (and many of which he would also recycle again after the war). Among his more persistent claims was the idea that humans instinctively identify as ugly those very features that characterize domestic animals and "overcivilized" people: flabby muscles, bowlegs, pug face, protruding stomach, and so on. Once again he warned at length about the diverse ways that domestication disrupted the normal functioning of instincts. And here he stressed once more that the domestication-induced *widening* of inborn schemata, producing an organism that was less selective in its choices of mates, was particularly dangerous to the health of a race.

In a section of the monograph entitled "The Value of Race Purity" ("Der Wert der Reinrassigkeit"), he allowed that race mixing produced the same sort of deleterious effects as domestication. He cited the case of a female duck that was the product of a cross between a Bahama pintail (*Anas bahamensis*) and an *Anas castanea*. Her courting movements were comparable to those of the females of her two parent species, but she displayed "the most extreme widening" when it came to her innate releasing schemata. Rather than court a brother, or males of either of her parent species, she courted other species of ducks entirely, "above all a certain female domestic duck and a white mallard drake." Lorenz found it particularly grotesque how she "doggedly forced herself on the more-than-three-times-larger object of her love."[116]

As in his previous writings, Lorenz insisted that an awareness of domestication-induced deficits in instincts was of the greatest importance when it came to race hygiene. Spengler, he said, had attributed the decline of civilizations to "the logic of the time." Lorenz in contrast attributed the decline of civilization to the domestication-induced genetic changes in the race. Here, following both Schopenhauer and Eugen Fischer, he underscored that domestication-induced changes were profoundly visible in white, northern Europeans. He quoted Schopenhauer as saying that the racial characteristics of whites had nothing "natural" about them, that "blonde hair and blue eyes already constitute a variety, almost an anomaly, analogous to the white mouse or at least to white horses." He quoted Fischer in turn as saying with respect to the blue eyes of the Nordic race: "There is no single free-living animal that has a pigmentation distribution of the eyes like a European," but the identical pigmentation is to be found "in nearly all domestic animals, individuals, or breeds."[117]

In this part of his monograph, Lorenz for the most part repeated what he had said in his earlier writings about the perils of domestication. Later in his monograph, however, he introduced a new dimension into the discussion. He continued to represent domestication as profoundly dangerous for the health of a people, but he added that it also constituted a "precondition of humanization [*Menschwerdung*]." Citing Whitman's identification of domestication-induced plasticity in instinctive behavior as "the open door through which the great educator, experience, comes in and works every wonder of intelligence," he portrayed domestication as a necessary prerequisite for the development of human intelligent behavior.

Then Lorenz went even farther. He credited domestication with being responsible for the "whole complex of characteristics" that L. Bolk had described as the "foetalization" of humankind.[118] This, essentially, was the zoologist's idea of "neoteny" as applied to man. Just as the axolotl had apparently *lost* the more adult stages of development of its lung-breathing ancestors to become a gill-breathing creature as an adult, so too it seemed that a "developmental inhibition" had produced the persistent "juvenescence" (*Verjugendlichung*) of humankind. Lorenz simultaneously portrayed neoteny in humans as "a genuine symptom of domestication" and as "the immediate cause of the characteristic intellectual attributes of man."[119] In short, man's primary mental attributes—his curiosity, his openness to his surroundings, and his potential for continuous, lifelong learning (at least up until the onset of senility)—were all made possible by his "domestication." As if these benefits were not enough, domestication had also, by Lorenz's account, gave rise to the variation among individuals without which the division of labor so essential to society would not have been possible.[120]

Here then was a much more ambiguous view of domestication than Lorenz had set forth in his earlier writings. Domestication was no longer to be viewed exclusively as an evil that needed to be attacked "wholesale with race-hygienic measures." It was also to be viewed as the precondition of the genesis of humankind. Man's nature was thus in its very essence fraught with danger, paradox, and possibility. Lorenz cited approvingly Arnold Gehlen's assertion that "man by nature is a cultural being," together with Gehlen's observation that man is a "jeopardized creature" with "a constitutional prospect of meeting with accidents."[121]

As Lorenz drew near to the end of his monograph, he emphasized not so much the dangers of racial decay as the precariousness, yet still the possibilities, of the human condition. The freedom of humankind, he said, owed itself to a process, domestication, that bordered hard on the pathological. "Only a domesticated being could become man," but this meant that it was impossible for him to have a hereditary constitution and inborn drives and inhibitions that were as harmoniously balanced as were those of a free-living animal. However, he was also a being who was able to foresee the consequences of his actions. If his freedom of thought gave him the possibility of going in the wrong direction, it also gave him the possibility of finding the right way. Few of the ways that lay open to him were not "disastrous cul-de-sacs," but it remained conceivable that with his sense of responsibility, based upon his "structured, categorical thinking," he might nonetheless skirt the many dangers that surrounded him and find his way to reach a height that was as yet only barely imaginable.[122]

At the very end of his monograph Lorenz posed the question "To whom is man answerable, if not to the society of all living men?"[123] Here he did not take the route he had taken in 1940 when he allowed that the true biological unit was the race and not the species. Neither, however, did he let his question stand as a rhetorical one, with its answer implicit in its initial phrasing. He proceeded to link what humans instinctively understood as "good" with the general direction of organic evolution, that is, from the simple to the complex. In his words, "Man thinks and values nothing other than what all creation has forever done: to create [produce] what is new and better."[124] Man was not responsible for humanity as it existed at the present, he said, so much as for "the generation coming after us, in which slumbers the possibility of undreamed-of and unlimited creative higher development." [125] In his ultimate summing up of this conclusion he wrote: "The value judgment of man, which feels the organic creation process as such to be the final, axiomatic value, springs from a structure inherent in all organic life, from radiolarians up to human artists and philosophers combined, and is thus in the true sense necessarily thinkable a priori, and no doubt not only for mankind, but for all living organisms generally conceivable in the conventional sense. This knowledge permits us a clear answer to the question, for which value and to whom is man as a responsible being accountable for his actions: not humanity and its fortune as a static system." Man is responsible, Lorenz concluded, not to humanity in the form of a static system, but rather to "the possibility of further, creative organic becoming." [126]

Lorenz's final wartime monograph thus ended on a note of high philosophizing that might or might not have been seen as compatible with his earlier statements (in this same monograph as well as in earlier writings) about the dangers of domestication and race mixing. The final tone of this monograph appears qualitatively different from that of his 1940 paper "Domestication-Caused Disruptions of Species-Specific Behavior," where he ended by praising the Nordic movement and stressing the importance of combating even "the smallest decline" in the "goodness of the blood." [127] But whether this difference in tone in 1943 represented a change in valence with respect to his earlier pronouncements or merely a philosophical elevation above the political fray is difficult to say. Lorenz often recycled material in his writings, and most of what he said in 1943 about the negative features of domestication-induced disruptions in instincts was in fact repeated from his paper of 1940.

The elements in Lorenz's 1943 monograph may well have been diverse enough alternately to worry and to reassure those who looked to this monograph for a clear view of Lorenz's political position. But then, too, not everyone at the time would have looked to the monograph primarily for political reasons. Whatever Erwin Stresemann, Lorenz's longtime adviser, might have thought about individual points in the monograph, he seems to have congratulated Lorenz on the overall intellectual achievement, cheerfully allowing that with this work published it would no longer matter if a bomb should fall on Lorenz's head. At least this seems to be the way to interpret what Lorenz wrote to Stresemann after the war, in 1948, when he finally returned to Austria safe and sound after spending three and a half years as a prisoner of war of the Russians. As he put it at the time, "The bomb, of which Erwin so charmingly spoke, that could calmly fall on my head after the publication of the 'Inborn Forms of Possible Experience,' has missed the mark by several meters (which he by the way also predicted, as I must mention by way of fairness)."[128]   .

CONCLUDING REMARKS

Fritz Ringer has written that in Germany in the early twentieth century, "the young were taught a dangerous respect for their own vital urges."[129] Lorenz, in focusing his research on animal instincts and in insisting that in humans the instinctive ability to recognize the moral worth of others was truer than any reasoned response, offered his own special contribution to this line of thought. He did not view his ideas as an accommodation to Nazi ideology. His thinking about domestication and race mixing was prior in origin, if not in elaboration, to his attempts to demonstrate his ideological worthiness to the Third Reich. The similarities between his views and those of Nazi writers—that is to say, *other* Nazi writers—may have been more a function of their descent from a common stock of eugenic thinking than of one party's borrowing ideas from the other. But this way of thinking about similarities (comparable to Lorenz's own identification of homologies in the course of comparative studies of behavior) should not be privileged to the exclusion of all others. Lorenz's views can also be considered from a more "ecological" perspective. He himself had something of a blind spot when it came to thinking about the influence of ecological contexts on the evolution of animal behavior displays. The same may be said of his thinking about his own ideas. Never feeling, and perhaps understandably so, that he had gotten his most important ideas *from* the Nazis, he was insensitive to the ways in which his own thinking might nonetheless have reflected the influence of the environment in which he was operating. His interests in epistemology notwithstanding, he was not very critical when it came to thinking about the origins of his own assumptions. He believed he was advancing universal truths when his views in fact displayed clear signs of having developed in the soil and climate of his own particular historical situation.

Within the population of biologists in the Third Reich different degrees of accommodation to the regime were displayed. We may ask where Lorenz fell along this spectrum (even if the idea of "spectrum" does not adequately characterize the multiplicity of positions a single individual might take over time in situating himself in the heterogeneous and evolving landscape of Nazi biology). What did he do, and what did he not do? More aggressively than many biologists (but still less so than others), he pronounced upon the lessons that biology and animal behavior could offer the

Reich. He did so in ways that were broadly consistent with the Nazi goal of creating a biologized German body politic. In the pages of the Nazi biology teacher's journal, *Der Biologe,* he argued that Darwinism, properly understood, led not to communism or socialism, but instead to national socialism. But he never cited Nazi political authorities in support of his views (unlike Ferdinand Rossner, for example). And he never branded a particular biological idea as distinctively Jewish (as Fritz Lenz did with the Lamarckian idea of the inheritance of acquired characters). Indeed, in none of his published writings did he make derogatory remarks about Jews. In private, however, just after the *Anschluss,* in the hopes of securing the professorship of psychology at the University of Vienna, he represented himself as someone who could develop a truly "German" psychology in place of the existing, flawed science that Jewish thinkers had been largely responsible for constructing.

In print Lorenz underscored the virtues and importance of race purity. He also expressed race-political concerns in his private correspondence. Writing to Heinroth after Britain declared war on Germany, Lorenz observed that from a "purely race-biological standpoint," it was a shame to have the two best "German peoples" of the world at war with each other while all the "nonwhite, black, yellow, Jewish and mixed races" stood by, rubbing their hands with glee.[130] When it came to his own scientific work, Lorenz was happy to intimate that his hybridization studies of diverse duck species had a bearing on broader issues of racial hygiene. And yet he never explicitly articulated why he thought the study of *interspecific* hybridity was a way of illuminating the dangers of *interracial* mixing. Equating the two of these seems highly problematic, but we have no record of him pondering this as an issue. He seems to have simply taken it for granted that the hybridizing of different duck species and the crossing of human races produced the same sorts of problems.

As for the connection between domestication and race mixing, Lorenz's line of reasoning here was simple enough, though it did have two parts. The first part was essentially *intra*racial, and this was where his warnings were usually targeted. He wrote mostly about the dangers appearing *within* a race when overly domesticated types threatened to spread like cancer cells through the racial body. The second step of his argument was to associate domestication with indiscriminate promiscuity. What took place in the barnyard, he thought, was analogous to what took place in big cities. Domestication and civilization led to a degradation of the innate schemata of animal species and human races, respectively. The result in each case was the increased likelihood that an individual organism would not make the racially appropriate choice of a mate. Degeneration within a race thus led to the additionally deleterious problem of race mixing. He argued that dangerous defectives had to be rigorously weeded out if the cascading consequences of degeneration were to be checked. The idea of eugenics as a state religion appealed strongly to him.

All this said, however, it still bears repeating that Lorenz's warnings against the genetic dangers of "domestication" did not constitute an effort that he regarded as a distinctively National Socialist project. In the immediate postwar period, while disavowing he had ever harbored any genuine Nazi sympathies, he continued to rail about the genetic and moral dangers of domestication. Indeed, he continued to do so for the rest of his career. His book *Civilized Man's Eight Deadly Sins,* published in 1973, contains a chapter on "genetic decay." There he once more set forth the analogy between the barnyard and civilized society and warned that the phenomena of domestication in animals were paralleled by the decay of genetically determined social behavior patterns in humans.[131]

In the Third Reich Lorenz achieved the kind of professional advancement that had been denied him in pre-*Anschluss* Austria. His appointment to a professorship at Königsberg in 1940 indicates that government authorities found him to be both politically and scientifically worthy. It is difficult to determine whether he reduced his political profile after attaining that post. In one new research direction that he pursued at Königsberg, his analysis of Kant's concept of the a priori in the light of modern biology, he made no mention of race or politics. The same was true of his comprehensive analysis of the phylogenetic affinities of various species of ducks and geese. In that tour de force he demonstrated convincingly that animal behavior patterns could be used for phylogenetic purposes. He also underscored the unique virtues of the particular scientific practices he had been cultivating. But he postponed there any presentation of the data he had collected in his multiyear study of the effects of hybridization on his birds' innate behavior patterns, and he raised no warning flags about the perils of race mixing.

Still, if Lorenz's paper on Kant and his monograph on the Anatinae were not burdened with political or race-purity rhetoric, the same cannot be said of his 1943 monograph on the innate forms of possible experience. There he voiced again his earlier concerns about domestication and racial degeneration. He worried once more about "parasites" and "cancerous" elements in the social body. He urged upon his readers the importance of race hygiene and the necessity of having a scientifically grounded race policy. In short, upon gaining his professorship at Königsberg he did not proceed to jettison the political baggage he had been carrying with him. But he also elaborated at some length in his 1943 monograph on a positive side of domestication, the side that created the preconditions for man's becoming human in the first place and that provided humankind with the possibilities of higher goals yet to be imagined and attained, however fraught with danger the attempt to reach those goals might be.

Lorenz did not see himself as toeing a Nazi line. This was no doubt in part because there was no single line to toe, and in part because he felt he had come to his ideas independently. In addition he believed himself to more knowledgeable about scientific matters than any party authorities, and certainly cleverer than the regular party hacks. But if he persuaded himself that his own ideas distinguished him from Nazi ideologues, he was nonetheless part of the broad circle of Nazi biologists. He was confident that biology could flourish under the Third Reich. He furthermore believed that he could personally contribute to the construction of a scientifically valid National Socialist biology. And he lent his scientific authority to the broad eugenic enterprise of Nazi race policy. He never explicitly endorsed the notion of Nordic racial *supremacy*. He never published derogatory statements about Jews. Nonetheless, in his endorsement of the idea of race purity, and in praising the Nordic movement for its efforts on that score, he signaled to others in the Nazi flock that, when it came to the subject of racial hygiene, he was essentially a "companion."

Later in his life Lorenz never really faced up to the possibility that his own behavior as a scientist under the Third Reich might have helped sanction the ideas and actions of others whose work he may not have fully endorsed himself. As we have seen, Erich Jaensch and Rudolph Hippius both cited their association with Lorenz. Both drew upon Lorenz's scientific standing to add authority to their own projects. But Lorenz does not seem to have viewed this as a problem. His stance, when he was later confronted about his past political behavior, was that whatever political gestures he made in the service of his biology were understandable and excusable as such. This, together with the explanation that he was "naive" about the intentions of the Nazis, constituted the essence of the "apology" he finally offered to the public, long after the war's end, on the occasion of his receipt of the Nobel Prize in 1973. He did not specifically acknowledge that as a scientist in the Third Reich, in promoting ideas of racial hygiene and using a language of "elimination," he had possibly made an indirect or inadvertent contribution to a program that resulted in genocide.

After serving as a military psychiatrist in Poznan from 1942 to 1944, Lorenz was called up in April 1944 to the Eastern Front as a physician.[132] In

June, near Vitebsk, the Russians took him prisoner. The word that initially came back from the front to his colleagues, friends, and family was that he was missing in action and presumed dead. He was not dead but instead interned in a Russian POW camp. He remained a Russian prisoner for more than three and a half years. It was not until February 1948 that he managed to return to Austria.

While a captive of the Russians, Lorenz served as a physician for his fellow prisoners. He also lectured to some of them on the science of animal behavior. He wrote these lectures on paper salvaged from old cement bags. He assembled them in a long manuscript that he described as "the first attempt to provide a cohesive account of a very young branch of biological research," the field of "comparative behavior study."[133] The "Russian manuscript," as it has come to be called, was a characteristically ambitious attempt on Lorenz's part to synthesize the natural sciences and the humanities. There he elaborated on themes he had introduced in his 1941 paper "Kant's Theory of the *A Priori* in the Light of Modern Biology," and his 1943 paper "The Innate Forms of Possible Experience." He pursued questions of evolutionary epistemology. He also called attention, once again, to the dangers of "domestication phenomena" in animals and humans, maintaining that "manifestations of ethical decay" in modern civilization were a function of "deficits in innate species-specific *social* behavior patterns" induced by domestication.[134]

Missing from this new synthesis, however, was any direct reference to race hygiene, the third pillar of the synthesis he had constructed in 1943.[135] Whether or not Lorenz's ideas on this score had changed, his immediate audience was certainly a different one. At the very least, a Russian censor would have to approve the manuscript before Lorenz could be allowed to take it with him back to Austria. This was not a time to tout eugenics or matters of racial purity. The Russians viewed eugenics as a fascist enterprise. Lorenz managed to impress his captors as having good antifascist tendencies himself.[136] In his manuscript he quoted from Karl Marx and made additional bows to the philosophical predilections of the Soviets. The weltanschauung fundamental to "all true natural scientific research," he wrote, "was clearly formulated long ago in the philosophy of dialectical materialism."[137] Once again Lorenz had refashioned himself for a new environment.

This latest refashioning, to be sure, was only superficial. And, of course, respectful references to Karl Marx were not going to help Lorenz when he returned to Austria. Back home, he would be confronted with new circumstances of time and place. It would not benefit him to look like either a former communist or a former Nazi. There was little danger of his being suspected of the former. For reasons this chapter has made evident, however, the issue of his relations with the Nazis was much more problematic. What differences this made for his own career and for the further development of ethology will be considered in the following chapters.

SIX

THE POSTWAR RECONSTRUCTION

OF ETHOLOGY

We all have a great longing for international interchange of ideas. I do not intend to cut off all relations with German scientific men. But first I must not see them for a long time, so as to overcome the psychological aversion resulting from the incredible German terror we underwent.

NIKO TINBERGEN IN A LETTER TO JULIAN HUXLEY,

PUBLISHED IN *NATURE* IN NOVEMBER 1945 [1]

It is my firm conviction that this congress is going to be of the utmost importance. Comparative Ethology, Vergleichende Verhaltensforschung, or however you chose to call it, is quite doubtlessly developing into a school at least as original and important as Behaviorism and Pavlov's Reflexology and quite certainly a much nearer approach to an exact natural science than both. It is certainly high time to come together and give it a name!

KONRAD LORENZ IN A LETTER TO W. H. THORPE OF 10 AUGUST 1948,

COMMENTING ON THE SYMPOSIUM OF THE SOCIETY FOR EXPERIMENTAL

BIOLOGY TO BE HELD IN CAMBRIDGE IN JULY 1949 [2]

At the close of the Second World War, the prospects of the young science of ethology were far from certain. Konrad Lorenz, who had taken the lead in laying the field's conceptual foundations in the 1930s, was missing in action and presumed dead. Niko Tinbergen, whose experimental and analytical talents had provided an invaluable complement to Lorenz's early theory building, had survived two years in a German prison camp and was free again, but his hopes of establishing ethology as a scientific discipline faced serious difficulties. Ethology perched uncomfortably on the boundary between biology and psychology, lacking secure institutional bases of its own. Though Tinbergen could plan to resume at the University of Leiden the program of behavioral studies he had started there before the war, the prospects of ethology's becoming an international discipline were unclear. What is more, there was no longer a journal to represent the fledgling science. The *Zeitschrift für Tierpsychologie* had been one of the war's many casualties. Beyond these difficulties, the material conditions of war-torn Europe made the immediate resumption of particular kinds of research impossible. Had Tinbergen been able to get a bicycle to ride to the site of his prewar fieldwork, his herring gull research would still have been thwarted by the live land mines left in the Dutch dunes by the retreating enemy. Added to these physical obstacles were the psychological wounds that the war had left in its wake. As Tinbergen set out to rebuild ethology, he looked first to British, Swiss, Finnish, and American colleagues for assistance. It would take some time, he felt, before he could work with German biologists again.

The precarious condition of ethology circa 1945 contrasts so markedly with the discipline's success over the next two decades that it is tempting to think of the war serving only as a temporary interruption of the field's development. Tinbergen, who was loath to think ill of most anyone, soon made his peace with his German colleagues, including Lorenz, who had not died in the war after all but had instead been captured and interned by the Russians. Tinbergen and Lorenz would continue to be ethology's leading figures well into the 1960s. Nevertheless, the precise course of ethology's development was far from preordained. The two decades after the war were for ethology a period of adaptive evolution. In this period, ethology's institutional bases were almost wholly restructured, its conceptual foundations were significantly reshaped, and its material practices were multiplied and refined. Furthermore, while the ethologists as a group sought to maintain coherence in their evolving discipline, the individual courses they fashioned for themselves were far from identical. And if the practitioners of the new field by and large succeeded in putting the experiences of war behind them for the good of their emerging discipline, memories of these experiences did not fade away entirely.

In the present chapter we will consider ethology's reconstitution and reinvigoration in the years immediately after the war. Making extensive use of manuscript correspondence of the major actors in this story, we will seek to reconstruct the interplay of material, disciplinary, institutional, political, and personal factors that were constitutive of ethology's reformation in the postwar period. We will focus first on Tinbergen's resumption of activities at Leiden, his efforts to reestablish an international ethological community, and his decision to leave his professorship at Leiden for a new position at Oxford. We will then look at Lorenz's return to Austria and his renewed efforts to find support for his research program. Finally we will consider the special symposium "Physiological Mechanisms in Animal Behavior," which was held in Cambridge in 1949. This was the first major scientific conference in the postwar period to address and call into question several of ethology's fundamental concepts. It was also the occasion for the first meeting of Tinbergen and Lorenz since the war had separated them.

RECOVERING FROM THE WAR: THE CASE OF TINBERGEN

As soon as the war ended, Niko Tinbergen set about reviving ethology internationally. For the previous two and a half years, he had essentially been cut off from other ethologists, and he craved renewed intellectual contact. Specifically, he wanted to be back in touch with scientists from the United States, Britain, and other non-German countries. As he explained to Margaret Morse Nice in June 1945, he was not ready to renew relations with the Germans:

> I know absolutely nothing about our German colleagues. Lorenz was in the Army Dept. of "Heerespyschologie" since 1941. He was rather nazi-infected, though I always considered him a honest and good fellow. But it is impossible for me to resume contact with him or his fellow-countrymen, I mean it is psychologically impossible. The "wounds of our soul" must heal, and that will take time. Originally, soon after the outbreak of the war, I had planned to personally organize the resuming of cooperation in scientific affairs. Now I see, that we will have to wait at least a year, first to see whether a German science will revive at all, secondly to wait till the Germans give proof of their goodwill in this respect.[3]

He expressed the same sentiments to Julian Huxley: "We all have a great longing for international interchange of ideas. I do not intend to cut off all relations with German scientific men. But first I must not see them for a long time, so as to overcome the psychological aversion resulting from the incredible German terror we underwent."[4]

Thinking about organizing a postwar symposium on animal behavior, Tinbergen decided it would have to be a small conference, involving ten to fifteen investigators who were personal friends: Pontus Palmgren from Finland, Monika Meyer-Holzapfel from Switzerland, Julian Huxley and David Lack from England, and Margaret Nice, Ernst Mayr, and Bill Vogt from the United States. He thought too of Lorenz, Erwin Stresemann, and Bernhard Rensch, but he put the notion aside. It was still too early to for him to work with them.[5]

In August of 1945 Tinbergen traveled to Switzerland, where he met with the Swiss animal behavior specialists Heini Hediger and Monika

Meyer-Holzapfel. Discussing with them the reestablishment of international ties, he concluded that what ethology needed even more than a postwar conference was a new journal to fill the gap left by the demise of the *Zeitschrift für Tierpsychologie.* That journal, together with the British *Journal of Animal Ecology,* had convinced Tinbergen just how crucial for a field a central journal could be. He wanted to establish a journal for ethology similar to the two just mentioned, but more international in character. He began laying plans for a *European Journal of Animal Behaviour.* Seeking international representation on his editorial board, he succeeded in lining up Hediger from Switzerland, Palmgren from Finland, and W. H. Thorpe from Britain (after Huxley and Lack successively begged off).[6]

Tinbergen took his journal plans to the Leiden publisher E. J. Brill. Brill had previously published the papers of the 1936 Leiden instinct conference in its series entitled *Folia Biotheoretica.* Brill had also published Tinbergen's long, programmatic monograph, "An Objectivistic Study of the Innate Behaviour of Animals," in 1942 in the series called *Bibliotheca Biotheoretica.* The publisher suggested for Tinbergen's new journal yet another ponderous Latin title, *Acta Ethologica.* Tinbergen was comfortable enough with this, but Thorpe expressed doubts. The problem, Thorpe said, was not that English readers would be put off by a Latin title, but rather that the word "ethology" was not much used in English-speaking countries. As an alternative, Thorpe suggested calling the new journal, *International Journal of Animal Behaviour.* Ultimately, the journal was simply entitled *Behaviour.*[7]

By January of 1946, Tinbergen was ready to start advertising the new journal. The advertisement went out over the names of the four editors on a flyer, printed by Brill, explaining the special nature of ethology and the special need, in the absence of the *Zeitschrift für Tierpsychologie,* for a "European journal devoted exclusively to the publication of modern behaviour studies." Defining ethology as "the study of animal behaviour," the flyer stated: "Interest in the study of animal behaviour (Ethology) is growing rapidly. The adoption of new methods of observation in the study of animals in the field, the development of suitable techniques for the experimental analysis of instinctive actions and recent advances in the study of the complex functions of the central nervous system have all tended to emphasize the great significance to general Biology of the scientific study of animal behaviour." It was not sufficient, the flyer went on to say, to study animal function from the standpoints of nerve and sense physiology. Other problems of animal behavior required the special methods of comparative ethology. Noting how the *Zeitschrift für Tierpsychologie* had served both to stimulate and to promote the study of animal behavior, the editors of the new journal concluded: "It appears to the undersigned that what is most urgently required is a truly international journal for comparative Ethology. We believe that there is now a unique opportunity to promote international scientific co-operation in this way, and we have accordingly decided to found a new international journal of comparative Ethology to be called 'Behaviour.'"[8]

Significantly enough, Tinbergen's certainty that ethology needed a journal of its own was not matched by any certainty on his part about where he should continue his own career. In September 1945 he expressed to Ernst Mayr an interest in leaving Holland, even though his position at Leiden was in the process of being upgraded. If the opportunity presented itself, he told Mayr, he might prefer a job in the United States. For that matter, he indicated, he might also like to spend some time in the Dutch East Indies.[9]

Two months later, Tinbergen reported to Mayr that he would soon have "a really good position" at Leiden. Still, he remained ambivalent about it. The only thing better, he admitted, would be the chance to be a pure researcher, working collaboratively in some kind of scientific institute with an endocrinologist and a physiologist, both interested in behavior, and also with "one or more ethologists." He acknowledged, however, that this was only a dream: an "absolute air castle." He conceded: "I think it would be wise to stay where I am and to work at the more humble but nonetheless attractive task of getting students interested in ethology."[10]

But even as he accustomed himself to the idea of a better position at Leiden, Tinbergen retained the idea of traveling abroad. The prospect of a trip to the United States especially appealed to him. This would not be possible before the fall of 1946, he told Mayr, because he first had to lecture for a year, start a student on thesis work, get his summer field camp up and running again, and make sure his international ethology journal got off the ground. After that, though, a visit to the States would be both possible and very useful: "I ought to see all the better ethologists and also more psychologists and I ought to get some Lashley-Beach [Karl Lashley and Frank Beach] people interested in our type of work, in order to come to [some] kind of cooperation." Such people, he believed, might be able to provide ethologists good advice.[11]

At the same time that the Dutch ethologist was pursuing the idea of a trip to the United States, he was also developing his contacts with British scientists. He had corresponded with David Lack as early as 1940, after Lack sent him some reprints. The two met face to face for the first time in February of 1946, when Lack came, at Tinbergen's invitation, to a symposium on ecology at Leiden. Following Lack's two-week stay in Holland, Tinbergen traveled with him back to England. At Oxford Tinbergen had the chance to meet and talk with Charles Elton, Bernard Tucker, and H. N. Southern; in Cambridge he made the acquaintance of W. H. Thorpe, James Gray, and others. In London he met Huxley and gave a general lecture to the Zoological Society. He was surprised to find, as he later reported to Mayr, "that the knowledge of the continental work (let us say, the work of the Lorenz-school) was very slight, but that many people were immensely interested." Characteristically, he was quick to acknowledge his own deficiencies: "At the same time, I saw that my own knowledge of the Anglo-American work in the same direction was just as bad if not worse." [12]

Within nine months of the war's end, Tinbergen had thus made major strides in rebuilding his international contacts. He had met with those whom he regarded as his Swiss, British, and Finnish counterparts, and he was also back in correspondence with American colleagues. There remained the question of reestablishing relations with the Germans. During the war, as Tinbergen afterward told Robert Yerkes, he had "experienced intensities of emotions as I never had before during quiet peace-time." These had given him "opportunities for many psychological observations on human nature, of my own and of others, often of a not very encouraging nature." [13] The war left him harboring intense, anti-German feelings. Nonetheless, he was taken aback, on his visit to England, when the zoologist Harold Munro Fox showed him a letter in which Otto Koehler asked Fox, "And how is Tinbergen? Does he still hate us?" [14]

Tinbergen's sentiments were particularly mixed with respect to Lorenz. He had become a great friend of Lorenz when he visited him in Altenberg in 1937. Bound together in the common project of establishing ethology as a science, the two had carried on an active scientific correspondence up until Tinbergen's imprisonment by the Germans in 1942. But as Tinbergen told Margaret Nice in June 1945, he could imagine reestablishing scientific relations only "with men whom we can trust, whom we know personally, and who have given proof of honest intentions." Two categories of German zoologists were clear to him. On the one hand were those whose wartime activities made them wholly unacceptable for any future relations. On the other hand were those who had done nothing to further the Nazi regime, and with whom he knew he would want to renew contact eventually. In the first category he placed the ornithologist Günther Niethammer, who he knew had served in the SS. In the category of those with whom he expected eventually to renew contact he identified

"Stresemann, Rensch, Von Frisch and, I hope, Laven." As for Lorenz and Koehler, they, in Tinbergen's mind, fell in a gray area somewhere in between. He was not sure what stance ought to be taken in their regard, but as he told Nice: "Personally I should regret if Lorenz and Koehler would be expelled."[15]

Tinbergen's ambivalence toward Lorenz found expression again in a letter he wrote to Joseph Hickey in September of 1945: "Lorenz has been more or less nazi; but knowing him so well personally I cannot believe that he agreed with the methods that Nazis adopted, or rather began to show, for their methods were criminal from the first day of their *debut*. He has written some very clever papers, a. o. [e.g.] one about the courtship movements of Anatinae, which is real pioneer work. He is undoubtedly *by far* the best man in our domain and I am very sorry that we will miss his leadership in future."[16]

In his letter to Huxley that *Nature* published in November 1945, Tinbergen included Lorenz with Stresemann, Rensch, von Frisch, and Laven in his list of "honest and reliable men" with whom he would, in time, want to renew contact.[17] Readers of *Nature* learned immediately, however, that renewing contact with Lorenz was unlikely. In addition to passing on the news from Tinbergen, Huxley added a letter from Stresemann describing the condition of German ornithology. Stresemann reported that Oskar Heinroth had died of pneumonia on 31 May 1945, "consequent upon many nights spent in an air-raid shelter during the heavy raids of March and April—an irreplaceable loss." Stresemann went on to say: "That great pioneer of modern animal psychology, Prof. Dr. Konrad Lorenz, was reported missing last summer near Witebsk, presumably killed in action. He was working as a physician in a field-hospital."[18]

Tinbergen first heard the news in late November or early December. He lamented to Margaret Nice: "This is a disastrous loss and I am very sad about it because we had become very good friends in 1937 and were continuously cooperating, that is to say he was continuously urging us and leading our work, ever since that time."[19] He repeated these sentiments a month later, calling Lorenz "one of my very best friends, a very good character."[20]

At the beginning of February 1946 word came that Lorenz was not dead after all. Stresemann heard the news from Oskar Heinroth's widow, Katharina, who had heard it from Lorenz's wife, who had just received a postcard from her husband. Stresemann passed the word immediately on to Margaret Nice: "He is a Russian prisoner, not wounded, and employed in a camp as a physician. Isn't that simply marvelous news? Please do spread it at once, and don't forget to inform Dr. Tinbergen and other friends abroad!"[21]

Nice, whose efforts on behalf of the ornithologists of war-torn Europe had led Stresemann to call her by the title of "the mother of international ornithology," relayed the news to Tinbergen.[22] He did not know what to believe.[23] He had already heard one mistaken report that Lorenz was alive. In this case it had been Lorenz's brother, not Lorenz himself, who had been seen in Vienna. Tinbergen wrote to Mayr about the uncertainty, observing: "You know Lorenz was a very good friend of mine and I am very anxious to know his fate and that of his family, but till now it was impossible for me to get in touch with them. Though he was a little nazi-infected I can not help to consider him as a friend to be helped."[24]

By May 1946 Tinbergen was in direct contact with Stresemann and Koehler, and the news was confirmed: Lorenz had made it through the war safely. Tinbergen exclaimed to Margaret Nice: "You will understand how glad I am that Konrad Lorenz is alive!" He added: "It is to be hoped that he will be released before long; but I am afraid the prisoners of war will be kept for a very long time." At the same time, Tinbergen responded sharply to Nice's positive appraisal of the German ornithologist Ludwig Schuster, whom Tinbergen knew to have occupied a high ministerial position in the German government:

> Do not forget that every German whom we knew as a nazi and who cheered heartily as long as things went well for Germany in the war, is very plaintive now and all at once claims he never had anything to do with the nazis. Do you understand how a high official on one of the government's bureaus could be "strongly opposed" to the régime? And yet stay and be left upon his post during the whole war? and make repeated promotions!? That is simply ridiculous, you know. We know what happened to people who were not even strongly but mildly opposed to the régime. I won't propagate hatred against the Germans in general, but we are shocked very much indeed by the naive attitude taken by the Americans and the British against nazis whose only argument is a winning smile and the lie that they always were opposed to it.[25]

Tinbergen's attitude toward Koehler had softened because Koehler had apologized straightforwardly to Tinbergen for his support of the Third Reich. What the Dutch demanded from their German colleagues, as Tinbergen told Nice, was an acknowledgment that what Germany had done was wrong, plus the willingness to "at least *once* say 'sorry brother, we are very sorry about it all and we will never support such a gang again.'" This was precisely what Koehler had conveyed in a letter to Tinbergen. The Dutch naturalist wrote back to Koehler, thanking him for acknowledging and lamenting that the Germans had (in Tinbergen's words) "committed a terrible crime against us and other peoples, so dreadful that is not to be described and not to be understood."[26] As Tinbergen explained in his letter to Nice, "If a man really *shows* he is sorry about it, I think we will forgive and forget and help them to make a new start."[27]

By the summer of 1946, the Germans with whom Tinbergen was back in touch consisted of Stresemann, Koehler, Gustav Kramer (who was together with Holst in Heidelberg), Bernhard Rensch, Hannes Laven, and Hans Peters. In addition Tinbergen had written to Lorenz's wife, Margarethe.[28] But Lorenz himself was still a prisoner of war, and there was no indication that the Russians were about to release him. To Koehler Tinbergen wrote: "If only Konrad could return! I have often to think of him and the beautiful pre-war period, when we exchanged ideas and results in a lively correspondence."[29] As it was, Tinbergen felt that the responsibility for ethology as a field continued to rest on his own shoulders. It was primarily up to him, he believed, to interpret ethology to the rest of the scientific world and to make the arrangements for the field's future development.[30]

Tinbergen had concluded his first letter to Huxley after the war with the comment "You may wonder whether we will have time to resume 'pure' scientific research, now that all kinds of reconstruction work will demand so much of our energy. I am sure that we will succeed in keeping part of our time for research."[31] In the course of the task of reconstruction, however, time was not the only problem. The simplest necessities were lacking. A year after the war's end, writing paper was still in short supply. It was a great boon to Tinbergen in the spring of 1946 when David Lack sent blocks of paper from England so Tinbergen's students could take laboratory notes.[32] Transportation into the field was also a problem. Before the war, Tinbergen and his students had had an easy hour's bicycle ride to the herring gull colony that was the site of their bird studies. During the war, however, German soldiers stole Tinbergen's bicycle, and after the war there were simply no bicycles to be had. Furthermore, as mentioned previously, even had he had a bicycle to ride to the site of his prewar gull studies, he would have had great difficulty doing research, for the retreating enemy had left the Dutch dunes booby-trapped with land mines.[33]

As for academic life at the University of Leiden, the university had not held classes since 1940. It quickly found itself flooded with students. Returnees were joined by other students arriving for the first time. Many needed extra help. The spring of 1946 found seven hundred students enrolled to study medicine, as contrasted with a prewar figure of two hundred.[34] Amid this turmoil, Tinbergen set about reestablishing the three kinds of instruction that had characterized his teaching program before the war: the laboratory block practical, the summer field studies, and the direction of doctoral research.

The laboratory work began first. Late in March 1946, back from his trip to England and a second postwar trip to Switzerland, Tinbergen took up his stickleback research again. Five years had elapsed since his last stickleback work. As he returned to the laboratory, he felt the loss of his former student, Joost ter Pelkwijk, who had been killed in the Dutch East Indies. He had designed the original dummy experiments with the sticklebacks. Tinbergen greatly regretted his death.[35] Still, Tinbergen felt glad to be back at work in the lab. As he began training a new crop of students, he decided such training should combine behavioral studies with knowledge and techniques from other fields. One of the students was prepared to study behavior in conjunction with endocrinology. Tinbergen hoped others would link behavioral studies with neurophysiology. In April he expressed his enthusiasm for the new work to David Lack: "I think it is one of the enjoyable trends in our work that we are not only digging deeper but at the same time joining up again with ecology, endocrinology, neurophysiology." Reporting further on the ongoing stickleback research he remarked: "It is amazing how many new things can be discovered each new season by very simple, if only consistent, methods! And how often one gets new individuals slightly different from all observed before. . . . We are analyzing habitat selection in sticklebacks, offering them different types of 'landscape' and comparing their effect on the fish and its responses to them."[36] In May he indicated to Margaret Nice that he and his students had been "working hard with our sticklebacks, trying to do a frontal attack on the 'internal causes'" of their behavior.[37]

The spring laboratory resulted in a paper by Tinbergen and his student J. J. A. van Iersel on "displacement reactions" in the stickleback. The two naturalists found that crowding several territorial male sticklebacks into one aquarium caused the fish to spend most of their time either fighting or threatening each other. Included in the threatening behavior was an activity the investigators identified as "displacement digging." The stickleback males, in the course of their boundary disputes, dug deep depressions at or near their nests. The action was more conspicuous than normal nest digging in that it involved the male's displaying his red ventral parts and erected ventral spines and performing brisk, shaking motions. Significantly, the males did not display this behavior when they were put in separate tanks and each was allowed to build a new nest without another male nearby.

Tinbergen and van Iersel interpreted displacement digging as serving the function of threatening other males—and having evolved through ritualization to do so. They did not believe this was true of every kind of displacement activity. "Pushing," "fanning," and "gluing" could all be elicited by thwarting the male's mating drive, and they evidently functioned as outlets for the thwarted drive, but they had not become ritualized. They elicited no special behavior on the part of the female. On the basis of this evidence, the investigators concluded there was a direct correlation between ritualization and signal function. Ritualization occurred when displacement reactions, which were initially outlets for thwarted drives, were subsequently adapted to the secondary function of serving as a signal or a releaser.[38]

While his spring 1946 stickleback work was progressing, Tinbergen was also preparing for his first summer field camp in five years. Among other things, he was planning to test the ability of birds to recognize camouflaged insects as prey. To this end, he hand reared a brood of jays so they would be accustomed to aviary life but naive with respect to the particular types of insect food he would later present to them.[39] Shortly before going to Hulshorst in July he wrote Margaret Nice, saying, "We will start our insect work again with some of the older students and with some representatives of the post-war student generation. . . . I wish you would be able to be with us there, for such camps are the highlights of our work."[40] Though the weather that summer proved abnormally cold and cloudy and hampered the research his team was able to conduct, this did not dampen Tinbergen's spirits. He told Nice when it was over: "It was delightful to be living in the tents again after 6 years' lapse, to have the long-tailed tits and the jays around during the day, and the nightjars and tawny owls at night."[41] Tinbergen, in sum, was finally back in the business of doing research. Both in the lab and in the field, he had reestablished his former practices.

When Tinbergen returned from his 1946 summer field camp, the most important event on his professional horizon was a chance to travel to the United States for a three-month visit. He had known about this since March, when Ernst Mayr, who had been at work on Tinbergen's behalf, reported: "It is all arranged that you will give a course of lectures in New York under the joint auspices of the American Museum of Natural History and Columbia University, and will receive the sum of four hundred dollars for your efforts.[42] Tinbergen began to wonder whether the lectures at Columbia might not serve "as a starting point for a little book, a kind of introduction to our mode of approach."[43]

Tinbergen set about planning how he would come to the States in November, work in the library of the American Museum, brush up on his English, and deliver individual lectures lined up for him by Mayr at various sites. He would then give the main series of lectures at Columbia in January before returning to Holland in mid-February.[44] Eight years earlier, on his previous trip to the States, financial considerations had confined him to the East Coast. This time he wanted to travel more widely. Ultimately his itinerary took him all the way to California, where he visited the University of California, Berkeley, and the California Institute of Technology. He also participated in the annual meeting of the Wilson Ornithological Club, held in Omaha. In addition, he lectured at the University of Chicago, the University of Wisconsin, and the University of Alberta. He stayed with Margaret Nice and her family in Chicago and with Ernst Mayr and his family in Tenafly, New Jersey. At Mayr's home he ended up sharing attic space with David Lack, who was also visiting, and who continued to impress Tinbergen as "a very stimulating man." In New York he had the opportunity to interact with the animal behavior people at the American Museum of Natural History, including Frank Beach, who had taken G. K. Noble's place there after Noble had died unexpectedly. Tinbergen succeeded in lining up Beach as an American editor for *Behaviour*.[45]

The Dutch ethologist gave his six lectures under the title "The Study of Innate Behavior in Animals" at the American Museum and Columbia in January and February 1947. He emphasized the external and internal causation of innate behavior in animals and the hierarchical organization of behavior systems. These had been the central themes of his *Acta Biotheoretica* paper of 1942. They continued to represent the areas, in his opinion, where ethology's claims were the strongest and about which his Anglo-American audience needed most to hear. Oxford University Press agreed to publish the lectures in extended form. The lectures thus served as the starting point for the first major text of ethology, Tinbergen's *The Study of Instinct*.[46]

Before he had finished the series, Tinbergen described in a letter to Otto Koehler his impressions of "how strong the interest in our way of working is." As he put it, "The psychologists, always working with rats and mazes, had up to now not much idea of what we are doing, but are ardently interested when they hear about it. The zoologists are similarly interested but they too up to now do not work 'on instinct.' "[47]

Upon returning to Holland, Tinbergen reported back to his American hosts, thanking them for their hospitality and noting what he had learned from the trip. To Margaret Nice he wrote: "This stay in the U.S. has been of much value to me. The American attitude is so different from ours; it takes a long time to come to understand it a little better than would have been possible from a distance, but I feel I have made some progress."[48] He also let Ernst Mayr know that Mayr's prodding of him in at least two respects had not been in vain. Begging the evolutionary biologist not to laugh, Tinbergen reported that upon returning to Leiden he had lectured to his older students on evolutionary problems. He also announced that he was planning to follow Mayr's suggestion of beginning a research program on the behavior of *Drosophila*.[49]

The spring and summer of 1947 proved to be a heady time for Tinbergen. In April he was promoted to a professorial chair at Leiden, a chair of experimental zoology created especially for him.[50] He was also made head of the Zoology Department. Then in June he and his students began observations at what for them was a new site for fieldwork, the bird sanctuary of Terschelling. In what he described at that time as "the most beautiful and gripping landscape I know," he and his students studied how herring gulls and oystercatchers recognize their eggs and young. They tested the stimuli releasing the pecking response of the herring gull chick. They also experimented on supernormal stimuli in the oystercatcher's recognition of its egg, roaring with laughter when the birds tried to incubate huge dummy eggs in preference to eggs of a normal size. Then in July they headed back to Hulshorst for their regular summer field camp.[51]

Yet even with the encouragement of the new professorship and the invigorating experience of a new field site, Tinbergen felt ill at ease. By summer's end, 1947, he was once more thinking about leaving Holland. The standard account of his eventual move was that he felt the need to take the message of Continental ethology to the Anglo-American scientific world and he decided the best way to do this was by taking up residence in an English-speaking country. Yet there were other, more personal, reasons as well.

His correspondence shows that he found life in postwar Holland psychologically trying. The experience of the German occupation of Holland continued to weigh upon him, and he felt dissatisfied with the social and political attitudes of his countrymen.[52] In August 1947, he explained his unhappiness to Mayr in the following terms: "One important thing is that the general lack of vitality of our people is disappointing me, and also their self-satisfied conservatism."[53] He asked Mayr whether there might be any chance in one or two years of finding in the States "a job that would allow me to continue what I consider my life work, viz., research on animal behavior and perhaps some teaching in zoology, especially animal behavior and related fields." He requested that inquiries be made discreetly, so that his colleagues at home would not learn immediately of his thoughts of moving. Mayr began testing the waters, enlisting Frank Beach and T. C. Schneirla in the process.

Tinbergen was not just discouraged by the political attitudes of his countrymen. He also felt oppressed by the various administrative and teaching duties that were falling upon his shoulders. The new professorship and the department headship were not unmitigated blessings. As professor of experimental biology, and one of only three professors in the department, he was responsible for elementary teaching in anatomy, sensory physiology, ethology, developmental physiology, and sometimes ecology. In addition he gave a short course on evolution. Only his graduate teaching focused on ethology.[54]

Tinbergen would have preferred to have more time to direct his graduate students, who by 1947 were engaged in a host of different researches in the field and in the lab. In the field they studied birds and insects, working with herring gulls and oystercatchers, three species of grasshoppers, and one species of butterfly. Among other things, they paid special attention to the way cryptic coloration protected insects from birds. In the lab, they studied fish: sticklebacks, cichlids, and pike. With the sticklebacks they investigated the influence of hormones on behavior and at the origin of social releaser movements. With the cichlids they conducted comparative studies of reproductive behavior. With the pike, they studied how young pike responded to sticklebacks.[55]

Mayr, as he pursued a position for Tinbergen in the States, felt that the ideal place for the Dutch ethologist "would be in an up-and-coming large department where they want to add a field that is not yet taken care of in American zoology." The problem, he told Tinbergen, was that "in every case known to me where there is a department head who is trying to do that, he is a physiologist and has no comprehension whatsoever of the value and importance of animal psychology."[56]

Tinbergen was grateful for Mayr's efforts. It was thus with some embarrassment that he wrote to Mayr in February 1948 to say that David Lack was also at work on his behalf, trying to get him a job at Oxford. Tinbergen had told Lack of his idea of moving to the States. Lack had warned him in reply that an exile in the United States could "lose his soul" there. This had not happened in the cases of the German-born Mayr and the Russian-born evolutionary geneticist Theodosius Dobzhansky, Lack admitted, but Mayr and Dobzhansky, he said, were exceptions. Tinbergen was not scared away. He explained to Mayr:

> In some respects I would prefer a job in the US, on the other hand, if Oxford would succeed in creating a position similar to that of David himself, I would seriously consider it, as you will doubtless understand. Of course, although Prof. Hardy, head of the zoology Department at Oxford, says: "we simply must get him" (as David wrote me), I have made it clear that I would only bite if any definite and satisfactory offer were made and could be weighed against eventual other possibilities. Also, it is by no means sure that Hardy will succeed in raising funds for an ethological institute. So that is the situation; I can quite understand that this might cause you to give up any attempts on my behalf, although I still think I would prefer a similar position in America.[57]

The spring of 1948 dragged on without any further news of a job for Tinbergen either at Oxford or in the States. Part of the delay at Oxford may have been because W. H. Thorpe, unbeknownst to Tinbergen, was angling for the Hope Professorship of Entomology there, and Thorpe had suggested to Hardy that Oxford might not be big enough to have two researchers in animal behavior.[58] Tinbergen began to feel discouraged. In addition to the uncertainties regarding a position for him at Oxford or in the States, he was not finding time to finish his book, and he did not think the papers being submitted to *Behaviour* were of as high a quality as hoped they would be. It also seemed to him that the world political scene was worsening. The main good news, as he reported to Mayr in May, 1948, was that Lorenz was back in Altenberg: "He seems to be in good shape, and is flowing over of energy and plans." Tinbergen hoped to see Lorenz in the fall of 1949 in Cambridge, where a major international conference on animal behavior was being planned.[59]

In June 1948 Tinbergen returned to Terschelling with thirty students to continue the previous year's work on herring gulls and oystercatchers.[60] In July they were back again in Hulshorst for more insect studies. At Hulshorst he was particularly pleased by Leen de Ruiter's work on the concealing coloration of insects, including countershading, and the demonstration that jays and finches are highly critical "selectors." But the most exciting event occurred late in the summer when David Lack visited camp, carrying with him a letter from Alister Hardy. Hardy was offering Tinbergen a post at Oxford. Tinbergen wrote Mayr: "Prof. Hardy, head of the Zoology Dept. there, wants to establish a subdepartment of animal behaviour organized along the lines of David's Institute and wants to have me there." Tinbergen promised Hardy that if the details of the offer could be firmed up soon, he would come to a decision on it by 1 January 1949. However, as he explained to Mayr: "The plans are still provisional. They also agreed that I would write to you now in order to tell you that I would still consider any offer that might possibly come from the U.S. before Christmas."[61]

Mayr reported that a possibility of a job for Tinbergen was developing at the California Institute of Technology. The school already had three Dutch biologists on its faculty, and Tinbergen found the prospect of joining them attractive. As he told Mayr, "When I was there Christmas 1946 I saw that it was a very active centre of research, and the presence of at least three Dutch biologists means something to me, in spite of the fact that I am trying to leave Holland."[62]

By this time, however, the pace of Tinbergen's postwar activities, following upon the long-term strain of the war years, was taking a toll on him. Since the war's end, he had traveled to Britain, Switzerland, and the United States in his efforts to revitalize ethology internationally. He had gotten his laboratory and field research up and going again and was directing graduate students in both kinds of work. He had published a number of short articles plus a long article on the experimental study of releasers (written in response to criticisms of the releaser concept by the American ornithologist A. L. Rand).[63] And he was hard at work trying to turn his American lectures into a book. Early in 1948 he had admitted to Mayr that he was exhausted: "I am at the end of my strength and lying down now and then." His efforts to finish his book, the demands of editing *Behaviour,* and his new job as director of the laboratory at Leiden were all weighing heavily upon him: "[The book] approaches completion, but the last parts will still take much energy. Also, Behaviour costs a lot of time, with criticizing manuscripts, keeping up contact with all the editors, urging the printers, etc. etc. My new job as director of the lab also costs me a great amount of energy, all those odd little jobs smashing up one's time into innumerable parts. I am no good for these things."[64]

Tinbergen completed his book manuscript and mailed it off to Oxford University Press just before Christmas, 1948. Then he took some time off to rest. Mentally and physically drained, he suffered something of a breakdown.[65] He canceled a lecture he had been scheduled to give at the University of London in January. By February, however, he was well enough to travel to Oxford. There, after long discussions, he accepted Hardy's offer to join Hardy's department.

Tinbergen's decision to leave Leiden for Oxford provoked ill feelings on the part of several of his Leiden colleagues. Professor van der Klaauw in particular was furious. He reproached both Tinbergen and Hardy—Tinbergen for his ingratitude for all that had been done for him at Leiden, and Hardy for stealing Tinbergen away.[66] But Tinbergen managed to get most of his colleagues to understand his move by emphasizing the importance of establishing a foothold for ethology in the English-speaking scientific world. And he was confident that he was leaving ethology at Leiden in good hands—the hands of his talented head assistant, J. J. van Iersel. Tinbergen also expected that close cooperation could be established between the programs at Oxford and Leiden. He told Mayr: "Most probably I will keep a minor appointment here [in Leiden] for part of the year to do some additional teaching and to organize field work by both Oxford and Leiden students. As it stands now the future for a further development for work in my field looks fine."[67]

THE RETURN OF LORENZ

On 18 February 1948, after nearly four years as a Russian prisoner of war, Konrad Lorenz returned to Austria and his home in Altenberg. He arrived carrying two caged birds—a crested lark and a starling—and a large manuscript on epistemology. Had he been willing to leave the manuscript behind him, he could have returned sooner. Other Austrian prisoners from the camp in Armenia where he had been interned had begun arriving in Vienna the previous November. When his fellow prisoners left the internment camp and boarded transport trains for the west, however, Lorenz headed instead to Moscow, seeking permission to take his manuscript out of the Soviet Union with him. The manuscript, written on crude paper from cement bags, constituted in Lorenz's words "the first attempt to provide a cohesive account of a very young branch of biological research," the field of "comparative behavior study." The Russian authorities ultimately gave him the permission he sought, upon the condition that he make a copy of the manuscript and leave the copy behind in Russia. Then, bound for Austria at last, he was delayed once more in his final homecoming, this time not by the Russian authorities but by flooding of the lower Danube.[68]

Some weeks after his arrival home, Lorenz reported to Erwin Stresemann his condition on his return: "a little gray, but entirely unbroken, and with a fat book manuscript under my arm." The Russians had treated him well, he said; he had been able to finish writing and to bring home with him the first volume of a book manuscript on comparative behavior study. As for his family back in Altenberg, he had found his children "rather big, of course, but thoroughly delightful" and his wife Gretl even more beautiful and sweet than he had remembered. There was no money, but there was plenty to eat, thanks to his wife's successful running of a small farm. Everyone laughed, he said, at how fat he had become in the few weeks since his return.[69]

Lorenz's immediate plans, he told Stresemann, were to bring out his book as soon as possible. He intended to publish German and English versions simultaneously, with the latter specially adjusted to the interests of an Anglo-American readership. Once published, the book was bound to enhance his claims on a professorship. But there was also the appealing alternative of teaming up with Erich von Holst and Gustav Kramer at the brand-new Kaiser-Wilhelm-Institut (soon to be renamed Max-Planck-Institute) für Meeresbiologie (marine biology) in Wilhelmshaven. It did not matter that the institute's name featured marine biology. The institute was under Holst's direction. He additionally directed the division of "nerve and flight physiology," while Kramer directed a division devoted to the "orientation and home-finding abilities of birds." Lorenz longed to join them: "We 3 in one institute, that would really be worth the effort!"[70]

For the timing being, however, Lorenz's ambitions far outstripped the realities of his situation. He was without a position. The professorship he had held at Königsberg no longer existed—Königsberg was now Kaliningrad, in Russian hands. Nor did he have the money to rebuild a research center for himself at Altenberg. The family financial resources were virtually nonexistent, having been depleted by the devaluation of the Austrian schilling and by costly repairs to the water pipes. As an alternative to an academic position, or a share in a Max Planck Institute, Lorenz contemplated a research institute of his own funded by the Austrian government. And then too there was the possibility (once again) of the directorship of the Schönbrunn zoo. Antonius was dead, having committed suicide at the end of the war. For the time being Lorenz found himself dependent on his wife's earnings, just as he had been in his early years as a *Privatdozent*. His most immediate opportunity to do research or to interact with students was to take part in the activities of an unfunded animal behavior research station that had been set up on the outskirts of Vienna by a young couple named Otto and Lili Koenig.

The Koenigs' story was an appealing one. At the end of the war this young couple—he an ornithologist and photographer and she a talented artist—moved onto the site of an abandoned antiaircraft battery and posted signs declaring: "Biologische Station Wilhelminenberg—Eintritt verboten!" The one-time military installation lent itself readily to the new purpose the Koenigs envisioned for it. The site consisted of six square kilometers or so of land that had once been the property of an Austrian archduke. Meadows, an old forest, an area of low bushes, two ponds, and two brooks graced the property. Since the whole of it was already fenced in, it was in effect an instant nature preserve. Seven vacant army barracks stood ready to be converted into living quarters, aviaries, and laboratories. With the help of a half dozen or so students from the University of Vienna, the Koenigs undertook the task of conversion. Eventually they legitimized their territorial acquisition by leasing the land for a nominal sum.[71]

Koenig had known Lorenz before the war and been inspired by him. He was engaged in a comparative study of the behavior of different species of herons. He was also studying the bearded tit, *Panurus biarmicus russicus.* The university students, including Irenäus Eibl-Eibesfeldt, Wolfgang Schleidt, Ilse Gilles, and Heinz Prechtl, added other animals to the station and began developing ethograms of the different species represented. Eibl, for example, studied the breeding behavior of the common toad, the play behavior of the badger, and the social behavior of mice (the latter study arising contingently when mice invaded the barracks with the onset of winter).[72]

Thus, when Lorenz appeared on Koenig's doorstep, Koenig could offer him not only space at the Biologische Station Wilhelminenberg but also a group of young ethologists eager to listen to the master. Lorenz welcomed the opportunity. He lectured from his book manuscript, and he shared his knowledge in less formal ways as well. But the situation at the station was inherently unstable. The enthusiasm and the pioneering spirit of the participants were not sufficient to overcome the problems brought on by meager resources, questions of authority, and difficult temperaments.

The Koenigs had gotten the Biologische Station Wilhelminenberg started in the first place by sinking into it money they had made by publishing a book on Sicily, where Koenig had been stationed during the war. The book was a popular success, but the approximately thirty-five thousand Austrian schillings of royalties that it generated were soon exhausted, and it took two thousand schillings a month to keep the station in operation. When Julian Huxley visited Vienna in the spring of 1948 in connection with a mission for UNESCO, he learned not only of the station's work but also of its financial problems. Huxley suggested to Lorenz that Lorenz write to David Lack and the ornithologist James Fisher in England inquiring about the possibility of getting help for the station from the British Trust for Ornithology.

Lorenz did just that. He wrote Lack describing the history of the sta-

6.1 Konrad Lorenz at the Wilhelminenberg in April 1948, lecturing on bird behavior to Otto Koenig, Lili Koenig, and four students. (Courtesy of I. Eibl-Eibesfeldt.)

tion and the excellent work being done there on a mere shoestring budget. He praised in particular Koenig's work on the bearded tit, saying, "It is one of the most thorough investigations on the system of actions of one species that I know of." He told Lack he valued Koenig's unpublished paper on the bearded tit "about as highly, as that of my dear friend Margaret Morse Nice on the American Song Sparrow, *Melospiza melodia.*" He also mentioned the work of Ilse Prechtl (the former Ilse Gilles) on the ethology of the house sparrow the gaping reactions of young passerines. (He was confident that Lack, the author of *The Life of the Robin,* would be interested to hear that the young robins at Wilhelminenberg did not seem to have an innate response of gaping at the red breast of the parent.) There were in addition, he said, "at least four equally important and interesting investigations" also underway at the station. But financial problems were threatening to undermine everything. Koenig himself, Lorenz noted, had "not done a stroke of productive work for months, because he was busy writing popular articles for newspapers as fast as he could, just to earn enough money to go on feeding the most important animals."[73]

Lorenz readily acknowledged to Lack that he had a personal stake in the fate of the Wilhelminenberg. Once he finished in the next year or so his big textbook on the comparative study of behavior, he said, he would return to original research, and then he would be "entirely dependent on the Biologische Station Wihelminenberg for a place where to work!" Exclamation point followed exclamation point as he linked his plight with the station's: "If it is smashed up I shall not know where to turn to! and there are a lot of interesting questions that I intend to investigate in the near future! The problem of releasers and innate releasing mechanism will require years of intensive work yet!"[74]

Lack passed Lorenz's letter on to Thorpe at Cambridge. Thorpe only a month earlier had decided to stay at Cambridge and develop animal behavior studies there rather than try his luck at Oxford. Always a hard-nosed calculator, Thorpe viewed the difficulties at the Wilhelminenberg as a potential opportunity for Cambridge. As he described things to Huxley, it would be difficult to raise in Britain the funds needed to underwrite the Wilhelminenberg station's annual budget (twenty-four thousand Austrian schillings was the equivalent of about six hundred pounds sterling), and even if these funds were raised, there would be problems transferring them to Austria. On the other hand, if the Koenigs were capable, cooperative, and adaptive people, doing excellent work on bird behavior, they might be just the people Thorpe needed to staff the Madingley Ornithological Field Station.[75]

Lorenz had written to Lack for the express purpose of securing British support for the Wilhelminenberg station. Ultimately, however, his letter served to bring his own situation, as much as the Koenigs', to the attention of others. Thorpe responded to Lorenz's letter by proposing that he (Thorpe) pay a brief visit to Austria not only to meet the Koenigs but also to make the personal acquaintance of Lorenz and von Frisch, both of whose work he much admired. Writing to Lorenz also gave Thorpe the chance to call attention to the symposium on animal behaviour that the Society for Experimental Biology was planning for July 1949 and to express the hope that Lorenz had already received the society's formal invitation to participate. He told Lorenz just what Lorenz wanted to hear: "It is most important to get the leading workers on all aspects of the subject together for such a meeting, and quite obviously the gathering would be incomplete without you."[76]

Lorenz responded with gusto, saying he would come to the symposium "even if I have to sell the piano to do so." He was convinced, he said, "that this congress is going to be of the utmost importance. Comparative Ethology, Vergleichende Verhaltensforschung, or however you chose to call it, is quite doubtlessly developing into a school at least as original and important as Behaviorism and Pavlov's Reflexology and quite certainly a much nearer approach to an exact natural science than both. It is certainly high time to come together and give it a name!"[77]

Responding to questions Thorpe had asked about the Biologische Station Wilhelminenberg and Lorenz's book, Lorenz reported that the station would be able to last a few more months without assistance, but no longer. Koenig had recently sold his kinematographic camera to pay for the station's expenses. Lorenz's book plans included proceeding simultaneously with separate German and English versions because of the different interests of the different audiences: "In German, the fight against idealistic-vitalistic ideology is all-important, in English it is much more necessary to stress the methodological errors of atomistic Behaviorism etc. etc."[78]

Huxley, who had already seen Lorenz in Austria earlier in the year, had had a chance to read a version of the introduction of Lorenz's book manuscript. His reaction was decidedly positive. He praised Lorenz's general approach and said it was obvious "how extremely important" the book was going to be. But he also offered constructive criticism. He recommended that Lorenz discuss additional species and cite additional authors. At the same time, he observed that Lorenz's presentation was weak with respect to modern genetics and the modern understanding of natural selection. And he challenged Lorenz in particular on the subject of domestication: "I must say I think you have over-estimated the importance of degenerative change in domestication. A modern racehorse is actually an improvement in certain respects on the original horse, and not, I think, in any sense degenerate. Furthermore, not all mutations in domestic animals are loss mutations; some of them are dominant." (To help set Lorenz straight, Huxley enclosed "a copy of my book on evolution, where I hope you will find some useful matter on this.") Huxley likewise doubted several of Lorenz's assertions about human nature. He criticized Lorenz's theorizing about the genetic bases of human behavioral deficiencies: "I find it difficult to believe that much of what you ascribe to loss-mutations in human behaviour really has a genetic base at all. I think you ought, at any rate, to mention the opposite possibility."[79]

Lorenz responded with typical enthusiasm, telling Huxley he was planning to follow virtually all of Huxley's suggestions. At the same time, he insisted that he was not focusing solely on the negative side of domestication, despite the impression his introduction might have given. "In later parts of the book," he maintained, "I am putting great stress upon the fact that 'domestication' is in no way to be confounded with what is generally called 'degeneration.' In many respects the domestic duck is a much more hardy and vital being than the mallard! Moreover, the rudimentation of innate behaviour, especially of innate releasing mechanisms, very often means a decided advance in mental development, as already Whitman knew very well. It is even my thesis, that the specific freedom from the bonds of innate reactions, characteristic of human behaviour, is largely due to just this consequence of man's domestication." Nonetheless, Lorenz had not changed his mind with respect to the main ideas he had expressed so vigorously in the late 1930s and early 1940s concerning human "domestication" and morality. He told Huxley: "I think I can show rather convincingly—and I still hope to convince you—that there are certain 'moral' defects of human beings which very probably have exactly the same genetical basis as some analogous phenomena in domesticated animals. I feel this sounds rather incredible in the form of this bald statement and I shall therefore change what I am saying about it in the introduction accordingly. Also I shall stress the fact that this explanation of human behaviour is, of course, still purely hypothetical." The problem, as he went on to note, was that "it is very difficult to condense one's 'thesis' in the way necessary for an introduction without seeming dogmatic and onesided." [80]

With respect to his search for gainful employment, Lorenz reported, his hopes for the directorship of Schönbrunn had been thwarted. Someone else—a man who "does not know a thing about animals and how to keep them"—had been appointed to the position. However, the Austrian Academy of Sciences, with the aid of a new source of funding, was planning on supporting a few small research institutes. One of these was to be an institute for comparative behavior study for Lorenz at Altenberg. Being able to stay in his own country, doing research at his own home, was just what Lorenz wanted. He had feared that without funding for Altenberg, he would have had to sell the family property. But he also worried that establishing an institute for him would jeopardize the chances of survival of the Wilhelminenberg station. Joining the two institutes would not work, he believed, because Koenig was "a very independent character," and "it would break his heart to work under anybody else, even with me." The best prospect, Lorenz felt, would be to have the two institutes funded independently. [81]

Thorpe traveled to Austria late in September and visited Lorenz, von Frisch, and the Koenigs, as he had planned. Upon his return to Cambridge, he wrote a letter to Richard Meister, now president of the Austrian Academy of Sciences, recommending that the Austrian academy support Lorenz's work. Establishing an institute for Lorenz at Altenberg would not be expensive, he said, and the rewards would be considerable. It "would be a most important step in the task of coordinating and advancing the sciences of zoology, physiology, psychology and social anthropology and in leading to a more realistic and comprehensive philosophy of biology than we yet possess." [82]

To Lorenz, Thorpe was proving to be a very welcome friend indeed. Thorpe surprised and pleased Lorenz by showing great interest in the philosophical part of Lorenz's book. Lorenz had planned to downplay the epistemological side of things in the Anglo-American version of the book, but with Thorpe's encouragement, he decided not to suppress any of the "Erkenntnistheorie" after all, saying "in my heart of hearts I too think it the most important part of it, only I do not quite dare to show it." [83] And Thorpe was providing more than just moral support. Before he left for Austria, he had been alerted by the ornithologist and environmentalist Max Nicholson to the fact that the author J. B. Priestley had royalty money tied up in Austria that could not be used elsewhere. Thorpe had thereupon written to Priestley's wife describing the situation of the Koenigs and Lorenz. As of 19 October 1948, Lorenz was able to tell Thorpe: "I have just got a letter from Mrs. Priestley, telling me that her husband is dedicating his Austrian royalties to Altenberg and to the Wilhelminenberg. I am tremendously grateful. . . . Of course I can feel your hand behind this happy miracle." [84]

While Lorenz's prospects were looking up, Thorpe's plans for the Koenigs were not. On the basis of his visit to Austria Thorpe concluded that Otto Koenig would not be an ideal colleague: Koenig was an excellent independent researcher, but he was unlikely to be a cooperative co-worker. [85] Thorpe agreed with Meister, furthermore, that if Priestley's royalties were not sufficient to support both Koenig and Lorenz, then Lorenz, who was in an entirely different scientific class from Koenig, had to come first. [86]

Help for Lorenz from the Priestleys, via the Austrian academy, materialized just before Christmas, 1948. Meister presented Lorenz with 3,770 Austrian schillings and the promise of more to come. Lorenz could now look forward to repairing the roof (damaged by the jackdaws) and to buying coke to heat his study (thus enabling him to get on with writing his book). He told Thorpe: "The jackdaws are really very expensive because of the damage they do to the roof and eaves. . . . If I were a hard hearted scientist I should do away with the jackdaws but they are so much a symbol of Altenberg, being the only birds who lived through the war, that I simply cannot do it. *You couldn't either!*" [87]

With funds from the Academy of Sciences (including a five-hundred-schilling-per-month stipend), Lorenz proceeded to reestablish Altenberg as a research center. Late in January 1949, he described developments to Thorpe as follows:

> Regarding Altenberg, things begin to substantiate. Stejskal, the fish man, is already living with us in Altenberg and is bringing a great number of big aquaria which he is putting at my disposal until his greenhouse (which is going to be built in our garden) is ready. So this spring I shall be able to keep some interesting fishes, I intend to so some observational work on Cottus, a group not very closely related to Percidae and totally unknown as yet. Furthermore the Prechtls are coming to live with us on March first, which will be a great improvement because Ilse is going to do some secretarial work for me. She has other work to do for her living, but even if she helps only a little, it will mean much. I am going to ask Meister wether the Academy of Science could not give Altenberg an official standing as an "Institute," even if the actual money supply is not substantially increased. I should find it easier to do some propaganda work and collect help, if Altenberg is an official institute and not just a private house.[88]

It soon looked as if the Austrian Academy of Sciences would endow his operation with the title of "Institute for Comparative Behavior Study, under the Direction of the Austrian Academy of Sciences" (Institut für Vergleichende Verhaltensforschung unter der Leitung der Oesterreichischen Akademie der Wissenschaften). This would give Lorenz's operation the official standing he so much desired for it.[89] For this to happen, however, he needed to be reinstalled at the University of Vienna with at least a docentship. Here troubles began to surface. Writing Thorpe at the end of February, Lorenz acknowledged "some most disagreeable difficulties" had arisen. Not until May, when the crisis had passed, did Lorenz offer Thorpe an account of what had happened:

> My re-installation at the Vienna University has progressed well enough, the accusations raised against me were happily so exaggerated that their falseness was evident and it was comparatively easy to show that I could never have been an old and active member of the Nazi party. If the fellow who seems to have wished me so well, had thought of something less serious, matters would have been much worse, because appearances were against me, the mere fact that I had been made a professor at a famous German university was dangerous, because so many people really *were* made professor at that time for merely political reasons. Who would be-

lieve me, if I told them that I was called to Koenigsberg because my studies of the I. R. Mechanisms threw an important light on Kant's teachings?
Well, it is happily over now, but I really have grown some more grey hairs
in the interval.[90]

Lorenz's denial of his Nazi past was successful with the Austrian authorities as well as with Thorpe. By the end of June, he was able to tell
Thorpe that his reinstallation at Vienna was finally "perfect" and Altenberg
was able officially to be an institute belonging to the Austrian Academy of
Sciences. The scene at Altenberg was one of prodigious activity: "There is
a lot of building going on everywhere, an aviary for small birds, terraria for
Schleidt's mice, Aquaria for my Cichlids are being built and got ready,
young sparrows are being reared for Ilse Prechtl's work and my bearded
tits, caught in the general fever of productive work, have successfully reared
one young (which afterwards perished through a stupid accident, but must
be counted as a great success nevertheless) and are incubating in a new and
better nest on four new and better eggs." [91] As Thorpe had recently learned
from Meister, thanks to the funds provided by Priestley's royalties, Lorenz
was to receive a total of some twenty thousand Austrian schillings for setting up his Institute. Three thousand Austrian schillings would be left over
for Koenig at the Biologische Station Wilhelminenberg.[92]

Lorenz was still without much of an income. The five hundred schillings per month from the Austrian academy was certainly a help, but it was
not a proper salary. Nonetheless, Lorenz had reason to be cheerful. For the
immediate future he could look forward to traveling to England for the
Cambridge symposium and a whole series of other lectures, film presentations, and visits that Thorpe and others were lining up for him.[93]

THE 1949 SYMPOSIUM OF THE
SOCIETY FOR EXPERIMENTAL BIOLOGY

The Society for Experimental Biology's symposium "Physiological Mechanisms in Animal Behaviour" originated as the brainchild of Tinbergen and
Thorpe. From the very beginning of the war, Tinbergen was looking ahead
to organizing a postwar symposium on animal behavior. After the war, he
found Thorpe eager to collaborate on such a venture. Thorpe was already
a convert to Lorenzian ethology when Tinbergen invited him late in 1945 to
serve as a coeditor of the new journal *Behaviour*. The two men first met in
February 1946, during Tinbergen's first postwar trip to England. Thorpe arranged for Tinbergen to stay in Cambridge at Jesus College, where Thorpe

was a don. Then Tinbergen hosted Thorpe at Tinbergen's home when Thorpe visited Leiden two months later. When Dutch biologists and members of the British Society for Experimental Biology (SEB) jointly staged a successful conference in Utrecht, this encouraged Tinbergen, Thorpe, and several members of the British Association for the Study of Animal Behaviour (ASAB) to think of having the SEB and the ASAB co-sponsor an international symposium on animal behavior.[94]

Thorpe drafted a list of potential topics for the symposium that was essentially a recital of the major themes of Lorenzian behavior theory: the "stereotyped nature of instincts"; the relation of instincts to reflexes, chain reflexes, and taxes; "instincts as endogenous stimulus production phenomena"; "reaction specific energy & specific exhaustibility"; the "Taxonomic & Phylogenetic importance of instinct;" "the releaser concept;" "learning & instinct;" and "substitute activities." The subtopics included innate releasing mechanisms, the experimental study of releasers, appetitive behavior, imprinting, and Thorpe's own favorite topic, latent learning.[95]

Thorpe and Tinbergen recognized from the beginning, however, that a symposium sponsored by the SEB could not be just the ethologists' own show. The SEB included physiologists whose views differed dramatically from the ethologists' with respect to the way motor patterns are controlled. Among them were James Gray, the head of Thorpe's department, and Hans Lissmann, a Russian-born German who had emigrated to Cambridge in the 1930s. They believed motor patterns were peripherally rather than centrally controlled, contrary to the view of Holst and Lorenz. Thorpe included Gray and Lissmann as well as Holst and Lorenz on his list of potential participants. For his own part, Thorpe suspected that Lorenz's instinct concept could probably survive the critiques of the Cambridge physiologists but Lorenz's identification of Holst's locomotory movements with pure instincts might not.[96]

When it came to recommending to the SEB's council that "the analysis of behavior" be the SEB's symposium topic for 1949, Thorpe did not commit himself to one side or the other in the dispute over central versus peripheral control. He simply identified five themes that the symposium might profitably address: "(1.) Rhythm: Central or Peripheral; (2.) Instincts, Chain reflexes and Release mechanisms; (3.) Orientation and Taxes; (4.) Instinct and Learning; and (5.) Latent learning and Insight." If time permitted, he suggested, these five topics could "lead on naturally" to a sixth topic, social behavior.[97]

The SEB council agreed to make the analysis of behavior the topic of its 1949 symposium. As the planning of the program progressed, social behavior fell by the wayside, and latent learning and insight were placed with habituation, conditioning, and imprinting under the general heading of "Instinct and Learning." This left four main areas for discussion:

(1) Rhythm: Central or Peripheral.
(2) Instincts, Chain Reflexes, Release Mechanisms and Substitute Activities.
(3) Orientation and Taxes.
(4) Instinct and Learning.[98]

Deciding that it would be possible to have twenty main speakers and a total of twenty hours for the formal papers and the discussion, the organizers set about identifying whom to invite.

The SEB symposium "Physiological Mechanisms in Animal Behaviour" was held in Cambridge from 18 to 22 July 1949. Four days of sessions were held over the five-day period. The symposium was a major event in the history of ethology for several reasons. In the first place, it figured prominently in the resumption of personal contacts that the war had disrupted. In the second place, the symposium provided a highly visible forum for the ethologists to set forth the main theoretical and methodological claims of their work, at least as these related to physiology. Thirdly, the symposium constituted the first major forum at which the ethologists found themselves cheek by jowl with physiologists whose approaches to the study of animal behavior were significantly different from those of the ethologists (despite Tinbergen's earlier efforts to speak of ethology as a branch of physiology).

The reuniting of Lorenz and Tinbergen in Thorpe's home in 1949 was a significant moment. The two men had already reestablished contact with one another through letters. They had furthermore made clear to each other their desire to pick up their prewar friendship where they had left off. Tinbergen, for example, had invited Lorenz to stay with him for a week or more in Holland after the conference in Cambridge was over. Nonetheless, it was at Thorpe's that the two men saw each other for the first time since before the war began.

Thorpe, years later, recalled the reunion of Lorenz and Tinbergen as "a moving event."[99] Lorenz, likewise years later, recalled of the reunion: "Though [Niko] had spent years in a German concentration camp and I even longer in a Soviet prisoner of war camp, we found that this had made no difference whatsoever, which Niko put in a nutshell by saying: 'We have won.'"[100]

Although Lorenz's recollection of Tinbergen's words rings true enough, his "no difference whatsoever" comment does not. Tinbergen and Tinbergen's country had suffered severely at the hands of the Germans. Tinbergen knew that Lorenz, to at least some degree, had been a supporter of the Nazis and German militarism. That Lorenz himself had subsequently been a longtime prisoner of war in a Russian camp seems to have made a real psychological difference to Tinbergen. It seems to have served, in a general sort of way, to balance things out. It made it easier for Tinbergen to do what he very much believed he should do, which was to forgive Lorenz, put the experience of the war behind them, and get on with the scientific work that the war had interrupted.

Whatever Tinbergen's personal feelings and reasoning in this regard may have been, the crucial thing for ethology's subsequent development was that Tinbergen's forgiving stance toward Lorenz was visible for all to see. As of the 1949 Cambridge symposium, Tinbergen and Lorenz were once again enjoying each other's company, much as they had done before the war intervened. A good example of their friendly interplay at the Cambridge meeting has been recounted by Robert Hinde: "We were walking down Jesus Lane in Cambridge, and Tinbergen and Lorenz were discussing how often you had to see an animal do something before you could say that the species did it. Konrad said he had never made such a claim unless he had seen the behaviour at least five times. Niko laughed and clapped him on the back and said 'Don't be silly, Konrad, you know you have often said it when you have only seen it once!' Konrad laughed even louder, acknowledging the point and enjoying the joke at his own expense."[101]

While the SEB symposium was the occasion for the personal reunion of Tinbergen and Lorenz, and it furthermore enabled them and others to establish important new contacts while also renewing old ones, the meeting was not harmonious in all respects. Bringing contrasting conceptual frameworks and diverse personalities into the same arena inevitably led to some overt clashes and various misunderstandings. Although Thorpe had originally envisioned the symposium as an opportunity for promoting Lorenzian instinct theory,[102] the symposium ended up as a meeting where the ethologists were in the minority, ethological papers constituted only a quarter to a third of the papers given, the study of instinct was one theme out of four, and the ethologists had to confine their focus to matters of physiological causation. To be sure, the ethologists were not displeased with this focus. On the subject of the physiological causation of behavior, they felt they had a major contribution to make. But this was not a subject on which their views would go unchallenged.[103]

Four ethologists came to the meeting from the continent: Lorenz from Altenberg, Tinbergen from Leiden, Baerends from Groningen, and Koehler from Freiburg. Holst in Wilhelmshaven prepared for the trip, but three days before the conference was to be begin he canceled his travel plans, sidelined by the recurrence of a heart problem. The majority of the participants in the symposium were from the departments and laboratories of zoology and physiology at the University of Cambridge: Gray, Lissmann, E. D. Adrian, J. S. Kennedy, C. F. A. Pantin, R. J. Pumphrey, J. E. Smith, and Thorpe himself. Also from Cambridge was the Reverend E. A. Armstrong, an amateur ornithologist who had been Thorpe's close friend since their school days. From University College, London came the physiologist G. P. Wells. From the Department of Zoological Field Studies at Oxford came the ornithologist P. H. Hartley. Finally, the physiological behaviorist Karl Lashley and the physiologist Paul Weiss traveled to the symposium from the United States.

The ethologists' cause was not helped when Holst, the only real physiologist among them, was unable to attend the meetings. Not only did they miss his expertise in their discussions with the physiologists, but he made the situation worse by mailing in to the symposium, at the last minute, comments sharply critical of Gray's and Lissmann's position on reflexes and peripheral control. Holst may have supposed that Lorenz, an excellent speaker of English, would be entrusted with the job of presenting the communication to the assembly, and that Lorenz would represent Holst's position diplomatically. Instead, Holst's paper was turned over to Lissmann for translation and presentation. The results were unfortunate.

Lissmann was a firm supporter of the idea of peripheral control—the idea that Holst was attacking.[104] On the conference's first day, which was devoted to the range of capabilities of sense organs, Lissmann gave a paper on proprioceptors and showed his hostility to Holst's idea of a central coordination independent of these sense organs. On the second day of the sessions, devoted to central and peripheral control of behaviour patterns, it was left to Lissmann to translate Holst's critique of the Gray-Lissmann position. Lissmann translated the critique word for word, apparently to everyone's discomfort but his own. As both Koehler and Lorenz reported to Holst after the event, the scene was an awkward one. Holst had not chosen a style suitable for the occasion, and the tone of his commentary, especially as delivered by Lissmann, appeared arrogant, disparaging, and offensive. Lissmann exacerbated the effect, Koehler told Holst, by smirking in the most embarrassing places. The paper was greeted with silence in the formal session, but it generated private discussions throughout the rest of the meeting. Koehler and Lorenz laid much of the blame on Lissmann. Tinbergen, for his part, had to explain to Koehler and Lorenz just how much the Germanic tone of Holst's paper grated on non-German ears so soon after the war.[105] Lissmann, who within two years would make the important discovery that certain fish can sense weak electric currents in their vicinity, was himself sensitive to the political as well as scientific currents that charged the atmosphere of the Cambridge symposium. He had refused to live in Germany after Hitler came to power there. It was his understanding that Lorenz, in contrast, had been sympathetic to the Nazis before Germany took over Austria.[106]

Holst's critique of Gray and Lissmann followed upon the papers presented by Adrian, Weiss, Gray, and Wells, all of which related to the theme of central versus peripheral control. The next day the conference theme was instinctive behavior. Lorenz, Tinbergen, Baerends, Koehler, and four additional participants gave papers."[107] The ethologists surveyed the sorts of topics that had come to characterize their field. These included the concept of the releaser (Baerends), the taxis component of instinctive behavior (Koehler), displacement activities (Armstrong), the mobbing of owl dummies by wild birds (Hartley), the hierarchical arrangement of instincts (Tinbergen), and the comparative method of studying animal behavior (Lorenz). What most piqued the physiologists' attention was a pair of diagrams. One was Lorenz's representation of his hydraulic model of instinctive behavior. The other was Tinbergen's sketch of the hierarchical arrangement of instinctive behavior systems.

In the published proceedings of the symposium, Lorenz's paper is by far the longest.[108] Indeed, it is fully twice as long as most of the other contributions. He was not able to present it as such at the symposium, since he, like the others, was given only an hour. Recognizing his paper was too long before he arrived in England, he thought of delivering part of it to the SEB Symposium and part of it to the Oxford Society for Experimental Psychology, to which he had also been invited to speak. In the end, though, he simply abbreviated the SEB paper and presented an entirely different paper to the experimental psychologists.[109]

Lorenz began his SEB paper by rehearsing the kinds of claims he had developed in his instinct papers of the 1930s.[110] Primary among these was his contention that the long-standing dispute between vitalists and mechanists had served to block a proper understanding of the nature of organic systems, the directedness of behavior, the apparent spontaneity of behavior, and the true nature of innate or instinctive behavior. The great strength of the comparative ethologists, he allowed, was that in contrast to the vitalistic and mechanistic schools of behavior study, the ethologists had erected their science on a secure inductive foundation. In characteristically sweeping fashion, he explained:

> It is an inviolable law of inductive natural science that it has to begin with pure observation, totally devoid of any preconceived theory and even working hypothesis. This law has been broken by one and all of the great schools of behaviour study, and this one fundamental, methodical fault, is at the bottom of all the errors of which we accuse vitalists and mechanists. To put it crudely in two examples. If William McDougall had known all H. Elliot Howard knew about 'reactions incomplete through lack of intensity,' he would never have confounded survival value and purpose. If J. B. Watson had only once reared a young bird in isolation, he would never have asserted that all complicated behaviour patterns were conditioned.[111]

That the pioneers of comparative ethology had managed to assemble a sound and thorough knowledge of animal behavior, Lorenz insisted, was primarily due to the fact that they were, first and foremost, animal lovers. Whitman's passion was pigeons; Heinroth's was ducks and geese. As Lorenz wanted to have it, "Happily ignorant of the great battle waged by vitalists and mechanists on the field of animal behaviour, happily free from even a working hypothesis, two 'simple zoologists' were just observing the pigeons and ducks they loved, and thus kept to the only way which leads to the accumulation of a sound, unbiased basis of induction, without which no natural science can arise."[112]

These men had also been successful, Lorenz suggested, because they were not only animal lovers, but they were also animal *keepers*. As opposed to observers of animals in the wild, he explained, keepers of animals were much more likely to notice when behavior patterns "miscarried." When this happened, when behavior patterns were activated by the "wrong" stimuli, it was possible to recognize the *particulate* nature of instinctive behavior, whereas in the wild instinctive behavior was more likely to impress the observer by its apparent adaptiveness or purposefulness. Animal keepers also recognized that innate behaviour patterns were not something that could be varied by the animal according to the requirements of the situation. Rather, innate behavior patterns were "something which animals of a species 'have got,' exactly in the same manner as they 'have got' claws or teeth of a definite morphological structure."[113] Such motor patterns, in other words, were like organs, and they could thus be used, like morphological structures, in *comparing* related species. It was on this basis, Lorenz said, that Whitman had written in 1898: "Instincts and organs are to be studied from the common viewpoint of phyletic descent." This insight, Lorenz allowed, marked "the birth of comparative ethology."

Lorenz went on to describe how one goes about constructing a "comparative anatomy of behavior." As the most thorough examples of such research up to that time, he cited Whitman's work on pigeons, Heinroth's work on ducks and geese, his own work on ducks, and the different contributions of A. Seitz, G. P. Baerends and J. Baerends van Roon, and A. Steiner on cichlid fishes. He also outlined his notion of releasers and how releasers had evolved. But the evolution of instinctive behavior patterns was not the subject of the SEB symposium, any more than was the mechanist-vitalist debate or the philosophy of science. At issue instead were the physiological mechanisms in behavior.

Lorenz devoted the second half of his paper to the ethologist's approach to physiology. He began by acknowledging that Whitman and Heinroth "were phylogenists and not physiologists."[114] He might appropriately enough have said the same of himself. He was prepared, nonetheless, to launch into a discussion of the physiology of behavior, beginning with an account of "the accumulation of action-specific energy." This was not an arbitrary notion, he insisted, but rather a concept that the facts forced upon the unbiased observer. Whitman, Heinroth, and Craig, he claimed, had all been inductivists. Without self-consciously developing any hypotheses about underlying physiological mechanisms, they had been led by the evidence to use words like "damming up," "discharging," and so forth in their discussions of instinctive behavior. The facts of animal behavior brought them unerringly to the implicit assumption *"that some sort of energy, specific to one definite activity, is stored up while this activity remains quiescent, and is consumed in its discharge"* (Lorenz's italics).[115]

The kinds of facts Lorenz had in mind consisted especially of what he for the time being called "energy accumulation activity" (following the suggestion of E. A. Armstrong) and "threshold lowering." The threshold for releasing the behavior in question is gradually lowered, he explained, to the point where "the activity in question will finally go off *in vacuo*, with an effect somewhat suggestive of the explosion of a boiler whose safety valve fails to function."

How was one to study such phenomena? Lorenz offered his auditors a set of procedures with a fancy name: "the method of dual quantification." This "method" corresponded to what Lorenz presumed to be the two causal factors at work in producing instinctive actions: "(1) the level attained by the accumulated action-specific energy at the moment and (2) the effectiveness of external stimulation."[116] But it was not possible to measure either of these factors directly. What were visible to the observer were "(a) the stimulus situation we are putting before the animal, and (b) the discharge of a specific activity thereby released."[117] The task for the investigator was to determine how much of the intensity of a discharged reaction was due to the internal accumulation of action-specific energy and how much was due to external stimuli.

Despite Lorenz's references to "quantification" and "measurement," what he actually offered was not a mass of painstakingly acquired quantitative data but instead a qualitative "hydro-mechanical model." He acknowledged the model's "extreme crudeness and simplicity," but he allowed that it was nonetheless able "to symbolize a surprising wealth of facts really encountered in the reactions of animals."[118]

His model and the diagram he used to represent it are reproduced below. As he explained, in the diagram, the tap T is constantly supplying the reservoir R with an endogenously produced liquid, corresponding to the "action-specific energy." The liquid's discharge through the cone valve V (a part of the releasing mechanism) is prevented by the spring S (the inhibition provided by the higher organizing centers). The rest of the releasing mechanism (the perceptual side of it) is represented by the scale pan SP and the string over the pulley connecting the scale pan to the cone valve. The *weight* on the scale-pan represents the impinging stimuli. In this setup, both the accumulation of liquid in the reservoir and the weight provided by the external stimuli tend to open the valve. The jet of liquid corresponds to the instinctive reaction itself. The distance the jet carries corresponds to the reaction's intensity (as measured on the scale G). The trough Tr beneath the scale has a series of outlets, representing how a series of different activities may be elicited depending on the strength of the reaction (as, for example, in cichlid fish, where the different behavior patterns displayed in fighting have different thresholds).

Lorenz wanted to have the best of both worlds with this model. He wanted to be able to appeal to it without at the same time being called to task for its deficiencies. He stated, "This contraption is, of course, still a very crude simplification of the real processes it is symbolizing, but experience has taught us that even the crudest simplisms often prove a valuable stimulus to investigation."[119] With the model as a guide, he maintained, one could proceed to conduct revealing experiments. Specifically, one could test which stimuli activate an innate releasing mechanism under which circumstances. This could be achieved by presenting the experimental subject

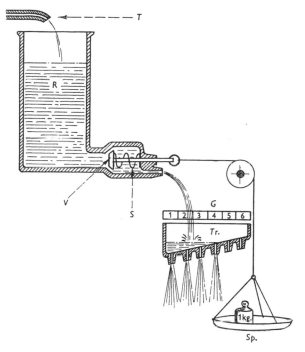

6.2 Konrad Lorenz's psycho hydraulic model of instinctive action (1950).

with a series of dummies, each successively improved upon, while the subject's action-specific energy reservoir was progressively "pumped out" by repeated discharges.

There was much more to Lorenz's paper than his hydromechanical model of instinctive action. Among other things, and as in so many of his other writings, he concluded with a warning about the predicament of humankind, accompanied by the suggestion that biological knowledge, rightly applied, could perhaps save the day. In this case the danger he identified was human aggression. "It is high time," he wrote, "that the collective human intellect got some control on the necessary outlets for certain endogenously generated drives, for instance 'aggression,' and some knowledge of human innate releasing mechanisms, especially those activating aggression. Hitherto it is only demagogues who seem to have a certain working knowledge of these matters and who, by devising surprisingly simple 'dummies,' are able to elicit fighting responses in human beings with about the same predictability as Tinbergen does in sticklebacks."[120] A decade and a half later, when he published his best-selling book *On Aggression*, Lorenz's understanding of aggression became a highly visible subject of contention.

In Cambridge in 1949, however, what the physiologists found most interesting—and most disputable—was his hydromechanical model and the
hypothetical "action-specific energy" that allegedly made the system go.

Tinbergen's symposium paper was as characteristic of Tinbergen as
Lorenz's paper was characteristic of Lorenz.[121] In contrast to Lorenz's mix
of bold assertions, wide-ranging examples, and engaging asides, Tinbergen's paper, "The Hierarchical Organization of Nervous Mechanisms Underlying Instinctive Behavior," was a more sober and more sharply focused
production. Unlike Lorenz's paper, furthermore, it was designed to fit the
time allotted for it. Yet it was not narrow in its aims. It was an effort at disciplinary bridge building. Tinbergen emphasized points of contact and co-
operation between ethology and neurophysiology. Although he, like Lorenz, set forth a model to account for specific aspects of the physiological
causation of behavior, his model, significantly and symbolically enough,
involved the incorporation of his domain of interest into a general scheme
of physiological causation previously proposed by the physiologist Paul
Weiss.

Tinbergen began his paper with an account of the kinds of ethological
evidence that had contributed to the notion of a hierarchical arrangement
of nervous mechanisms in the central nervous system. He had come to the
idea initially, he explained, through his work on the reproductive behavior
of the three-spined stickleback. In the case of the stickleback, as he described it, the awakening of the male's reproductive instinct in the spring
leads the fish to migrate to shallow freshwater. In this first stage, the highest, most general level of the reproductive instinct is activated, and this results in appetitive, seeking behavior of a generalized kind. In order for the
next level of instinctive behavior to be activated, the fish has to encounter
specific stimuli, namely, a rise in temperature and the visual stimulus of
vegetation. These serve to elicit territorial behavior. Additional stimuli, the
visual stimuli either from specific plants or from an intruding male displaying breeding coloration, elicit nesting or fighting, respectively. Fighting
in turn takes the form of different sorts of movements—biting, chasing,
and so on—each of which is dependent upon specific stimuli received
from the intruder.

Tinbergen described this behavioral progression as a successive nar-
rowing-down of the potential responses on the male stickleback's part.
Moving from the more general to the more specialized level of activation,
the animal becomes increasingly ready to perform certain responses, but
not others. This could only mean, Tinbergen said, that the hierarchical sys-

tem was "a system of nervous 'centres,' the higher centres controlling a number of centres of a next lower level, each of these in their turn controlling a number of lower centres, etc." [122] He made no specific claim about the actual physiological basis of these "centers" beyond saying that what was involved was "a hierarchical system in the central nervous system." Provisionally, he said, one ought simply to think of the centers as "functionally characterized systems."

Tinbergen explained his system further by calling on Wallace Craig's distinction between appetitive behavior and consummatory actions. Most ethological studies, Tinbergen noted, had focused on the consummatory actions. Less attention had been given to the more complicated, appetitive behavior leading up to these actions. Baerends's work on the digger wasp, Tinbergen allowed, was an important exception that showed how the analysis of appetitive behavior could be carried further. For his auditors in Cambridge, Tinbergen used the hunting behavior of the peregrine falcon to illustrate his understanding of how appetitive behavior is organized. He summarized this understanding as follows: "Activation of the centre of the highest level results in appetitive behaviour of a generalized kind. This is carried on until a new stimulus with a more restricted effect releases a subordinated type of appetitive behaviour. This again is continued until the next stimulus releases a still more restricted type of appetitive behaviour and this is carried on until the consummatory act is released." [123]

Tinbergen went on to explain that according to this model the releasing stimuli do not actually call forth reflexive responses; they instead serve to disinhibit already activated centers. As he put it, "The instinctive centres seem to be in a state of readiness, they are constantly being loaded from within, but their discharge is prevented by a block. . . . The adequate sign stimuli act upon a reflex-like 'innate releasing mechanism' (I. R. M.), and this mechanism, upon stimulation, removes the block." [124]

Tinbergen proceeded to set forth a "graphic model" of the instinctive behavior with which he was most familiar, the reproductive instinct of the male three-spined stickleback. With circles representing "centres," elongated rectangles representing inhibitory blocks, arrows representing causal factors, schematic pictures of nerve cells representing innate releasing mechanisms, and drawn lines representing motor paths, he pictured the organization of an "instinctive centre" (see fig. 6.3).

Tinbergen's discussion of the stickleback's behavior at the "territorial center" level and lower illustrates nicely his conception of how "hierarchical centers" function:

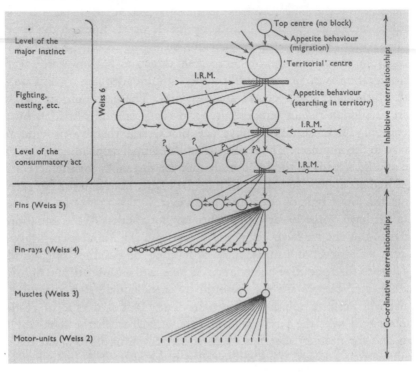

6.3 Niko Tinbergen's model of the hierarchical organization of drives (1950).

As long as the blocks of the fighting, building, etc., centers are not removed by the adequate sign stimuli, the animal will perform the appetitive behaviour. This is a type of restless swimming all over the territory, while the animal is on the look-out for rivals and for nesting materials. This is carried on until, for instance, an intruding male appears. This removes the block of the fighting centre; all impulses flow to this centre, and the animal stops its aimless wandering over the territory and attacks. The type of attack, that is to say, the activation of one of the five next lower centers, is decided by the behaviour of the intruder; his movements finally provide the stimuli which call forth (or rather enable the fish to perform) one special type of fighting which is a consummatory act.[125]

Keen to join the ethologists' cause with that of the neurophysiologists, Tinbergen allowed that "the existence of internal, intrinsic central nervous mechanisms responsible for and controlling coordinated motor patterns of an order of complexity of the consummatory acts" had been made "highly probable" by the work of neurophysiologists. He named in this regard four particular individuals, all of whom had been invited to the symposium: Holst, Gray, Lissmann, and Weiss. Ignoring for his immediate purposes the differences among these researchers, Tinbergen credited them with showing "that the relatively low centers, of the level of the consummatory act and locomotion, have their anatomical basis somewhere in the spinal cord."[126]

Tinbergen went on to explain that after coming to his own ideas on the hierarchical organization of instinct, he had read Paul Weiss's 1940 paper on self-differentiation of central nervous patterns. There Weiss also stressed the hierarchical organization of central nervous mechanisms. Weiss's focus, however, was *below* that of Tinbergen's lowest level (i.e., below the level of the consummatory act). Weiss's system had six levels. Tinbergen happily announced that the whole complex of systems he himself was considering fitted "without any trouble" into Weiss's system at the sixth and highest level, the level of the behavior of the animal as a whole.[127]

Remarking on the progress instinct studies had made in the past two decades, Tinbergen observed that whereas earlier writers had distinguished instincts by the special ends to which they were directed (reproduction, feeding, escape, etc.), it was now possible to distinguish between instincts on a neurophysiological basis. "We are justified in concluding," he said, "that in any definition of an instinct its neurophysiological foundation will have to be mentioned just as well as the end toward which it is directed."[128]

In emphasizing the ties between ethology and neurophysiology, Tinbergen was not simply trying to bring about the rapprochement of the different interests represented at the Cambridge symposium. He was pursuing the trajectory he had followed ever since the 1930s, when he distinguished his "objective" approach from that of his animal psychologist countrymen Portielje and Bierens de Haan. He was still concerned with confirming ethology's independence from a subjectivistic animal psychology. Acknowledging that there was a subjective side to the phenomena he was studying—one knew from introspection, he said, that instinctive activities have emotions attached to them—he maintained that psychological results did not necessarily conflict with the results of the ethologists' objectivistic approach. What would be "of the greatest importance for future ethological research," he insisted, was "to clarify the relations between these two types of research."[129] But there can be no doubt about where Tinbergen stood at this time with respect to whom he wanted to have as allies. To make ethology a credible science, he believed, it was necessary to make it more like neurophysiology and less like subjectivistic psychology.

On the last day of the conference, the theme was learning. Among the contributions were a presentation by Thorpe on concepts of learning and their relation to concepts of instinct, plus Lashley's now-classic paper "In Search of the Engram," where Lashley surveyed his thirty years of trying to find specific memory traces in the brain.

The Cambridge symposium featured not only scientific papers but also a number of films (Lorenz presented his film on the ethology of the greylag goose) and a series of roundtable discussions. Organized by Thorpe, the roundtable discussions were designed to bring some uniformity to the terms used to describe and account for animal behavior. Thorpe was concerned about the way familiar terms were taking on new meanings and leading to various misunderstandings as animal behavior studies expanded. Others shared this concern.[130]

Not everyone gets excited about sessions dealing with the definition of terms, and not everyone attended Thorpe's nomenclature sessions, but it was certainly true that different understandings of the same word—"instinct," for example—had been a source of major confusion in the past. It was also widely agreed at the symposium that Lorenz's phrase "action-specific energy" was a phrase that raised a great many problems.[131] In the informal exchanges among the participants, the topic of "action-specific energy," like the "fluid" it was intended to represent, kept bubbling up. Lorenz left the conference acknowledging that "energy" was perhaps a misleading word. He had agreed, at least in principle, with the suggestion that the more neutral phrase "specific action potential" (SAP) be used as a simple description of the potentiality of an animal to display a certain type of behavior at a given time. At the same time, he was convinced that what he had characterized as "action-specific energy," in the way it corresponded to Holst's "automatic-rhythmic stimulus production," remained the key issue regarding the physiological mechanisms constitutive of instinctive behavior.[132]

At the end of the symposium, the ethologists felt generally pleased with the results, notwithstanding the fiasco of Holst's communiqué, the lack of consensus regarding central versus peripheral control, and the physiologists' doubts about Lorenz's concept of action-specific energy. Tinbergen wrote to Lack immediately afterward, saying: "The Cambridge symposium was a most remarkable success and I am more convinced than ever that I did the right thing in finally seizing the opportunity created by you, to settle in the so highly stimulating English environment. Meeting people like the Cambridge group, Young, Lashley, Weiss and so many other world class workers has been a tremendous spur to us, and I believe that several of them begin to see that we too are not altogether fools."[133]

Lorenz prepared a report on the conference for the Austrian Academy of Sciences. He noted how interested both Lashley and Weiss had been in his own findings and Tinbergen's. The two Americans, he said, had been particularly intrigued by the idea of innate releasing mechanisms, and they had urged him not only to write a short book on ethology in English but also to come give lectures in the United States in the following year.[134]

Lorenz reported more informally to Holst on the fruitfulness of the conference, particularly with respect to how the different participants kept coming back to the same phenomena involving the central nervous system. He had been impressed, he said, by the way the complementary approaches of Weiss and Tinbergen to behavioral phenomena at different levels of integration promised to bridge the gap between physiologists and ethologists. However, it was "A great SHAME!," he exclaimed (capitalizing and accentuating the word "shame" [*Jammer*] with letters three times as large as the rest of his writing), "that you weren't there!" Lorenz allowed that all the participants (except for Lissmann, who had manifestly enjoyed the consternation caused by Holst's absence) had regretted Holst's inability to attend. The ethologists' cause would have been greatly strengthened, he said, had Holst been present to defend the idea of endogenous stimulus production.[135]

Koehler also wrote to Holst, describing the conference as "most taxing but really rewarding." Even in Holst's absence, Koehler said, "an essential step forward" had been taken, and the prospects for the future looked good: "Thorpe is wholly committed to us, Tinbergen and Lorenz worked superbly, and Tinbergen's call to Oxford, which was acclaimed most warmly, is the best guarantee of a long-lasting effect."[136]

Tinbergen reported on the conference to Mayr in similarly favorable terms, saying "several first-rate neurophysiologists attended and joined the discussion and told us where we were wrong (as well as heard where we thought they were wrong). Adrian, Gray, Weiss and Young were present and gave addresses. Lashley was there too; it was the first time I met him, and he made quite an impression on us all." Tinbergen added: "It was an extremely useful and stimulating conference; I have never worked so intensely as during those days." He also noted that the conference had afforded him the pleasure of seeing Lorenz, Lorenz's wife, and Koehler again, and that Lorenz was coming to stay with him in Leiden "for about a fortnight."[137]

Lorenz's postconference travels began with a selective tour of England, including a week's visit to the Priestleys on the Isle of Wight, a stay at the ornithologist Peter Scott's at Slimbridge, and a trip to Oxford.[138] From

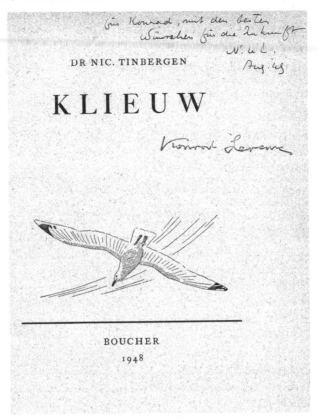

6.4 The title page of the copy of *Klieuw* that Niko Tinbergen gave to Konrad Lorenz in August 1949 with the inscription "Für Konrad, mit den besten Wünschen für die Zukunft" (For Konrad, with best wishes for the future). The title page also bears Lorenz's signature stamp. (Illustration courtesy of the Konrad Lorenz Institute for Evolution and Cognition Research.)

there he went to Holland to stay for a week or so with Tinbergen in Leiden, during which time he also visited Hulshorst and talked with Jan van Iersel. He took home from his visit with the Tinbergen family a gift that was small in size but large in symbolic significance—an inscribed copy of Tinbergen's recently published book, *Klieuw,* one of the two stories Tinbergen had composed for his children while he was a prisoner of war of the Germans. The Dutch naturalist's interest in putting wartime experiences behind him was signaled by the inscription: "Für Konrad, mit den besten Wünschen für die Zukunft" (For Konrad, with best wishes for the future).[139]

Lorenz did not get back to Altenberg until the beginning of September, at which point he sat down to write various letters. One of these was to

Wallace Craig, to whom he painted a rosy picture of the Cambridge congress, saying: "The 'Symposium on Physiological Mechanisms in Animal Behaviour' really was altogether a success. I do not think that ever before so much of a real understanding between physiologists and behaviour students sensu stricto has been reached." [140]

Lorenz told Craig that what had impressed him most at the Cambridge conference was Tinbergen's paper on the hierarchical organization of instincts. Visiting Tinbergen in Leiden had evidently reinforced this impression, for when he wrote to Thorpe early in September he was thinking hard about the consequences of Tinbergen's hierarchical model for his own notion of action-specific energy. Following a long discussion with Tinbergen in Leiden, the issue, as Lorenz saw it, was "on what level of integration the exhaustibility and accumulation of A.S.E. is a real thing." As he explained to Thorpe:

On the highest levels of integration, for instance when the stickleback is appetitively searching for the stimulus-situation in which he can turn red and territorial, the appetitive behaviour is evidently not, or not to an appreciable degree, caused by the state of accumulation of the several consummatory activities at the lowest end of the hierarchy. Quite certainly it is the hormones which are causally responsible for this generalized appetitive behaviour. But even one rung down the ladder, when the fish has reached the situation, in which his releasing mechanism responds to the right habitat for breeding, and shunts his appetites to the next-more-special step, making him beautifully red and territorial, the picture changes! Now, it is quite certainly of the greatest influence upon his subsequent appetitive activities, whether his fighting, or his courting reactions are more accumulated or more exhausted. On these intermediate niveaux of integration, all I have said about A.S.E. holds doubtlessly true! But one rung still farther down the ladder, it evidently does not: What we call "one" activity, fighting, for instance, consists of 5 (in Astatotilapia) or even 8 (in Betta) very different endogenous movements, all of which are activated by *the same* A.S.E., only at different thresholds of specific excitation. The high level (searching for habitat) and the low level (fin-spreading gill-membrane expansion etc.) both are not activated by an energy specific of this particular item! This is not so very clear, but I am not so very clear about it myself! On the other hand, my very long talk with Jan van Iersel in Hulshorst has led me to one definite conclusion: The phenomena of displacement are, in themselves, the most coercing argument for the specificity and spontaneous generation of some sort of "energy." (I think that the phenomena in question are so very far from elementary physics that it can hardly lead to misconception, if we continue to use the term energy! I said so to Niko and, come home, found a letter from Erich von Holst, in which he says exactly the same.)[141]

To Lorenz, the course of action he needed to follow seemed clear: "What I am going to do next, and personally, is to do some good quantitative studies upon the mutual interaction of thresholds, say, of fighting and courting in some cichlid. I felt quite ashamed when, talking with Kennedy, I was not able to prove what I actually am sure I *know,* by giving protocolls of exact experiments!" In point of fact, however, experimentation was not Lorenz's forte, and as he settled back in at Altenberg after his nearly two months away, there were other tasks that needed his attention. Finishing the book manuscript was one. Establishing a stronger institutional base for himself was another.

The Cambridge experience whetted the ethologists' appetites for further international meetings. But they were not primarily interested in settings where they would have to explain and defend their views and approach. They were thinking instead of small gatherings where they could come together to discuss with one another their newest ideas. Holst made a suggestion of this sort just before the Cambridge meeting took place. He invited Lorenz to stop in Wilhelmshaven on his way back from England, and he expressed the hope that Koehler might do the same. He thought too of Tinbergen, "who lives so close by that it would be almost insulting if he didn't come here once sooner or later." Holst raised the possibility of holding a small-scale symposium in Wilhelmshaven the following summer.[142]

Lorenz could not visit Wilhelmshaven on his way back from England because he had already planned to spend time in Leiden with Tinbergen, but he liked the idea of periodic, small-scale gatherings. A variant of the idea showed up in a letter he wrote to Thorpe upon returning to Altenberg. Commenting on "what a good friend you have been to me during this last year," Lorenz told the English biologist: "I admit that it is, on principle, a good plan, to have different representatives of the same branch of science in different localities, and not huddled together into one institute. But you, Niko and I must in some way develop a fixed habit of meeting about once a year for at least a fortnight or so. I felt like a lost greylag gosling after I had parted from you and I quite exactly repeated the experience when parting from Niko in Leiden."[143] How such get-togethers might be organized and where they would take place remained for the time being unspecified.

In the meantime, Tinbergen felt he had done the right thing in deciding to move to Oxford. Lorenz, in contrast, recognized that he himself was not yet adequately settled. He had just seen Thorpe's base of operations at Cambridge, where plans for the Madingley Ornithological Field Station were proceeding apace. He knew that Tinbergen would soon be securely established at Oxford, a great scientific center and indeed a wonderful place for anyone interested in biological field studies. He recognized that his German friends Holst and Kramer were well ensconced at Holst's new Max Planck Institute in Wilhelmshaven. As he compared his colleagues' institutional settings with his own grandly titled but poorly funded institute at Altenberg, he was willing to admit to pangs of envy.[144]

Much had been accomplished in the four years since the war had ended. A new international journal had been established to help define the field and serve as an outlet for new work. Important new research programs had been set up or were in the process of being set up at Oxford, Cambridge, Groningen, and Wilhelmshaven. Within the international community of students of animal behavior, the fractures caused by the war were on their way to healing, as exemplified by the happy reunion of Tinbergen and Lorenz. The ethologists had basically held their own in a major international conference on the physiology of behavior, and they were looking forward to establishing regular international meetings of their own. It was clear that the more controversial of the ethologists' major postulates, particularly Lorenz's idea of action-specific energy, would need to be subjected to further scrutiny. What was less easy to predict was just how ethology itself would change in the face of new challenges, new institutional settings, and an influx of bright new recruits into the field.

SEVEN

ETHOLOGY'S NEW SETTINGS

I am at a critical phase now having to begin something new under new conditions of fauna, opportunities, and university organisation, and it will take some years to get my program smoothly running again.

NIKO TINBERGEN IN A LETTER TO ERNST MAYR, 4 SEPTEMBER 1950 [1]

The "Max Planck Gesellschaft" has decided to *build an institute for me!* I shall have a *full professor's fee* and be able to employ *three assistants!* Technically my institute will be part of Wilhelmshaven, but practically independent, situated in an incredibly picturesque castle near Münster in Westfalen. The castle is named Buldern and is surrounded by a gigantic moat and any number of big ponds, waiting for ducks and geese, a big empty greenhouse is waiting for aquaria.

KONRAD LORENZ IN A LETTER TO W. H. THORPE, 1 NOVEMBER 1950 [2]

Niko Tinbergen's appointment at Oxford began on the first of October 1949. He came ahead of his family, not bringing them from Holland until the beginning of the New Year. He had been determined to make the move, but he was nonetheless pleased to find, after some months in Oxford, that it still seemed like a good choice. "We feel that we will be able to root here," he told his friend Ernst Mayr in April 1950.[3]

The year 1950 was an important one for developments at Cambridge as well. That was the year that the Ornithological Field Station at Madingley, located just a short bicycle ride away from Cambridge, began operations under the direction of W. H. Thorpe, a recent convert to Lorenzian ethology. The young Robert Hinde became curator of the station, and Peter Marler and G. V. T. Matthews soon arrived as graduate students.

The same year proved decisive for Konrad Lorenz as well. Early in the year his hopes for a secure institutional base of his own soared when he learned that he was the frontrunner for the professorship of zoology at the University of Graz. Political considerations of a variety of sorts intervened to dash his hopes at Graz, but diverse negotiations by his friends helped him ultimately gain the result he had always craved most: a research station created by the Max-Planck-Gesellschaft (MPG) especially for him.

In each of the above cases—Tinbergen's appointment to the new lectureship in animal behavior at Oxford, Thorpe's inaugurating of the new Ornithological Field Station at Madingley, and Lorenz's relocation to an MPG research station of his own in Buldern, Westphalia—the possibilities were ripe for ethology's adaptive radiation. Lorenz and Tinbergen, the leading figures of the new science, were moving into new settings with new possibilities. So too in effect was Thorpe, for although he had spent most of his career at Cambridge, most of that had been as an entomologist. He had only recently decided to refashion himself as an ethologist studying birds, and with the new Ornithological Field Station at Madingley he had the chance to construct a center that would play a major role in ethology's subsequent development.

In the meantime there were other centers as well where researchers were devoting themselves to the study of animal behavior. One of these was in Holland at the University of Groningen, where Gerard Baerends had taken the position of professor of zoology in 1946 and was building a program featuring behavioral and ecological studies of the animal as a whole. Another was in the United States at the American Museum of Natural History in New York. There, in the Department of Experimental Biology founded by Gladwyn Kingsley Noble in the late 1920s, flourished a tradition of animal behavior research that was distinct from the mainstream of American comparative psychology and also from Continental ethology.

At each of these various sites—Oxford, Cambridge, Buldern, Groningen, and the American Museum in New York—investigators of animal behavior were confronted with their own peculiar opportunities and challenges. These local differences, combined with possibilities ranging from competition to division of labor to collaboration between sites, inevitably complicated the enterprise of ethology. At the same time they provided the new science with a heterogeneity that enhanced its potential for continued growth.

In this chapter our emphasis will be on the mutual construction of individual careers and particular research centers and the ways that different centers came to be identified with different sets of issues, methodologies, and material trappings. We will look here in particular at the approach to natural history championed by Alister Hardy at Oxford, and the context Hardy's Oxford provided for Niko Tinbergen's further development as a field naturalist; W. H. Thorpe's path to becoming an ethologist, and his efforts, once he became an ethologist, to divide up the territory of ethology in such a way as to protect certain lines of research at Cambridge; the academic politics that led to Lorenz's finally being appointed to a Max Planck Institute in Germany; and the animal behavior program at the American Museum of Natural History.[4]

ALISTER HARDY AND OXFORD ZOOLOGY

When Niko Tinbergen left his professorship at Leiden to take up the more junior post of lecturer in animal behavior at Oxford, he did so, as we have seen, for a variety of reasons. Not the least of these was that Oxford promised to be an exceptionally vibrant place for a field biologist. The man most responsible for Oxford's recent reputation in this regard was Alister Claverling Hardy. Hardy had been appointed Linacre Professor of Zoology and Comparative Anatomy at Oxford in 1945. In submitting his application for the post, he had made it perfectly clear that his goal was to reinvigorate Oxford zoology by giving special prominence to field studies of animal ecology, behavior, and evolution.

Hardy had not always envisioned himself in such a role. Near the beginning of his career, in the midst of an intellectual crisis of sorts, he had told Julian Huxley, his former zoology tutor at Oxford, that a life in academic zoology did not appeal to him. Hardy had reached this conclusion while working at the Zoological Station in Naples, a year after he had graduated from Oxford with distinction in zoology. He had received an Oxford Biological Scholarship to do research in Naples, but the morphological studies he was conducting there were failing to sustain his interest. As he explained to his former tutor: "I frankly do not care whether Priapulus or any other animal excretes by solenocytes, open nephria or coelomoducts anymore than I care whether a motorcycle has one or other type of valve, carbureter or magneto. I admit that collectively all these are of great interest to the student of evolution—but isolated I find them in themselves dull and I find I have not the temperament for such work."[5]

The reason he had been drawn to biology in the first place, Hardy told Huxley, was "the desire to understand better the meaning of life." Studying morphology was not bringing him any closer to this goal. Hardy recognized, furthermore, that if he were to become an academic zoologist and teach at a university, he would be expected to be an expert in morphology. Alone in Naples, and very uncertain about his qualifications for an academic career, Hardy decided that "wider" interests—interests of a political, social, and religious sort—needed to play a larger role in his future. If he were to enjoy and be good at "any research of a purely intellectual nature," he explained to Huxley, it would have to be as a hobby. When it came to having a *profession*, what he needed was "to be working—really working at something directly useful to the community." He had learned of a post that was opening up in the fisheries, and this seemed to him just the thing he was looking for. He saw it as "a hundred times" more appealing than that of demonstrating and lecturing at a university. The work in the fisheries would certainly be useful. On top of this, he was attracted to oceanography and to sea life in general. A university position, in contrast, had the disadvantages of the "petty jealousy and backbiting that goes on in Donnish circles." Hardy also disliked, he said, "the unreality of the whole life . . . a life which during term one never gets away from. Senior Common rooms would drive me mad in a very short time!"[6]

Twenty-five years later Hardy would become Linacre Professor of Zoology at Oxford, a fellow of Merton College, and an important addition to the Merton Senior Common Room. In 1921, however, he began his career as a marine ecologist, based at the Lowestoft Fisheries Laboratory with an appointment as assistant naturalist to the Ministry of Agriculture and Fisheries. The post attracted him for intellectual as well as practical reasons. The director of scientific investigations of the Ministry of Fisheries was E. S. Russell, who was an articulate advocate of a "functional view" of animal life. Russell rejected the mechanistic interpretation of animal life, insisting instead on the essentially purposive nature of animal activities. Hardy found Russell's understanding of the nature of animal life to be highly congenial with his own thinking.

In 1924 Hardy was appointed chief zoologist on the RRS *Discovery* expeditions to the Antarctic (1925–1927). The ecological work of this expedition delighted him. The work was utilitarian. It also allowed him to think about man's place in the biological world. In July 1926 he wrote Huxley: "I believe one of the important contributions zoology can make to the world is working out ecological methods that can be turned in time to the study of man." Interestingly enough, in addition to expressing his grand hopes for ecology, he also expressed the hope of returning one day to Oxford. E. S. Goodrich would not keep his chair at Oxford much longer, Hardy recognized, and though it was almost certain that Gavin de Beer would be Goodrich's successor, Hardy considered trying for the position himself. He told Huxley, "I should love to have a shot at linking up with Elton and building with him a great Ecological Institution.[7]

It would take Hardy another two decades to get back to Oxford. In the meantime, he was appointed Professor of Zoology at University College, Hull, in 1928 and then Regius Professor of Natural History at the University of Aberdeen in 1942. In both of these posts he brought the study of marine ecology to bear on the problems of the fishing industry. He was also able to give considerable thought to what a suitably modernized natural history should look like. He made this last topic the main theme of his inaugural address as Regius Professor at Aberdeen in 1942.

Hardy entitled his Aberdeen inaugural address "Natural History—Old and New."[8] In his address he extolled the old style of natural history, characterizing it as "magnificent descriptive Natural History," and then proceeded to identify the quantitative and experimental methods of the new natural history, particularly as exemplified in the new science of ecology. He wanted laboratory studies and fieldwork to be more closely coordinated with one another, particularly in the study of evolution. Throughout the previous quarter of a century, he maintained, evolutionary work had tended to emphasize laboratory studies "of a little fly kept in bottles." The time had now come, he said, "to leave the laboratory for a bit of fresh air—to bring these concepts [of the gene and the gene complex] to the test in the living world outside." He called for "more and more research into the working of natural selection in the wild." At the same time, he expressed doubt that the natural selection of genetic differences was all that was needed to account for evolutionary change. "There is a pressing need," he wrote, "for the objective study of animal behaviour in the field: of an animal's power of choice." It appeared to him that there was still more to evolution than modern biologists were allowing: "It would seem that choice must be governed in part by the gene complex, but is it entirely? Have we in this quarter of a century discovered the whole or the real essence of the evolutionary mechanism? I think it unlikely. The hypothesis of racial habit has for the time being been eclipsed by the brilliance of gene analysis and by inherent difficulties in putting it to the test of experiment. We must recognise that a test of this hypothesis of the development of racial habit can only be valid if carried on for a long period of time."

Hardy recognized that such long-term investigations required special institutional structures to support them. He concluded his lecture by indicating how beneficial it would be to have "an institute of real experimental Natural History dealing with the ecology, behaviour and evolution of living things in nature or under as far as possible natural conditions." He soon enough found himself with the opportunity to develop such an institution, but at Oxford instead of Aberdeen.

In May 1945, the University of Oxford announced a call for applications for the Linacre Professorship of Zoology and Comparative Anatomy. Hardy applied. He acknowledged in his application that the directions of his own research had been "far removed from the grand tradition of Oxonian Comparative Anatomy." He was not a morphologist, he admitted, and he was not going to claim to be one. What he proposed instead was to "help to maintain the spirit of Oxford scholarship . . . and at the same time throw myself into the development of the department as a centre for the study of the living animal under natural conditions." He stated exactly what he wanted to do for Oxford zoology: "If I could join my marine ecological interests with Elton's mammalian work, Ford's entomological evolutionary studies, Baker's interest in climate and breeding seasons, Tucker's field bird observations, and Professor Hale Carpenter's work at the Hope Department, I believe we could, working together, build up a strong school of experimental and observational research; one which would help to place the study of living animals in the field on the road to gaining equality with physiology in biological prestige."

Hardy had an eye not only for the special things that might be accomplished at Oxford but also for how zoology at Oxford should relate to zoology at Cambridge. It made sense, he believed, for Oxford and Cambridge not to try to duplicate each other's efforts. He wrote: "I believe that since Cambridge has developed so much on the side of laboratory experiment, it would be well for Oxford to follow for a time a different but equally important line." He had no intention of doing away with physiological research at Oxford. He wanted, he said, "to keep a proper balance." But he planned nonetheless to develop field studies. He was also continuing to think along utilitarian lines. He felt there should be a teaching as well as a research side of the enterprise, because the empire needed well-trained marine and terrestrial ecologists "fired with the spirit of the old field naturalists, yet equipped with the training for a quantitative and experimental attack."[9]

In July 1945, within a fortnight of submitting his application, Hardy was chosen as Linacre Professor of Zoology and Comparative Anatomy at Oxford. The appointment was not to begin officially until January of 1946, but he visited his new department in September 1945. It was then, if not sooner, that he began to learn of the different constituencies who hoped that his coming might be the occasion for restructuring zoology at Oxford in ways that would be to their own advantages.

A continuing issue for Oxford zoology was the existence of two independent units, the Edward Grey Institute of Field Ornithology and the Bureau of Animal Populations (BAP), both of which were loosely tied to the Department of Zoology. The university registrar was concerned about the structural problems created by these two independent units, and he urged Hardy to take both units into the Department of Zoology. Hardy was anx-

ious, however, about the prospect of adding as many as seven senior research people (four from the BAP and three from the Edward Grey Institute) to his department. He worried that this might jeopardize his chances
of appointing any more researchers in zoology.[10]

Another person who hoped that Hardy's arrival at Oxford might be
an opportunity for restructuring Oxford zoology was Charles Elton, the
director of the BAP. The BAP had grown during the war as the result of
practical work conducted under the Agricultural Research Council. Elton
wanted the BAP upgraded to the status of a subdepartment. He likewise
wanted to have all his assistants upgraded to the rank of university demonstrator (his own rank at the time). This plan astonished and displeased the
other demonstrators in zoology (John R. Baker, E. B. Ford, Peter Medawar,
H. K. Pusey, and Bernard Tucker). They were outraged to hear that the university was being asked to spend almost as much on the subdepartment of
animal population as on "the whole of the rest of zoology." Tucker, the ornithologist of the group, had himself been hoping for the establishment of
a subdepartment of ornithology, but Ford, Medawar, Pusey, and Baker
persuaded him to withdraw his proposal and not to ask the university for
more than a small annual subsidy of two hundred pounds for ornithological studies. Tucker then agreed to go along with the others in opposing the
foundation of a subdepartment of animal population. Following Hardy's
September visit to the department, Baker, a friend of Hardy's since their
undergraduate days together, reported to Hardy how high the local sentiments were running: "In the quarter of a century that I have been in the department, I have never known such intensity of feeling."[11]

Hardy himself did not help matters when he sent to the zoological staff
a circular letter in which he stressed the importance of ecological field
studies. The nonecologists on the staff read this as an indication that Hardy
viewed the work of Elton, Carpenter, and Lack as the most important zoological work being conducted at Oxford. Likewise galling to the unhappy
zoologists was Hardy's use of the phrases "my department" and "captain
and crew." Both of these ideas, Baker later explained to Hardy, were "quite
unheard of."[12] The aggrieved staff protested openly about the direction in
which Oxford zoology seemed to be going.

It was a testimony to Hardy's administrative and diplomatic skills that
he was able to calm things down, construct a plan that satisfied almost
everyone, and still promote the field studies to which he was committed.[13]
His solution was to unite the BAP and the Edward Grey Institute in a new
department, the Department of Zoological Field Studies. Established by
statute in July 1947, the new department took as its focus "the study of an-

imals in nature: their ecology, behaviour and evolution." As Linacre Professor, Hardy headed both the new Department of Zoological Field Studies and the old Department of Zoology and Comparative Anatomy. This way, he believed, field zoology and laboratory zoology could be kept in close communion with each other.

"Zoology outside the laboratory" was the theme Hardy chose for his lecture as president of the zoology section of the British Association for the Advancement of Science in 1949. There he happily expressed his conviction that zoology outside the laboratory was "at last coming into its own." He noted that the Royal Society's 1946 "Report on the Needs of Research in Fundamental Science after the War" had emphasized the importance of research "directed towards the study of living organisms and their relation with their natural surroundings." He also pointed with pride to his new Department of Zoological Field Studies, which he assumed to be the first department in the world with this title. In addition he noted Oxford's recent decision to devote the woodlands of the Wytham estate to biological field studies, Charles Elton's institution of a student course in terrestrial ecology, and, most significant of all, "an event of great importance for Zoology in this country: Professor Tinbergen, formerly of Leiden University, is coming to Oxford next month to take up an appointment specially created for him: University Lecturer in Animal Behaviour."[14] Tinbergen's appointment, interestingly enough, was made in the Department of Zoology rather than the Department of Zoological Field Studies.

Hardy had his own thoughts about what a field ethologist might be able to say about evolution. He believed that behavioral choices had a role to play in the evolutionary process. He supposed that the "Baldwin effect" or "organic selection" effectively allowed a quasi-Lamarckian kind of organic development. But Tinbergen was never attracted to this idea. Nor did he ever express sympathy for Hardy's enthusiasm about telepathy. At least once, nonetheless, movie camera in hand, the transplanted Dutchman ventured out in the field with Hardy to film the synchronized movements of bird flocks in flight, a phenomenon that Hardy, like Selous before him, thought might involve extrasensory perception. Oxford's new lecturer was surely joining Hardy for the pleasure of being out in the field, and perhaps also to be politic, not because he thought that making films of bird flight would provide evidence either for or against ESP. Indeed, to others Tinbergen out-and-out deprecated the whole idea. As for Hardy's sense of the need for a new synthesis of science and religion, this too was entirely foreign to Tinbergen's way of thinking. But such differences never outweighed the bond the two men felt as field naturalists. Although Tinbergen at times

lamented that Oxford University was not earmarking any funds for his research group, he always regarded Hardy to be personally supportive of the
group's work.[15]

TINBERGEN AT OXFORD

Tinbergen's transplantation from Leiden to Oxford was of immense importance both for Tinbergen's own career and for ethology in general. Reflecting back years later on the move, he acknowledged that it had been a
life-changing—and research-altering event.

The move required adaptation on Tinbergen's part. This became clear
to him very quickly as he set about living in this new country and working
within an educational system that struck him as "entirely alien." At Leiden,
the centerpieces of his undergraduate ethological teaching were the spring
fieldwork and the block practical. These enabled him to direct students in
the early stages of serious but achievable fieldwork and/or laboratory research. In contrast, as he soon discovered, Oxford's system of "condensed
three year courses and three terms per year" left little or no room for introducing Oxford undergraduates to research. Tinbergen's lecturing responsibilities included a yearly set of sixteen lectures on animal behavior
for advanced students and a stint every other year in his department's "Animal Kingdom" course. The responsibility that fell to him in the latter regard was a set of nine lectures and nine practicals on the molluscs! Hardy
believed that all the lecturers in the department should take part in the
"Animal Kingdom" course. Tinbergen was prepared to do his share, even
though the Mollusca were hardly his specialty.[16]

As for the *graduate* students at Oxford, they were expected to be ready
to start on their theses right away, even if they had no prior research experience. Tinbergen found most of them insufficiently prepared to do so. After three and a half years experience with the Oxford system, Tinbergen
told Huxley he found it "absolutely necessary for me to spend many weeks
with my D.phil. people in the field during the first season, in order to train
them while together kind-of-wrestling with the object and its problems."
He had to coach the students at length, he said, "before leaving them to
their own wits." The key here, he said, was getting students to have a feeling for biology as a whole complex of interrelated problems. He elaborated:

> Because in behaviour it proves so very difficult to acquire the "biological"
> outlook, (much more difficult than in most branches of physiology for
> instance), the training takes such a long time; only exceptional people

grasp the type of approach within a year, and most of then need two years to develop it. The elementary lectures I give do help a little, but without confrontation with the object itself they are of relatively little use. This all has forced me to keep the number of D. phil people down (although it is very high as it is: 9) and to spend most of my time closely collaborating with them: sitting in front of the aquarium or on the cliffs, and discussing difficulties during special seminars. I believe it is now working well. But a constant watch on strict thinking and sharp, even pedantic formulation of each problem and each conclusion is, I believe, essential now.[17]

Tinbergen's promotion of "the 'biological' outlook," his insistence on precise thinking, and his collaboration with his students were hallmarks of his teaching. The locales where the serious teaching was done, as indicated above, were the lab (the aquarium), the field (the cliffs—or dunes), and special seminars (especially the Friday evening seminars he held at his home). What he demanded of his students he demanded of himself, and vice versa. The seminars contributed not only to the intellectual formation of Tinbergen's individual students but also to the social bonding of successive cohorts. The same was true of shared research experiences in the laboratory or the field. Tinbergen's students felt themselves to be members of a "privileged elite" under the direction of "the maestro."[18]

Tinbergen's research as well as his teaching underwent changes after his move to Oxford. In part this was because he felt it was time for his research to enter a new phase anyway. But it was also because of new opportunities that his new setting provided. When, after a year in Oxford, he received an invitation from Ernst Mayr to come to the United States to give a field course on animal behavior, Tinbergen begged off, saying: "I am at a critical phase now having to begin something new under new conditions of fauna, opportunities, and university organisation, and it will take some years to get my program smoothly running again."[19]

When Tinbergen made this statement (in September 1950) it had been more than a year since he had had the chance to do any serious fieldwork. In the summer of 1949 he had been occupied with the move to Oxford and the Society for Experimental Biology symposium at Cambridge. In the spring of 1950 conferences in Germany, Sweden, and Holland had likewise left him no time to do field studies. Explaining further to Mayr why he could not come just then to the States, he wrote: "I think you know such critical phases in your own work just as I do: after a period of detailed research and subsequent theoretical digestion you need time and attention to plan new work that can carry you a little further; one feels that it is essential that one does not stop at the front reached, just carrying on with the same type of work, but that one devises work that can throw light on the familiar problems from another angle, and I find this stage of reconnaissance, extremely difficult. I think I ought to return to detailed research after a period in which I was rather a 'tradesman in ethology!' " [20]

How does one decide how to structure a research program? In 1962 the Association for the Study of Animal Behavior met in Oxford, and Tinbergen took the opportunity to reflect back upon the development of his animal behavior research program at Oxford. The program had been shaped, he acknowledged, by a variety of choices. These choices, in turn, had been influenced by local conditions and by more general considerations about where the field of ethology should be heading. He had wanted to have a small group of researchers who were interested in a range of problems and also a range of animal species, but who did not spread their efforts so widely that they were unable to study selected problems and selected animal species in depth. He had been convinced, furthermore, "that ethology would be doomed if it did not attempt to enter certain 'no-man's-lands' between it and sister disciplines." The problem that remained, however, was which sister disciplines to choose, since there were so many possibilities: "e.g., neurophysiology and sensory physiology, endocrinology; human psychology; ecology, genetics and evolution; anatomy; experimental embryology." Tinbergen described the results of the selection process as follows: "We considered that some work on neurophysiological aspects of behaviour ought to be done in any such group, and that for the rest our choice should be guided by opportunity, such as the possibility of close contact with specialists. For this reason we decided to pay special attention to problems of ecology, genetics and evolution, fields in which Oxford can boast to have flourishing schools under Dr. A. J. Cain. C. S. Elton, Dr. E. B. Ford, Professor A. C. Hardy and Dr. D. Lack."

Beyond this, Tinbergen acknowledged, there had also been the question of being economical with limited resources. Sticklebacks and *Drosophila* could be maintained and studied cheaply in the laboratory. Colonial seabirds could be the focus of research of the group's fieldwork. As Tinbergen concluded: "These considerations determined the character of the research programme, which consequently is a compromise between broadness of approach and penetration in depth." [21]

Funding had also been an issue. When Tinbergen first arrived in Oxford, the only funds for his research group took the form of a five-hundred-pound start-up grant from the Agricultural Research Council. Things approved dramatically in 1951 when Tinbergen received a five-year grant of ten thousand pounds from the Nuffield Foundation. He regarded this as quite munificent for a small group working in a little-known area.[22] In 1956 this grant was renewed for another five years, something of a rarity as far as Nuffield funds were concerned.[23] When that grant expired in 1961, Tinbergen received a major, new grant from the Nature Conservancy for "a permanent research unit with emphasis on field work" that would extend up until the time of his retirement. It gave him the opportunity to appoint two senior scientists and more junior scientists as well.[24] Other sources of funding for Tinbergen's group came from the Ford Foundation and the United States Air Force Office of Aerospace Research. As of 1962, Tinbergen could report that his group consisted of the reader in animal behaviour (i.e., Tinbergen himself) plus "three post-doctoral research workers in various grades, a Secretary, and nine graduates working for the degree of D.Phil."[25] The program had come a long way in a dozen years.

W. H. THORPE AND ETHOLOGY AT CAMBRIDGE

Whereas the key figure in establishing ethology at Oxford was Tinbergen, the key figure in establishing ethology at Cambridge was William Homan Thorpe. Thorpe, unlike Tinbergen, was not brought in from the outside to establish a new field of study. He was already at Cambridge when he decided to push the cause of ethology. However, he had not started his career as an ethologist. He began his career as an entomologist and later transformed himself into an ethologically oriented ornithologist. He was not a transplant but a self-fashioned retread.

Thorpe, like his Continental counterparts, had been devoted to natural history as a youth.[26] As a schoolboy, though, he had not imagined that his interest in natural history would lead him to a career as a zoologist. Periods of ill health prevented him for performing particularly well at the boarding school he attended, and he was convinced that he would not be able to hold his own in university studies. News reached him, however, of the need for economic entomologists. This led him to the idea of pursuing an agricultural degree at Cambridge.[27] He went up to Cambridge in 1921. There he attended lectures in zoology as well as agriculture. After receiving his degree in agriculture, he stayed on to write a thesis on the biological control of insects. The thesis earned him his diploma in agricultural science.

At Cambridge Thorpe engaged in the sorts of extracurricular activities one might well expect of someone who was later to become an ethologist. He was a founding member of the Cambridge Bird Club. He also published several short papers on field ornithology in *British Birds* and the *Ibis*.

But his route to the study of animal behavior was not the same as Lorenz's or Tinbergen's. In the first place, his research, reflecting the agricultural orientation of his studies, had a decidedly utilitarian component to it. In the second place, his research engaged him directly with the question of the mechanisms by which evolution works. Lorenz and Tinbergen, in contrast, had not worried much about the mechanisms of evolution early their own careers. They seem to have simply taken it for granted that Darwinian natural selection was the primary agent of evolutionary change. Thorpe on the other hand was attracted to the Lamarckian idea of the inheritance of acquired characters. What pointed him in this direction was his own research on the isolation of biological races of insects by food preferences, plus the influence of some Lamarckian thinkers with whom he came in contact.

Thorpe was a student in the audience on 30 April 1923, when the Austrian experimental zoologist Paul Kammerer delivered to the Cambridge Natural History Society a lecture entitled "Breeding Experiments on the Inheritance of Acquired Characters." Years later Thorpe told Arthur Koestler that the lecture "had remained one of the strongest impressions of my undergraduate career."[28] The charismatic Kammerer did not convert Thorpe to Lamarckism on the spot. Thorpe at the time did not feel sufficiently qualified to judge the issue.[29] Nevertheless, Kammerer may well have struck a responsive chord with him when he stated: "The necessities of life have almost compelled me to abandon all hope of pursuing ever again my proper work—the work of experimental research. I hope and wish with all my heart that this hospitable land may offer opportunity to many workers to test what has already been achieved and to bring to a satisfactory conclusion what has been begun."[30] Kammerer's plea was seconded by J. Stanley Gardiner, the professor of zoology at Cambridge. According to an eyewitness account, Gardiner "urged the youth of this generation to take up K's work without delay and either to prove or disprove it."[31]

Kammerer committed suicide on 23 September 1926, six weeks after the American biologist G. K. Noble, who was then best known as a herpetologist, announced in the journal *Nature* that Kammerer's evidence for the inheritance of acquired characters appeared to have been doctored. Noble reported that Kammerer's only remaining specimen of the midwife toad, *Alytes obstetricans,* lacked the nuptial pads that Kammerer claimed to have induced in it. Under microscopic investigation, the alleged nuptial pads seemed to be nothing more than India ink or another such substance injected under the specimen's skin where the pads were supposed to be. Many scientists interpreted Kammerer's suicide as an admission that his evidence was fraudulent. Others were not convinced, or at least did not believe that the inheritance of acquired characters was thereby disproved. Kammerer's suicide did nothing, for example, to change the views of the ardent neo-Lamarckian William MacBride. Nor did it discourage the efforts of J. W. Heslop Harrison, who, in a 1927 paper on the egg-laying instincts of the sawfly (*Pontania salicis*), claimed to have demonstrated the genetic transference of changes in food insect food habits.[32]

Thorpe, in his own study of the formation of biological races in insects, was more critical and cautious than Heslop Harrison had been when it came to deciding what had been proven and what had not. He opted not to commit himself completely. He nonetheless made clear what direction he was leaning. Reviewing the experimental work on biological races, he concluded: "Many of these experiments are easily explained on some form of Lamarckian theory, but extremely difficult to account for on any other lines." He acknowledged that the experiments he had surveyed were not extensive enough "to carry complete conviction on this point," but he allowed that "they do, however, suggest most profitable fields for further work of this nature and, taken together, they provide a quite considerable amount of the ever-growing body of circumstantial evidence for the theory."[33]

At the time he published this review, Thorpe was working under the direction of W. R. Thompson at the Imperial Bureau of Entomology's Farnham Royal Parasite Laboratory. Thompson himself was positively disposed toward Lamarckism. He warned Thorpe, however, that embracing Lamarckism could associate Thorpe with the wrong scientists and thus hurt his career.[34] In fact, things seem to have worked the other way around. Gardiner, who had been so enthusiastic about Kammerer's work, invited Thorpe to return to Cambridge to take up the post of lecturer in entomology.

Back in Cambridge with this academic post and as appointment as fellow and tutor at Jesus College, Thorpe continued to study the phenomenon of insect host preferences and their importance in the formation of biological races. As his researches continued, however, his expectations of demonstrating a Lamarckian influence in evolution waned. By 1937 he was suggesting an alternative mechanism — "pre-imaginal olfactory conditioning" — as being of at least "possible importance in the early stages of evolutionary divergence." He acknowledged this to be a version of the "host selection principle" initially advanced by the American entomologist B. D. Walsh in the previous century. According to this view, a polyphagous insect species might split into isolated populations, restricted to different host species, if the adult females were attracted to and laid their eggs upon the particular host species they had fed on as larvae. Explaining insect host preferences in this way made it unnecessary, Thorpe noted, to call upon Lamarckian theory.[35]

Thorpe had conducted his work on olfactory conditioning using a parasitic wasp, *Nemeritis canescens*, which usually parasitizes two species of flour moth, *Ephestia kühniella* and *E. elutella*. He found he could also raise *Nemeritis* on another moth, *Meliphora grisella*, by contaminating *Meliphora* larvae with the scent of a normal host. Trying to produce a Lamarckian effect, he raised *Nemeritis* on *Meliphora* for a series of successive generations. He hoped to get an increase in the percentage of insects choosing *Meliphora*, but after eleven generations no increase was to be seen, and he decided to drop this line of experimentation.[36] In 1940 he stated with respect to his "biological races" of a decade earlier, "no work that has since been carried out either on the natural occurrence or the experimental production of biological races in insects has given any conclusive evidence of any Lamarckian effect."[37]

While Thorpe's studies of the conditioning of insects to particular hosts offered no support for the idea of the inheritance of acquired characters, they led him in the late 1930s to consult other work on conditioning in insects, including Karl von Frisch's experiments on bees and Charles Henry Turner's work on the cockroach. Early in the 1940s, he went on to publish a three-part article on learning in insects and other arthropods.[38] There he allowed that it was not possible to discuss learning satisfactorily until one offered an adequate definition of "instinct." There too he first cited the work of Konrad Lorenz, along with the work of other instinct theorists including Wallace Craig, Erich von Holst, J. A. Bierens de Haan, and the Reverend Edward A. Armstrong.[39] It may well have been Armstrong who introduced Thorpe to Lorenz's writings.[40]

Thorpe decided that Lorenz's ideas of imprinting could be brought into fruitful conjunction with his own earlier studies on habitat selection in insects. Contrary to the view of Ernst Mayr (and also David Lack) that geographical isolation is primary in the process of speciation, Thorpe argued that ecological factors can be just as important as geographical ones. He suggested that a kind of "locality imprinting," based on olfactory cues, could over time "set the direction for the selective processes tending to bring about genotypic isolation." The result would "closely simulate a Lamarckian effect."[41] As Thorpe now saw it, Lorenzian imprinting made it easier to understand the "Baldwin effect," by which habits that were at first only phenotypic could become genetically fixed in a race.[42]

Thorpe at this point decided that the scientific subjects that interested him most were learning and instinct in animals. He nurtured other broad interests as well, in particular the role of behavior in evolution and the relations of science and religion, but issues of learning and instinct took center stage in his research. Most important for his research, he decided, as he later put it, "that at all costs I must attempt to switch over from entomology to ornithology."[43] Birds seemed to him to be ideal subjects for studying both instinct and learning, and as an avid bird-watcher he had long taken an interest in them.

In the summer of 1945, Thorpe suggested that the Institute for the Study of Animal Behaviour promote a program of researches on animal learning. He pointed to the advantages of studying birds and to the concomitant importance of establishing aviaries or an ornithological field station. Soon he was promoting the idea with James Gray, who had succeeded Gardiner as professor of zoology at Cambridge. Gray suggested forming a steering committee to consider the establishment of an ornithological field station. Thorpe constituted a committee consisting of Gray, David Lack, Edward Armstrong, and himself. From Gray, he wanted institutional and financial support. From Lack, who had recently been appointed director of the Edward Grey Institute for Ornithology at Oxford, Thorpe wanted an understanding about a division of labor between Thorpe's enterprise and Lack's. Specifically, he wanted Lack to agree that if Cambridge committed resources to building aviaries, similar facilities and a competing research program would not be developed at Oxford.[44]

As negotiations regarding funding and a site for the field station dragged on into 1948, Thorpe had reason to rethink his recent decision about switching from entomology to ornithology. In June 1948 he entertained the possibility of becoming the Hope Professor of Entomology at Oxford. He explored the idea with Alister Hardy. Hardy's cooperation and support, as Linacre Professor, would be critical in determining the attractiveness of the position. Thorpe asked Hardy directly: "Would you be prepared to go all out to help me develop the Hope Department as a department of Comparative Behaviour Studies, primarily concerned of course with Insects but by no means confined to them?"

Thorpe worried that this plan might be compromised if Hardy succeeded in attracting Tinbergen to Hardy's department. As Thorpe put it, "Would not the coming of Tinbergen to your department make the development of the Hope along the lines I want, more rather than less difficult? I can see the argument being put forward that this subject was already being catered for in your dept." Thorpe wondered whether the proposed demonstratorship for Tinbergen might be transferred to the Hope Department, and likewise whether any behavior research students could also be transferred to the Hope, if it appeared they would fit better under Thorpe than under anyone on Hardy's staff.[45]

Hardy promised that he would "most definitely" do all he could to help Thorpe develop the Hope Department as a department for comparative behavior studies. However, he said, the university would not allow the transfer of a demonstratorship from one department to another. He was eager, he said, to have the Department of Zoology and Comparative Anatomy, the Department of Zoological Field Studies, and the Hope Department closely linked to each other, but there were limits to what he could do.

Thorpe mulled things over and then decided that the chances that the Hope Department would develop in the way he wanted it to do remained "too distant and nebulous" for him to accept the professorship. Besides, as he explained to Hardy, he had talked things over with James Gray, and the prospects of developing animal behavior facilities at Cambridge were looking up. The university had recently acquired the Madingley estate, which appeared to be a promising site for an experimental aviary and ornithological field station. Thorpe chose to stay at Cambridge.

It was fortunate for the future of Cambridge ethology that plans for the Ornithological Field Station at Madingley were well in place a year later when the symposium "Physiological Mechanisms in Animal Behaviour" convened in Cambridge. Holst's attack on Gray's work on the peripheral control of nervous processes gave Gray no reason to feel kindly toward Lorenzian ethology, and, personal feelings aside, he was inclined as an experimental physiologist to see the work of the ethologists as rather "woolly." Later, the Ornithological Field Station would annoy him as a drain on his budget.[46] From Robert Hinde's perspective as curator of the operation, Gray seemed disposed to put obstacles in the way whenever the researchers at Madingley tried to take a step, but Thorpe's administrative skills were up to the challenge.[47] Madingley would become one of the preeminent centers for animal behavior studies in the world.

When the Ornithological Field Station began operations in 1950, Thorpe was its director and Hinde was its curator. Thorpe had thought of others for the role of curator, including R. E. Moreau of Oxford, the Koenigs from Vienna, and apparently even Lorenz himself, but Hinde was the one who ultimately got the job.[48] The station was situated on an L-shaped piece of land consisting of about an acre and a half of deciduous woods and two and a half acres of field. A wire fence was erected around the site to keep out foxes and rats. An existing hut, sixty feet in length, was equipped

with cages for birds, and more than sixty aviaries of various sizes were constructed or purchased and located around the site. For an office and work-
shop, Thorpe made use of the old village blacksmith's shop, which was located just off the main site.

During the station's first year of operation, Thorpe and Hinde initiated
a number of research projects for which Madingley would later be especially well known. These included studies of song learning, imprinting, and
"insight learning." The first year also witnessed the appearance on the
scene of the first graduate students at the station, Peter Marler and G. V. T.
Matthews, and the introduction of a variety of different bird species for
study. At the same time, a decision was made about one species, which, as
Thorpe put it, was "likely to be our chief experimental species for some
time to come." This was the chaffinch (*Fringilla coelebs*).

Song learning headed the list of Madingley research topics from the
beginning. Thorpe set about establishing experimental protocols to test
the nature of the song-learning process. He was keen to know whether
chaffinches innately recognized the sound patterns of their species. To test
this, he needed to be able to isolate birds from each other. He also needed
good recording equipment. He turned to the BBC Engineering Depart
ment and the General Post Office Engineering Research Station at Dollis
Hill for technical advice and help with matters of soundproofing, recording, and reproduction. From the Royal Society he secured a grant that
helped him buy a special, semiportable recorder.[49]

That was only the beginning of his search for better and better equipment. He learned of a new invention, the sound spectrograph, that he supposed could be put to use analyzing bird vocalizations. In 1951 he took a
collection of birdsong recordings to the Bell Telephone Laboratories in
New York for conversion into spectrographic form. Very impressed by the
results, he asked whether any sound spectrograph machines existed yet in
Britain. He was told there was one such machine in use in a government
laboratory. Thorpe tracked it down at the Admiralty Research Laboratory
at Teddington. At the time it was the only sound spectrograph in use in the
country and it was top secret. The Admiralty was using it to analyze the
noises made by submarines. Thorpe nonetheless was permitted to bring
his disc recordings to Teddington once a month to use the Admiralty's
sound spectrograph to study birdsong. His confidential report to the University regarding work at Madingley from 1951 to 1952 acknowledged the
debt: "The process of development of the song in isolated and hand-reared
birds has been followed in detail by sound recording and the results are being further analysed and compared by means of the sound spectrograph,

for the use of which we are greatly indebted to a Government Department which wishes at present to be nameless." In his annual report for 1952–1953, Thorpe was able not only to credit the Admiralty Research Laboratory by name but also to record his success in securing a sound spectrograph for Madingley's full-time use. A $2,700 grant from the Rockefeller Foundation enabled him to purchase a machine in the United States.[50]

Thorpe's work on birdsong development was the origin of a whole series of researches at Madingley, including important studies by Peter Marler and then further work by Marler's students and others.[51] Thorpe and Hinde also took up the study of imprinting, and this work likewise developed into a long and fruitful research tradition, highlighted by the studies of Patrick Bateson, who joined the field station in 1959–1960.

Thorpe in 1950 additionally undertook studies of "string pulling" in birds, hoping that this might "become a standard technique for the investigator of the higher types of learning and that it [might] open the way to an investigation of the existence of true imitative behaviour in birds." It was already known that a number of different species had the ability to secure food suspended on a string by pulling up the string with their beaks and holding on to the pulled-up string with a foot. In the first year of Madingley's operations Thorpe tested the string-pulling abilities of some eleven different bird species. This work was later continued by Margaret Vince, working both at Madingley and in the laboratory of the Cambridge Psychology Department.

Robert Hinde's other early research projects at Madingley were likewise of considerable consequence for the future development of animal behavior studies. He continued the line of studies he had undertaken at Oxford under David Lack and Niko Tinbergen on the behavior of great tits. He also began analyzing the mobbing of owls by chaffinches as a means of testing experimentally Lorenz's idea of action-specific energy.

As for Peter Marler, he had come to Cambridge in 1951 from a job at the newly founded Nature Conservancy. He already had a Ph.D. in botany from the University of London, but he had become interested in the development of chaffinch song. The Nature Conservancy arranged for him to go to Cambridge with a research fellowship. He worked closely with Hinde as he developed a second doctoral dissertation, this one on the behavior of chaffinches in the Madingley woods. The report of Madingley's first year of operation mentions Marler's having two projects: one with Hinde on the courtship and copulation of the chaffinch in captivity; the other, officially his own, on behavior during the annual cycle of the chaffinch. Over the next several years Marler worked with Thorpe making recordings of bird-song. Together they learned how to use the sound spectrograph as well as other, more primitive, recording devices. But Marler's opportunities to work on birdsong were not unlimited. Thorpe's anxieties about research turf were made clear to his students as well as to workers at other insti-tutions. As Marler has recounted it, "I felt no inhibitions about working on calls, but Bill, as the boss, exercised territorial rights when it came to song."[52]

The work at the Madingley facility developed steadily, as did the facil-ity itself. The Nuffield Foundation promised to fund the Ornithological Field Station from 1951 to 1954 if the university agreed to provide compa-rable support for two years afterward. Notwithstanding Gray's lukewarm view of the station, the university agreed to round out the five-year pack-age. As of the fall of 1952, seven researchers were regularly using the station: William Thorpe, Robert Hinde, Peter Marler, Margaret Vince, G. V. T. Matthews (studying bird navigation), R. J. Andrew (studying the compar-ative behavior of buntings), and A. R. Jennings (studying bird diseases). These numbers increased through the decade. By 1960 Thorpe could list nearly twenty individuals who had made use of the field station that year. He could also record with pleasure that the university had established, within the Department of Zoology, a Sub-department of Animal Behav-iour with Thorpe as its director and the Madingley field station as its nucleus.

SETTLING LORENZ

Having described how ethology was institutionalized at Oxford and Cam-bridge, let us return to Lorenz's story where we left it. In September of 1949 Lorenz was back in Altenberg after nearly two months away in England and Holland. His situation was better in two respects than it had been at the be-ginning of the year: he had been officially reinstated as a lecturer at the Uni-versity of Vienna and his research station at his home in Altenberg had been officially taken under the wing of the Austrian Academy of Sciences and designated the Institute for Comparative Behavior Study. Four stu-dents—Ilse Prechtl, Heinz Prechtl, Wolfgang Schleidt, and Irenäus Eibl-Eibesfeldt—had come to live at Altenberg and were helping him carry on research. He felt very fortunate to have them, not only because they were excellent researchers but also because they were willing to work essentially without pay. Financially, Lorenz was still just scraping by. He wrote to Wallace Craig much the same thing Craig had written to C. C. Adams some four decades earlier: "I scarcely know how to buy shoes and clothes for my family."[53]

To make money, Lorenz published a popular book recounting his experiences with animals around—and in—the house. Entitled *Er redete mit dem Vieh, den Vögeln und den Fischen* (literally *He Talks with the Beasts, the Birds, and the Fish* but subsequently published in English as *King Solomon's Ring*), the book was a great success. Just as the book's sales began to bring in fresh revenues for him, however, the Austrian tax authorities insisted on taxing Lorenz's home not as an institute, with a low tax burden, but instead as a single-family dwelling, at a much higher rate. The amount levied exceeded Lorenz's modest income as a lecturer at the University of Vienna. Lorenz enlisted Richard Meister, the rector of the university and vice president of the Austrian Academy of Sciences, to help argue the case for the lower, institutional tax rate, but in November, 1949, the outcome remained in doubt.[54]

At this point Lorenz began contemplating other sources of support. In particular he began to think about raising funds through a lecture tour to the United States. At the SEB symposium in Cambridge, both Karl Lashley and Paul Weiss had encouraged him to come to the States to talk about his work. Lorenz wondered in a letter to Ernst Mayr whether an American lecture tour might be one way of encouraging the Rockefeller Foundation to support Lorenz's institute at Altenberg.

Lorenz also inquired of Mayr about the level of anti-German sentiment in the States. Would this be a problem for Lorenz as a lecturer? There was also, Lorenz confided, a more personal matter to mention. Karl von Frisch had already tested the Rockefeller waters on Lorenz's behalf, only to be told, as Lorenz put it to Mayr, "that I had been a member of the German Army, Psychological Division, and therefore a dangerous Nazi." Deciding he had better give Mayr his version of the story, Lorenz explained: "I was indeed for two months working at a personnel examination board [*Personalprüfstelle*] in Posen as an apprentice, so to speak, but a promotion to personnel consultant [*Personalbegutachter*] never occurred, and I was never elevated into the Army Civil Service because the entire Army Psychological Division was completely dissolved in June of 1943."[55] Lorenz continued: "I consider it potentially useful if I let you know the following as well: I was not a member of the NSDAP (by the way, again not owing to my own merits, but due to a lucky coincidence, i.e. invited to join I clashed in time with the party organization) and am presently newly appointed at the University of Vienna."[56]

Sorting truth from fiction in this statement is not entirely simple. It was certainly true that Lorenz felt it was "potentially useful" to tell Mayr what he did. His veracity about his party membership, on the other hand, was more nebulous. In 1938 he applied for membership in the NSDAP and was accepted. In 1949, however—and here indeed he was lucky—the Austrian authorities chose to say that he had not received his party membership card and thus had never officially become a party member. Lorenz could thus allow to Mayr that he had never been a party member without bothering to mention that in saying this he was reporting a fortunate ruling rather than an unambiguous matter of fact. What his statement makes perfectly clear, in any case, is that he wanted his recent rehabilitation at Vienna to be taken as a sign that the issue of his having a Nazi past was now finally officially closed.

In addition to testing the water, figuratively speaking, with regard to a lecture tour to the United States, Lorenz was also thinking at this time about real water—and about filming ducks. Specifically, he was planning in the spring to go to Slimbridge, England, on Peter Scott's invitation. There he would meet up with his friend and former assistant, Alfred Seitz, and the two of them would spend a month or more filming the behavior of different species. They would film all the behavior patterns of all the species Lorenz had described in his anatid monograph of 1941, plus the behavior of additional species at Slimbridge. As Lorenz saw it, the filming project and the lecture tour were related. He had had great success previously when he had lectured and at the same time shown his film on the ethology of the greylag goose. He imagined that new footage from Slimbridge, together with the old greylag film, would provide an excellent foundation for his lecturing tour to the States, whenever that took place.[57]

In the meantime, Lorenz had other lectures to present. In January 1950, he traveled to Germany to lecture in both Munich and Freiburg. The drawing card in Munich was a sociological conference convened by Professor Ernesto Grassi. Lorenz went there expecting and no doubt also hoping to be the enfant terrible of an evening debate on the concepts of *Gestalt* and *Ganzheit*.[58] The other participants included Grassi, August Thienemann, Adolf Portmann, and Thure von Uexküll (the son of Jakob von Uexküll). The concepts of *Gestalt* and *Ganzheit* may have been the intended focus of the discussion, but questions concerning evolution also came to the forefront. The Munich paper *Neue Zeitung* reported on this under the headline "Das Ende der Abstammungslehre" ("The End of the Theory of Evolution"). Lorenz was not mentioned in the account (a disappointment, one would imagine, for the would-be enfant terrible), but

Portmann was represented as saying it was a mistake to draw parallels between animal and human societies. His views were seconded by the author of the *Neue Zeitung* piece, who noted that the word "cell" had been used not only for the "cells" of the beehive but also for the "cells" or sections of German towns overseen by the Nazi officials called *Zellenleiter*. Otto Koehler wrote to Margaret Nice about the article, expressing his dismay at how Darwinian biology was routinely discredited by such implausible and unfair associations.[59]

This postwar linking of Darwinism and Nazism was not new. Two years earlier Koehler had complained to Nice of the same problem: "There are many people now who say that zoology is irreligious, and that nature is from the devil, and Darwin was the preface of Hitler and nonsense like that. And there are zoologists f.i. in Switzerland who seem a little to compromise with such formulations."[60] Koehler may well have had the Swiss zoologist Portmann in mind. Behind closed doors, Portmann was disposed to identify Lorenz with Darwinism and Nazism. An entry dated 19 November 1950 in the diary of the associate director of the Humanities Division of the Rockefeller Foundation indicates that Portmann had told him that Lorenz "had been the most outstanding Nazi in Austria." The note in the diary continued: "He [Lorenz] is undoubtedly a very able biologist, but Portman would hope that he would not receive a position where he would be brought in touch with students."[61]

From Munich, Lorenz went to Freiburg, where he stayed a week with Koehler and lectured to students five times. The lectures elicited extended discussions afterward. Koehler reported to Nice that their mutual friend was "quite the old Lorenz, full of spirits, wits, just as you know him."[62] While Lorenz was still in Freiburg, Holst came for a visit and he, Lorenz, and Koehler spent a full day together. Lorenz later told Thorpe of his trip:

> My short lecturing tour to Germany was most exciting. That poor country is at present visited by an epidemy of anti-Darwinism and existantial philosophy and I had the most wonderful rows in some discussions. Once I found myself moved to say that it had been a complete mistake to lecture on comparative ethology at all to such an audience, the correct thing to do would have been to teach them the main facts about the origin of species, or perhaps, better still, about why the astronomy of Kopernikus was preferable to that of Ptolemy. After Cambridge it was as if I had stepped into H. G. Wells's time machine and travelled back a few hundred years! They would not believe that the question of Darwinism was never even mentioned in our congress in Cambridge. Poor Ger-

many! I was seriously depressed on my coming back, mainly, because even the young students are fanatically anti-Darwin, because they identify him with national socialism. One more cause for hating the latter, damn it![63]

Lorenz hated to see Germans engaged in Darwin bashing. On the other hand, he was impressed by the vitality of Germany and the vigor of the rebuilding efforts in Munich and Freiburg, and he was stunned to find that professors in Germany were earning six times more salary than he was.[64] But he soon learned that his own fortunes were likely to change for the better. In mid-February, shortly after returning from Freiburg, he received a letter from Holst saying: "Perhaps you know it already—that Frisch is going back to Munich, and . . . wants to have you as his successor at Graz." Lorenz was ecstatic. He decided to put his Slimbridge plans and his idea of an American lecture tour on hold.[65]

Frisch, for his part, had decided to return to his earlier post at Munich only after extended negotiations with the university administration there. The details of that story, which included a contest with the chemist Heinrich Wieland over the rights to Frisch's previous lab space, need not detain us here. What was important for Lorenz's situation was that Frisch was returning to Munich and leaving vacant his position at Graz.

Seven candidates were being considered for the post. Frisch viewed Lorenz as the clear front-runner. Frisch alerted Holst, however, to the possibility that Lorenz might be opposed. He asked Holst to provide an official review of the candidates. Holst responded quickly, agreeing that Lorenz was without doubt the top candidate. He ranked Lorenz "by far in first place" and identified him as a "multifaceted zoologist with outstanding knowledge in anatomy, systematics, ecology, and at the same time the founder of a new intellectual and research direction in behavior theory or animal psychology." Lorenz and his school, Holst wrote, were liberating the study of animal behavior from its "materialistic-behavioristic straitjacket" and elevating it to "a recognized branch of the broad discipline of zoology." Lorenz was "recognized at home and abroad," he said, as "the leader of his field." Already, Holst point out, several of Lorenz's disciples (most notably Tinbergen, who had just been called from Leiden to Oxford) possessed important positions. Any university faculty, Holst insisted, would be honored to be able to claim among its members such a "pathbreaking researcher" and "most stimulating teacher" as Lorenz.[66]

Holst let Lorenz know he had sent a glowing recommendation to Graz on Lorenz's behalf. But he also gave Lorenz a warning. "Strong counter-currents, Holst explained," threatened Lorenz's candidacy. "Your enemies," he told his friend, "are surely with pleasure gathering together everything that proves (apart from earlier statements that can be used politically) that you advocate an unethical, pro-Darwinist, materialistic, etc. conception of human nature and human origins and especially of the nature of the human spirit." He advised Lorenz to lie low for as long as the appointment remained undecided. Specifically he asked him "not to give any dust-stirring lectures on, e.g., the animal dimensions of human behavior, human phylogeny, scientific attempts to explain religiosity, etc., even if it's difficult." [67]

How seriously Lorenz took Holst's concerns is hard to tell. What mattered to him most was the main news, about which he was thrilled. He reported enthusiastically to Thorpe (who, as it turns out, had already written to Frisch on Lorenz's behalf): "Karl von Frisch is leaving his Institute in Graz and going back to his old big institute in Munich and has appointed me *primo et unico loco* for his successor in Graz. I shall have to take over his lecture in Graz provisionally about April 15th and I am even now starting to learn comparative anatomy of vertebrates. The probability that I shall definitely be appointed as professor in Graz is about 95%." [68]

Lorenz went on to tell Thorpe that the rector of the University of Vienna (Richard Meister) and the professor of zoology (Otto Storch) were trying to keep Lorenz in Vienna. They had gone so far as to suggest to Frisch that he take Lorenz's friend Wilhelm Marinelli instead of Lorenz for the Graz position so that Lorenz could have Marinelli's position as *Extraordinariat* (associate professor) in Vienna. Lorenz allowed that this prospect was also appealing to him. His research facility in Altenberg made an associate professor's post at neighboring Vienna almost as attractive as a full professorship at the more distant Graz. Lorenz visited Frisch to talk this scenario over, but Frisch was unreceptive to it. Lorenz explained to Thorpe: "He succeeded in convincing me that it is more or less my duty to take on the professorship in Graz in the interests of Comparative Ethology which hitherto is not represented on the continent by a whole institute with a comparative ethologist for its director except in Groningen (Baerends). Anyhow, he would have none of Marinelli." [69]

Everything at this point seemed to be going swimmingly for Lorenz. Institutions were finally competing with each other to give him what he wanted and needed most: "an official, paid job." In addition, his popular book was selling "unexpectedly well," and Julian Huxley had just offered to write an introduction for the English version. At Huxley's instigation, furthermore, Lorenz had agreed to write two monographs for the New Naturalist series, one on the jackdaw and one on the greylag goose.[70]

As late as the end of March 1950, Lorenz expected to start substituting for Frisch in Graz by the middle of April. He exulted to Huxley about how honored he was to be Frisch's first choice and how splendid the facilities at Graz were: "The Zoological department in Graz is by far the most modern, largest and best-staffed in Austria, a fine building with large sunny rooms, very suitable for aquaria. Even now I am collecting and breeding Cichlids as fast as I can."[71] By the end of April, however, the position was still not finalized. Koehler began to worry. He wrote Holst, "I fear the Austrians are perhaps no less scheming than the Bavarians." Lorenz was more positive, but a note of caution intruded in a letter to Thorpe: "It is still soon to crow about Graz, I have not got it yet."[72]

In the meantime Lorenz wanted to be sure that Thorpe was coming to the small conference Holst was organizing in Wilhelmshaven. "There are going to be the best 'animal-people' in the world," he said, and meeting Holst himself was "worth travelling any distance." Lorenz assured Thorpe that Holst was "an entirely different person from what you must think judging from his belligerent little paper read by Lissmann at the congress." Lorenz himself was planning to come "in the hope of getting a bit clearer about what action-specific energy might really be!"[73]

The Wilhelmshaven conference was held from 30 May to 5 June 1950. Thorpe and Tinbergen were among those in attendance. Perhaps Lorenz at this time expressed to his friends his increasing anxieties about Graz. Whatever the case, by the end of June he recognized the cause was lost. He wrote Thorpe: "I am sorry to say that I shall not get the chair of Graz. Although I have expected it, it nevertheless is rather a blow, mainly because it means that I shall, in all probability, *never* get a good position in Austria." The *faculty* had put him forward as its first choice, he said, but the *ministry* was unwilling to have him. It was time, he believed, to try to leave Austria. He could no longer continue "spending 75% of my energies on writing popular booklets and worrying about financial cares." Some day, he told Thorpe, he might still be awarded the Max Planck Institute he dreamed of sharing with Holst, but in the meantime he needed to think of something else. "I am 46 years old," he wrote, "and time is running away without my doing anything really valuable, except writing my big book, but that means only using up my old capital of knowledge without acquiring any new." Lorenz asked both Thorpe and Tinbergen whether either of them could find him a temporary job in England. "What I am dreaming of," he told

Thorpe, "is a small job in Oxford or Cambridge which enables me to do some work on cichlids and leaves me time to work on the Slimbridge water-fowl collection during the critical months in spring."[74]

Thorpe and Tinbergen set about seeing what they could do. Thorpe asked Gray whether anything was possible at Cambridge. Tinbergen came to Cambridge at Thorpe's request to talk over Lorenz's situation.[75] Thorpe then contacted Peter Scott (who had already invited Lorenz to come to Slimbridge for two months the following spring) and Max Nicholson (who had impressed Lorenz as someone with "pull"). Thorpe asked Scott about the possibility of a research post for Lorenz at Slimbridge, combined with an academic position at Cambridge or elsewhere. It would be a real coup, Thorpe suggested, to work out something for Lorenz in Britain: "I am sure you realise that he is perhaps the most outstanding and original student of the behaviour of the higher animals living today. He has partly completed an immense work on animal behaviour which is obviously going to be a landmark in the subject, and he is one of the most stimulating and original characters I have ever come across. It will be an immense asset to British zoology to get him to this country."[76]

Lorenz's book actually remained a good ways from being done. Thorpe had encouraged Lorenz to develop the epistemological side of the book in its English as well as its German version. Lorenz's German friends, on the other hand, were inclined to grow anxious when Lorenz turned philo-sophical. Holst, upon reading the large, manuscript "introduction" to Lo-renz's book, felt his initial pleasure turn to alarm. As he told Koehler, Lo-renz in epistemological matters was "like the elephant in the china shop." It was "a true shame," Holst said, for such a "genius at observation" to step out of his natural element and handicap from its very beginning a book that otherwise could have been outstanding. Holst hoped he could per-suade Lorenz to make some changes.

Koehler understood well Holst's anxieties. Koehler's wife, a philoso-pher, had tried previously, but to no avail, to explain to Lorenz that his un-derstanding of Kant was shaky. Holst proceeded to write Lorenz a twenty-page "critical and cautionary letter" about Lorenz's philosophizing in the draft "introduction." Receiving no response, he concluded he had not been of the least influence. Gustav Kramer in the meantime also provided Lo-renz with critical comments on the same material. Whether it was because of his friends' comments or because other things were demanding his at-tention, Lorenz put his book project aside for the time being.[77]

In the middle of August 1950, Lorenz decided that if he were to have a chance at a position in Great Britain, he had better provide his British friends with a fuller account of why he had not gotten the professorship at Graz.[78] As indicated above in chapter 5, he wrote to Thorpe and Huxley explaining to them how he had come to write his 1940 paper in *Der Biologe*. To both men he stressed his troubles with the Austrian educational authorities before the war as well as afterward. To Thorpe he wrote:

> It is now quite evident that the Austrian authorities, meaning the Unterrichtsministerium, will have none of comparative ethology. I think it advisable to inform you about all the tedious details, excuse the long letter. You know that in the year 1935 when I tried to become Dozent for "Tierpsychologie," meaning comparative ethology, this was impossible, because the catholic ministery denied the poor animals the right to have a psyche. I had to apply again and the[n] became "Dozent für Zoologie mit besonderer Berücksichtigung der vergleichenden Anatomie und Psychologie," the last two words being smuggled in, hoping nobody would read as far as that.

Lorenz then told Thorpe how his article in *Der Biologe* was essentially an attack on the antievolutionist Ernst Krieck, and not really a pro-Nazi piece at all, but the article was now being used against him:

> Some time ago, Meister told me that this little paper had been unearthed by some dear friend, and the ministery's attention drawn to it, also that the authorities particularly objected to one passage, where I wrote that the Schuschnigg ministery regarded everybody who believed in Darwin as a nazi and (quite thruthfullx) that this had been what had driven me into the Nazi camp.[79]

Lorenz thereupon decided, as he recounted it to Thorpe, to tell the whole story to the minister himself. However, he then learned from his friend Marinelli, who had a friend who knew the minister, that the minister was not really concerned with the paper in *Der Biologe* but objected instead to Lorenz's field of study, namely, animal psychology. It thus seemed, to Lorenz's dismay, that the situation was no different from what it had been in 1935. The minister did not want future gymnasium teachers to be taught Lorenz's "dangerous science."[80]

Lorenz railed further to Huxley:

> The minister does not care three whoops in hell about that article and what I wrote in it, but welcomes it only as an excuse. What he actually ob-

jects to is that "Tierpsychologie" should be preached from a chair at an university where all prospective middle school teachers of all Styria are bound to hear it. He regards this science as dangerous for exactly the same reasons for which old Krieck objected to it: He thinks that it is detremental to the true recognition of the uniqueness of Man and his "dignity." The animal has no soul and therefore Tierpsychologie is objectionable in itself. You may remember that the same objection was raised by the Schuschniggian minister of education in 1935 when I first tried to get the Dozentur for that science! Confound Nazism, Catholicism—and Marxism too, since it has given birth to Mitchurin and Lysenko!"[81]

Self-serving as Lorenz's account to Thorpe and Huxley may have been, it may also have been not so very far off the mark with respect to the failure of Lorenz's candidacy at Graz. It is entirely plausible *both* that the 1940 article from *Der Biologe* was called up to discredit him *and* that the fundamental objection to him was his evolutionary view of human psychology. It is impossible to say whether Lorenz really believed that if Thorpe and Huxley went back to his 1940 article it would be obvious to them that he had never been a Nazi at all. In any case, there is no evidence that either Thorpe or Huxley actually made any effort to check Lorenz's claim. They valued Lorenz for his pathbreaking science and his intelligence. It no doubt also made a difference to them that Germany was no longer the enemy. The primary threat to world peace and modern science, as viewed from the West, was now the Soviet Union.

In his letter to Huxley, Lorenz described what Thorpe, Tinbergen, and Scott were trying to arrange for him so he could come to work in England. He would conduct research at Slimbridge, he explained, and one or more universities would pay him to give guest lectures. Although this would be only a temporary solution to his job problem, it had two attractions. The first was scientific. As he cheerfully put it, "Regard it this way: Here am I, an ethologist, particularly specialized on Anatidae, without any of these birds, except some domestic geese that my wife keeps and breeds for food. In Slimbridge, there is an unique collection of anatides without an ethologist to work on it. These two facts alone do, in my opinion justify what we are trying to do." The second attraction involved the strategic dimensions of such a move: "Nemo propheta in patria, and a year or two of absence will certainly not diminish my chances of getting a position in Austria after all. If I stay on, struggling along here in Altenberg, as the authorities certainly expect me to do, I shall never get an adequate position, if I go away for a time, there is just a chance that I do."[82]

Lorenz was soon proved right in thinking that going away would make him look more attractive to his countrymen. When Lorenz's German friends learned that some kind of position was being cobbled together for him in England, they recognized this might inspire the Max Planck Society to construct a better alternative for him and thereby induce him to refrain from going. Holst, Lorenz, and Kramer had long shared a dream of building up an institute for behavioral physiology in the vicinity of some university. As Holst explained to Koehler, a site like Freiburg (where Koehler was teaching) would be best. Holst did not know what the chances were. As he put it, "favorite dreams tend not to come true, their value is rather that they float in the distance, where the leanness of reality cannot hurt them." He set off for the Max Planck headquarters in Göttingen to talk with Ernst Telschow, director of the Max Planck general administration, to see if anything could be done to keep Lorenz from emigrating.[83]

Holst urged Telschow that it was more important than ever to keep outstanding German scientists from leaving Germany and taking positions abroad. Telschow agreed. The result of the interview, as Holst reported to Lorenz at the beginning of September, was not yet the guarantee of an institute, but it was a promise of DM 6,000 per year. What Holst needed to know immediately from Lorenz, before writing to MPG President Otto Hahn on the matter, was whether Lorenz, on the basis of this support, would be willing to decline for some years "das Oxford-Angebot" (the Oxford offer)."[84]

Had Lorenz told Holst that an offer was coming from *Oxford?* Or had Holtz perhaps mistakenly constructed this out of the information that Tinbergen at Oxford and others of Lorenz's friends in Britain were putting *something* together there for Lorenz? Or was "the offer from Oxford" a fiction recognized by Holst and Lorenz alike, perhaps not so very far from the truth, and certainly strategically advantageous when it came to promoting Lorenz's cause to the Max Planck Society? Whatever the case, ten days later Holst had an even brighter picture to paint for Lorenz than he had been able to offer previously. In the new picture there was not only research money but on top of it a whole new site, including a castle. The name of the place was Buldern. It was located in Westphalia some twenty kilometers south of Münster.

Buldern and its castle were the property of Baron Gisbert von Romberg, a science enthusiast eager and willing to put a workplace and scientific resources at Lorenz's disposal. As Holst represented the man to Lorenz, Romberg was "one of that dying breed of people who promote research unselfishly out of idealism." Inside the castle would be working space for

Lorenz. Outside the castle, the waters already teemed with fish, and wild geese and other birds flourished among the reeds. The overall setup at Buldern, Holst allowed, seemed "nothing less than ideal" for Lorenz's kind of research. Furthermore, as testimony to their personal eagerness to make this work, Holst and Kramer were prepared to divert some of their own Max Planck funding to help with Lorenz's start-up costs—DM 3,000 year on top of the DM 6,000 Telschow had already verbally guaranteed.

Holst noted additional attractions of Buldern, such as its proximity to the University of Münster, where there would be good colleagues. Holst and Kramer also wanted Lorenz to realize how much start-up time he would require if he went to Oxford. At Buldern, Holst said, no time would be lost, because Holst would immediately furnish Lorenz with aquaria, and Lorenz from the beginning would have a well-furnished workplace for himself and his co-workers. As icing on the cake, Holst noted that Buldern, by Romberg's account, was a fine place for raising geese. The castle grounds, furthermore, were beautiful. Holst made arrangements to take Lorenz to Buldern to see things for himself.[85]

The British were now, without knowing it, a step behind the Germans, but they were nonetheless making progress in finding Lorenz a position. Thorpe, Tinbergen, Hardy, Scott, and Nicholson jointly formulated a plan: the Nature Conservancy would channel a salary for Lorenz through a university. The university affiliation would be Bristol—not Oxford, as Holst understood it, or Cambridge, as Thorpe had initially dreamed. Bristol's attraction was that it was close enough to Slimbridge so that Lorenz could have a reasonable commute to either site from somewhere in between. The plan proved agreeable to J. E. Harris, the professor of zoology at Bristol, and to the university administration. J. B. S. Haldane offered one hundred pounds of his own (for personal reasons, and without Lorenz's knowing it) to enhance the package. Tinbergen and Thorpe discussed with each other how soon—and how—to let Lorenz know about how the way plans were developing. Vividly remembering what life in Holland had been like under the Nazis, Tinbergen worried what the Russian authorities who controlled Altenberg might do if they learned Lorenz had a job offer in Britain. Tinbergen explained to Thorpe that such anxieties should not be taken lightly: "Most of my English friends have not the slightest idea of the attitude of mind and of the complete immorality and ruthlessness of civil servants of totalitarian states."[86]

As the British plans took shape, the Max Planck Society was moving more speedily still. Lorenz met MPG President Otto Hahn at a conference in Munich; Hahn invited Lorenz to come to talk to him in Göttingen. The

Göttingen meeting took place on October 30. Two days later, on his way back to Altenberg, Lorenz began writing to tell Thorpe the news: "The 'Max Planck Gesellschaft' has decided to *build an institute for me!* I shall have a *full professor's fee* and be able to employ *three assistants!* Technically my institute will be part of Wilhelmshaven, but practically independent, situated in an incredibly picturesque castle near Münster in Westfalen. The castle is named Buldern and is surrounded by a gigantic moat and any number of big ponds, waiting for ducks and geese, a big empty greenhouse is waiting for aquaria."[87]

Lorenz had additionally learned, via a letter from his wife in Altenberg, that Peter Scott was also making him an offer: nine hundred pounds per year, plus use of Scott's boat the *Beatrice,* to work at Slimbridge. Buldern, however, was even more attractive than this. It was not a matter of salary, for in that regard the Buldern and Slimbridge offers were about the same. From the standpoint of research, Lorenz judged the facilities in Buldern were better for studying cichlids but not for studying anatids. What decisively tipped the scale in Buldern's favor, Lorenz explained, was that "the Max Planck Ges. offers me a permanent employment, with a pension when I get old and stupid; and also that I shall have *three* permanent jobs for Ilse Prechtl, Schleidt and Eibel."

Even after learning that a Max Planck appointment for Lorenz was in the works, Lorenz's advocates in Britain continued to pursue an official offer to him from Bristol and the Nature Conservancy. This was not unreasonable on their part. Lorenz had indicated that the Max Planck plan for Buldern was not a fait accompli (it still needed to be scrutinized by a commission and then approved by the MPG senate).[88]

In the meantime Lorenz told Huxley the good news about his meeting with Hahn (though Huxley had in fact already heard the news from Warren Weaver of the Rockefeller Foundation, who had heard it from von Frisch).[89] Thanks to Holst's efforts, Lorenz reported, Hahn was now promising Lorenz DM 20,000 for the first year, with DM 9,000 in addition to be contributed by Holst and Kramer from their Wilhelmshaven funds. Lorenz expressed his delight: "My old dream of a Max Planck institute seems to become realized at last. At a most conservative estimate my salary will be twice that of a professor in Graz and I shall be able to offer my Altenbergian collaborators reliable positions which I should only be able to do for one of them if I had got the professorship in Graz."

Lorenz professed to Huxley that he still intended to come to Slimbridge in the spring of 1951 for two months. The relative proximity of Buldern to Slimbridge, he said, as compared to the much greater distance between Altenberg and Slimbridge, would enable him to return to Slimbridge frequently. The question was how to organize his research. Tempting as it was, he did not want to build up a big collection of anatids at Buldern. The problem, he said, was "not to let our new institute grow into a Zoo!" This, he acknowledged, was a real danger, "as all of us are exactly the same kind of animal lovers and you know how one is apt to buy 'just one More' beastie until all one's time is taken up with feeding and cleaning cages and none left for scientific observation." He therefore swore he would "conscientiously concentrate on *geese* in Buldern, starting out from the Greylag as my best-known species, and slowly extending my investigations on the closest allied, Bean, Whitefront, Snow etc." In addition he would continue his comparative study of cichlids.[90]

Lorenz apologized to Huxley for having put his friends in England to so much effort on his behalf. He had never expected, he said, that Holst would be able to achieve anything for him through the Max Planck Society. As it was, he acknowledged, it was largely because of the English activity that the Max Planck Society had come through for him at all: "This paradox is explained by the fact that it is a special point in the programme of the Max Planck people to prevent the emigration of German-speaking scientists. They would not have given a damn if I had starved in Austria (well, that is in injustice, they would have given one) but they reacted at once when Holst told them that I was going to England for good." As it was, it was the Austrians who were left to bemoan the fact that they had not done enough to keep top scientists like Frisch and Lorenz in their own country.[91]

Lorenz was grateful to hear that Tinbergen understood his choice of Buldern. He thanked Tinbergen for this, writing to him indeed from Buldern, where, at the beginning of December, he was already at work setting things up. He advised Tinbergen not to discontinue efforts in England until a Max Planck contract was actually finalized, but he indicated that the Buldern operation had become a sure thing. A sign of Baron Romberg's helpfulness, Lorenz said, was that while Lorenz was sitting in the castle writing to Tinbergen, Romberg was driving Lorenz's wife and Wolfgang Schleidt to Münster to buy angle irons and glass for the new aquaria (with funds that Holst and Kramer had made available). Romberg was, in Lorenz's words, "absolutely unbelievable." He was a skilled engineer and a clever physiologist. Most of all, he had stepped in as deus ex machina to provide Lorenz with the research station of his dreams. If one were to read such a thing in a novel, Lorenz told Tinbergen, one would dismiss it as "far too unrealistic." On top of everything, Romberg was "personally the nicest fellow I have become acquainted with in a long time."[92]

The Max Planck deliberations proceeded more quickly than Lorenz expected they would, hastened by the worry that Lorenz would go to England and expedited by the fact that Holst chaired the commission that reviewed the proposal to situate Lorenz at Buldern under the aegis of Holst's Max Planck Institute at Wilhelmshaven. Hahn phoned Hans Bauer, the distinguished *Drosophila* geneticist who directed the unit for cytogenetics there. Bauer confirmed having neither scientific nor personal objections to Lorenz. Hahn reported this to Boris Rajewsky, Director of the Max Planck Institute for Biophysics, who was responsible for presenting the report of Holst's commission to the MPG senate. Hahn told Rajewsky he had heard "the same favorable opinion" of Lorenz from "the most different quarters."[93]

Some MPG members had complaints about the proposal when they first heard of it. One objection was to the way Holst back in September had gone directly to Telschow instead of consulting colleagues. Another objection involved the inappropriateness of creating an inland research institute under the wing of what was by title an Institute for Marine Biology. A third concern had to do with Romberg, who nurtured hopes of gaining an MPG affiliation of his own. Incidental to this grumbling, but a potential factor in the senate vote nonetheless, was the readiness of one or more MPG members to disavow Hahn, the aging president of the society. When the senators met in official session on 19 December 1950, they were presented with a proposal to create research opportunities for Lorenz, "an animal psychologist of great renown," within the framework of Holst's Max Planck Institute of Marine Biology at Wilhelmshaven. Baron Romberg's offer to provide accommodations for Lorenz at Buldern was understood as part of the arrangement. President Hahn strongly supported the plan, pointing out that Lorenz otherwise would accept a position in England. The senate endorsed the plan without a dissenting vote, earmarking DM 40,000 for Lorenz's operation.[94]

Whether these arrangements would actually work out remained very much in question. All of Europe was anxious about the tense situation in world politics. Lorenz allowed to Huxley that he was worried about his apparent good luck: "I am conditioned by the fact that a world war broke out as soon as the K. W. G. had decided to build me an institute in Altenberg in 1938." Koehler was afraid that the division of Germany and Europe would inevitably lead to war, and that under the circumstances the Lorenzes might prefer to be *farther* from Altenberg, that is, in Bristol rather than Buldern. Holst, however, assured Koehler that Lorenz appeared to be "in his element" at Buldern. "Everything so far," Holst acknowledged, had gone better than he "had really expected," except for the behavior of some of his Max Planck colleagues. As for the political scene, he hoped the god of war would be merciful and spare a soil that was already "bloody enough."[95]

Lorenz truly felt very much in his element as he settled in at Buldern in December 1950. On the last day of the year he wrote to Thorpe apologizing for all the trouble he had put Thorpe and Tinbergen through on his behalf. Of his current situation he stated: "At present, Schleidt and I are sitting here in Buldern, building Aquaria and generally planning and adapting things. Our aquaria-greenhouse is going to be quite wonderful and the goose-territory will be really unique, better even than Peter Scott's. Erich von Holst has given us, viz. our institute, his little DKW-car as a Christmas gift. . . . It is absolutely like a dream, to have money enough and to move about freely in one's own car, buying aquaria and birds and everything with easy nonchalance, as if it were a matter of course."[96]

The building of the aquaria and the preparation of the goose territory proceeded apace. Lorenz sent out orders for cichlids and for greylag goose eggs. He purchased all the interesting cichlid species he could find. For the greylag eggs he ultimately wrote to some thirteen different suppliers, hoping to get enough eggs to produce fifty greylag goslings. If half these goslings were subsequently lost, as he expected they would be, this would still leave him a suitable number for starting a colony.[97]

The Buldern arrangement was conceived from the beginning as a temporary arrangement for Lorenz, prior to establishing something better. It remains the case, however, that it was there that he set up his first MPG operation. And it was there that he was joined by his students Schleidt, Eibl-Eibesfeldt, and the Prechtls; there that numerous other visitors and students came to interact with him; and there that the first official international meeting of ethologists was held early in the spring of 1952.

All in all, Lorenz's operation at Buldern lasted some six years. The idyllic part of the stay was much shorter than that. Baron Romberg died early in the summer of 1952, His heirs proved much less inclined than he had been to view with favor Lorenz's extended family of animals and researchers. Late in June 1952, amid squabbles with the new baron about what was to be done with the old baron's scientific equipment, Holst wrote Lorenz to suggest that the time had come to look elsewhere. More precisely, it was time to find the right place for the three-person institute of which he, Lorenz, and Kramer had long dreamed. He and Kramer, Holst explained, favored a southerly climate, something in Bavaria or the Black Forest region.

7.1 Lorenz, ducks, and geese at Buldern. (Photo by Alfred Seitz. Copy provided by the Konrad Lorenz Institute for Evolution and Cognition Research.)

Within a matter of weeks, Lorenz was off inspecting possible sites for a new, more permanent Max Planck installation.[98]

ANIMAL BEHAVIOR STUDIES AT THE AMERICAN MUSEUM OF NATURAL HISTORY

As they busied themselves institutionalizing ethology in their own (or neighboring) countries in the postwar period, the Continental and British ethologists paid relatively little attention to animal behavior studies across the ocean in America. Behavior studies in the States had long been dominated by comparative psychologists, most of whom were behaviorists of one sort or another. From the ethologists' perspective, American comparative psychology exhibited a handful of conspicuous, interconnected failings. It focused on one, single animal species (the white rat) instead of a wide array of different organisms. It was based entirely in the laboratory and neglected to consider how animals behaved under natural conditions. Finally, it concerned itself almost exclusively with learned behavior and did not take instincts into account. The ethologists confidently assured them-

selves that running white rats through mazes was not an enterprise that deserved the title "comparative."

This view of American animal behavior studies was not without justi
fication. In 1932, only partly tongue in cheek, the American comparative
psychologist Edward Chace Tolman dedicated his book, *Purposive Behavior in Animals and Men,* to "M. N. A."—that is, *Mus norvegicus albinus,* the
white rat. His reason for doing so, he said, was to indicate "where perhaps,
most of all, the final credit or discredit belongs." [99] Tolman's gesture was appropriate. In the years between the two world wars, American animal psychologists devoted an extraordinary percentage of their studies to this
single mammalian species. Just prior to the United States' entry into the
Second World War, in the four-year period from 1938 through 1941, some
two-thirds of the papers published in the *Journal of Comparative Psychology* and more than half the papers delivered at national meetings were
based on laboratory studies of the white rat. In no other country did the
study of animal behavior have such a profile. [100]

But American comparative psychology was not monolithic. Some
practitioners of the discipline by the 1940s were insisting on the importance of studying more than one animal species. Two such scientists were
Frank Beach and T. C. Schneirla at the American Museum of Natural History in New York, working in the Department of Experimental Biology
founded by Gladwyn Kingsley Noble. Both Beach and Schneirla lamented
how the discipline of comparative psychology had narrowed from an earlier, broader base to become little more than a science of "rat learning."
Their own researches were designed to provide a more genuinely comparative and biological understanding of animal actions. [101]

Frank Beach had done his doctoral research at the University of Chicago in the 1930s, studying the influence of neocortical lesions on maternal
care in rats. He structured his work along lines suggested by Karl Lashley,
who had recently left Chicago for Harvard. After finishing his doctoral thesis at Chicago in 1936, Beach spent a year at Harvard in Lashley's lab. When
Lashley learned that a position had been created for an assistant curator in
the Department of Experimental Biology at the American Museum of Natural History, he recommended Beach for it. Noble offered Beach the job
and Beach took it. He moved to New York in the fall of 1937.

The Department of Experimental Biology at the American Museum of
Natural History was established by Noble in 1928 as a center where natural
history and experimental biology would come together in the common
study of the evolution of the neural and hormonal bases of vertebrate social behavior. Noble's own research focused on the social behavior of the

lower vertebrates—fishes, amphibians, reptiles, and birds. He hired Beach to study the social behavior of mammals. Beach was intrigued by the research orientation of Noble's department and of the American Museum as a whole. The attention given at the museum to evolution, adaptation, and natural populations afforded him a much broader view of biology than his lab experiences at Chicago and Harvard had provided him previously. As for Noble, the department head, Beach found him to be brilliant, energetic, and "the only man I ever met with whom I simply could not get along."[102]

When Noble died suddenly in 1940 at the age of forty-seven from a streptococcus infection known as "Ludwig's quinsy," it looked as if his department might die with him. The department had been very much the expression of Noble's own distinctive interests and entrepreneurial abilities. Roy Chapman Andrews, the American Museum's director, doubted that the department could continue without its founder. Beach managed to save the unit from being disbanded. He became its curator and reconstituted it under a new name: the Department of Animal Behavior. Three years later, in 1943, he invited the animal psychologist T. C. Schneirla to join the department as associate curator. Four years later still, when Beach accepted a professorship of psychology at Yale, Schneirla took over Beach's post as curator.[103]

Schneirla had studied comparative psychology at the University of Michigan, writing a doctoral dissertation on learning and orientation in ants. After teaching at New York University from 1927 to 1930, he spent a year (1930–1931) as a National Research Council fellow with Karl Lashley at Chicago. This brought him into contact with Norman Maier, with whom he coauthored in 1935 the classic textbook *Principles of Animal Psychology*. He returned to NYU in the fall of 1931 and continued on the faculty there until his death in 1968.[104]

Preserved among Schneirla's papers is a manuscript identifying the aims and methods of the Department of Animal Behavior at the American Museum in 1943, the year Schneirla became associate curator there. The document identifies the "primary, over-all objective" of the department as being "to discover, develop, and study the interrelationships of basic, scientific theories or principles which explain the behavior of animal life." Restated, the department was concerned with the "broad, psychobiological principles governing behavior."

Three "secondary objectives" characterized the department's nature further. The first was *"To concentrate research upon behavior patterns possessing fundamental biological significance,"* that is, patterns that were "directly important to the organism's existence and to the perpetuation of the species." Along these lines, the organism was to be studied "as a living re-action whole." Its "inter-individual relationships and even inter-group functions" were understood to "possess fundamental biological signifi-cance." The second of the department's objectives at this level was "*To study the same types of behavior in different animals at distinctive levels in the phy-letic series,*" that is, to investigate "phylogenetically representative species" and to look at "general reaction patterns persisting throughout many phyla." Finally, the department was "*To employ every available investiga-tional technique and approach necessary to the realization of the primary ob-jective.*" In other words, observations and experiments in the field would be used to complement more controlled sorts of investigations in the lab-oratory. What is more, biological and psychological points of view would be pursued simultaneously. From the department's perspective, clearly, the study of animal behavior was not a marginal area of specialization but in-stead an enterprise that belonged at the very heart of the life sciences. In the self-confident and self-justifying words of the document, "The field of animal-behavior study is centrally related to branches of zoology such as ecology, morphology, physiology and systematics, and is an important meeting ground and area of synthesis of these subjects."

Reproductive behavior was the department's primary research focus. Its attractions as a subject of study included "its roots deep in the animal's inherited constitution" and the fact that it was "subject to modification through experience." The department worked on insects, fish, amphibia, birds, and mammals with the expectation that "a knowledge of inter-phyletic similarities and differences will yield information as to the prob-able course of the evolution of the behavior patterns under investigation." It recognized that some aspects of reproductive behavior were best studied under field conditions but laboratory work was crucial for getting at "the animal's internal milieu" and the neurophysiological and endocrinological dimensions of behavior. To the latter end, the staff of the department was "very profitably" making use of "techniques borrowed from the disciplines of experimental neurology, endocrinology, physiology and psychology." Progress was being made in the area of "the modifiability of innately orga-nized behavior patterns," the result being "an increased understanding of the processes through which behavior is altered as a result of experience." [105]

The interests of the Department of Animal Behavior at the American Museum of Natural History were thus a far cry from those of rat-running learning theorists. They were in fact more akin to the interests of the Con-tinental ethologists. Researchers at the American Museum focused on be-havior of "fundamental biological significance" at different levels of the

evolutionary scale. They were interested especially in questions of reproductive and parental behavior. Among their concerns were the stereotyped
forms of behavior generally referred to as "instinctive." And they appreciated fieldwork as well as lab work.

However, there were differences as well as similarities between the
American Museum's researchers and the ethologists. The Americans
tended to do more endocrinological work. They were also disposed to
put quotation marks around the word "instinctive" and to believe that
"innately organized behavior patterns" were subject to modification. They
were particularly intent on analyzing the *development* of behavior in the individual. The ethologists, in contrast, paid more attention to the biological
function of behavior patterns and to behavioral evolution. They were inclined to compare closely related species (while the American Museum scientists prided themselves on their thinking about the differences between
organisms from different phyletic levels). The ethologists (especially Lorenz) were also committed to the idea that instinctive actions, properly
identified, were elementary and unmodifiable, even if the larger behavior
patterns of which they were a part were subject to modification.

Schneirla conducted field as well as laboratory studies on the behavior
of army ants. Beach did not do fieldwork himself, but he took an interest
in the work of people who did, as, for example, C. R. Carpenter. In 1946
Beach told Carpenter it was "important that everything possible be done"
to facilitate Carpenter's field studies of primate behavior, given the quality
of Carpenter's previous work and "the fact that no other contemporary
psychologist in this country appears to be interested in extending this type
of work." [106] When Tinbergen asked him in 1947 to serve as American editor of Tinbergen's new journal, *Behaviour*, Beach readily agreed. The following year Beach collaborated with Ernst Mayr in trying to find Tinbergen a position in the United States. In the 1950s Beach played a major role
in building bridges between psychology and ethology. Schneirla too was
interested, as he put it in 1944, in "a rapprochment of psychologists and zoologists interested in the more naturalistic aspects of animal study." To this
end he explored the possibility of reviving Yerkes' old *Journal of Animal
Behavior*.[107]

In 1946 Schneirla published a critique of American comparative psychology's overreliance on the white rat. There he allowed that "while intensive concentration upon the capacities of a caged animal may have great
advantages for experimental control, it is a rather artificial way of studying
animal psychology in general and does not lead to a *comparative* psychology." [108] He cited Tinbergen's critique of American psychology to the same

effect, and he noted that while European investigators would benefit from increased exposure to American laboratory and statistical procedures, Americans in turn had much to gain from the "naturalistic movement" in European animal psychology. Where American psychology was particularly strong, he suggested, was in having "an underlying tendency . . . to give ontogeny its due."

In January of 1947 Schneirla had the opportunity to meet Tinbergen and hear him deliver at the American Museum and Columbia University his six-lecture series on the study of instinct. Three months later Schneirla outlined a "provisional sketch" for a research program at the Museum under the title: "Studies on Learning Capacity and Adaptive Behavior as Factors in Relation to Problems of Heredity and Evolution." Schneirla's plan was to seek a three-year grant from the Macy Foundation in immediate support of the ornithological part of the program. He imagined that the National Research Council would be able to help fund the program beginning in 1948. Schneirla's sketch makes clear that he was thinking, among other things, about the ideas of the Continental ethologists. Under the subheading "Study the development of early behavior in the animal: role of learning," he noted in parentheses: "Here Lehrman's study fits in very nicely. It will be desirable to subject the entire thesis of Lorenz and Tinbergen to a careful experimental examination." [109] Three years later, reporting to Robert Yerkes on work in the Department of Animal Behavior at the American Museum, Schneirla remarked: "Lehrman expects to finish his main study on parent-young feeding relations in the ring-neck dove, this year. He is getting results, as you may recall, that are rather sharply at odds with the Lorenz interpretations of such patterns." [110] The researcher to whom Schneirla referred was Daniel S. Lehrman, a doctoral candidate working under Schneirla's direction at New York University. He was soon to burst on the scene as Lorenzian ethology's sharpest and most serious critic.

Born in 1919, Daniel S. Lehrman grew up in the Bronx with a passion for bird-watching. While a student at City College, he worked as a volunteer under G. K. Noble at the American Museum of Natural History. He made good friends among the New York ornithologists, especially William Vogt. Ernst Mayr recognized the young man's talents and put him on the editorial board of the Linnaean Society of New York. When Niko Tinbergen made his first trip to the United States in 1938, he met "Danny" at the American Museum. The Dutch naturalist was impressed by the young American. After Tinbergen returned to Holland, the two corresponded regularly with each other.

This was not a happy period in Lehrman's life, however. Family bickering at home upset him. His attendance and his academic record at City College were spotty. And he found working with Noble at the Museum to be extremely frustrating. Coauthoring a paper with Noble on egg recognition in the laughing gull proved especially difficult. Lehrman conducted most of the experiments for the study, but Noble was disinclined to treat Lehrman as anything like a worthy partner in the enterprise. When their interpretations of the data differed, Noble shouted the younger man down. There were a few places in the manuscript where Noble appeared to be willing to let Lehrman express his views, but Noble reneged and reinserted his own ideas without telling Lehrman he was doing so and sent the manuscript off for publication.[111]

Lehrman's first published comments on Lorenzian ethology appeared in a review that he wrote for *Bird-Banding* in 1941—a dozen years before his epoch-making *Quarterly Review of Biology* critique of 1953. *Bird-Banding* had become a major source of information about Continental ethology for American readers, thanks to the efforts of Margaret Morse Nice. Nice invited Lehrman to write a review Lorenz's 1939 paper "Comparative Behavior Studies." Lehrman's account laid out the basic principles of Lorenzian ethology and lauded Lorenz's paper, saying: "This is certainly one of the most important comprehensive papers on animal behavior that is known to the present reviewer. Whatever may be the eventual status of the individual aspects of Lorenz's theories, he has provided both a theoretical attitude and a methodological approach that bids fair to become essential for the investigation and understanding of behaviour." The young Lehrman's response to Lorenz's work was thus quite the opposite of that of the ornithologist A. L. Rand, also of the American Museum of Natural History, whose sharply critical comments on Lorenz's ideas appeared in the same year.[112]

Yet Lehrman was not prepared to swallow Lorenz's theorizing hook, line, and sinker. There was "at least one respect," the young man noted, "in which the sweeping nature of Lorenz's conclusions hardly seems justified by the evidence." Lorenz had sought to understand instinctive behavior in terms of a single type of process operating in the central nervous system. He had drawn his physiological evidence primarily from work that the distinguished neurophysiologist Charles S. Sherrington and Holst had independently conducted on fishes. Lehrman called attention to "the great differences in organization of the nervous system at different evolutionary levels and for different types of behavior" (one of Noble's major themes). He questioned whether what was true for the swimming movements of fish would necessarily apply to the egg-rolling behavior of birds. Overall, however, he remained greatly impressed by Lorenz's work. Regardless of what the physiological foundation of Lorenz's principles might prove to be, he wrote, these principles were "powerful tools" for investigating behavior. The field observer of bird behavior, he said, was likely to gain from Lorenz's views "an insight into the causes of the behavior" that was "superior and more fruitful than any that can be obtained otherwise." Lehrman announced he had translated Lorenz's paper into English and was prepared to loan copies of the translation to whoever might be interested.[113]

Lehrman joined the army in 1942. During the war, he served as a translator of German and also as a cryptographer. After the war, he returned to City College and received his bachelor of science degree in 1946. He thereupon enrolled in graduate studies in psychology at New York University. When Tinbergen turned his manuscript of *The Study of Instinct* over to Oxford University Press, New York, for publication, it was Lehrman who got the job of rendering the manuscript into fluent (American) English.

Given his early association with Tinbergen, his early enthusiasm for Lorenz's work, and the biological as well as psychological interests of the American Museum's animal behavior people, Lehrman might not seem to have been a likely candidate to write a sharp attack on Continental ethology in the early 1950s. There were a variety of factors, however, that contributed to his doing so. The idea of doing the paper was suggested to him by his teacher, T. C. Schneirla. From the standpoint of theory, Schneirla believed strongly that Lorenz's sharp dichotomy between the "innate" and the "acquired" was the kind of thing that hindered rather than helped the study of behavior—particularly the study of behavioral development. He also shared Noble's view (and Lehrman's too, as we have seen) that Lorenz's writings on instinct failed to consider how different the physiological mechanisms that operated in the behavior of animals of different phyletic levels had to be. On top of this, Schneirla was active on the political Left. He had been a supporter of the anti-Franco forces in Spain; he had participated in the organization of the New York University staff. Whether or not he was responsible for calling Lehrman's attention to various pro-Nazi passages in Lorenz's wartime publications, Schneirla was not disposed to regard Lorenz's career in the Third Reich as easily forgettable.[114]

As it was, even before Lehrman's blast of Continental ethology in the pages of the *Quarterly Review of Biology*, Tinbergen had a sense of Schneirla's unhappiness with ethology. Tinbergen noted this in a letter to Ernst Mayr in April 1953: "I feel that in the States two attitudes towards our work are developing: zoologists and some psychologists are getting interested in our type of work, although healthily critical, other psychologists, among them our friend Schneirla, are becoming more and more emotionally opposed. This is partly our fault, partly theirs." [115] Tinbergen apparently did not know it at the time, but Lehrman's critique of Continental ethology was soon to follow.

EIGHT

ATTRACTING ATTENTION

I have always thought I was a psychologist. I am not too sure now. Last summer, when I visited Dr. Tinbergen, he introduced me to the professor of zoology at Oxford and said, in a very apologetic way, "Dr. Lehrman is a professor of psychology but he is really more of a zoologist." I took this as a compliment.

DANIEL LEHRMAN AT THE 1954 MACY
CONFERENCE "GROUP PROCESSES"

I do *not* agree with Thorpe that we should keep the conference small. First, . . . there are now already too many first-class ethologists. Second, just such a policy would in my opinion isolate and impoverish ethology. We must dissolve the somewhat artificial borders between ethology and other biological research areas such as neurophysiology, endocrinology, evolution-research and genetics, [and] human psychology.

NIKO TINBERGEN IN A LETTER OF 28 DECEMBER 1960
TO OTTO KOEHLER REGARDING PLANNING FOR THE
1961 INTERNATIONAL ETHOLOGICAL CONGRESS

Ethology came into its own in the 1950s. With Lorenz and Tinbergen as its most conspicuous champions, it attracted increasing attention from biologists, psychologists, and a wider public. New disciples were drawn to the field. Critics attacked. Clearly, the ethologists had established an identity for themselves. But this was not an identity that could remain fixed for long. If the field were to have a robust future, it would have to be able to grow, develop, and adapt to new and changing circumstances.

Of the two cofounders of ethology, it was Tinbergen who worried more about what the field needed in order to grow and develop in a coordinated and productive fashion. It was he, furthermore, who pointed most clearly toward the future with his development of a whole new program of researches in the area of behavioral ecology. Lorenz for his part continued to be a major, charismatic figure, promoting his ideas and his approach with gusto and charm, reaching ever wider audiences through his books, lectures, television appearances, and so forth. But his days of productive new research were largely over, even though he finally had the institutional support for which he had dreamt for decades. His Max Planck institutes first at Buldern and then at Seewiesen enhanced his ability to attract and teach new disciples, but the ideas he promulgated and defended in the 1950s were by and large ideas he had formulated back in the 1930s. The exceptions, most notably his concept of the "innate schoolmarm," he generated in the course of responding to critics, not as the result of new empirical or experimental work.

Lorenz, in sum, remained a giant in the field, but Tinbergen was better suited to guide ethology in new directions. He was less invested than Lorenz was in defending ethology's early conceptual foundations. His research program at Oxford continued to develop in new and interesting ways. His successes there reinforced his sense of the importance of fieldwork for ethology and, more generally, for biology. He constructed a vision of ethology's aims and methods to which ethologists have been happy to turn for self-identification and guidance ever since.

The heterogeneity of ethologists' projects and perspectives in the 1950s and 1960s makes it impossible to encompass in a single chapter all that went on in the field at this time. Tinbergen's story is by no means the only story to tell. Nevertheless, if we keep Tinbergen's activities and concerns close to the center of our narrative, we will not stray too far from the central problems of the discipline. His book *The Study of Instinct* defined the field at the beginning of the 1950s. His emerging sense of the "four questions" of ethology became the roadmap for ethology's development in the 1960s and beyond.

THE STUDY OF INSTINCT

Graduate research programs, special research centers, biennial international meetings, scientific publications, and presentations to the public served mutually to constitute ethology's identity in the 1950s.[1] If there was one single thing to which an outsider could turn to begin to understand ethology, however, it was Niko Tinbergen's book *The Study of Instinct*. Published finally in 1951, after having first been sent off to press three years earlier, *The Study of Instinct* was ethology's first major text.[2] It provided a benchmark for how far ethology had come. At the same time it testified to how far the field still needed to go. Tinbergen's book thus provides an excellent starting point for a consideration of ethology's growth and development in the second half of the twentieth century.

The Study of Instinct was first and foremost about the physiological causation of instinctive behavior. This corresponded to what Tinbergen,

Lorenz, and their colleagues had emphasized up to that time. Of the book's eight chapters, the first five were devoted to questions of causation. The last three chapters dealt with the topics of ontogeny, adaptiveness (function), and evolution, respectively. Tinbergen was not especially proud of these last three chapters. By his own account, he only added them "after considerable hesitation."[3] Although he wanted to be able to portray ethology as a coherent whole, the only coherence to which he could point was in the ethologists' understanding of "the causes underlying instinctive behaviour."

Understanding the causes of instinctive behavior meant for Tinbergen identifying the "physiological mechanisms" underlying instinctive behavior and the causal factors, both "internal" and "external," associated with these mechanisms. The internal causal factors were "hormones, internal sensory stimuli, and, perhaps, intrinsic or automatic nervous impulses generated by the central nervous system itself." These controlled the "motivation" of the animal, both qualitatively and quantitatively, by serving constantly to prime the instinctive mechanisms. They roused the animal to engage in appetitive behavior and determined the threshold needed for an instinctive action to be released. But they did not by themselves elicit the *performance* of the instinctive actions (the "fixed patterns"). Innate releasing mechanisms, each corresponding to a specific instinctive action, had first to be "unblocked."

This is where external factors came in. The first of these were the releasers or "sign stimuli." Operating according to the "rule of heterogeneous summation," they triggered the innate releasing mechanisms with which they were associated and thereby unblocked the performance of the associated instinctive motor pattern—the "consummatory act." In addition to releasing stimuli there were directing stimuli, "enabling or forcing the animal to orient itself in relation to the environment."[4]

According to this very mechanical view, each instinctive action is paired with its own specific innate releasing mechanism. To account for the *coordination* of behavior, Tinbergen explained that the mechanisms underlying instinctive actions were related to each other by belonging to, and being hierarchically arranged within, separate overarching systems. When one system is strongly activated, this prevents the simultaneous performance of instinctive behavior patterns belonging to another system. "Displacement activities" occur when conflicting drives such as fighting or fleeing are activated simultaneously and each prevents the discharge of the other. Within a "major instinct," Tinbergen went on to explain, "impulses" flow successively from higher to lower centers as "appetitive behavior" brings the animal into contact with the appropriate "sign stimuli" and the impulses are finally used up.[5]

Such, in brief, was Tinbergen's theory of instinctive causation as of the beginning of the 1950s. The picture was an elegant one. It combined Craig's distinction between appetitive behavior and consummatory acts, Lorenz's notion of releasers and innate releasing mechanisms, Holst's ideas on the endogenous production of motor impulses, and Tinbergen's account of the interrelations of instinctive actions and their hierarchical organization within the central nervous system. In doing so, it made sense of a vast array of evidence. This scheme was not entirely identical to Lorenz's. Where Lorenz talked about "action-specific energy," Tinbergen offered "impulses" (though these too "flowed," or "drained away," or were "used up").

Tinbergen, much more than Lorenz, was disposed to qualify his claims. He insisted that the ethologist's evidence was still "very fragmentary" and his generalizations were correspondingly of "a very tentative nature." He offered his diagrams of the operation and hierarchical arrangement of instinctive centers as "no more than a working hypothesis of a type that helps to put our thoughts in order." Still, this model of instinctive action seemed to him "to cover the reality better than any theory thus far advanced." As he put it: "Its concreteness gives it a high heuristic value, and it is to be hoped that continued research in the near future will follow these lines and fill in, change, and adapt the sketchy frame."[6]

At the same time that Tinbergen acknowledged that the ethologist's understanding of behavioral causation was still only provisional, he insisted that his understanding of the development, adaptiveness, and evolution of behavior was sketchier still. A synthetic treatment of behavioral development, he allowed, was "not yet possible." He offered his chapter on the adaptiveness of behavior not as "an exhaustive treatise," he said, but "merely" to show "the necessity as well as the possibility of an objective study of directiveness of behaviour." He introduced his chapter on behavioral evolution with a discussion of "why our knowledge of the evolution of behaviour is still so backward."[7] In short, as Tinbergen represented things in *The Study of Instinct*, the young science of ethology still had a great deal more to do.

INTERNATIONAL MEETINGS AND
EARLY CRITICISMS OF ETHOLOGICAL THEORY

In 1950, the ethologists initiated a series of international meetings that were to play a key role in establishing the collective identify of the field. Prior to

that date, there had been just two international gatherings of real consequence for the study of animal behavior. The first was the Leiden instinct
conference of 1936. The second was the Cambridge "Physiological Mechanisms in Animal Behavior" conference, held in 1949. The Leiden meeting
brought Lorenz and Tinbergen together for the first time and united them
in opposition to the older generation of subjective animal psychologists.
The Cambridge conference gave ethology's cofounders their first opportunity to regroup after the war. It also allowed them and their followers to
stake out their claims in a broader scientific arena. As we have seen, however, the Cambridge conference was not a clear triumph for ethology. The
physiologists not only outnumbered them, but also remained relatively
dubious about the ethologists' models of behavioral causation.

After the Cambridge experience, the leaders decided to meet again
soon to talk informally among themselves about their latest researches. At
the same time, they were willing to participate in other international conferences and share the stage there with other students of animal behavior.
Both the "in-house" and the multiparty conferences became important occasions for the critical evaluation of the concepts and practices of the field.

The immediate precursor of the international ethological congresses
was the small meeting Erich von Holst convened in the spring of 1950
(30 May–5 June) at his new Max Planck Institute in Wilhelmshaven. There
half a dozen visitors joined an equal number of members of Holst's institute for a roundtable devoted to *Verhaltensphysiologie* (behavioral physiology). Only two of the twelve participants—Thorpe and Tinbergen—were
not German or Austrian. The two met in Holland and made the trip into
Germany together by train. For Tinbergen, it was five years after the German occupation of his homeland and his first trip to Germany. When the
train first stopped inside Germany and he heard German being spoken by
everyone outside his train car window, his body shook uncontrollably.[8]

The other symposium participants at the roundtable symposium were
Lorenz from Altenberg, Koehler from Freiburg, Herbert Böhm from Er-
langen, and Wolfgang Metzger from Münster. The Wilhelmshaven contingent consisted of Holst himself, Bernhard Hassenstein, Gustav Kramer,
Horst Mittelstaedt, Lore Schoen, and Ursula von Saint-Paul.

Holst's idea for the gathering was to have informal presentations of
work in progress —"only half-laid eggs," as he liked to put it. There was
no preestablished program. Not until they arrived in Wilhelmshaven, apparently, were the participants told to talk about the aspects of their researches that they felt would most attract the interest of the group as a
whole.[9] This approach may have encouraged informality, but it seems not

to have inspired the participants to try out their very latest ideas. Most of the speakers resorted to topics they had been thinking about for some time. Tinbergen spoke on how the reproductive behavior of the stickleback demonstrated the hierarchical organization of instincts. Thorpe discussed different types of learning. Lorenz rehearsed the old mechanism-vitalism debate before discussing the concepts of *Ganzheit* and *Gestalt*. Koehler presented a film illustrating the counting abilities of birds. Kramer discussed his researches on the flight orientation of starlings.

The most novel idea introduced at the meeting came from Holst himself and his student Mittelstaedt. They called it the "reafference" principle. Challenging the thinking of reflex physiologists, Holst and Mittelstaedt focused not on the relation between stimulus and response but instead on what then happens back in the central nervous system upon the arrival of *afferent* impulses evoked in the effectors and receptors by *efferent* impulses previously transmitted to the periphery. The reafference principle in effect introduced into ethology the cybernetic notion of negative feedback. Its target was reflex physiology and the latter's impoverished view of the functioning of the central nervous system. Eventually, it came to be seen as also undermining Lorenz's notion of how the consummatory act ends a behavior sequence by releasing action-specific energy. At the 1950 meeting in Wilhelmshaven, however, this threat to Lorenzian theory was apparently not perceived.[10]

The last day of the Wilhelmshaven symposium, like the last day of the Cambridge meeting the previous year, was devoted to efforts to clarify the terminology of animal behavior studies. At the end, the conferees agreed that it would be good to hold another such roundtable discussion the following year.[11] As it happened, however, it was almost two years before the next major gathering of ethologists occurred.

The site of the next meeting was Lorenz's new Max Planck Institute at Schloss Buldern in Westphalia. It took place in March 1952. "*Vergleichende Verhaltensforschung*"—comparative behavior study—was what Lorenz still preferred to call his field, and he grandly entitled the meeting the "I. Internationales Symposion der Vergleichenden Verhaltensforscher" ("First International Symposium of Comparative Behavior Researchers").

The meeting was a major event—a ten-day affair, with thirty individuals named on the program and twenty-five different presentations scheduled. The majority of the participants came from Germany, but animal behavior researchers from other parts of Europe and Britain were also well represented. Tinbergen came from Oxford, bringing his students W. M. S. Russell and L. de Ruiter with him. Thorpe did not make the trip from

Cambridge, but Robert Hinde did, and he delivered one of the most significant papers of the conference. Dutch ethology was represented by Baerends from Groningen and Jan van Iersel from Leiden, Swiss ethology by Heini Hediger from Basel and Monika Meyer-Holzapfel from Bern. Three Scandinavian scientists took part in the meeting: E. Fabricius from Sweden, L. von Haartman from Finland, and H. Poulsen from Denmark. Rémy Chauvin came to the conference from France (where twentieth-century animal behavior studies had focused on the behavior of social insects and had proceeded essentially unaffected by the work of Lorenz and Tinbergen).[12] From within Germany itself, Lorenz's Buldern team was the largest contingent. It included Lorenz himself, Irenäus Eibl-Eibesfeldt, Heinz Prechtl, Wolfgang Schleidt, and the young Swiss ethologist Uli Weidmann. Other participants were Otto Koehler from Freiburg, Bernhard Rensch and Wolfgang Metzger from nearby Munster, Lorenz's former student Paul Leyhausen from Bonn, Katharina Heinroth from Berlin, Heinz Sielmann from Munich, and Gustav Kramer and F. Goethe (but not Erich von Holst) from Wilhelmshaven.

Typically two (but never more than four) papers were given on a single day, thus allowing ample time for discussion. Films shown in the evenings also played a prominent role in the conference. Baerends, Fabricius, Koehler, Leyhausen, Lorenz, Schleidt, Sielmann, and Tinbergen all presented films.

The topics of the papers were diverse. The more senior scientists tended to address broad themes, such as the problem of expressive movements in higher animals (Lorenz), fight and threat in mating behavior (Tinbergen), and stimulus and schema (Metzger). The more junior scientists were inclined to offer papers more closely tied to particular species or particular experiments, such as discussions of the reproductive biology of the squirrel (Eibl-Eibesfeldt), pairing behavior and sex hormones in *Xenopus* (Russell), and the responses of chaffinches to owls (Hinde). But there was not a perfect correlation between an investigator's seniority in the field and the generality of his paper's title. The more junior researcher L. de Ruiter entitled his talk "Some Ethological Aspects of Protective Coloration in Animals," while the more senior researcher Gerard Baerends entitled his "Analysis of the Dance of the Male *Lipistes reticulatus*." Nor did the breadth of a paper's title necessarily indicate the full dimensions of the paper's significance. If there was a bombshell at the meeting, it was dropped by Robert Hinde in his paper on how chaffinches react to owls.

Three years previously, Lorenz had left the Cambridge "Physiological

8.1 Participants at the "First International Symposium of Comparative Behavior Researchers," held at Buldern in 1952. Niko Tinbergen and Jan van Iersel are followed by Konrad Lorenz, L. de Ruiter, and others. (Courtesy of I. Eibl-Eibesfeldt.)

Mechanisms" symposium convinced of the need of doing experimental work on action-specific energy. Hinde's study of the mobbing behavior of the chaffinch responded to this need.[13] Lorenz was excited by Hinde and Hinde's work, and some weeks after the Buldern meeting he wrote Thorpe to tell him so:

> I think you ought to be congratulated on your choice of Robert Hinde as an assistant! He really has become a first-rate ethologist and his paper was one of the most interesting—if most disturbing—that were read at our congress. Do you quite realize what an amazing fact it is, that that mobbing reaction, after having been released ONCE, just fades to half its original intensity and NEVER again reaches 100% of what it did at first, but only just a little above 50%????? I think I can show definitely that the reaction of the greylag gosling to the warning call behaves *quite* exactly like Robert Hinde's mobbing! And mark, he gets this drop in intensity even when he does not use a weak dummy, but the actual, live predator, and even if the owl frightens the chaffinch by flying about in its cage and, to all intents and purposes, flies *after* the finch.[14]

Lorenz wondered whether reinforcement might be "necessary to keep up the effect of the IRM." He thought of taking a new batch of goslings and testing to see if he could reinforce their reactions to the warning cry "by letting an eagle dummy swoop at them every time I warn them." Perhaps under these circumstances the reaction would not fade as it had with his previous goslings and with Hinde's chaffinches. "What irks me," Lorenz continued, "is that this fading cannot be natural! It would jeopardize the whole survival value of the mobbing, would it not, if the whole thing could function only just *once* with its full intensity!?" He concluded: "It certainly is one of the most intriguing problems, and one that I subconsciously know for years, but did not realize before Hinde put it in clear words. Good for him!!!"

Lorenz did not stay this happy about Hinde's work for long. Hinde was soon to emerge as the leading critic of the energy and drive models that Lorenz and Tinbergen had offered to explain how instinctive behavior works.[15] Hinde would likewise side with the American psychologist Daniel Lehrman when Lehrman attacked Lorenz for being insufficiently analytical on matters of behavioral development. Only a month after the Buldern meeting, however, all this remained in the future, and Lorenz was in the best of moods. He had hosted the first international ethological congress at his own new Max Planck Institute and things had gone very well indeed. The intellectual fecundity of the congress combined with the biological fecundity of his immediate animal and human associates had him in high spirits. He told Thorpe: "Buldern is at present a paradise of flowers, lilacs and innumerable ducklings and goslings. Ilse Prechtl is due to hatch a prechtling next week. The Bahama pintail duck is sitting on 9 eggs, the teals, handreared last year, produced one tealing. . . . Barheaded geese are about to lay, 7 graylags have hatched, another batch of 6 is due next week but one. Two pairs of Canadas are sitting on 6 eggs each. Isn't it wonderful? I am writing to you at exactly 6 o'clock in the morning, *after* having done my first feeding and watering round through the baby cages!"[16]

The first cracks that had begun to show in ethology's theoretical structure became increasingly evident to the new generation of students, who indeed were responsible for introducing some of the cracks in the first place. Their readiness to challenge the thinking of the field's founders was noted by the young American psychologist Bill Verplanck, who traveled to Britain and the Continent early in 1953 to observe Tinbergen's program and to visit other ethological centers as well. As he wrote in one of the personal "Newsletters" he composed for his family back in the States, "All the Tinbergen people are now no longer under the illusion that they are finding

8.2 Lorenz at Buldern. (Photo by Alfred Seitz. Copy provided by the Konrad Lorenz Institute for Evolution and Cognition Research.)

anything out about the nervous system, even Niko (who now looks pained if people press him about his 'hierarchy' notion). But still Niko thinks he's talking about something when he says 'drive.' Some of his students aren't sure even of this. And the data if nothing else, seem to be driving them all over to a position not unlike an American one." With respect to Lorenz and his students, Verplanck observed, "they still like to think that they're studying 'centers,' and are quick to use von Holst and Hess as justifications for thinking so. But they are more open minded than I had expected them to be—that is, Uli [Weidmann] proved to be."[17]

Verplanck's additional observations of Lorenz are equally instructive. Almost as soon as he arrived in England in February 1953, he had the chance to hear and meet Lorenz, who had come to Oxford for a week. He recorded this impression: "He is a very remarkable man. A sort of Freud for animals. That is, he makes very sound (ie reproducible) observations in a more or less intuitive way. He talks about them in a language mostly all his own, that is shot though with Germanic poetry. He has the knack of pointing at what seems most important. And it takes a whole lot of other people to come along afterwards and clean up after him. Tinbergen told me this morning (Perhaps before I got a chance to tell him) that Lorenz cannot do a clean experiment. Always leaves holes."[18]

When Lorenz went on from Oxford to Slimbridge, in order to make movies of ducks and geese, Verplanck tagged along with Tinbergen and had the chance to see Lorenz in an informal, outdoor setting. "Seeing Lorenz in action," he recorded, "and getting to know him in a small group was fine. He is quite a character. Big, bearded, late-fiftiesh, jovial, enthusiastic. One can see easily how it is that everybody likes him. He is a man who really enjoys himself, and I think that he is a little taken aback at the great interest in what he is doing. Neither does he seem to give a damn. There is none of the sound and the fury of some European academics about him." That said, Verplanck insightfully added, "On the other hand, it is not unlikely that the situation was not such as to put him on his dignity."[19] A year or so later, the sound and fury would be released.

Before that occurred, the Second International Ethological Congress was held, twenty-one months after the first. It met in Oxford from 9 to 20 December 1953. Thirty-five papers were scheduled, and a total of sixty-two individuals were registered as "members" of the congress. The research programs at Buldern, Wilhelmshaven, Groningen, Leiden, Cambridge, and Oxford were each represented by at least three scientists.[20] Changes of address since the previous meeting signaled an interweaving of personnel and places within the growing ethological community. Leen de Ruiter, who had earlier come from Holland to Oxford to work with Tinbergen on a D.Phil. degree, was now back in Holland as a member of the faculty at Groningen. Uli Weidmann, a Swiss zoologist who two years earlier had been at Buldern, was now in Oxford on a postdoctoral fellowship and married to the former Rita White, another Tinbergen student. Under Tinbergen's direction the Weidmanns were both working on black-headed gulls.

Lorenz delivered a conference paper on the comparative study of the behavior of ducks. In it he previewed what the conferees might see the following day on a special excursion to Peter Scott's wildlife refuge at Slimbridge. Tinbergen gave two papers, one on comparative studies of the behavior of gulls, the other on the pecking behavior of the black-headed gull. Meanwhile, some of the younger participants continued the critical examination of the concepts of drive and hierarchical organization that Lorenz and Tinbergen had used to account for the physiological causation of behavior. Margaret Bastock discussed "some problems of 'drive'" as part of her paper on the courtship behavior of *Drosophila melanogaster,* and Robert Hinde addressed the concept of "specific action potential" and the hierarchical organization of behavior mechanisms.

Other papers published in this period suggest what some of the comments at Oxford must have involved. Bastock was already the coauthor with her fellow Oxford students, Desmond Morris and Martin Moynihan, of a 1953 paper on displacement activities. Among other things, the paper addressed some recently discovered cases of displacement activities that did not fit Lorenz's hypothesis of the way a drive's energy was consumed when the motor acts appropriate to that drive were performed. The young Oxford authors proposed an alternative to Lorenz's mechanism along the lines of Holst's and Mittelstaedt's reafference principle. They suggested that what discharged the energy of a black-headed gull's incubation drive was not the act of sitting on eggs but rather the sensory stimuli received as a result of that act. In other words, what was necessary to release the accumulated nervous energy of a drive was not the performance of the consummatory act but rather the feedback from "consummatory stimuli." More evidence was needed, the authors allowed, to decide whether their suggestion should be regarded as "an addition to or a replacement of Lorenz's idea." In any case, they had not felt inhibited about challenging Lorenz's model. Indeed, they went on to suggest that what Lorenz called "vacuum activities" might often be responses to suboptimal stimuli and probably fit a "feed-back" explanation better than they fitted Lorenz's model.[21]

If Bastock, Morris, and Moynihan pointed to problems with Lorenz's model, they nonetheless continued to couch their discussion in terms of "drive," "nervous energy," and "spark-over." Hinde went further. By the time of the Oxford conference of 1953 he was expressing dissatisfaction with "analogies of anything flowing" (nerve impulses excepted).[22]

Hinde had not always been this averse to drive models of behavior. To the contrary, he had relied upon such models in his earliest researches. He had learned how to analyze behavior from Tinbergen, who had come to Oxford when Hinde was beginning his second year of D.Phil. research at the Edward Grey Institute under David Lack.[23] Tinbergen explained various animal postures and motor patterns as the result of conflicting drives, such as fighting and fleeing or fighting and courting. By Tinbergen's account, displacement or derived activities functioned as outlets for the surplus of drive produced when conflicting instincts, simultaneously activated, were unable to be released through their appropriate consummatory acts.

Hinde initially followed Tinbergen's lead. Benefiting from what Tinbergen told him about yet-to-be-published researches on the herring gull and the three-spined stickleback, Hinde explained in Tinbergian fashion the display behavior of great tits (*Parus major*) and related species. He proposed that the "head-up" display of the great tit was developed from "an inhibited fleeing movement." "Displacement pecking," he suggested, "is also probably caused by inter-action between the fighting and fleeing drives—apparently when the two drives are of such intensity that the internal conflict can be resolved in no other way except by a dissipation of the energy through a totally different channel." [24]

But Hinde's contentment with "drive" explanations of instinctive behavior did not last. By 1953 at the ethological congress in Oxford his critique of ethology's chief models of behavioral causation was underway. A few years later he would state baldly that the "useful life" of Lorenz's "hydraulic reservoir" model and Tinbergen's "hierarchical system of centers" was near its limit. [25]

The critique of the ethologists' models, which Lorenz and Tinbergen were always ready to identify as "provisional," took a variety of forms. Always averse to pretense, Tinbergen decided a few months before the Oxford congress that it would be good to have a light-hearted evening as part of the proceedings. He approached Desmond Morris with the idea. "I am considering," he told Morris, "the possibility of an ethological nonsense evening, kind of cabaret or variety. For instance a review of origin and adaptive radiation of imaginary derived movements-plus-morphological releasers in man. And something like demonstration of movements, of which the audience must guess what they mean (head bobbing of plover, zigzagdance, head flagging etc.). Or terminological oddities. Or a little play about the ethologist at work and at play (and at home). Let's think of as much nonsense as we can collect." [26]

Morris rose to the challenge. Ultimately he hit upon a scheme that differed from any of those Tinbergen had suggested. With the help of fellow graduate students Aubrey Manning and David Blest, Morris assembled an extraordinary Rube Goldberg type of machine ostensibly ridiculing a German theorist who had taken Lorenz's flush-toilet model too far by trying to explain the workings of the brain in terms of water flowing through pipes.

The students' machine was "designed to demonstrate every popular etho-logical principle then in vogue, and to do it by means of colored liquids." In addition to a conglomeration of taps and tubes the contraption fea-tured "a row of very large balloons which slowly filled with colored water until they dangled obscenely, high over the front row of senior delegates." "Mock giant sperm" were then fired along wires to the balloons. The bal-loons exploded, drenching the delegates in the first three rows of the audi-ence. A standard toilet with tank was thereupon brought on the stage to signify a desire to return to simpler models. Lorenz found the demonstra-tion highly amusing and roared with laughter as the whole machine self-destructed.[27]

Thus, through well-aimed humor as well as more serious critical anal-ysis, the ethologists acknowledged that the early theoretical structures of their field should not be taken as sacred. By 1953, key features of the theo-retical framework that Lorenz and Tinbergen had presented to the physi-ologists in Cambridge just four years earlier were being challenged from within the ethological community itself.

EXTERNAL CRITICS

Criticism of ethological theory came from outside the ethological commu-nity as well. Two early critiques came from the Canadian comparative psy-chologist D. O. Hebb and the Cambridge physiologist J. S. Kennedy. The papers appeared in *The British Journal of Animal Behaviour* in 1953 and 1954. Hebb's paper was a reworking of an address he had given at a meet-ing of the British Association for the Study of Animal Behaviour in 1952. His main target was the way ethologists juxtaposed instinctive and learned behavior. That distinction, he insisted, was a false one: "We cannot di-chotomize behaviour into learned and unlearned, environmentally deter-mined and hereditarily determined." It was thus logically impossible, he said, to follow Tinbergen's stated program of studying innate behaviour be-fore studying learning. Hebb's basic claim was that "the behaviour one can actually observe and experiment with is an inextricable tangle of the two influences, and one of them is nothing without the other."[28]

J. S. Kennedy launched his attack from a very different angle in a pa-per entitled "Is Modern Ethology Objective?" Kennedy's answer was no. Despite all its mechanistic trappings, he argued, modern ethology was fun-damentally subjective. Lorenzian theory was dualistic like Freud's, Kennedy maintained, in its postulation of an accumulable energy, "inside and quite distinct from reflexes," serving as a prime mover of the animal. The ethologists' rejection of reflex explanations of behavior, Kennedy insisted, was based on an outmoded "telephone-exchange" view of reflexes and failed to take into account advances made by modern reflex physiologists.[29]

The ethologists did not feel especially threatened by the criticisms from Hebb and Kennedy.[30] Having already weathered the critiques of the subjectivist psychologists (such as Bierens de Haan) and the peripheral-control physiologists (notably Gray and Lissmann), they basically shrugged off the latest complaints. However, the attack they soon received from the American comparative psychologist Daniel Lehrman cut deeper. Published in the December 1953 issue of the *Quarterly Review of Biology*, Lehrman's critique initiated a vigorous, continuing debate between American comparative psychologists on the one hand and Continental ethologists on the other.

DANIEL LEHRMAN'S CRITIQUE OF LORENZIAN ETHOLOGY

The American comparative psychologists were a group the ethologists found hard to understand. The ethologists shared the American psychologists' objectivistic approach. They were inclined, however, to equate American comparative psychology with Watsonian behaviorism, which they believed manifested a series of interrelated and conspicuous deficiencies. These included a too exclusive focus on one single animal species (the white rat), a failure to study animals in their natural settings, a failure to learn the entire range of behavioral repertoires an animal species had at its disposal, an underestimation of the importance of instinct, and a too facile assumption that studies of learning in the white rat could be readily extrapolated to humans and other higher animals. The ethologists confidently assured themselves that running white rats through mazes was an enterprise that did not deserve the title "comparative."

The ethologists and the comparative psychologists differed not only in their research practices but also in the primary questions that attracted them. Where the ethologists were interested primarily in the areas of behavioral evolution, causation, and function, the comparative psychologists were more interested in behavioral development. The ethologists, to be sure, were the ones who had first called attention to the phenomenon of imprinting. They had also distinguished between innate behavior and learned behavior. But this distinction, as they constructed, seemed all too uncritical to their American counterparts. And this was the point that took center stage in Daniel Lehrman's *Quarterly Review of Biology* (*QRB*) paper entitled "A Critique of Konrad Lorenz's Theory of Instinctive Behavior."[31]

Lehrman's foremost complaint was that Lorenz's definitions of instinct and innateness left unexplored the whole question of behavioral development. He leveled two other major charges as well. The first was that the ethologists had no basis for assuming, as they did, that the same physiological mechanisms were at work in the behavior of animals representing widely disparate taxonomic groups and levels organization. The second was that the ethologists had moved too readily and uncritically in their discussions from the behavior of animals to the behavior of humans.

Lehrman's paper also raised the question of the political dimensions of Lorenz's thinking, but it did not do so as aggressively as Lehrman had originally intended. According to Jay S. Rosenblatt, the draft of the paper that Lehrman initially sent to the *QRB* built toward a final section highlighting Lorenz's endorsement of Nazi ideas on race purity. It identified these ideas, furthermore, as the ideological consequences of Lorenz's scientific theorizing. However, on the advice of a number of influential scientists who read the draft, and in response to objections from the *QRB* editors as well, Lehrman was persuaded to restructure his paper and tone it down. He acquiesced to the argument that too much attention to the political dimensions of Lorenz's thinking would distract readers from the paper's main scientific arguments against Lorenzian instinct theory. Lehrman still mentioned Lorenz's politics, but he did so in a less emotional fashion, and in the middle of the paper, where the discussion was less conspicuous than it would have been at the end.[32] It bears noting that in 1949, when the question of finding American funding to support Lorenz's work arose, W. C. Allee reported to Karl von Frisch, "I find a strong, or shall I say bitter, attitude on the part of all Jews with whom I have discussed Dr. Lorenz's problems."[33] Lehrman's 1953 observations on Lorenz's wartime writings indicate that such feelings had not disappeared.

Lehrman's paper sent a shock wave through the ethological community. It raised immediate questions about the author. Who was he? What warrant did he have to say the things he was saying? Was he a typical American behaviorist, that is, an ultraenvironmentalist? Many of the ethologists imagined this to be the case, judging from the favorable attention Lehrman gave to Z. Y. Kuo's remarkable claim that the species-specific pecking motions of the chick are not "innate" but instead develop in the egg as the chick's head is raised and lowered by the beating of the chick's heart.[34] And what were his political motivations?

Lehrman began his paper with the observation that American investigators were beginning to show increasing interest in Lorenz's system and school. The interest was coming not only from ornithologists, zoologists, and ecologists, but also from comparative and experimental psychologists. The attractions of Lorenz's thinking, Lehrman said, were to be found in its "diagrammatic simplicity," its employment of concepts from neurophysiology, and its attention to the life cycles of animals in natural situations rather than the artificial setting of the laboratory. All of these factors, he said, helped explain the appeal of Lorenz's views in Europe, where students of animal behavior were more likely to be zoologists, zookeepers, naturalists, or physiologists than psychologists (the reverse being the case in America).

As a means of introducing Lorenz's basic approach, Lehrman reviewed Lorenz's and Tinbergen's joint study of the egg-rolling behavior of the greylag goose. He described Lorenz's concept of "the instinctive act" (including Lorenz's idea of action-specific energy accumulated in some kind of reservoir), the innate releasing mechanism, the innate releasing pattern, the taxis, and appetitive behavior. He explained further that according to Lorenz, the performance of the act, not its biological result, is the goal of the appetitive behavior. Then he itemized what he identified as the problems common to instinct theories, namely, "(1) the problem of innateness and the maturation of behavior; (2) the problem of levels of organization in an organism; (3) the nature of evolutionary levels of behavioral organization . . . ; and (4) the manner in which physiological concepts may be properly used in behavioral analysis."[35]

Lehrman objected to Lorenz's and Tinbergen's use of the word "innate" on the grounds that it sidestepped any serious attention to the problem of development. It diverted attention from "investigating developmental processes in order to gain insight into the actual mechanisms of behavior and their interrelations." Lehrman credited Lorenz with having provided a "brilliant approach to the taxonomic analysis of behavior characteristics" in Lorenz's 1941 study of ducks and geese. He insisted, however, that "the fact that a characteristic is a good taxonomic character does not mean that it developed autonomously."

Lehrman went on to insist that Lorenz's simple classification of behavior as either innate or not similarly obscured the question of the different levels of organization possessed by organisms at different places on the evolutionary scale.[36] His critique of Lorenz's ideas on behavioral evolution was much the same. He maintained: "Lorenz's application of the concept of evolutionary change does not consist of analyzing the different ways in which behavior patterns at different evolutionary levels depend on the structure and life of the organism. It consists rather of abstracting aspects of behavior, reifying them as specific autonomous mechanisms, and then citing them as demonstrations of 'evolution' in a purely descriptive taxo-nomic sense."[37]

What may have stung the ethologists most about Lehrman's attack was his suggestion that ethologists, proud as they were of studying many different animals and bringing an evolutionary perspective to their subject, were, when all was said and done, not all that comparative or evolutionary in their work. Lorenz's reification of instinct, Lehrman insisted, "leads to a 'comparative' psychology which consists of comparing levels in terms of *resemblances* between them, without that careful consideration of *differences* in organization which is essential to an understanding of evolutionary change, and of the historical emergence of new capacities."[38] What the ethologists ultimately offered, Lehrman said, was a "preformationist" rather than a developmental view of instinctive acts.[39]

Lorenz and his followers were particularly careless, Lehrman indicated, in their use of analogies between animal and human behavior. Here, he said, their confusion of levels was particularly obvious. He took Tinbergen to task for suggesting that a human's pursuit of sports or science depended on the same kind of internal factors that led a dog to hunt even when the dog was well fed. He found even more problems in Lorenz's writing. Most troubling to him was Lorenz's equating of "the effects of civilization in humans and the effects of domestication in animals." Tackling Lorenz's discussion of degenerative mutations that are not eliminated by natural selection under the conditions of modern civilization, Lehrman wrote: "He presents this as a scientific reason for societies to erect social prohibitions to take the place of the degenerated releaser mechanisms which originally kept races from interbreeding. This is presented by Lorenz in the context of a discussion of the scientific justification for the then existing (1940) German legal restrictions against marriage between Germans and non-Germans." Lehrman also quoted Lorenz's observation that "the face of an Asiatic is enigmatic to us because the physiognomic characteristics to which our innate perceptual patterns respond are not connected with the same behavioral characteristics as in our race." Lehrman countered: "Social psychologists will all agree that the various degrees of difficulty which different people have in learning to recognize and respond to facial expressions in a culture different from their own is at least partly dependent upon the attitude with which they approach the strange culture to begin with."[40]

Lehrman thus attacked Lorenzian ethology from a variety of angles. As he represented the situation, Lorenz had relied on preconceived, preformationist notions of instinct, ignored the issues of development and the different levels of organization on which behavior depended, and promoted ideas that were politically reprehensible.

The December 1953 issue of the *Quarterly Review of Biology* began making its way to biologists early in 1954.[41] Not all American biologists were pleased with the critique. The American entomologist Alfred E. Emerson wrote Lehrman's mentor, Schneirla: "I think the recent article by Lehrman on 'instinct' is open to severe criticism. And from what I hear, it will receive such criticism. I do not envy you for your prominent association with this supercritical demolition without a constructive replacement of discarded theory by a better theory to explain the facts."[42]

On the ethologists' side, Lorenz was furious. Otto Koehler took the paper and threw it against the wall. Tinbergen proved much more receptive, though he admitted to Koehler that his own first reading of the paper had given him "an adrenalin attack." It seemed to him then that Lehrman was "emotionally excited," "did not know the latest things," "cited falsely and selectively," "had not correctly understood" ethology's questions, and "on the whole was not mature enough to be allowed to publish such a multifaceted and fundamental general critique."[43]

Before long, however, Tinbergen, began to see some merit in Lehrman's critique, even if he continued to see it on the whole as "not very well grounded." He told Ernst Mayr in March 1954: "I have now carefully read Dany Lehrman's 'blast.' I think it is a very creditable piece of criticism, particularly where the ontogeny of behaviour is concerned. He really puts his finger on several weak spots in our work, weak sometimes in the actual content of our conclusions, sometimes in our semantics." Mayr had told Tinbergen that the ethologists needed to reply to Lehrman's attack. Tinbergen wanted to be diplomatic rather than belligerent in his response: "Since we do not only want to reveal the other fellow's faults but above all to bring what is valuable in American psychology in better contact with our work, I do not want to be unduly aggressive. After all I think, quiet arguing, presentation of pertinent facts and arguments, and accepting criticism where it is correct, act best in the long run." Tinbergen was happy to interpret Lehrman's critique as evidence of the inroads that ethology was making in America and of "the tremendous influence Lorenz has had on the organisation of [Lehrman's] thoughts," however much Lehrman considered himself opposed to Lorenz.[44]

The first Continental ethologists to confront Lehrman personally after the *QRB* attack were the Dutch ethologists Jan van Iersel and Gerard Baerends, both former Tinbergen students. They had been invited by Frank Beach to take part, along with Lehrman, in the Fourteenth International

8.3 Daniel Lehrman, bird-watcher. Lehrman's standing in the eyes of the ethologists was greatly enhanced when they learned of his enthusiasm for bird watching. Here he is shown on a bird-watching outing at the southern tip of Sweden in 1961, seven years after the publication of his famous critique of Lorenzian ethology. The American psychologist (*on the right*) is accompanied by Holgar Paulsen, director of the Copenhagen zoo. (Photo courtesy of Colin Beer.)

Psychological Congress, held in Montreal in June 1954.[45] In the course of their first day with Lehrman, the Dutch scientists found they had more in common with him than they had imagined. What broke the ice was their discovery that he had the same passion for bird-watching that they did. Baerends later recalled:

> The passing of some North American birds, such as nighthawks that Jan and I had not seen before, helped to reveal that Danny also was an enthusiastic field ornithologist and that he fully shared our interest in the species-specificity of behavior. It further became clear to the three of us that a considerable part of Danny's aversion to various ethological concepts was due to misunderstandings. Frequently he had read more into

the descriptions of concepts than had actually been intended, since he was looking at them from a background different than the one from which they had arisen. It was a surprise to Danny that we at least considered the concepts only as temporary constructs that should be tested, amended, changed, or replaced as soon as they were no longer sufficiently supported by the available evidence. At the end of the day I do not think that any major essential difference remained between us, and we certainly had become very good friends.[46]

THE PARIS INSTINCT CONFERENCE

For Lorenz, who met Lehrman for the first time only a matter of days after the Dutch ethologists did, the sense of difference would never really disappear, even if he and Lehrman came to identify each other (though still with ambivalence) as friends. Lorenz's first meeting with Lehrman (and Schneirla) took place in Paris in June 1954. The occasion was an elegant conference on instinct convened in Paris by the French zoologist Pierre-Paul Grassé and sponsored by the Singer-Polignac Foundation. Lorenz, Schneirla, and Lehrman were all invited to give papers at the conference. So too was Tinbergen, but he declined the opportunity and arranged to have his student Desmond Morris go in his place. Among the other speakers were Karl von Frisch, Otto Koehler, J. B. S. Haldane, and Haldane's wife, Helen Spurway. As the conference proceeded in the elegant salon of the Singer-Polignac building, the different participants had multiple and different things on their minds. Grassé was piqued at the way that French animal behavior studies and his own work had been essentially ignored by Lorenz and Tinbergen (the "objectivist school"). Lorenz viewed Schneirla and Lehrman as both scientifically and ideologically suspect, and they felt the same of him. Lorenz's relations with Haldane and Spurway involved not only scientific and ideological differences but the emotional reverberations of a love affair that had taken place between Lorenz and Spurway, beginning with Lorenz's trip to England in 1949 and lasting for fully a year. The affair had gone on, it seems, with Haldane's awareness and even encouragement. Not long after the affair was over, Spurway wrote a review of Lorenz's *King Solomon's Ring* entitled "Behold, My Child, the Nordic Dog." In 1954, however, she still retained a keen interest in Lorenz and his ideas. She reminded Lehrman at the Paris symposium that physiologists continued to support Lorenz, and furthermore that "The greatness of Lorenz is that he has provided concepts applicable in fields very different from those in which he made his observations."[47]

8.4 Participants at the 1954 Paris "Instinct" conference. *Front row:* Karl von Frisch, J. B. S. Haldane, Helen Spurway, and Pierre Grassé. Among those in the second row are Otto Koehler (*far left*) and T. C. Schneirla (*far right*). The back row includes Daniel Lehrman (*second from left*), Konrad Lorenz (*third from left*), and Desmond Morris (*fifth from left*). (Photo courtesy of Desmond Morris.)

With respect to the scientific skirmishing in Paris, Lorenz had the chance for the first thrust. Near the beginning of his talk he identified Lehrman's *QRB* critique as one of the several recent attacks that had by and large suppressed "the most important facts" on which ethological theory was based.[48] He then quickly backpedaled to say that he would not dream of accusing his critics of consciously or even unconsciously suppressing facts, and that the problem was thus the ethologists' own: they had failed to make clear the facts that were most fundamental to their theories. But as he reviewed the series of facts that he felt deserved special attention, his aggressive juices began welling up again (to use an analogy akin to his own model of instinctive action). Back in an accusatory mode, he complained: "It is the wildest caricature of scientific truth to reject all these facts, for purely dogmatic reasons, and then to accuse the men who found them out, of being dogmatists, as Dr. Lehrman has indubitably done." Lorenz asked his auditors to compare the substantial weight of evidence that pointed to "preformation of motor coordinations within the central nervous system" against "Dr. Kuo's theory, desperately supported by Dr. Lehrman, that the domestic chick learns to peck by having its head moved up and down passively by the action of the heart." [49]

The Austrian naturalist insisted that his critics were wrong to assume that he really believed in his "world famous model of the flushing reservoir." "Please believe me," he said, "that none of our youngest students has ever believed for a moment that this object of ridicule is more than a parable or should be regarded as an attempt at explanation of facts." [50] He likewise rejected the notion that ethologists believed that identical behavioral mechanisms were responsible for the behavior of animals of very different levels of organic complexity. He allowed, nonetheless, that the prospects of being able to explain "the characteristic properties of instinctive movements" according to processes found at "lower levels of the central nervous system" were well worth pursuing. In his words, "it seems to me to be the most obvious kind of commonsense to seek for explanations in this particular direction and not in that of learning, conditioned responses or, still worse, 'chain reflexes.' And this last sentence comprises pretty much all the 'theory' we have developed concerning instinctive movements." [51]

Schneirla and Lehrman had their opportunities to respond. Schneirla explained that his objections to Lorenz's system centered on its "deficient attention to the need for analytical investigation of ontogenetic processes." The issue, he insisted, was not one of determining the relative importance of heredity versus the environment in the production of any particular trait. "There seems to be no possible gain," he allowed, "in reviving that controversy." A more realistic approach, he said, would be to discourage the habit of sharply opposing the concepts of "learned" and "unlearned" when trying to understand development. [52]

Lorenz failed to understand what the Americans were driving at. He happily announced that the ethologists had not neglected ontogeny at all. After all, Heinroth had spent years raising young birds in isolation from adults to see how their behavior would develop. Lehrman replied that Lorenz did not appreciate what Schneirla meant by ontogeny. Heinroth's studies, Lehrman acknowledged, were ontogenetic "in a sense," but as he went on to explain: "I mean no disrespect for these contributions of Heinroth when I say that they are not a substitute for the analysis of the physiological conditions which are changing during the emergence of the behavior elements, and which constitute the development from one stage to the next, which accounts for the behavioral changes." [53] Lehrman did not like the whole Lorenzian schematics of behavior patterns emanating from centers, he said, because this was not conducive to the sort of epigenetic analyses needed to better understand the causes of behavior.

Temporarily in a more conciliatory frame of mind, Lorenz suggested that the preceding discussion had brought both sides "very near to a mutual understanding of what each of us means when he talks of 'Ontogeny.'" He continued, "Of course, I entirely agree that we must study the individual development of behaviour much more thoroughly and with a very similar technique as the one used in experimental embryology. Our excuse for not having done more in this direction as yet is that we had to begin in a purely descriptive way before proceeding to physiological analysis." In response to his critics' worry that students exposed to Lorenz's "nativistic theories" would imbibe the wrong scientific attitudes, Lorenz allowed that there were already "many cases in which the experimental works of young ethologists have forced us to abandon our earlier positions." The problem was that others, outside the field of ethology, did not understand the provisional nature of ethology's concepts and models: "We have forever to fight against their tendency to regard our very hypothetical and very provisional concepts and explanations for gospel truth."[54]

Schneirla and Lehrman were not convinced that Lorenz held his concepts as provisionally as he claimed. Earlier in the discussion, with respect to what Lorenz had just been saying about the innate workings of the nervous system, Schneirla commented: "It is good to learn that Dr. Lorenz is not altogether adamant in these matters, although he still seems largely consistent with his earlier positions as to how the nervous system works innately. Even now, I am not really too sure whether he takes the drainage hypothesis more or less seriously than do his students."[55]

At the conference's end, the privilege of summing up was given to the doyen of French animal psychology, Henri Piéron. The job, Piéron acknowledged, was not an easy one. The conference had explored the behavior of animals from protozoans and daphnia all the way to humans. One issue, nonetheless, had dominated the meetings "and stirred up the most heated debates." This was the question of the roles played by "hereditary transmissions and individual experiences" in determining behavior. On this score, Piéron genially remarked that "Lorenz, who has known so well how to make his ducklings follow him," had been less successful in getting some of his colleagues to do the same. He had in particular run up against "the gentle obstinacy of M. Schneirla," a man who had shown himself to be "very resolved in staying with his opinions."[56]

What had happened at the Paris meeting was interpreted differently in different quarters, as shown in correspondence between Tinbergen (who had not been there) and Koehler (who had). The historical interest of this exchange merits our attention to it in some detail.

Tinbergen reported to Koehler in a letter of August 1954 that he had heard much from Desmond Morris about the Paris conference. He mentioned in addition that Lehrman had visited Tinbergen's group in Oxford. It seemed, Tinbergen said, that Lehrman was "beginning to understand that we are not such fools as he perhaps believed." On the other hand, Tinbergen acknowledged, he was inclined to agree with Lehrman that "we must do more about the ontogeny of behavior patterns, and that even the 'inborn' behavior patterns could, by means of exact study, still yield additional unexpected results." [57]

Koehler doubted that Lehrman had anything of value to offer. What is more, he was already committed to saying this in print. Since before the war, the German zoologist had commanded the review section of the *Zeitschrift für Tierpsychologie* like an artillery captain at the top of a high hill. He had hoisted the flag of Lorenzian ethology and then blasted away at anyone on the horizon who looked like an enemy. By the time Tinbergen's mention of the positive elements of Lehrman's *QRB* article reached him, Koehler had already sent a review of Lehrman's article off to the *Zeitschrift für Tierpsychologie* for publication. "It is rather sharp," he admitted to Tinbergen, "but Konrad thought it should be." [58]

Lorenz was nonetheless of the opinion, Koehler reported, that Lehrman (though not Schneirla) could be won over to the ethologists' side. Koehler for his part had a hard time envisioning Lehrman as someone likely to "swim against the current" of American psychology and swing the Americans around to the ethologists' way of thinking. Koehler left the Paris meetings impressed not by Lehrman's promise but by Lorenz's patience. Indeed, as he insisted after Tinbergen pressed him further on the subject, "I admired Konrad's patience, in the face of both the frightful Mrs. Haldane and also Lehrman, and I can only characterize the role played by Schneirla in particular and his follower Lehrman as deplorable." [59]

Five years earlier in Cambridge Tinbergen had found it necessary to spell out to both Koehler and Lorenz why the British physiologists were upset by the particularly Germanic tone of Erich von Holst's attack on the work of James Gray and Hans Lissmann. Once again he felt the need to set Koehler straight about political sensitivities:

> One must not forget that [Lehrman] and his other Jewish colleagues have understandably taken it badly of Konrad that he at the time in his domestication work rather strongly supported the Nazi conviction that mixing of Aryans and Jews was undesirable. They have felt themselves offended and injured by his statements. We can, knowing Konrad and lov-

ing him as a person, easily say: we must now finally forget that, but for them, as Jews, who moreover do not know Konrad personally, the thing naturally stands differently. Danny Lehrman is now however convinced that he must bury that, and also will really do it; he is really a nice fellow.[60]

Tinbergen went on to allow that he felt, having read a copy of Lorenz's Paris lecture, that the lecture was "much too sharp" and that Lorenz had been wrong to assume Lehrman was a typical American psychologist who knew little about animals. Lehrman, Tinbergen said, "is yet not an ethologist, but stands very close to us, and his doctoral thesis on the influence of prolactin on the parental behavior of doves, which is going to appear in *Behaviour,* is really world-class." There was merit, he again stated, in Lehrman's objection to the assumptions the ethologists had brought to their discussions of behavioral development, even if there was also much in Lehrman's critique was wrong. Overall, he considered Lehrman to be open to ethology's concerns and expected him "to develop into an extremely valuable intermediary between the two groups."

As Tinbergen saw it, a constructive interaction between the American psychologists and the European ethologists was both necessary and likely. Baerends, van Iersel, and Hinde had all been recently to America, he noted, and they had all remarked on the way that there were many Americans who were open to ethological work (as many, Tinbergen suggested, as there were "dogmatic ethologists among us!"). It was time to think of tactics, he believed. Baerends had written to him saying that Lorenz's lecture in Paris had seemed tactically wrong. As Tinbergen put it to Koehler, "Konrad reminded him of an angry lion, who shakes the dust from his fur and gives a furious roar." It would have been wiser, Tinbergen said, if Lorenz had presented the fundamental facts calmly and shown "how unavoidable *many* of his key conclusions are." Planning for the future, Tinbergen noted, "I will be together with Konrad for a whole day before the Ithaca Congress and would like to discuss with him entirely calmly what our tactics ought to be."[61]

Koehler himself was not calmed by Tinbergen's letter. To the contrary, it seems to have made him angrier, and he fired off a letter to Tinbergen in return. Lehrman's critique struck Koehler as an example of "insolence." "Everything has its limits," Koehler told Tinbergen, "this was mad impudence." A man who was not really launched on his career, who was yet to publish his doctoral thesis, "ought not to begin with such a pamphlet." Koehler allowed that he himself had become "perhaps rather sensitized" by the political contexts in which science had been discussed for some time.

For nine years, he said, he had had to read in the daily West German press that Darwin had invented the genetics of the Nazis. In the meantime, the East German press called Darwin a hero, but when they said "Darwin" they really meant "Lamarck and Pavlov." He told Tinbergen he was "fed up" with being presented this "stew" of ideologies and religious intolerance. "When one is permitted to write that genetics ceases with the zygote, everything else is development, and ontogeny is further nothing but pre-formation, Mr. Tinbergen, that goes too far, that does not belong in a de-mocracy, otherwise we are lost, that is advocacy, vote-gathering, but not science." [62]

Koehler saw no reason to excuse Lehrman's sensitivities about the past: "It is certain that Mr. Lehrman through his political attack on Konrad has for his part done no good service. We have now these 9 years of denazifica-tion, reeducation, etc. behind us and are perhaps with this hindsight also sensitized." In Koehler's mind, it was time to even the score. He had no in-tention of letting Lehrman off gently in reviewing Lehrman's QRB piece for the *Zeitschrift für Tierpsychologie*.

Tinbergen's perception of the situation was very different. The etholo-gists' "first and foremost goal," he told Koehler, needed to be awakening other researchers to what ethology was about. When it came to scientific exchanges, he counseled, ethologists should observe "the utmost emo-tional moderation," so as "to try to understand the opponent's mistakes and to proceed in a way that prevents him from emotionally isolating him-self." At the same time, the opponent needed to be exposed "again and again to our facts and chains of reasoning." Offering an optimistic view of the recent critiques of ethology, Tinbergen exclaimed that Lorenz in any event was now "being taken seriously!" "This," Tinbergen continued, "is the first, and perhaps the most important step toward full recognition!" Misguided as the critiques of Hebb, Kennedy, and Lehrman might have been, Tinbergen said, it seemed evident to him that ethology was on the verge of making inroads in America. "And that," said Tinbergen, was to him "extraordinarily important," given that he had set for himself, as his "life task, working as an apostle for ethology in the English-speaking world."

Tinbergen agreed that there was no hope of winning Schneirla over. On the other hand, "Lehrman, Beach and others" looked like "potential strongholds." "In any case, Tinbergen insisted, "Lehrman in *one* thing is right. We use the word 'inborn' sloppily." After pursuing this point, and then acknowledging other ways in which Lehrman was *not* right, Tinber-gen returned to the question of Lorenz's lecture in Paris. Though neither he nor Baerends had been at the conference, Tinbergen said, they both had *read* Lorenz's lecture. It seemed clear to Tinbergen that Lorenz's anger was apparent throughout the lecture, and had led him to say "things that are not exactly true." Thinking about the tactics to cultivate in America, Tinbergen took exception to what he knew of the behavior of Eckhard Hess, who as a strong supporter of Lorenz was in fact "so hyper-ethological" (and unwilling to consider any criticism of ethology) that he was doing harm to ethology's cause.

What ethologists needed to think about, Tinbergen believed, was how to win over American psychologists like Frank Beach "who seriously take the trouble to understand us." This would not be simple, he said, because "first they have no zoological insight and knowledge; second because they above all think only about causation, not about species-preserving function or evolution, and in their causal researches they think atomistically, very rarely synthetically." His own experiences with young students in England, he noted, had impressed on him just how hard it was to bring them around to thinking in terms "that seem so simple and self-evident to us." A "long growth process" was required—basically two years. The few people like Robert Hinde who caught on right away were, as Tinbergen put it "really ethologists from the beginning." To bring students around to the ethological way of thinking required skill and great patience. "How much harder must it then be," as he put it to Koehler, "if one has to deal with people who first are not so young any more; second already carry factual and theoretical baggage!" One cause of hope, as it seemed to Tinbergen, was that one found more open-mindedness in America than in Europe. He also noted with satisfaction that five European ethologists had been invited to the States in a single year and that a special session on ethology was being organized for the American Ornithological Society's meetings in Madison in the fall.[63]

Koehler's next reply was gentler in tone than the previous one had been. His review of Lehrman was already in print, he told Tinbergen, so that was that. "But," he said, "when you see Lehrman, greet him for me and say that I am an old rough fellow, and have said that for his roughness he must pay for it. But despite this, I like this fat fellow in his way, and I am glad that you see him not as an opportunist but as an honest seeker. That he is bright, I have observed precisely." Koehler this time around observed that "Lehrman makes exactly the same mistake as we do in the opposite direction" (that is, he analyzed what he called "learned" no more carefully than the ethologists analyzed what they called "innate"). No doubt remembering that the review he had written was not this evenhanded, he ended his letter to Tinbergen by observing, "Mr. Lehrman must not take it badly, that as one treats others, so one should expect to be treated."[64]

THE 1954 MACY CONFERENCE

Three months after the Paris conference on instinct, the European ethologists and American comparative psychologists met again, this time at the conference "Group Processes" sponsored by the Josiah Macy, Jr. Foundation and held in Ithaca, New York, late in September 1954. There, amid a remarkable assembly of thirty or so biologists, psychologists, psychotherapists, and others, the ethologists and the comparative psychologists had another chance to lay out and explore their similarities and differences— and to interact with the other participants who had come to discuss the interrelations between biology and psychology. Konrad Lorenz, T. C. Schneirla, and Daniel Lehrman were again present. This time Niko Tinbergen and Frank Beach also took part. Of these five, all but Schneirla were featured speakers. Also speaking were two additional biologists: L. Thomas Evans, from Yeshiva University College of Medicine in New York, and Helen Blauvelt, from the Cornell Behavior Farm Laboratory.

The Macy Foundation prided itself on having developed a conference format that encouraged communication across disciplines and discouraged deference to authority. Group discussion took precedence over formal presentations, and interruptions of the presentations were encouraged "at any time." This was eminently feasible in this case because five days were allotted for the six talks. It was also *advisable* in this case because among the other attendees were individuals who were certain to have things to say. They included the anthropologist Margaret Mead, the psychologist Jerome Bruner, the psychoanalyst Erik Erikson, the evolutionary biologist Ernst Mayr, the biochemist and psychoanalyst I. Arthur Mirsky, and the Medical Director of the Macy Foundation, Frank Fremont-Smith.

Amid such an array of intellects, interests, and expertise there were multiple opportunities for participants to position and reposition themselves and, in the end, to take different lessons away from the overall experience. From the beginning, when the participants introduced themselves, an enthusiasm for interdisciplinary thinking prevailed. Lorenz identified himself as "still mainly a phylogeneticist of behavior, though ethology is at present being integrated into true physiology." Lehrman stated: "I have always thought I was a psychologist. I am not too sure now. Last summer, when I visited Dr. Tinbergen, he introduced me to the professor of zoology at Oxford and said, in a very apologetic way, 'Dr. Lehrman is a professor of psychology but he is really more of a zoologist.' I took this as a compliment." Tinbergen was similarly playful and congenial. He described himself as "a hunter at heart" who had become a zoologist because his Dutch ancestors had killed off all the worthwhile game in Holland. He explained: "I like to be in the open. I am perhaps also very lazy; I like to watch animals, not to kill them. I have remained in the appetitive stage of hunting behavior. I stalk animals and look at them." He went on to describe how he had come in contact with Lorenz and how more recently he had been stimulated by Lehrman's attack on ethology in the *Quarterly Review of Biology*. He had come to the conference, he said, not only to find out "where we had made mistakes, where we had gone too far," but also to defend some of ethology's facts and conclusions. He hoped that interacting with the American psychologists might help generate "a common research program." [65]

A detailed analysis of the various positions adopted, points made, and evolving dynamics of the 1954 Macy conference cannot be offered here. Space does not permit it. For our purposes it will be enough to direct our attention to several features of Tinbergen's session, which Tinbergen diplomatically entitled "Psychology and Ethology as Supplementary Parts of a Science of Behavior." [66]

Tinbergen quickly found his session exemplifying the Macy principle of interruptions "at any time." He had not spoken for more than three minutes before he was interrupted by a question on the origins of the word "ethology." The delay proved to be a long one. Before it was over, he, Lorenz, Mayr, Mead, Bruner, Schneirla and others had a say on what previous writers had meant when they had used the word "ethology." Eager to get on with his paper, Tinbergen stated that when he now used the word "ethology," he meant "the biological study of behavior." Lorenz butted in again, saying, "Don't give a definition. I would just say, historically, it is that branch of research started by Oskar Heinroth." Tinbergen protested: "It has certain characteristics." Lorenz responded: "It has but you cannot define them." [67]

Tinbergen was irritated, but he kept his irritation to himself. He was convinced that one *could* define ethology's characteristics. After a few additional remarks by the conference participants, he proceeded to spell out the more salient of these features. He allowed that while Lehrman had criticized "Lorenz's *Theory* of Instinctive Behavior," what really distinguished ethologists from psychologists was the *approach* Lorenz had developed. Ethologists, Tinbergen explained, were trained as zoologists, and their concerns went beyond the psychologists' concern with questions of immediate causation. Zoologists were additionally concerned with the problems of survival value and evolution, which were problems that psychologists were inclined to neglect.

Tinbergen's mention of survival value and evolution was the occasion of another interruption. Lorenz launched into a discourse on the different sorts of questions that biologists ask and a defense against the charge that ethologists were "teleologists." This gave Ernst Mayr the opportunity to offer a slightly different version of "how?" and "why?" questions in biology. Schneirla got into the act, noting that American psychologists had come to be properly critical of certain uses of the word "purposive." After an interchange with Mayr about the schools of thought to which particular psychologists belonged, Schneirla suggested that William McDougall's purposive psychology had perhaps influenced Lorenz in ways that Lorenz himself did not recognize. When Lorenz offered that he shared Schneirla's critical view of Edward Armstrong's and McDougall's teleological thinking, Schneirla agreed this his position and Lorenz's were "much the same, on this point." Schneirla and Lorenz then lectured each other back and forth on teleology until the psychologist John Spiegel allowed that he did not understand why the discussion was becoming so heated. Fremont-Smith seized the moment and gave the floor back to Tinbergen. The foundation's encouragement of interruptions had been carried far enough. Twenty minutes or so had elapsed since Tinbergen had last spoken.[68]

Back on track, Tinbergen returned to the pair of concerns he associated above all with ethologists: "How does the behavior contribute toward survival?" And "How has the behavioral equipment of each species developed in evolution?" He then identified additional differences between ethologists and comparative psychologists. Ethologists studied a greater number of animal species than psychologists did. Furthermore, with each species they studied, they tried to learn the animal's whole behavioral repertoire, not just parts of it. Finally, Tinbergen observed, ethology was clearly the more junior of the two sciences. With only a few trained workers, "very limited financial resources," and a commitment to studying a broader range of species and phenomena, it was relatively speaking "still in its infancy."

Lorenz took up the theme of the relative size and support of ethological and psychological research teams, saying: "I consider myself very lucky if I have about six people working with me; Dr. Tinbergen has about seven, and they do co-ordinated work. In the American schools there is very fine co-ordinated, organized collaboration, the co-working of an army of single specialists. That makes them much more exact." Lehrman immediately chimed in, "It only looks that way."[69]

Lehrman had visited Tinbergen in Oxford a few months earlier. The two had thus already had a chance to discuss their differences. These previous discussions helped convince Tinbergen that the ethologists' use of the term "innate" was not viable. As he explained to the other attendees of the Macy conference, "in using too sweeping a term, we tended to smother our inclination to investigate fully all the problems of ontogeny. We were aware of more problems than our use of the word 'innate' might suggest, but dropping the term helped me in restating the problems more fully." Nonetheless, Tinbergen insisted, the ethologists' use of the word "innate" had made sense in its original context. As he explained, "stressing of the unconditioned nature of such responses was in itself a justified reaction to American overemphasis on conditioning, which was very striking when we began our work."[70]

As for Lehrman's objection to the ethologists' use of the term "innate releasing mechanism" to describe the behavior of animals vastly different from one another, Tinbergen denied that ethologists believed that "the physiological mechanism must be the same in all cases." This, he declared, was something he and his colleagues had never believed and never implied. They had used the term provisionally and functionally. They had not prejudiced their thinking with respect to how the behavior was actually achieved. They did not suppose that IRMs in greatly different animals were in fact homologous structures. "Here," Tinbergen protested, "I honestly feel that our critics merely show a lack of zoological training."[71]

Tinbergen's paper and the conference as a whole provided the occasion for more jousting but also for some reconciliation between the ethologists and the comparative psychologists. Lorenz reported back to Erich von Holst several weeks after the conference that Lehrman was, "upon closer examination and after the dismantling of all mutual hostilities, a clever, analytical fellow."[72] Later still, after concluding what had been a very long tour for him in the United States, Lorenz confidently told Bill Thorpe: "My American trip was very tiring but altogether a success. The best of the learning theorists like Beach and others are just beginning to see for themselves that learning cannot explain 'everything.' Curiously enough, Lehrmanns objectionable paper seems to have helped considerably to the recognition of this fact. The best means to convince people that there is such a thing as instinctive movements is the film. I played duck films to Frank Beach until he nearly fainted, he got seriouser and seriouser

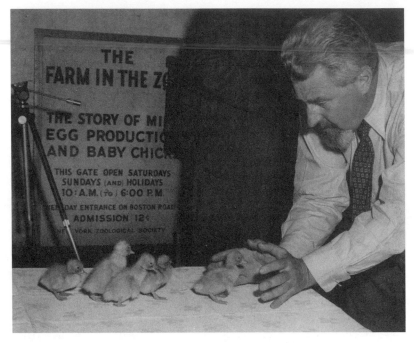

8.5 Lorenz's fame spread in the 1950s through his publication of popular books on animals, his public lectures, and his appearance on American television. This picture appeared on the front page of the *New York Herald Tribune* on 23 January 1955 with an account that began: "Dr. Konrad Z. Lorenz, the tall bearded dignified Austrian who has been giving invitation lectures on animal behavior at American universities and scientific institutions, became the foster mother of five baby ducks yesterday at the Bronx Zoo." The story continued: "In less than two hours Dr. Lorenz, who speaks both duck and goose, had the ducklings following him around and coming to his call." (Photo courtesy of the Konrad Lorenz Institute for Evolution and Cognition Research.)

and in the end he said in a small voice: 'You know I did not believe a word of it and now I believe everything.'"[73]

Significantly, while the Macy conference brought into focus the differences between the ethologists and the comparative psychologists, Tinbergen left the meeting thinking not only about these differences but also about the differences between himself and Lorenz. It bothered him that Lorenz had been so cavalier about how ethology should be defined, or whether it could be defined at all. To Tinbergen, defining ethology as "the biology of behavior" was of the utmost importance. This, he felt, was not just a convenient kind of shorthand. Instead, it provided critical guidelines for the field's future development. He was not convinced Lorenz fully understood how important this was.[74]

Indeed, it can be argued that Lorenz's successes at this point were less in advancing the field's research front than in bringing it broader popular attention. His appearance at the Macy conference was but one stop in the midst of a three-month lecture tour of the United States that included the Messenger Lectures at Cornell, the Dunham Lectures at Harvard Medical School, and an appearance on the American Museum of Natural History's television show *Adventure,* where he was able to charm a large viewing audience by displaying five baby ducklings from the Bronx zoo that he had imprinted on himself only a day earlier.

FURTHER INTERNATIONAL CONGRESSES
AND THE PROBLEM OF CONTROLLING
ETHOLOGY'S GROWTH AND DIVERSIFICATION

The ethologists met again at a conference of their own in Groningen in September 1955. The Third International Ethological Congress lasted twelve days and featured forty talks and five film sessions. Germany, Britain, and Holland provided the largest contingents. Swiss, Scandinavian, and French ethologists also participated. Here too, for the first time, was a contingent from the United States: Frank Beach, Daniel Lehrman, Eckhard Hess (University of Chicago), and D. Davenport (University of California, Santa Barbara). Beach chaired a day's worth of sessions devoted to ontogeny of behavior and social behavior, Hess gave a paper on imprinting, and Lehrman gave a presentation on incubation behavior in ringdoves. Lehrman and Beach would become stalwarts at the international ethological congresses, both serving as members of the organizing committee that determined who should be invited to the meetings.

The chairs for the nine days of sessions effectively constituted a Who's Who of the growing ethological community: Lorenz, Tinbergen, Thorpe, Baerends, Koehler, Kramer, Hediger, Hinde, and Beach. Among those conspicuously absent was Karl von Frisch. Despite his brilliant contributions to the study of animal behavior and his development of a strong cohort of researchers, he did not take an active role in the development of the ethological community in the postwar period. He never attended an international ethological congress. Likewise missing from the Third International Ethological Congress and all the others was T. C. Schneirla. As for the brilliant Erich von Holst, he appeared on a congress program only once after hosting the 1950 symposium at Wilhelmshaven that was the precursor of the later meetings. He skipped the assemblies at Buldern, Oxford, Groningen, and Freiburg and then showed up in Cambridge in 1959. There he chaired a session and gave an unprecedented special series of four evening lectures on experiments he and his co-workers had been pursuing on the release of drive behavior in chickens by means of brain stem stimulation.

From the Groningen conference onward, the congresses were characterized by a steady growth in numbers of participants, an increasing diversity of topics and approaches, and more and more scientists from the United States. At Freiburg in 1957, out of some 115 participants, only 7 Americans attended, with 5 (Beach, Lehrman, Hess, Vincent Dethier, and Kenneth Roeder) giving papers. Two years later in Cambridge, 22 of the 128 participants were Americans, and 15 of them gave presentations.[75] In Starnberg, Germany in 1961, 31 Americans participated and 18 of them gave papers (nearly 35% of the total of fifty-two papers).

Not all the Americans were comparative psychologists like Beach and Lehrman. They included the zoologist Donald Griffin, famous for his work on echolocation in bats, and the neurophysiologist Kenneth Roeder, similarly noted for his work on the neurophysiology of the behavior of insects. Among them too were biologists who had received at least some ethological training in Europe. Such was the case with Martin Moynihan, who took a degree with Tinbergen at Oxford (1950–1953); George Barlow, who spent two years with Lorenz in Seewiesen (1958–1960); Gordon Burghardt, who studied with Lorenz in Seewiesen (1963); and John Fentress, who took his doctoral degree at Cambridge with Thorpe (1961–1965). By Donald Dewsbury's calculation, eighty-three Americans traveled to Europe between 1949 and 1974 to work with European ethologists. The vast majority went to one or another of five major centers: Cambridge, Seewiesen, Leiden, Groningen, and Oxford. In addition there were scientists from other countries than the United States who took up careers in the States after studying ethology first in Europe. Among the earliest of these were Peter Marler, R. J. Andrew, Colin Beer, Franz Sauer, Frank McKinney, and Wolfgang Schleidt. Ethology was successfully transplanting itself across the Atlantic.[76]

The international congresses remained "invitation-only" affairs for years, but even then it was not easy to keep down the meeting size. Thorpe wanted to keep the meeting size small, but Tinbergen saw the drawbacks to this. As he told his fellow conference organizer Koehler, "I don't agree with Thorpe that we should keep the Conference small. First, . . . there are now already too many first-class ethologists. Second, just such a policy would in my opinion isolate and impoverish ethology. We must dissolve the somewhat artificial borders between ethology and other biological research areas such as neurophysiology, endocrinology, evolution-research and genetics, [and] human psychology." Tinbergen suggested that the conferences continue to be held every two years, but on a cycle whereby once every four years there would be "an abominable monster-congress" and then in between, again once every four years, a much smaller conference of "institute directors with their best students."[77]

Tinbergen's idea about alternating "monster-congresses" and small ones was not endorsed by the organizing committee. The biennial meetings congresses continued to grow. Up until the 1967 congress in Stockholm the organizers were able to avoid scheduling simultaneous sessions, but in Stockholm they decided the change finally had to be made.[78] The printed program from this meeting lists 264 participants plus 34 spouses. Yet even in Stockholm a sense of collective enterprise remained. As Colin Beer recollected several years later, "Lehrman gave one of the plenary session lectures at the 1967 Ethological Conference, and those who were there will remember how he called on Lorenz to help him hold up the blackboard at one stage of the proceedings. The spectacle of those two rival titans standing like supporters of an escutcheon was so eloquent that it drew loud applause. In his speech at the closing dinner Lorenz remarked on the symbolic significance that all were apparently wont to see in the incident. That conference was notable, in my opinion, for its spirit of harmony, enthusiasm and optimism for ethology's future."[79] Reporting on the conference to Tinbergen, who was unable to attend because of illness, Lorenz wrote: "I believe Danny Lehrman has now finally understood me, and we have both in many discussions astonishingly come to exactly the same thing."[80]

Harmonious as the spirit of the Stockholm conference may have been, the ethologists were finding it difficult to maintain their field's identity. And this was the case independently of the question of the size of the meetings or the fact that Lehrman and Lorenz were still far apart from each in their thinking, the symbol of cooperation notwithstanding. Tinbergen had already signaled a warning at the Freiburg congress in 1957. On a positive note, he allowed that ethology was going through "a very fascinating phase, a phase of rapid development of the conceptual framework." Some of ethology's major concepts, he said, were showing simultaneously their usefulness and their provisional nature. He described the conference as something of a "milestone" for himself "after a period of confusion of several years." Thanks to criticism "from many parts," the ethologists had begun to see "what was wrong with our conceptual framework." A new approach was beginning to emerge. As Tinbergen put it, "I feel that I have gone, like a pupating caterpillar, through a phase of breakdown and reorganization, and that I am now more or less like the butterfly hatching something new

8.6 Lorenz and Tinbergen at the International Ethological Congress in Starnberg, 1961.

and something viable is going to emerge; it will live happily for some time. Of course, it will die, but it will have laid its eggs, and a new period of growth (and perhaps one of confusion and reorganization) will follow."

But if Tinbergen was happy to speak of his own metamorphosis, he did not fail to remark on what he identified as the "darker sides" of the Freiburg conference. Although he had been pleased to see an increasing rapprochement between ethologists and neurophysiologists, he was troubled by the way that the physiological analysis of behavioral patterns seemed to be dominating ethological studies. Research on problems of biological function, ontogeny, and evolution were simply not keeping pace with the study of physiological causation. The problem, as Tinbergen interpreted it, was not so much a matter of theory as it was of not recruiting enough *naturalists* to the field: "We are attracting increasing numbers of physiologically-minded young workers; a different type than we attracted twenty years ago, when our group consisted almost entirely of naturalists. I feel very strongly that we still need the naturalists, that we must make sure to attract them in numbers, because it is only through them that we can check whether we are explaining the entire behaviour, the phenomena in all their complexity. More important still: it is the naturalists who at every stage of our science *discover,* and call our attention to phenomena to be understood; they are in touch with the full riches of nature. We must not lose them or we will lose contact with reality." [81]

Others at subsequent meetings remarked on changes they had observed. In Starnberg in 1961 Kenneth Roeder remarked that at this meeting, in comparison to the meeting in Freiburg, there was "a great deal more straight-forward reporting" and not so much discussion of general ethological concepts. Also conspicuous at this meeting was the great diversity in the different approaches being brought to bear on behavioral questions. [82] Lorenz for his part in 1961 expressed his own anxieties about the directions in which ethology was going. He claimed not to lament the fact that some of his own theories were being thrown overboard. He objected strenuously, however, to the jettisoning of some of ethology's key *questions.* The 1959 ethological congress in Cambridge had left him "in a state of the deepest depression," he said, because the new generation seemed to be ignoring two of the subjects most central to ethology: the species-preserving function of behavior and the study of behavioral evolution. In particular, the comparative studies so dear to his heart were being dismissed as no more than inspired guessing. To add insult to injury, "English-speaking ethologists" were allowing this cavalier attitude to go unchallenged. But all was not lost, Lorenz said. He saw signs of hope. The pendulum seemed to be swinging back in the right direction. A prominent example of this was Niko Tinbergen's recent work identifying the competing selection pressures to which behavior patterns were subject. [83]

NINE

TINBERGEN'S VISION FOR ETHOLOGY

> Does it hit you hard when I move in to the attack? Or do you think I am just being dim? I feel so *convinced* I am (more or less) right! And you know, Erich [von Holst] is wrong when he says that I am no fighter and that I always "give in." If this were so, I would give in to you rather than to anybody else. I hope for a vigorous reply!
>
> NIKO TINBERGEN LETTER TO KONRAD LORENZ, 31 JANUARY 1961,
> CHALLENGING LORENZ'S VIEWS ON "INNATE" BEHAVIOR

> After long doubts I now know what I am going to write for Konrad's Festschrift, and I'm now sitting every free hour at the type-writer. . . . I want to write something for Konrad that he likes a lot. I am now writing some thoughts down about (don't panic) "What is ethology?"
>
> NIKO TINBERGEN LETTER TO OTTO KOEHLER, 6 FEBRUARY 1963,
> ANNOUNCING HIS PLANS FOR WHAT WAS TO BECOME HIS
> SINGLE MOST IMPORTANT SCIENTIFIC PAPER

TINBERGEN'S METAMORPHOSIS

What was this new work by Tinbergen to which Lorenz referred so optimistically at the Starnberg conference in 1961? And what happened to Tinbergen in the 1950s that led him to describe his development in terms of a metamorphosis from caterpillar to butterfly? To understand this change we need to review Tinbergen's development up to this point.

In the 1930s and 1940s Tinbergen's claims regarding ethology's legitimacy as a science were centered on his identification of ethology as an "objectivistic" approach to the study of animal behavior. What this meant for him, as we have seen, was that ethology was "applying physiological methods to the objects of animal Psychology" and focusing on the issue of behavioral *causation.* There were a number of reasons for him to define his enterprise in this way. One was his strong conviction that life processes are to be understood in terms of their underlying *mechanisms,* a conviction that he developed in opposition to subjectivists like Bierens de Haan and E. S. Russell. A second was his intuitive sense that animal displays must be the consequence of internal, physical "excitement" in the organism. Third, he recognized, at least subconsciously, that in the pecking order of the life sciences natural history ranked near the bottom while physiology was at the top. Making natural history more physiological would bolster its scientific respectability. But Tinbergen was not trained as a physiologist. The kind of "dissection" he preferred was analytical. It did not involve using a scalpel. Nor did it involve sophisticated quantitative analysis, for that matter. Analyzing behavioral phenomena meant breaking behavior patterns down into their component parts, into "special combinations of simple muscle contractions." It also meant studying the internal and external stimuli and mechanisms that motivated, released, and oriented the kinds of behavioral phenomena in which he was interested.[1] Examples of this were his early studies with Lorenz (on the egg-rolling behavior of the greylag goose), D. J. Kuenen (on the gaping response in thrushes), and A. C. Perdeck (on the begging response in herring gull chicks).[2]

Tinbergen carried such analyses of behavioral causation further in the late 1940s and early 1950s, most notably in his studies of displacement activities and other behavioral displays. His primary organisms of choice up to that point, the herring gull and the three-spined stickleback, continued to serve him well. To an audience of fellow ornithologists at the International Ornithological Congress in Uppsala in 1950, he explained that the herring gull has three different threat postures, each indicative of the intensity of the motivations underlying it.

In the first of these postures, the "upright threat posture," Tinbergen discerned the "three components of the intention movement of actual attack," namely, "trying to come on top of the opponent, pecking, [and] delivering wing blows." At the same time, he saw evidence of the tendency to withdraw in the way that the bird pulled back its neck. "The stronger the tendency to withdraw," he said, "the more the neck is brought back." To Tinbergen, the upright threat posture was "clearly recognizable as the combined intention movements of the two antagonistic drives: attack and withdrawal, or escape." When the intensity of the gull's conflicting drives increases, Tinbergen went on to explain, new behavior patterns are exhibited: first, "grass pulling" ("displacement nest building"), and then, at an even higher level of motivation, "choking." As Tinbergen summed this up, "All aggressive behaviour of the Herring Gull, at first sight a curious mixture of relevant and irrelevant movements, is caused by varying degrees of activation of two drives: attack and escape."[3]

Tinbergen used the same sort of analysis to account for the courtship behavior of the three-spined stickleback. In a paper with his former student Jan van Iersel, Tinbergen explained the "zig-zag dance" of the male stickleback as a combination of hostile and sexual movements on the fish's part. The male's "zig" in the direction of the female was a hostile, attack motion. The "zag" was a motion comparable to that of leading the female to the nest. Tinbergen and his student Martin Moynihan offered a similar explanation of courtship in the black-headed gull. They interpreted the "forward" display, the first movement in the species' mutual courtship, as an essentially hostile posture. They interpreted the next movement, "head flagging," as an "appeasement gesture."[4]

In the early 1950s Tinbergen continued to teach students how to analyze the stimulus situations that release "innate responses" and how to identify the multiple motivations that underlie threat and courtship displays. It made sense to him to call "conflict movements" those actions that seemed to arise as the result of the simultaneous activation of different "major functional systems." Thinking about "major functional systems" was in turn related to his conceptualization of "the hierarchical organization of the behaviour machinery." He underscored the importance of forging ties with the physiologists, who were pursuing the mechanisms of behavioral action at a finer level. But he was not disposed to become more of a physiologist himself. As he told the present writer in 1979: "Once I got as far as The Study of Instinct, I realized that the further analysis of the physiology of behaviour would have to dig into the nervous system. Here I balked; I was not really interested in poking inside a system that seemed much too complex to me, and I had not the slightest technical expertise, nor knowledge of nervous functions as then known. I also felt that the level of the firing neurone was too far away from that of the ultimate output as integrated behaviour. This was a clear example of non-rational shying-away from something I felt unable to tackle. Also, I remained drawn to the outdoors."[5]

Tinbergen continued to write in the early 1950s about the motivational basis of social signals, most notably in the paper on "derived activities" that he published in the *Quarterly Review of Biology* in 1952 and then again the following year in his book *The Herring Gull's World*. By the time these appeared, however, he was inclined to view them as representative of his research past. Making these results visible to a wider English-speaking audience had been part of the "missionary" work he felt obliged to do. As he told Otto Koehler in the fall of 1952, he had paid for the past three years with headaches and various painkillers, but it had been "worth the effort," for it appeared that his effort "to colonize England ethologically" had been successful. Now he was ready to broaden his research program. Specifically, he wanted to undertake comparative studies as a means of gaining insights into the evolution of behavior over time.[6]

Through the 1950s and beyond, other investigators would pursue issues deriving from Lorenz's and Tinbergen's early models of instinctive behavior. Problems such as the activation and deactivation of instinctive urges, the nature of the fixed action pattern, and behavioral motivation provided a rich field of study for those who sought to pursue ethology from the causal side of things. Tinbergen paid close attention to this work, but he came to see new lines of research as equally worthy of development and more satisfying to him personally. The remainder of this chapter is not an attempt to encompass the wealth of different contributions that a burgeoning population of investigators of animal behavior made to the field of ethology.[7] Nor does it track the considerable popular attention that Lorenz in particular (but others as well) drew to animal behavior studies through public lectures, television appearances, and so forth.[8] Rather, it is an examination of the evolution of Tinbergen's own research program and his vision for the enterprise that he had helped to cofound.

TINBERGEN AND THE COMPARATIVE STUDY OF BEHAVIOR

Tinbergen arrived in Oxford in the fall of 1949 already planning to undertake comparative studies of behavior. Although he had not done comparative work previously, he did not have to start from scratch. His previous work on the herring gull and the three-spined stickleback provided starting points for him. His new plan was simple: he would study additional species of gulls in the field and additional species of sticklebacks in the laboratory. "Comparison becomes more and more urgent for us," he wrote in September 1949 to his friend Ernst Mayr. Hoping to get his hands on some American sticklebacks, Tinbergen asked Mayr whether someone traveling by airplane from the United States might be able to transport live sticklebacks to England.[9]

In 1949 Tinbergen regarded Lorenz's 1941 paper on ducks as the best model of comparative behavior studies in existence. He hoped at Oxford to direct researches of comparable value. Nonetheless, he imagined his studies might end up qualitatively different from Lorenz's. He had never found much appeal in the enterprise of identifying and tracing homologies. As a field naturalist, he wanted to see how species are adapted to different ecological niches.[10] What he and his students discovered with respect to adaptive radiation ultimately surpassed his expectations.

Tinbergen began lab work at Oxford before fieldwork for the simple reason that in his first spring after arriving in England he was tied up attending conferences. Holst's small gathering in Germany, the International Ornithological Congress in Sweden, and another meeting in Holland all conflicted with the project of studying gulls in their breeding season. He determined, however, which species to watch once he got back in the field again. His choice was the black-headed gull, *Larus ridibundus*.[11] Meanwhile, back in the laboratory, he identified *Spinachia*, the sea stickleback, as the species to compare with its three-spined relative, *Gasterosteus aculeatus*.[12] As of 1951 Tinbergen had two new graduate students, Desmond Morris and Philip Guiton, in the lab studying sticklebacks. In a room in the basement of the Zoology Department, these two set up fifty or so fish tanks and began their researches. Morris took on the ten-spined stickleback (*Pygosteus pungitius*). Guiton continued the study of *Gasterosteus aculeatus*.[13]

Inevitably, though, it was in the field rather than the lab that the behavioral differences between species could best be examined in terms of the animals' ecology. In the field, furthermore, was where Tinbergen preferred to be. It was there that he felt most engaged as a scientist, most complete as a person, and most disposed to bond with students and colleagues as a "comrade-in-arms."[14] His field students undoubtedly felt privileged in this regard, although not when it came to the noisy bonhomie he cultivated in camp at the crack of dawn. As they saw it, his practice of awakening them at 5:00 a.m. by whistling or singing nursery rhymes was inclined to inspire thoughts of mayhem more than thoughts of comradeship.[15]

As indicated in chapter 4, Tinbergen's early enthusiasm for field natural history preceded any specific theoretical concerns on his part. When he went to Greenland in 1932, it was not *in order* to study the territorial behavior of the snow bunting in spring. Rather, the snow bunting project was the scientific excuse he constructed to justify going to Greenland. Later in his career, his fieldwork would guide his theorizing as much as theory guided his fieldwork.

Field naturalists learn to adjust to the seasonal rhythms of the animals they study. Because birds in the course of the breeding season pass from one part of the reproductive cycle to the next relatively quickly, the field ornithologist is rarely able to study any one topic for more than a few weeks at a time. Adapting to the birds' timetables, Tinbergen and his students took up different questions as the birds passed through the different stages of the reproductive cycle. As he explained in his book *Curious Naturalists* (1958), "We switch, with the birds, from one phase to the next, adding a little to our results every year."[16] Working under these and other constraints, it was necessary to be opportunistic. Nonetheless, Tinbergen became increasingly confident that fieldwork could address certain issues better than any other approach could. Over time he came to think of his gull studies not only as a means of understanding the lives of gulls and the sociology of gull communities but also as "a kind of test case of the possibilities of field work."[17]

The gull work of the Oxford program began in earnest in the spring of 1951.[18] Tinbergen set two doctoral students to work studying the behavior of black-headed gulls at Scolt Head Island on the coast of Norfolk. Martin Moynihan focused on hostility between individuals and on the formation of pairs; Rita White studied the relations between parents and chicks. Ultimately, neither the Scolt Head site nor a different site tried in 1952 proved satisfactory, but Tinbergen's enthusiasm was not dampened. In the spring of 1953 he transferred the base of the black-headed gull studies to the Ravenglass Peninsula in Cumberland. He put other students to work studying other birds in the Farne Islands in Northumberland. Now he had a total of six students (three doctoral students and three postdocs) engaged in comparative behavior studies of gulls and terns. In addition to Moynihan and White, the group included Michael Cullen, Esther Sager, Frank McKinney, and Uli Weidmann. Weidmann became another investigator of the black-headed gull (and also the husband of Rita White). Cullen, Sager, and McKinney worked on arctic terns, kittiwakes, and eider ducks, respectively.

Tinbergen could look forward not only to the findings of the six field ornithologists in his own research group but to reports from other European gull watchers as well. One of these was his former student Gerard Baerends, who was continuing work on the herring gull. Others studied the sandwich tern, the common tern, the little tern, the black tern, the Caspian tern, and the little gull.[19] Altogether they constituted an international cohort of gull researchers, each undertaking the kind of painstaking observations necessary for understanding the causation and evolution of behavior.

Tinbergen was pleased by the way this community of researchers was developing. He was also excited by the gulls themselves. The kittiwake in particular made an especially strong impression on him. In July 1952, he reported to Ernst Mayr: "I have just been to the Farne Islands where I saw, for the first time, breeding Kittiwakes and thus saw my next gull species' behaviour pattern. A most intriguing species! The most aggressive of all I have seen; there is severe competition for nest sites, quite different from the situation in the Herring gull."[20] Eleven months later, toward the end of a visit to the Farne Islands site, Tinbergen wrote Huxley: "The 'Farne boys' (and girl) are doing well I think. My stay is drawing to a close; I have made a 'complete' film record of the kittiwake's displays, and can now show 3 gull species in comparison."[21]

ESTHER (SAGER) CULLEN'S STUDY OF THE KITTIWAKE

By the time Tinbergen wrote the above words to Huxley in June 1953, his student Esther Sager, a young Swiss woman, was making excellent progress with the kittiwake (*Rissa tridactyla*). She had taken immediately to the birds on her first, brief visit to the Farne Islands in June 1952. In 1953 she embarked on her first full season of watching kittiwakes on the Inner Farne.

Two generations earlier Edmund Selous had described the "wretchedness, cold, and discomfort" of bird-watching in north temperate Europe during the breeding season. He had made it clear that bird-watchers have to be inventive if they are to have the slightest hope of staying warm and dry for any length of time under such conditions. The American psychologist Bill Verplanck, who visited Tinbergen's team on the Inner Farne in late February and early March of 1953 and then later in the season, described one of the beautiful days during his first stay: "Warm in the sun (Of course, warm with longhandled drawers, a heavy flannel shirt, two sweaters, a muffler, a duffle coat, sheepskin gloves, and a wool hat; Also heavy wool sox in the Marine boots)." He also described how the researchers' typical day began: "Up at 5:30. Shiver. Drink hot Nescafe, eat bread and marge and marmalade. Off to the top of the cliffs on the West side of the island. Sit immobile, behind a rock, or in a hide for three hours, watching."[22]

Esther Sager found that a sleeping bag and hot water bottle, in addition to the sorts of clothing mentioned by Verplanck, improved her chances of surviving her own bird-watching vigils. From her lookout position on the edge of one cliff, she observed the behavior of some thirty pairs of kittiwakes nesting on an opposing cliff. Equipped with field glasses and a notebook (plus a chart of wing tip markings she devised to help her keep track of individual birds), she documented in detail how the kittiwake's unique nesting habit correlated with its distinctive behavior patterns. All in all she observed kittiwakes on the Inner Farne for three full seasons (1953–1955), with shorter stints in 1952 and 1956.[23]

Tinbergen was thrilled with what she found. He was also pleased with his matchmaking: Esther Sager and Mike Cullen married in 1954 (following the example of Rita White and Uli Weidmann before them).[24] In the fall of 1955, as Esther Cullen was writing up her results for publication, Tinbergen reported enthusiastically to Huxley about her work. She was providing, Tinbergen said, "a detailed description of those characters (mainly

9.1 Watching kittiwakes from a shelter. (Photo by Niko Tinbergen. Courtesy of the Tinbergen family. Copy provided by The Bodleian Library, University of Oxford, MS. Eng. C. 3142, C. 224.)

behavioural, but also some others) in which Kittiwakes differ from other gulls, and an attempt to see whether these differences can be interpreted as adaptive." The kittiwake's cliff-breeding habit, Tinbergen explained, was obviously an antipredator device. Cullen had identified "a multitude of peculiarities" in the bird's behavior and morphology that were apparently linked to its cliff-breeding habit. Tinbergen delighted in the results: "The whole study is, I think, fascinating, extremely precise and critical, and a wonderful description of adaptive divergence of behaviour."[25]

Esther Cullen completed her paper and submitted it to the *Ibis* in the summer of 1956. It appeared in print the following year. The author's stated aims were to identify the ways in which the Kittiwake differs in its behavior from ground-nesting gulls and to show how these peculiarities related to the species' cliff-nesting habit.

For information on ground-nesting gulls, Cullen was able to draw on the expertise of Tinbergen and fellow students Martin Moynihan and Rita and Uli Weidmann. She could also call on a growing scientific literature on gull behavior, highlighted by Tinbergen's recent book, *The Herring Gull's*

World, and Moynihan's recent monograph on the black-headed gull. Other ground-nesting gulls had been studied as well, though not as thoroughly. All in all Cullen had ample material on which to base a comparison of her cliff-nesting species with its ground-nesting relatives.

Her discoveries were striking. She found that the kittiwake's mode of attacking opponents, for example, involved a whole series of interconnected peculiarities, all of which were linked to the bird's habit of nesting on very narrow ledges—sometimes as little as four inches wide—on steep cliffs. The attack of a herring or black-headed gull characteristically involves getting above the opponent to peck down on it or grasping a part of the opponent and then pulling. The kittiwake's attack, in contrast, is directed horizontally forward. The bird tries to grasp the opponent's beak and then twist it from side to side. As Cullen explained, the fighting methods of the ground-nesting gulls—pecking downward and pulling backward—are simply not suitable for encounters on narrow ledges.

Cullen went on to argue that the kittiwake's reliance on one specialized mode of fighting had repercussions for other aspects of its behavior as well. On the one hand, the kittiwake lacks the "upright" threat posture typical of ground-nesting gulls (the posture Tinbergen had interpreted as having developed from the intention movement of rising up to peck down at an opponent from above). On the other hand, the kittiwake uses its beak more prominently in its aggressive encounters than ground-nesting gulls use theirs. The kittiwake also has as an "appeasement movement" a hiding of the beak that is much more exaggerated than that of any ground-nesting species. Cullen concluded that the kittiwake's beak serves as both a directing and releasing stimulus for other kittiwakes.

Cullen also identified important differences in the behavior of young kittiwakes and young ground-nesting gulls. These too were correlated with the kittiwake's cliff-nesting habit. Whereas young ground-nesting gulls can normally run away from fights, young kittiwakes, confined to their narrow ledges, cannot. Instead of running away, young kittiwakes show head turning, which serves to "appease" opponents (i.e., it tends to stop their attacks). This head turning not only removes the beak from view, it exposes a special black band across the nape of the young bird's neck. Among gulls, this marking is unique to the kittiwake and its close relative *Rissa brevirostris* (which Cullen also included in her study). Cullen concluded that both the head turning and the black neck band of the young kittiwake are characters that developed as correlates of the species' cliff-nesting habit.

Cullen went on to show how a whole host of other characteristics of the kittiwake corresponded to the kittiwake's ancestral move from a ground-

9.2 Esther Cullen with kittiwakes. (Photo by Eric Hosking. Copy provided by Hans Kruuk and Jaap Tinbergen.)

nesting to a cliff-nesting habit. These included courtship displays, nest construction, nest sanitation, feeding of the young, the parent's lack of recognition of their own chicks, clutch size, and the infrequency of alarm calls. She closed her paper by listing the distinctive adaptations of the kittiwake and stating: "With the adaptations described in this paper . . . I hope to have shown how this one change to nesting on tiny ledges on steep cliffs has had repercussions in many aspects of the life of the species and has led to morphological changes as well as a great many alterations in behaviour. In many animals adaptive differences between species have been described but I know of no other case where one relatively simple change can be shown to have been responsible for so many alterations."[26]

These discoveries played a critical role in Tinbergen's sense of metamorphosis in the mid-1950s. If, at the International Ethological Congress in Freiburg in 1957, he felt like a butterfly emerging from a chrysalis, this was directly related to the recent successes of his research group's fieldwork, highlighted by Esther Cullen's work on the kittiwake. Tinbergen was

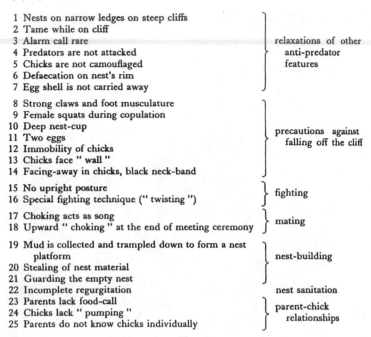

1 Nests on narrow ledges on steep cliffs	
2 Tame while on cliff	
3 Alarm call rare	relaxations of other
4 Predators are not attacked	anti-predator
5 Chicks are not camouflaged	features
6 Defaecation on nest's rim	
7 Egg shell is not carried away	
8 Strong claws and foot musculature	
9 Female squats during copulation	
10 Deep nest-cup	precautions against
11 Two eggs	falling off the cliff
12 Immobility of chicks	
13 Chicks face " wall "	
14 Facing-away in chicks, black neck-band	
15 No upright posture	fighting
16 Special fighting technique (" twisting ")	
17 Choking acts as song	mating
18 Upward " choking " at the end of meeting ceremony	
19 Mud is collected and trampled down to form a nest platform	nest-building
20 Stealing of nest material	
21 Guarding the empty nest	
22 Incomplete regurgitation	nest sanitation
23 Parents lack food-call	parent-chick
24 Chicks lack " pumping "	relationships
25 Parents do not know chicks individually	

9.3 Adaptations in the Kittiwake to cliff nesting: a chart constructed by Niko Tinbergen, based on the researches of Esther Cullen. (From Tinbergen, "Behaviour, systematics, and natural selection," *Ibis* 101 [1959]: 324.)

back in the field in earnest. He was building on his earlier experience as a field naturalist, but his horizons were widening. Comparative gull studies were opening up new ways of getting at the topic of behavioral function. This was new intellectual territory. He felt he was pioneering again, not simply serving as a missionary spreading the word about past achievements.

Tinbergen summarized Cullen's findings in a 1959 paper of his own in the *Ibis*. He rearranged her list of kittiwake adaptations slightly to present twenty-five "peculiarities of the Kittiwake, in which this species differs from the other gulls."[27] These he presented in the table shown in figure 9.3.

Tinbergen offered the kittiwake as a prime example not only of adaptive radiation but also of the interrelatedness and coherence of a species' adaptive characters within a whole adaptive system. If some of the peculiarities of the kittiwake did not seem adaptive at first glance, he said, a consideration of their functional contexts made their significance clear. This was true of all twenty-five of the distinctive kittiwake characters identified by Esther Cullen. Together they constituted a comprehensive adaptive system. Thinking about adapted features as *systems* led in turn, Tinbergen said, to recognizing that "selection pressures must often be in conflict with each other," leading inevitably to different sorts of "compromises." Characters that by themselves appeared to look like the results of random change proved on closer examination to show the indirect effects of system-wide selection.[28]

In Tinbergen's view, these conclusions bore not only on the evolution of behavioral, morphological, and physiological characters. They also had implications for scientific theory and practice. The study of function was "forcing the behaviour student to an attitude of healthy respect for the all-pervading power of selection." This, furthermore, was something the researcher in the field could see much better than the taxonomist back in the museum. Not being in such close touch with living animals, taxonomists were more disposed "to attribute characters to random change." Tinbergen predicted that disputes between the student of behavior and the museum taxonomist would fade if it were agreed that the real aim of comparative studies was not to classify animals but to arrive at "a better description of how and why species have diverged the way they have."[29]

At the same time that Tinbergen underscored the advantages the student of behavior has over the museum taxonomist, he also distinguished among different approaches to the study of behavior. Field observers, he maintained, were in a position to discover things that observers of animals in captivity were not. While the ethologists had often distinguished their own work from that of the American comparative psychologists with the too narrow focus of the latter on the laboratory rat, Tinbergen in this instance was thinking about the differences between his own work and Lorenz's.

As we saw in an earlier chapter, Lorenz's theorizing was informed by (and in turn was used to justify) his practice of watching tame, free-flying birds at and around his home research station in Altenberg. In promoting his work in the 1930s, he argued that when it comes to identifying homologies and reconstructing phylogenies, behavioral characters are often more reliable than morphological ones. He believed this to be particularly true of the special displays that serve as releasers of the fixed action patterns of members of the same species. These displays, in his view, needed to be understood above all as *historical* arrangements within the species. They were not closely linked to the ecology of the species, they thus were not subject to convergence, and this made them exceptionally valuable for taxonomic purposes.

Tinbergen by the late 1950s had come to see the situation differently. His theorizing, like Lorenz's, was informed by (and similarly used to justify) his own particular scientific practices. For him, fieldwork was the key to unlocking special insights about behavioral evolution. Fieldwork made it clear that the displays of the kittiwake are not arbitrary conventions but instead are intimately related to the species' cliff-dwelling habit. Fieldwork showed also that in similar functional contexts, convergences are indeed possible.

Tinbergen discussed behavioral convergences further in a major 1959 progress report on comparative studies of gull behavior. There he explained that a gull has only a limited number of movements it can make to signal its likelihood of attacking an opponent or retreating from it. All of the bird's postures associated with pecking (attacking) involve pointing the bill at the opponent. Conversely, all of the bird's threat or appeasement postures involve doing something that is the opposite of this, for which there are only a limited number of possibilities available. A bird can turn its head away horizontally, turn its head upward in the vertical plane, turn its head downward until its beak pointed backward, or pull its head and bill back while still facing the opponent. Given this restricted number of alternatives, Tinbergen declared, it is not surprising to find similar displays in very different species. "Bill-up" postures, for example, are found in terns, gannets, and the great tit. Cases of convergence such as these, Tinbergen allowed, called into question Lorenz's claims about the conservatism (and nonconvergence) of displays.[30]

Pleased as he was with how much the comparative study of behavior patterns seemed to illuminate about behavioral evolution, Tinbergen recognized that comparative work was only a first step. It had to be combined, he said, with studies of behavioral causation and function. As he and Robert Hinde put it in a coauthored paper of 1958: "The fertility of the comparative method is . . . enormously enhanced when it is coupled with studies of function and causation. These enable us to distinguish between homology and convergence, give us insight into the origin and later adaptation of 'derived' movements, and permit a more accurate description of the true innate differences between species." But if comparative study was an indispensable phase of research, it could not by itself get at "the ultimate problem" of "the dynamics of behavioral evolution."[31] "Mere observation," as Tinbergen had put it two years earlier, could only give "tentative conclusions" when it came to studying survival value."[32] Tinbergen had come to regard the *experimental* study of behavioral function as a necessary component of his group's research.

9.4 A comparison of the meeting ceremonies of the black-headed gull (*left*), the little gull (*center*), and the kittiwake (*right*). (From Tinbergen, "Comparative studies of the behaviour of gulls [Laridae]: a progress report," *Behaviour* 15 [1959]: 1–70.)

TINBERGEN AND THE FURTHER
ANALYSIS OF BEHAVIORAL FUNCTION

Tinbergen had given some attention to problems of behavioral function early in his career, most notably in his prewar study of the functions of sexual fighting, territory, and song in the snow bunting.[33] After the war, he began studying experimentally how color patterns function as camouflage, signals, or means of defense against predators. In Holland with L. de Ruiter he studied how countershading in caterpillars and the camouflaged character of stick insects helped the organisms escape discovery by jays.[34] Soon afterward, at Oxford, he directed David Blest's studies of how eyespot patterns in butterflies and moths intimidate would-be predators.[35] In 1955 Tinbergen teamed up with Bernard Kettlewell to demonstrate the importance of predation in the wild. Kettlewell needed support in countering an ornithologist who expressed doubt that birds could eat the number of moths suggested by Kettlewell's studies. Tinbergen spent three weeks with Kettlewell out in the field filming birds capturing moths.[36] Tinbergen also collaborated with R. Hoogland and Desmond Morris in an analysis of the way stickleback spines served as a defense against predators.[37] All of this work served in effect as background for Tinbergen's team's fieldwork on black-headed gulls. It was there that Tinbergen's study of behavioral function reached its fullest development.

From the first initiatives in 1951 on into the 1960s, the black-headed gull work of Tinbergen's research team reflected the close interplay between the annual rhythms of the animal subject and the emergent intellectual aims of the naturalists. The behavior of the gulls posed questions that led the researchers to new projects, and the new projects generated additional questions. In June 1958 Tinbergen wrote to Julian Huxley describing the multiplicity of projects in which his students were engaged. Colin Beer, Tinbergen said, had unraveled nicely for the black-headed gull "the internal fluctuation of the willingness to (a) brood on eggs, (b) roll in an egg on the nest's rim, and (c) carrying away the egg shell when the chick has hatched." In the meantime the Oxford team's comparative work was illuminating the differences in "choking" as it appeared in the kittiwake, the black-headed gull, and other species. On this score Tinbergen reported: "This is only one of the many occasions where we begin to see how even signal movements can diverge in speciation because of their dependence on other parts of the total behaviour pattern and not because of *direct* pressure on them themselves. It is curious how, for a better understanding of these things, one should always try and understand as much as possible about causation and function as well as about the way selection could press directly." [38]

Encouraged by Huxley to propose projects for which special funding might be found, Tinbergen explained his team's gull studies as follows: "Various members of our group (M. Moynihan, E. Cullen, G. Manley, C. Beer, R. Weidmann, U. Weidmann) have studied the behaviour of a number of representatives of the large gulls (Herring Gull and relatives); the hooded gulls (Black-headed Gull and relatives); and the Kittiwake. We would like to continue these studies on as broad a basis as possible, that is: investigating more species, studying the total behaviour pattern of each species, and giving equal attention to problems of causation, function and evolution. There is an obvious 'interfertility' between all these aspects: studies on motivation of postures helps in discovering their evolutionary origin; comparison between species throws light on ritualisation; adaptations in one sphere of behaviour may have repercussions on other behaviour elements (see Cullen's paper on the Kittiwake which shows the many ramifying effects of cliff breeding), etc." Tinbergen had not done much work on behavioral development, but he recognized its importance: "Another urgent task would be to study the ontogeny of behaviour by raising

9.5 Niko Tinbergen in 1959. (Photo by Colin Beer. Courtesy of Colin Beer.)

birds in isolation, and by interchanging young of two species, preferably Kittiwake and Black-headed Gull." [39]

Among the many features of the black-headed gull's behavior there was one in particular that Tinbergen found particularly intriguing. This was the bird's removal of eggshells from in or around the nest. Colin Beer attacked the problem from the standpoint of causation. [40] Comparative considerations then provided Tinbergen's group with clues on the survival value of this behavior. The black-headed gull carries shells away from its nest, but the kittiwake does not. This suggested to Tinbergen and his co-workers that the function of eggshell removal was unlikely to be that it removed the danger of injury from the shell itself (from the shell's sharp edges, for example, or from bacterial infection). It appeared more likely that eggshell removal served an antipredator function, specifically a camouflaging of the nest and brood.

Tinbergen's group designed experiments to gain a closer understanding of predation in the black-headed gull colony. They tested the role of egg color as camouflage. They also tested the importance of eggshell removal in protecting the brood. Tinbergen's letters from the field testify to how he pleased he was with the work. In April 1960 he described the results of the eggshell removal studies to Ernst Mayr:

> We have now proved, with mass field tests with the natural predators Carrion Crows and Herring Gulls, that the near presence (within 4 inches) of an egg shell greatly endangers the other eggs (and by implication the chicks), and we have indications, which we hope to follow up this season too, that freshly hatched egg shells attract mammalian predators by scent. The combination of an analysis of the stimulus situation and other aspects of causation, and experiments about the survival value begins to make a fascinating story. What astonished me most was the tremendous level of predation one can observe. . . . All our tests point to egg shell removal as a vital corollary to the camouflage of the brood.[41]

Ten days earlier Tinbergen had told Huxley about the camouflage studies:

> We put out equal numbers of 2 kinds of differently coloured eggs and watch predators take them. Unexpected differences between carrion crows, Herring Gulls and Bl.[ack] h.[eaded] gulls as predators! Crows see camouflaged eggs *much* more readily than H.[erring] Gulls. Black-headed gulls are uninterested until an egg has broken; Herring Gulls persistently follow crows about; and when crows find something they join them. All show the most amazing (and crows also precise) location-conditioning.[42]

Among the many things that Tinbergen's group observed over the course of the 1959 and 1960 seasons at Ravenglass was that eggshell removal did not take place immediately after a chick hatched. The naturalists identified a reason for this apparent delay: leaving newly hatched chicks alone exposed them to the danger of being eaten by neighboring gulls. Tinbergen's team reported: "In both years there were a number of Black-headed Gulls in the colony which preyed selectively on nearly hatched eggs and on wet chicks. Although we are certain that not all gulls engage in this 'cannibalism,' this type of predation is very common, particularly towards the end of the season." The phenomenon had actually thwarted some of the researchers' attempts to study ontogeny. As Tinbergen and his co-workers put it, "many of our efforts to observe the development of the be-haviour during the first few hours of a chick's life (which were usually done late in the season) were time and again frustrated by the wet chicks' being snatched away by such robber gulls immediately after hatching."[43]

With this discovery Tinbergen and his co-workers identified an addi-tional function of the gulls' territoriality: "reducing the likelihood of pre-dation by neighboring gulls." This in turn led to an appreciation of "the compromise character of colony density." Colonial nesting, Tinbergen ex-plained, has the benefit of making it possible for the gulls to attack en masse predators like the carrion crow. Simultaneously, spacing out of breed-ing pairs (achieved by territorial hostility) has the dual advantages of dis-couraging predators (by increasing their search time between finds) and reducing the likelihood of predation by neighboring gulls. The picture that emerged from the study of the gulls' antipredator system was "one of great complexity and beautiful adaptedness."[44]

The black-headed gull studies reinforced what the kittiwake work had shown earlier, namely, that conflicting selection pressures inevitably led to "compromises" within adaptive systems. The gulls' whole antipredator system represented a balancing of conflicting demands. There were con-flicts between the safety requirements of the parents and those of the chicks. Likewise, there was a tension between the two opposing modes of defense: crowding and spacing out. In addition, different predators elicited from the gulls different, often incongruent antipredator strategies. As Tin-bergen and his co-workers explained: "Herring Gulls and Crows might be prevented entirely from taking eggs and chicks if the gulls stayed on the nests, but this would expose them to the Foxes. While Herring Gulls and Crows exert pressure towards quick egg shell removal, neighboring gulls exert an opposite pressure; the timing of the response is a compromise."[45]

Tinbergen's researches received a substantial boost in 1961 when the Nature Conservancy awarded him a major, multiyear grant. Tinbergen wrote enthusiastically to Huxley about the help the grant was going to provide:

> One day you must come and visit us in Ravenglass! Until now, we worked from year to year, never knowing whether we could continue, but now, with the Nature Conservancy grant starting in October, we are start-ing on a long-range programme: a population study with colour ringed birds, experiments about the threat function of the brown face (follow-ing up what you and James [Fisher] did), experiments on the effect of Facing Away, continuation of the egg shell work, and a programme on

development of behaviour in the chicks. . . . I myself shall be in the camp from March 30 till April 13, from May 12 till May 30 and from June 20 till July 7.[46]

As the work of Tinbergen's team continued on these multiple fronts, Tinbergen came to feel that the work on survival value was what suited him best. He also concluded it was critical that he keep doing such work, because so few other investigators at the time were doing anything like it.[47] When Ernst Mayr in 1963 sent Tinbergen a copy of his magisterial new book, *Animal Species and Evolution,* Tinbergen wrote back saying the gesture encouraged him to continue his pursuit of "questions of survival value and selection pressures."[48] His group's latest work, he reported, was showing that not only eggshell removal and the spacing out of nests but also the synchronization of egg laying serves an antipredator function.[49]

Regarding himself as better at observing and analyzing things than at constructing new theories, Tinbergen allowed to Mayr that his studies of the interaction of selection pressures might not result in "exciting new ideas." He modestly stated, "I just want to understand in as much detail as possible, why my particular animals are as they are." But if his main strength was not in formulating general theories, he remained an able critic of the theories of others, especially when these theories were advanced without adequate empirical or conceptual foundations. He was not persuaded, for example, by V. C. Wynne-Edwards's argument that birds in times of straitened circumstances limit their clutch sizes for "the good of the species." Tinbergen took strong exception to this. He felt that the phenomena Wynne-Edwards described could be adequately accounted for by "selection pressures acting upon the individual and the family."[50]

"ON AIMS AND METHODS OF ETHOLOGY"

As his studies of behavioral function thrived, Tinbergen continued to keep an eye on the development of ethology as a whole. In October of 1962, he wrote confidentially to Otto Koehler to recommend the preparation of a Festschrift for Konrad Lorenz on the occasion of his sixtieth birthday the following year. The following February he announced the subject on which he had decided to write: "After long doubts I now know what I am going to write for Konrad's Festschrift, and I'm now sitting every free hour at the typewriter. . . . I want to write something for Konrad that he likes a lot. I am now writing some thoughts down about (don't panic) 'What is ethology?'"[51]

The result of Tinbergen's efforts was the paper he entitled "On Aims and Methods of Ethology." It was published with the other papers from the Festschrift in the *Zeitschrift für Tierpsychologie* in 1963. It has served ever since as the standard identification of what the biological study of behavior is (or should be) all about. Ethologists (and other biologists as well) routinely cite it today when they want to identify the range of questions in the biologist's purview.

What is routinely remembered about Tinbergen's paper is that it was there that he defined ethology as "the biology of behavior" and then went on to explain that to study behavior biologically is to ask four distinct kinds of questions about it: (1) What is its physiological causation? (2) What is its function or survival value? (3) How has it evolved over time? (4) How has it developed in the individual?[52] Less often remembered about Tinbergen's 1963 paper is that he also offered a historical thesis in it. His thesis was that Lorenz deserved the title of "father of modern ethology" because he was the man who "made us look at behaviour through the eyes of biologists." This, Tinbergen said, and not the great mass of novel facts that Lorenz had discovered, was what constituted Lorenz's greatest contribution to the study of behavior.

Praise flows freely in commemorative collections like Festschriften, and Tinbergen himself was a generous man, at times to the point of being self-effacing. In this case, in crediting Lorenz with bringing biological questions to bear on the study of animal behavior, Tinbergen was assigning Lorenz the role that many ethologists today might be more likely to ascribe to Tinbergen himself. We need to look more closely at what Tinbergen had to say in his 1963 paper.

The case for calling Lorenz "the father of ethology" was certainly clear enough. Lorenz had provided the field with its primary theoretical foundations back in the 1930s. He had called attention to (and made sense of) a host of behavioral phenomena, many of them essentially new to science. Furthermore, he was keen to insist that biological questions be brought to the forefront of studies of animal behavior. This was precisely the theme he chose in 1937 when he addressed the first public meeting of the new German Society for Animal Psychology.

Tinbergen's praise for Lorenz was both genuine and appropriate. At the same time, however, Tinbergen believed that the job of making ethology a truly unified biological science was not yet fully accomplished. What he described as Lorenz's main contribution to ethology was what he also regarded as ethology's greatest continuing challenge. Privately, he was not sure that Lorenz fully appreciated all the dimensions of this challenge.

Tinbergen's task in writing the Festschrift paper was made more delicate by the fact that Lorenz was angry with him. Tinbergen had recently disagreed with Lorenz on a number of points and in particular on the subjects of "innate behavior" and behavioral development. Lorenz did not take the criticism well. Tinbergen may have hoped that the many positive things he said about Lorenz in the Festschrift paper would in some measure mollify his friend. On the other hand, he felt obliged to spell out his own view of the best course for ethology's future. He also was not prepared to back down on the points regarding behavioral development where he thought Lorenz was clinging to outmoded ideas.

Before looking further at Tinbergen's "Aims and Methods of Ethology," let us examine the conceptual rift that had emerged between the two ethologists. In 1951, in his book *The Study of Instinct,* Tinbergen had been content to identify his subject as "innate behavior." This he defined as "behaviour that has not been changed by learning processes."[53] He continued to use the phrase "innate behaviour" in 1953 in his book *The Herring Gull's World.* In 1954, however, he took to heart Daniel Lehrman's claim that ethologists had been too uncritical in their thinking and writing about "instincts" and "innate behavior." At the Macy conference that year Tinbergen granted that the ethologists' use of the word "innate" had been too sweeping, and, furthermore, that it had worked against a more systematic and analytical investigation of development. Three years later Tinbergen (along with Hinde, Baerends, van Iersel, and several other ethologists, but not Lorenz), spent a month with Lehrman and other American comparative psychologists at a conference organized by Frank Beach in Palo Alto, California.[54] The ongoing interaction with Lehrman in particular led Tinbergen himself to stress how important it was to think carefully about the problem of development.

Be that as it may, most of the work of Tinbergen's research group was aimed at issues of causation, evolution, and function, not at issues of development. The Cambridge ethologists paid much more attention to issues of behavioral development. When Tinbergen's group took up the question of how the oystercatcher learns to feed, their emphasis was less on matters of theory than on what they could learn from careful observation. In 1963 Tinbergen felt justified in writing: "Systematic descriptions of behaviour ontogeny are still rare and fragmentary."[55]

Lorenz's response to Lehrman's critique was very different from Tinbergen's. He had a much greater investment than Tinbergen did in the concept of innate behavior patterns. Where Tinbergen acknowledged that at least some of Lehrman's critique had hit its mark, Lorenz dug in his heels and constructed a counterattack. This took the form of a lengthy mono-graph, published in German in 1961 in the *Zeitschrift für Tierpsychologie*, Lorenz entitled it "Phylogenetische Anpassung und adaptive Modifikation des Verhaltens" ("Phylogenetic Adaptation and Adaptive Modification of Behavior").[56] The original German version was subsequently revised and published in English in 1965 in a volume entitled *Evolution and Modification of Behavior*.[57]

Lorenz did not construct his countercritique with only Daniel Lehr-man and the American psychologists in mind. He complained also that Tinbergen and other "English-speaking ethologists" had either stopped us-ing the word "innate" or had decided to use it in a very restricted way. To Lorenz's mind, these "ethologists writing in English" had granted too much to their American critics. The concept of the "innate," as he saw it, remained valid, as did the deprivation experiment that helped one identify what was innate. Lorenz likewise defended his view that instinctive and learned components of behavior "come in chunks" that can be clearly sep-arated from each other.[58]

Lorenz argued that there are basically just two ways that an organism gains from its environment the kinds of information it needs to survive. The first is through evolution. The second is through individual develop-ment. In the first case, he explained, "it is the species which, by means of mutation and selection, achieves adaptedness insuring survival." In the second case, "it is the interaction between the individual and its surround-ings" that provides the information on the environment that is then incor-porated into the living organism.[59]

Rather than concluding that "learning" has a role to play in "every phylogenetically adapted behavior mechanism" (a view he ascribed to the American psychologists), Lorenz turned the tables—at least to his own satisfaction—on his critics. As he saw it, an organism's ability to learn cer-tain things is itself a function of mechanisms "built into the organic system in the course of its evolution."[60] He argued that "learning, like any other organic function regularly achieving survival value, is performed by or-ganic structures evolved in the course of phylogeny under the selection pressure of just that survival value."[61] Several years earlier, in responding to Lehrman, Lorenz had sarcastically observed that Lehrman's views re-quired one to postulate an "innate schoolmarm" that guided what an ani-mal could learn. The more Lorenz subsequently thought about this idea, the better he liked it. Innate mechanisms determined the kinds of learning possible in a species.[62]

Upon reading the first, German version of Lorenz's monograph in

January 1961, Tinbergen wrote Lorenz a long, four-page, single-spaced letter. He told Lorenz he was worried, because "for the first time in our life we are really of a different opinion!"[63] He was perfectly happy, he said, with Lorenz's formulation that phylogeny and ontogeny were two different modes of getting information into the organism. He liked also the way Lorenz had stressed "that 'predispositions to learn' special things are built-in" to different species. It was "easy to see," Tinbergen allowed, "why some behaviours must be largely built-in, and others must have the chance of individual adjustment: a small butterfly cannot afford to have to practice flight, and a young song bird cannot afford to have a very sharply defined IRM for food; its initial 'open-mindedness' and environment-controlled specialisation allows it to live in many more habitats than it could otherwise do."

But how was one to study how developmental change is achieved? On the subject of what the deprivation experiment could tell the experimenter, Tinbergen found his own position "diametrically opposed" to Lorenz's. As Tinbergen saw it, raising an organism in isolation from some normal factor in its environment (parents or nestmates, for example) and then finding that the organism developed a behavior independent of that factor's presence or absence, was not sufficient to justify saying that the behavior was innate. He willingly granted to Lorenz that many American psychologists had overemphasized the importance of learning or environmental induction in behavior. He also allowed that the ethologists' early emphasis on "innate" behaviour had been "a (healthy) reaction" to the psychologists' position. He was convinced, however, that this emphasis on "innate" behavior had outlived its usefulness. One could legitimately use the words "innate" and "learnt" when it came to talking about *processes,* he said, but not in talking about *characters.*

The problem, Tinbergen insisted to Lorenz, was not just semantic: "All this would be merely squibbling about words, if it had not an effect on research. It cannot be denied that our former attitude made us far too glib in assuming that species-specific behaviour characters were innate, and that systematic ontogeny studies began only under the spur of American criticism—your imprinting work being rather an exception." Much more work needed to be done, Tinbergen declared, in order to know how the "machinery of behavior" changed in the course of development. He suggested that a solution to their present differences of opinion might be found if the two of them recast their discussions about "innate behavior" in terms of survival value. "What we call 'innate' behaviour elements," he proposed, "are those that have to be 'ready for use' when first used."

Tinbergen concluded his letter to Lorenz with a paragraph that says much about the personalities of the two men: "Does it hit you hard when I move in to the attack? Or do you think I am just being dim? I feel so *convinced* I am (more or less) right! And you know, Erich [von Holst] is wrong when he says that I am no fighter and that I always 'give in.' If this were so, I would give in to you rather than to anybody else. I hope for a vigorous reply!"

It may be that Lorenz gave Tinbergen a vigorous reply, but the Tinbergen Papers at Oxford show no sign of it. Long after the fact, indeed a full decade and a half later, Tinbergen acknowledged to Ernst Mayr that there was a "sad side" to his friendship with Lorenz: "I know that I hurt Konrad whenever I disagree. Twice in the past I have criticised him on specific points, and both times he was really peeved."[64]

When it came to writing his paper for Lorenz's 1963 Festschrift, Tinbergen did not want to make his friend any unhappier. On the other hand, his preeminent concern was to ensure for ethology a healthy future. His biggest worry was that the field might split up "into seemingly unrelated sub-sciences" or become "an isolated '-ism.'"

The best way to guard against such dangers, Tinbergen believed, was to be clear about the kinds of questions one should be asking. Julian Huxley, Tinbergen explained, "likes to speak of 'the three major problems of biology': that of *causation,* that of *survival* value, and that of *evolution.*" Tinbergen added a fourth problem to Huxley's list, that of *ontogeny.* "There is, of course," he wrote, "overlap between the fields covered by these questions, yet I believe with Huxley that it is useful both to distinguish between them and to insist that a comprehensive, coherent science of Ethology has to give equal attention to each of them and to their integration."[65]

Tinbergen stressed the importance of maintaining other balances as well. There needed to be a balance, he said, between description and analysis. There also needed to be a balance in the kinds of people attracted to the field.

The heart of Tinbergen's paper was a sequential discussion of the problems of causation, function, ontogeny, and evolution. His observations on the study of causation gave due credit to Lorenz's pathbreaking attention to species-specific behavior patterns and to Lorenz's emphasis upon "internal causal factors" in the control of behavior. He acknowledged that the models he and Lorenz had constructed in their attempts to "physiologize" their thinking had been oversimplified. Neither Lorenz's psycho-hydraulic model nor his own model of the hierarchical organization of behavior systems had made any provision for negative feedback. He was happy to report, however, that new studies were advancing the field and that the boundaries between neurophysiology, physiological psychology, and ethology were disappearing. It appeared to him that a generalized "physiology of behavior" was becoming the domain for treating the broad problem of behavioral causation.[66]

Twelve years earlier in *The Study of Instinct* Tinbergen had given the lion's share of his attention to questions of causation. Behavioral causation studies continued to dominate the field, but Tinbergen was now more interested in redressing the imbalance than mirroring it. He thus devoted less space to work on causation and more to studies of survival value. By his account, studies of the function of behavior had suffered from relative neglect for quite some time. Facile hypotheses about the survival value of particular structures and behaviors, offered in the years after Darwin, had created a backlash, and questions of survival value tended to disappear altogether from the biologist's purview. Lorenz, said Tinbergen, was one of the few recent biologists to take an interest in survival value, and this indeed was one of the things that made his work so appealing to many naturalists. Naturalists liked what Lorenz had to say because they were "people who saw the whole animal in action in its natural surroundings, and who could not help seeing that every animal has to cope in numerous ways with a hostile, or at least uncooperative environment." Tinbergen noted in this regard that it was "regrettable" that Lorenz's new institute was named the Institut für Verhaltensphysiologie (Institute for Behavioral Physiology), since Lorenz's work and that of his institute went well beyond questions of physiology.

What Tinbergen refrained from mentioning was that Lorenz himself had never taken the study of survival value into the field. Lorenz had not concerned himself with the multiple ways in which animals cope with hostile environments. He had not displayed Tinbergen's sensitivity to matters of ecological context and to the possibility of specific behavior patterns aiding some members of the species and not others. To the contrary, he had spoken in the most general sort of terms about behavior patterns working "for the good of the species." But Tinbergen recognized that this was not the time or place to criticize Lorenz's understanding of evolution. Lorenz had promoted the idea of thinking about the survival value of specific instinctive behavior patterns, and Tinbergen was happy to credit him with having done so.

That said, the topic of survival value remained, in Tinbergen's view, seriously underdeveloped. Many doubted, in fact, that the topic was amenable to scientific study. Tinbergen argued to the contrary that the survival value of specific behavior patterns could be investigated not only by com-

paring the differences between closely related species but also by conducting experiments in the field. First, however, one needed to be aware that a
particular behavior pattern might have some sort of survival value. As he
put it, "It took me ten years of observation to realize that the removal of the
empty eggshell after hatching, which I had known all along the Black-
headed gulls to do, might have a definite function, and that even the length
of delay of the response, which varies with the circumstances . . . may be
adaptive." After one suspected a particular behavior pattern had a function, then one could devise experiments to test the hypothesis. Tinbergen
acknowledged fully the difficulties of constructing experiments in natural
situations. At the same time, he insisted on the value of studying survival
value as doggedly as possible. Only in this way could one gain a clearer
sense of the role of natural selection in evolution, and, more specifically, of
how animals survive in the environments in which they live.

Having dealt with behavioral function, Tinbergen moved on to the
subject of behavioral development. Here was an area, he admitted, where
there had been (and continued to be) "a real clash of opinion." Ethologists and psychologists had come to the subject of ontogeny from different angles, and different understandings of words such as "innate," "acquired," "instinct," and "learning" seemed only to complicate matters
further. He was prepared to admit that the ethologists in the past had been
insufficiently critical in the way they had identified behavior patterns as
"innate." He now believed it was not enough to eliminate one or several
possible environmental influences on behavioral development and then, if
the behavior manifested itself anyway, declare the behavior to be "innate."
Deprivation experiments could show which environmental influences
were *not* significant for the development of a behavior pattern. They could
not show that the pattern was "innate."

Pursuing this line of thought further, Tinbergen said, would mean "I
should have to cross swords with my friend Konrad Lorenz himself—both
a pleasure and a serious task requiring the most thorough preparation."
This, however, said Tinbergen, was not the occasion to do so. (He did not
acknowledge that he had already tried this with Lorenz in private and
things had not gone very well.) He concluded this section of his paper with
the claim that Lorenz, by insisting on a thoroughgoing biological approach, had influenced the study of behavioral ontogeny just as much as he
had influenced the other aspects of ethology, "even though his evaluation
of the part played by internal determinants may have been on the optimistic side."

The final question Tinbergen treated in his paper was that of behavioral evolution. He credited Lorenz here with promoting systematic comparative studies of behavior and appreciating that behavior patterns could, for purposes of systematics, be treated as "organs." He then described work that served to illuminate the dynamics as well as the history of evolutionary change. He reiterated what he had said in earlier papers about the diversity of the selection pressures operating on animals and how "the animal that survives best must be a compromise." He admitted that studies of selection pressures in the present could not demonstrate definitively the specific results for which selection was responsible in the past. He insisted, nonetheless, that ethology had a special contribution to make to understanding how evolution worked, and thus to biological thought in general.[67]

Summing up his paper, Tinbergen returned to the question of Lorenz's contribution to modern ethology. He proposed that Lorenz's major contribution was "his insistence that behaviour phenomena can, and indeed must, be studied in fundamentally the same way as other biological phenomena." As he put it: "The central point in Lorenz's life work . . . seems to me his clear recognition that behaviour is part and parcel of the adaptive equipment of animals; that, as such, its short-term causation can be studied in fundamentally the same way as that of other life processes; that its survival value can be studied just as systematically as its causation; that the study of its ontogeny is similar to that of the ontogeny of structure; and that the study of its evolution likewise follows the same lines as that of the evolution of form."[68]

Tinbergen had used the occasion of the Lorenz Festschrift to honor Lorenz's past contributions. At the same time he had crafted a vision for ethology's future, a vision of coordinated, balanced development encompassing all the fundamental questions of the life sciences as these applied to animal behavior.[69]

CULTIVATING THE FIELD

Were these in fact all the basic questions one might want to ask about animal behavior? Could one ask, after all, about an animal's subjective experience?[70] Tinbergen never thought the question belonged in the science of ethology. He strongly opposed devoting attention to questions of animal subjectivity, consciousness, and cognition because he thought they could not be studied objectively. He stated this explicitly in *The Study of Instinct,* and he did not change his position for the rest of his career, although colleagues such as Otto Koehler and Julian Huxley, questioned or challenged him on the matter. "No compromise possible!" he exclaimed to Koehler in

1951. He stressed that his methodological focus had "nothing to do with whether I think that animals experience something. I do think that, but it is in my opinion totally irrelevant. Experiences are not perceivable, and thereby not usable in animal ethology."[71]

In response to one of Huxley's later attempts to convert him, Tinbergen wrote back saying that "attempting to find out anything about subjective phenomena in animals" had not been a prerequisite for him when it came to making progress "with what I really wanted to understand about animal behaviour." He explained further: "It is so to speak enough of a job for me to try and understand the machinery of behaviour in the very same way any other life process is studied biologically (with reference to underlying causation, function, ontogeny and evolution), and I find it an extremely satisfying job; I feel that we are getting somewhere. Any attempt at synthesizing the data I find in this way with subjective phenomena in other subjects has always hindered my work." Having typed out this response to Huxley, Tinbergen evidently felt he had stated things too one-sidedly, for he added by hand: "Perhaps you agree with Bierens de Haan who once said: you ethologists are like people who look at a painting with colour filters in front of your eyes so as to deliberately to miss the most essential thing: colour."[72]

At other times Tinbergen's tone was not so conciliatory. At the Darwin centennial symposium at the University of Chicago in November 1959, on a panel with Huxley, Tinbergen spoke again of his commitment to the objective study of behavior as "a matter of principle with no compromise possible." He commented further: "Some may say our view is very narrow. All right, it is narrow; but we feel we must recognize that science *is* a limited occupation and is only one way of meeting nature."[73]

Huxley and Tinbergen continued to bat the question back and forth. In 1962 Huxley told Tinbergen: "As regards psychometabolism I think we are at semantic cross purposes. I am quite sure you are right in thinking that the actual technique of investigating animal behaviour must be a 'behaviourist' one; on the other hand, the recognition that their behaviour must be accompanied by subjective awareness may often suggest subjects for investigation, and in any case is in my opinion quite essential for the interpretation of the results in terms of any biological theory."[74] When Huxley brought the subject up again in 1965, Tinbergen replied: "Whether or not one can deduce anything useful about the subjective phenomena going on inside an animal—here we shall never agree. My conviction remains that our inclination to try and feel and say something about these things is one of the most serious obstacles to progress in human psychology!"[75]

Tinbergen in the mid-1960s had many things to think about besides animal subjective experience. He continued to direct his research group's multipronged researches on gull behavior. He also devoted considerable energy to making films about animal behavior. In addition, he was paying increasing attention to what ethology could contribute to the study of human behavior. He considered the matter sufficiently important that he gave time and energy to an extended campaign to establish at Oxford a course on the human sciences.[76] Another of his concerns was helping to establish the Serengeti Research Institute. In January of 1965, upon returning from six weeks in the Serengeti, he wrote enthusiastically to Huxley about the experience:

> Here is a unique set-up: a marvelous area for studies on population dynamics and behavioural ecology of the plains game and its many predators; a very competent team of young, devoted zoologists; an extremely knowledgeable leader, Phil Glover, who is himself an excellent plant- and "range"-ecologist; and an inspired, very active, and highly competent director of the National Parks, John Owen. . . . The research that is in progress seems to me pretty well unique in that it is really aiming at essential problems: how are numbers controlled; how does the behaviour of the many species make them fit into their particular niches, what habitats do they need in the different seasons, and why; what limits their success and keeps them all in balance; how will the habitat have to be managed, etc.

It seemed to Tinbergen that the research project stood to benefit simultaneously science, Tanzania, and conservation.[77]

Judging from the quality and importance of his "Aims and Methods" paper of 1963 and the range of activities Tinbergen pursued to good effect in the 1960s, one could well imagine that this was the period when Tinbergen reached the height of his abilities. If he did, he was not able to enjoy it for long. In April 1965, feeling he was on the verge of a nervous breakdown, he begged off from participating in a major symposium on behavioral ritualization that Julian Huxley was organizing under the auspices of the Royal Society.[78] The doctors diagnosed him as having "some unknown virus infection of the central nervous system."[79] His symptoms, as he later looked back on them, included "a curiously rhythmic recurrence of periods of pretty bad depressions." These tended to appear when he came under "even relatively light stress."[80] He tried to cut back on commitments. He continued only those activities he viewed as absolutely essential, such as providing guidance to his students in their researches. As it was, the students were receiving excellent guidance from Mike Cullen, who since the mid-1950s had acted as Tinbergen's "right-hand man" in the Oxford Animal Behaviour Research Group and who in fact was able to help the students with the quantitative analysis of data at a level well beyond Tinbergen's ken.[81] Working with them provided him a stimulation that enabled him to "rally more or less," at least enough to keep his research group going.[82]

The Fourteenth International Ornithological Congress was held in Oxford in July 1966. David Lack served as president of the congress, Tinbergen as secretary-general. By Lack's account Tinbergen had "the main responsibility" for the success of the congress, and he worked extremely hard to see that things went well. Lack recognized the strain this put on his friend and colleague: "Indeed at one time the burden seemed so heavy that I wondered whether it could be right for us, once every four years, to use a leading scientist for this post, to the detriment of his own research and hence of ornithological progress."[83] Tinbergen suffered both before and during the meeting, but he managed to present a masterly paper nonetheless. Designed as a companion piece to Lack's presidential address on interrelationships in breeding adaptations in marine birds, he spoke on the adaptive features of the black-headed gull.[84]

Tinbergen used his research group's gull work to illuminate how experiments as well as observations could be brought to bear on the effects of natural selection. His group's many discoveries gave him a wide range of examples for discussion. Ian J. Patterson had found that house-hunting pairs of black-headed gulls try to nest in the densest part of the colony. Hans Kruuk had established that outlying nests, because of predation, are subjected to the highest mortality. Other work showed that the coloration of the eggs and young acts as camouflage, and that the territorial spacing out of broods played a role in camouflaging them. Bob Mash's work revealed how the brown mask of the black-headed gull served to intimidate territorial rivals. And Patterson too was studying the deterring effect of displays. This work showed how the signaling system of the birds effects the spacing out of pairs. Taken altogether, these studies went a long way toward demonstrating the pressure favoring colonialism, the counterpressure promoting spacing out, and the mechanism by which spacing out is achieved.

Tinbergen went on to describe additional work on the bird's courtship (by Martin Moynihan and Gilbert Manley) and on the synchronization of the breeding cycle (by Colin Beer and Ian J. Patterson). The whole of this work illustrated the "astonishing finesse" in the ways birds were adapted to their environments. It reinforced all that Tinbergen had been saying for a decade about the way an organism's features should not be considered in isolation but needed to be understood as "parts of larger systems." Pressures in one functional sphere could lead to secondary adjustments in another. Conflicting pressures could lead to compromises. There could also be selection for interspecific distinctiveness.

Tinbergen found lessons in this for theory and practice alike. "If our field studies have convinced me of one thing," he wrote, "it is of the fact that the imagination of even the best field biologist falls far short of the reality; what has been found so far about the nature of selection pressures ought to make us realize how little we know of their true nature and of their variety." Once again he criticized Wynne-Edwards's argument for group selection. Much more needed to be learned, he said, before one had "more than an inkling of the extent and the details of adaptedness."

Tinbergen ended his paper with a general plea for more fieldwork and more studies of behavioral function. "Field craft," he felt, had "atrophied alarmingly even among biologists" and was "in urgent need of redevelopment." As for the balance of studies in his own field, he wrote: "I submit that a biological science which gives all its energies to the analysis of causal mechanisms underlying life processes and neglects to study, with equal thoroughness, how these mechanisms allow the animals to maintain themselves, is a deplorably lop-sided Biology."[85]

A little more than five months after the ornithological congress, Tinbergen felt "back to normal," full of energy, and mentally alert. He wrote Mayr, "I feel like having woken up from a years-long awful dream." In the flush of recovered health, he even worried about overtaxing his research group with an "explosion of activity." He also took Mayr to task for not showing as much interest in Tinbergen's studies of survival value as Tinbergen thought that he should."[86]

Mayr expressed pleasure at his old friend's restored health. He confessed that he took natural selection so much for granted that new evidence for selection's effectiveness did not particularly excite him. He agreed, nonetheless, that "the ecology of behavior" represented "one of the great frontiers." He also identified two other behavioral problems needing attention. The first was the variation of behavior within populations. The second was behavioral "maturation."[87]

Tinbergen wrote back immediately, describing in detail what he saw as the course of action before him. He had "to proceed on many fronts," he said, to keep ethology developing into a "respectable" science and to prevent it from deteriorating into a narrow "ism." He planned to promote his

9.6 Tinbergen in the field. (Photo by Lary Shaffer. Courtesy of Lary Shaffer.)

"own top-priority line," namely, behavioral ecology. As for behavior genetics, he intended to leave that to his "very able co-worker, Stella Cross-
ley, and to the Mannings [Aubrey Manning and Margaret Bastock]." He
would need also, he said, to continue with "the building of bridges to
neuro- and sensory physiology, and equally if not more important, the
propagating of applying ethological methods rather than results to the
study of Man." All this could be achieved, he felt, only by training "the best
possible youngsters," not as a self-contained group but instead within the
context of a full-fledged zoology department that was "balanced and
many-sided" itself.

He then continued his discussion with Mayr about behavioral ecology:

As you say, the work done so far on this behavioural ecology, and the
analysis of environmental pressures, is slight. But that is mainly because

so few modern ethologists are interested in it—they all turn to the more physiologically oriented aspects. And don't forget: this experimental work is very, very time-consuming! It has to be done well, but, for instance, in order to really get the decisive experiments showing what the brown facial mask of the Black-headed Gull does, and what Facing Away in courtship does, one man had to work for four years. I have merely summarised very briefly the results, but behind many of the short statements in say, my papers on adaptive radiation in gulls, and my paper given at the I.O.C., are often years of jolly difficult work. Yet all this simply has to be done, if we are not to get stuck in, admittedly exciting, hunches such as thrown out by brilliant Konrad. I have had to fight a hard battle to make people see what I am after. I use always your own remark: "we know so little of the natural environment." Well—to increase the knowledge is jolly hard work.[88]

Late in February 1968 Tinbergen delivered his inaugural lecture in a newly created position at Oxford, Professor of Animal Behavior. He chose as his subject "On War and Peace in Animals and Man."[89] This gave him the opportunity to explain publicly, yet diplomatically (or as he so hoped), the ways in which his own approach to the question of animal and human aggression differed from those expressed in two recent, best-selling books. One was Konrad Lorenz's *On Aggression*. The other was Desmond Morris's *The Naked Ape*. Tinbergen's critique of his two friends was not that they were talking, as ethologists, about human behavior. He fully agreed that ethologists needed to bring their insights to bear on serious problems of human behavior, especially the problem of human aggression. The problem with the books in question, he allowed, was they offered as certainties what were "no more than likely guesses."

What ethologists could offer the study of human behavior, Tinbergen urged, was not a set of results but rather a general approach, together with "a little simple common sense, and discipline."[90] The approach he had in mind was nothing less than ethology's "comprehensive, integrated attack" on the four problems of biology. The coordinated effort to understand the causation, development, evolution, and survival value of behavior, he maintained, needed to be brought to bear on human behavior and on specific problems like human aggression.[91]

Tinbergen's inaugural lecture was the kind of measured, thoughtful, and inclusive-of-others sort of presentation one would expect from him. Where many others would have been disposed to see fundamental oppositions, he looked for common ground. He argued, for example, that Robert Hinde and Konrad Lorenz were not really diametrically opposed in their views on the mechanics of aggression, and likewise that Lorenz and the American psychologist T. C. Schneirla each had important points to make with respect to the development of behavior. He argued further that Lorenz's *On Aggression,* despite its overassertiveness, its various factual mistakes, and its opportunities for misunderstanding, "must be taken more seriously as a positive contribution to our problem than many critics have done." He agreed with Lorenz "that elimination, through education, of the internal urge to fight will turn out to be very difficult, if not impossible." Like Lorenz, too, he ended up by suggesting scientific research as one of the best ways of sublimating aggression. In his words, "We scientists will have to sublimate our aggression into an all-out attack on the enemy within. For this the enemy must be recognized for what it is: our unknown selves, or deeper down, our refusal to admit that man is, to himself, unknown."[92]

Lorenz reacted to the lecture sharply. However constructively Tinbergen had tried to phrase his criticisms, however carefully he had tried to explain to Lorenz beforehand the broader context in which he was voicing them (a context which included his unwavering recognition of Lorenz as the founder of the field), the Austrian biologist took things badly. He felt unfairly treated — indeed deserted — on multiple fronts. He wrote Tinbergen a nine-page letter expressing his hurt and his anger, voicing too his general dismay about the directions in which ethology, as it seemed to him, was sliding. He accused Tinbergen of appropriating his ideas without acknowledgment; lumping him unfairly with Desmond Morris; taking his words and putting them into the mouth of the enemy, Schneirla; agreeing with Schneirla that "everything is acquired"; and attempting a synthesis of Lorenz's, Schneirla's, and Hinde's views when no such synthesis was possible. He bristled (as he had done many times before) at the notion, going all the way back to Lehrman's first attack on him, that the Americans had significant things to teach him about ontogeny. "I know of no ethological institute," he insisted, "that occupies itself more with ontogenesis than ours. Eibl does nothing else but study the ontogeny of learning. Helga Fischer, Irene Würdinger, [and] Bob Martin do ontogeny, Schutz occupies himself full time with it. I would like to know if that ant-Ideologue Schneirla has ever studied the ontogenesis of a vertebrate, e.g. a young dog." He warned against any compromise with the opponents. He feared, he said, that the ideas through which he and Tinbergen had advanced the study of animal behavior would "fall into darkness, even become

9.7 In the same period during which Lorenz complained bitterly about the treatment of his ideas by "English-speaking ethologists," he received an abundance of honorary degrees and other prestigious awards (from institutions in Great Britain and the United States as well as in Europe). He is shown here in 1969 being helped with sartorial adjustments by his wife, Gretl, prior to his receipt of Germany's highest scientific award, the Orden pour le Mérite. (Photo by H. Kacher. Copy provided by the Konrad Lorenz Institute for Evolution and Cognition Research.)

extinguished, if one compromises with the representatives of stimulus-response psychology, such as Hinde does." Hinde, he fumed, had "castrated" ethology.[93]

Tinbergen responded to Lorenz with what he identified as "the most serious letter I have ever written to you, and on about which I had to think a long time."[94] Observing that nothing must be allowed to impair "our long-standing and deep friendship," he set about trying to patch things up. He did so by trying not simply to mollify his friend but also to make him understand a variety of things he had never quite grasped. In the course of his letter he underscored Lorenz's leadership role in the field, spelled out why certain tactics that might make sense in Germany were inappropriate for the different mental climates in Britain and America, explained that he (Tinbergen) had not agreed with Schneirla that "everything is acquired," pointed out that people would be more impressed if he displayed independence of thought instead of slavishly following Lorenz's ideas, and told Lorenz other things as well. He was obviously glad he could cite his forthcoming article on the history of ethology, where he reminded the world that Lorenz was the one whose unifying vision and knowledge and intuition had made the field of ethology possible.[95] But he also made clear that Lorenz needed to look at him not as a wayward pupil or son who had deserted his father, but rather as a pupil or son who had grown into a full partner, and whose *differences* with the father in fact enriched the father's contribution.

Tinbergen emphasized how he had now spent nearly twenty years as a missionary for ethology to the English-speaking world, and how he had come to understand, much better than Lorenz could, the differences between German and British attitudes. "You and Erich," he wrote, often told me that I was not aggressive enough [as a tactician and mediator]. But you two really know the Anglo-American mind far less well than I do, and you must believe me when I simply report that you and Erich have applied the wrong, too assertative tactics with respect to this particular job. You cannot know how difficult the struggle has been and still is." With both Bill Thorpe and Robert Hinde, he said, he still had "to convince them that in their attitude they just are too narrow, and do not really bother about survival value and evolution." Earlier in his letter he had also noted with respect to differences between British and German thinking: "Germany still lives in the unhappy aftermath of the Nazi time, when Biology was discredited in the eyes of many people by, (simplified), the nazis preaching not only 'we must recognise the animals in our selves' but 'we must yield to the animal inside.' That, I am sure, is an important difference between the German and the British attitude nowadays: *you* have to define your science in quite a different mental climate than *we* have; many Germans are still traumatised by having been subjected to this misapplication of Biology."

Tinbergen allowed to his friend that he did understand "the heavy burden you must be carrying now; doing the utmost to preach the gospel, again and again, and then feeling that people are hostile, are stupid." Reminding Lorenz "we have to accept that it is the fate of pioneers to be understood," he also told him: "And in a sense I am also a pioneer, even though I play and with complete justification second fiddle to you whom I consider THE Father of Ethology. But please don't be a 'Father figure'—do enjoy honest clashes of opinion as you used to do. You must not let yourself be worn down by the resistance we meet—it's part of our job as missionaries. But we each do it in our own way, and while you may disagree with my tactics, you must not doubt my integrity and sense of fairness to you. I too have to carry the burden of a missionary, and by God I have to fight on many fronts too."

Tinbergen reminded Lorenz of how far they had come since the 1930s, when their careers were just beginning, and how much success they in fact had had. He concluded by saying: "Cheer up Konrad, do believe me in what I say about my motivation, and rest awhile and look back, with the satisfaction that is due to you, on what you have achieved so far. And even so we can truly say: this is only the beginning—the real explosion is still to come." It was a long time, however, before Lorenz responded. He did not write back to Tinbergen for nearly three months, leaving Tinbergen in the meantime to be worried about Lorenz's unhappiness.[96]

The year 1968 was important for Tinbergen not only in terms of his new professorship and his unwished-for clash with Lorenz. It was also a year that saw him achieve real success in presenting his ideas through film. In December of 1968, BBC television aired *Signals for Survival,* a film Tinbergen created in collaboration with the producer Hugh Falkus. The film elucidated the signals by which the interactions of lesser black-backed gulls are regulated. It also expressed Tinbergen's growing concerns about the health of the environment. In 1969 the film won the Prix Italia for best television documentary.[97]

Tinbergen's promotion of the ethological approach and his worries about the deterioration of the environment (and the interrelations of environmental and human problems) found expression again in the new introduction he wrote in 1969 for a reprinting of *The Study of Instinct.* There he argued once more that modern behavioral and biological studies were placing too much emphasis on causation and paying too little attention to questions relating to the ecological and evolutionary dimensions of behavior. He also identified what he believed to be the source of this imbalance. The problem was that the "knowledge of the causes underlying natural events provides us with the power to manipulate these events and bully them into subservience." This desire for power, this "urge to conquer nature," was disrupting the natural world. The goal of science, Tinbergen urged, should be understanding, not simply power. What needed to be studied in the future (and for the future) was "the relation between our behaviour and our environment."[98]

Although he agreed to reissuing *The Study of Instinct,* Tinbergen recognized the book was long out of date. He thought of writing up a new ethology text based on the set of sixteen animal behavior lectures he gave annually to undergraduates at Oxford. Outlines of these lectures remain among his papers at Oxford. They make clear that had he written the book, he would have highlighted once more the importance of the biological approach.[99] But he never finished the book. He ended up devoting himself to

9.8 Tinbergen the filmmaker. Tinbergen's success as a maker of educational nature films reached its peak with his 1968 BBC film, *Signals for Survival,* made with Hugh Falkus. The film won the coveted Italia Prize in 1969 for best television documentary. (Photo by Lary Shaffer. Courtesy of Lary Shaffer.)

other projects, most notably the study of childhood autism, an enterprise he undertook with his wife. At the same time, his health problems returned, preventing him from accomplishing as much as he hoped. At the International Ornithological Congress in Oxford in 1966 he had been elected president of the next congress, scheduled for The Hague in 1970. In the fall of 1969 he resigned the post, citing reasons of health. The following year, he did not attend the meeting.[100]

In March of 1972 Tinbergen traveled again to the United States. He lectured at Harvard, using the occasion as a dress rehearsal for the Croonian Lecture he would give at the Royal Society in May (a lecture entitled "Functional Ethology and the Human Sciences").[101] He also paid a visit to Daniel Lehrman's lab in Newark. It was nearly thirty-four years since Tinbergen's first trip to the United States and his first meeting with Lehrman. Since then, Lehrman had become a major figure among that "group of people" known as "ethologists," a group Lehrman characterized as having "a common interest in understanding the behavior of animals in relation to their natural environment (including fellow members of the species)."[102] Tinbergen found the visit to Lehrman's Newark lab very stimulating. The

9.9 Daniel Lehrman introducing Niko Tinbergen at a seminar at Rutgers University, March 1972. (Photo by Rae Silver. Courtesy of Rae Silver. Copy provided by Colin Beer.)

downside of it, however, as he reported to Mayr, was he "was worried about D's overweight and signs of strain. We have grown very fond of him over the years."[103]

Six months later Lehrman was dead of a heart attack at the age of fifty-three. The following year Lorenz reached the age of seventy and had to retire from his Max Planck Institute at Seewiesen. Tinbergen retired from Oxford a year later, in 1974, at the age of sixty-six. In the midst of these changes, the Nobel Prize for Physiology or Medicine for 1973 was awarded to Lorenz, Tinbergen, and Karl von Frisch. The prize was a testimony to how much the ethologists had accomplished. At the same time, though, an era in ethology was clearly over.

TEN

CONCLUSION: ETHOLOGY'S ECOLOGIES

Scientific truth is wrested from reality existing outside and independent of the human brain. Since this reality is the same for all human beings, all correct scientific results will always agree with each other, in whatever national or political surroundings they may be gained. Should a scientist, in the conscious or even unconscious wish to make his results agree with his political doctrine, falsify or color the results of his work, be it ever so slightly, reality will put in an insuperable veto: these particular results will simply fail on practical application.

KONRAD LORENZ, *ON AGGRESSION*, 1966[1]

The danger for the scientist is to not test the limits of his science, and thus of his knowledge. It's to mix what he believes and what he knows. And especially, it's the certainty of being right.

FRANÇOIS JACOB, *OF FLIES, MICE, AND MEN*, 1998

Was the Nobel Prize for Physiology or Medicine of 1973 awarded for what the ethologists Thorpe and Hinde proclaimed it was? That is to say, for the creation of a new science? This interpretation appealed to the ethologists themselves. It confirmed publicly and dramatically the value of all they had been doing. But founding an important new discipline is not in and of itself sufficient to attract the attention of the Nobel authorities. Furthermore, it was not all that clear how the work of Frisch, Lorenz, and Tinbergen fit within the Nobel category of "Physiology or Medicine." Frisch, Lorenz, and Tinbergen were naturalists who had focused their attentions on the behavior of bees, fishes, and birds. Wonderful as their discoveries had been, the most conspicuous of these had not been developed at the level of analysis characteristic of *physiological* studies, commonly understood. Instead, it seems to have been the potential bearing of the ethologists' studies on human health that convinced the awards committee of the legitimacy of their candidacy for a Nobel Prize.

Professor Börje Cronholm of the Royal Karolinska Institute took care to identify these connections at the award ceremonies in Stockholm in December 1973. The findings of ethology, he allowed, were illuminating "the importance of specific experiences during critical periods for the normal development of the individual." Ethological studies were also leading, he said, to the conclusion "that the psychosocial situation of an individual cannot be too adverse to its biological equipment without serious consequences." Speaking directly to the recipients about the discoveries they had made, Cronholm stated: "Aside from their value in themselves, your discoveries have had a far-reaching influence on such medical disciplines as social medicine, psychiatry and psychosomatic medicine. For that reason it was very much in agreement with the spirit of Alfred Nobel's will when the medical faculty of the Caroline Institute awarded you this year's Nobel Prize."

One suspects that Cronholm's words were not merely the window dressing needed to award the Nobel Prize to Frisch, Lorenz, and Tinbergen for founding an important new biological discipline. Had the ethologists devoted their entire careers to studying the behavior of "lower" animals, it is doubtful they would have received the prize. However, while Frisch continued to focus on his honeybee studies, Lorenz and Tinbergen both came to consider very seriously the bearing of ethological studies on matters of the human condition and, more specifically, human health. Characteristically, Lorenz offered his pronouncements on the human predicament with great authority. Tinbergen (likewise characteristically), was much more circumspect in what he had to say. In any case, each felt called upon to offer his views, as a student of animal behavior, on how the stresses of modern life were affecting human health and what could be done about it.

Lorenz and Tinbergen were not the only ethologists to suggest that either the findings or the techniques of ethology could be used to come to grips with the pathologies generated by life in the modern world. In the 1950s other ethologists (and nonethologists as well) began to pay attention to questions of animal behavior as these related to human health. The British psychoanalyst John Bowlby, introduced by Julian Huxley to ethology and to the writings of Lorenz and Tinbergen, decided that ethology provided "a wide range of new concepts to try out in our theorizing."[2] He believed that imprinting in animals offered parallels to the mother-infant relationship in humans, and he decided these parallels were worth pursuing. As his work progressed, he paid close attention to the latest studies on imprinting, especially the work of Patrick Bateson at Cambridge. Bowlby also had a long and mutually instructive interaction with Robert Hinde.[3]

Without access to the records of the Nobel Prize committee, one can simply surmise that it based its 1973 award for "Physiology or Medicine" not on any one, single thing but instead on a variety of different things that together, like Alfred Seitz's "law of heterogeneous summation," added up to a favorable conclusion. These included a wealth of discrete and striking discoveries about animal behavior, the establishment of a brand-new field of scientific inquiry, and the elaboration of ideas and techniques relevant to the understanding and treatment of human disease. At the same time, however, the ethologists' concern for the status of their discipline, the aspirations of different individuals to speak authoritatively about the human condition, and public perceptions of the field's utility did not fit seamlessly together. Whenever one ethologist—for example, Lorenz—hazarded publicly a controversial opinion about the relevance of animal behavior studies for understanding human behavior, other ethologists had the occasion to worry about how this might jeopardize their enterprise as a whole. Pronouncements about the human condition could easily be seen as distinct from, and perhaps even detrimental to, the basic job of getting on with the work of the discipline.

In this final chapter, we address a pair of topics related to ethology's popular and professional development in the 1960s and 1970s. The first is how the founders viewed the human condition and how they thought the study of animal behavior might help in improving the human condition. The second is the identity problem that ethology faced in the 1970s and how ethology's claims to leadership in animal behavior studies came to be challenged by a new contender. We then conclude by reflecting on the history of ethology, with special attention to the intertwined careers of Lorenz and Tinbergen.

LORENZ AND TINBERGEN ON THE HUMAN CONDITION

The twentieth century was certainly not the first period to witness scientists (and others) offering biological insights on the human condition. In the mid-eighteenth century the Swedish naturalist Linnaeus not only classified the human species as a part of nature, he also suggested that "man's" natural inclinations served to maintain the "economy of nature." When human populations grew too large, Linnaeus said, then "envy and malignancy toward neighbors" abounded, and it became "a war of all against all!" Half a century later, Thomas Robert Malthus insisted that the need for food and "the passion between the sexes" were biological requirements of the human species that constituted an insurmountable obstacle to the rosy visions of social equality and biological improvement that various Enlightenment thinkers had recently been promoting. In 1838, mulling privately over his new idea of evolution, the young Charles Darwin wrote in his "M" notebook, "He who understands baboon would do more toward metaphysics than Locke." In his *Origin of Species* of 1859, Darwin finally made his theory of evolution public, but he continued to keep his thoughts about human evolution to himself. It was not until 1871, in his *Descent of Man,* that he argued that the human race was the product of a long process of biological evolution and furthermore that "a severe struggle" would probably continue to be necessary if the race were to rise further.[4]

The *Origin of Species* and *Descent of Man* inspired a flood of treatises about the biological dimensions and determinants of human destiny. In the twentieth century, the shocks of the Great War and then World War II stimulated further thought on these matters. Biologists, social theorists, and others debated not only the broad topic of how evolutionary theory might illuminate social theory, they also addressed the more specific questions Why do animals fight? and What does animal behavior tell us about why humans fight?[5] In sum, when Konrad Lorenz began writing about such subjects, first in the 1930s and then with renewed vigor after the war, he was by no means the first to do so.

Two things, however, were novel about the postwar period of the 1950s and 1960s with respect to the discussion of human aggression. The first was the threat of imminent nuclear war. If biologists worrying about genetic deterioration had earlier in the century written about "mankind at the crossroads," in the 1950s it seemed that the human race was not so much at a crossroads as on the edge of a precipice. The atomic physicists set the hands of the "doomsday clock" only minutes before midnight. Bomb shelters were built in backyards by citizens, who then worried not only about how well these structures would protect them from blast and radiation but also about how to deal, when the time came, with neighbors who had not built shelters for themselves. The American television series *The Twilight Zone* offered an episode in which, as missiles were launched, humans fought at shelter doors "like naked animals."[6]

Also new in the years after the Second World War was the flowering of ethology as a branch of biological science. In Germany there was some distrust of Darwinian biology on the part of those who felt that biology had contributed to the crimes of the Third Reich, but a more general trend, in Europe and the United States alike, was to pay increasing attention to the views of scientific experts. Konrad Lorenz relished the role of scientific authority. From early in his career he had believed that the study of animal behavior could illuminate the wellsprings of human nature. His experiences in the war years did not check his readiness to pronounce upon the lessons of animal behavior for the human race. He did so in lectures, articles, and popular books. But his pronouncements did not go uncontested. Not only thinkers from other disciplines but also many ethologists felt a need to distinguish their own views from Lorenz's. Among these was Niko Tinbergen.

Lorenz and Tinbergen each came to the topic of human aggression in a way that was true to his own scientific style and personal temperament. Lorenz came impetuously; Tinbergen came cautiously. Still, both came. Seen from afar, their broad views on the animal roots of the human condition appear very much the same: each worried that the human species was ill equipped biologically to handle the consequences of its rapid cultural evolution. Seen from closer up, significant differences in their positions become evident.

Lorenz presented his postwar views on the human condition not as provisional hypotheses but rather as biological truths bearing on the perilous situation in which the human species now found itself. The human species, he was convinced, had gotten out of sync with its own biological heritage. He had an historical argument to offer here, namely, that the cultural evolution of humankind had outpaced the race's biological evolution, and that humankind had reached a critical moment in its history. Ironically, however, he proved conspicuously unreflective about the historically situated character of his own thinking.

Lorenz's postwar writings on the human condition involved a variety of themes. Among these were the biological basis of human aggression, the continuing dangers of genetic decay, the conflict between the generations, and the deterioration of the environment. Here we will concentrate on his ideas concerning human aggression. These tended to cluster about two main claims. The first was the idea that the human species is exceptional among higher animals in its lack of innate inhibitions about killing its own kind. The second was the notion that aggression is an instinct, and that like all other instincts it builds up internally, like a fluid in a reservoir, until it is eventually discharged.

Lorenz made the first claim in several places, most visibly perhaps in the conclusion to his book *King Solomon's Ring*.[7] As indicated in a previous chapter, he wrote this book when he came back from the war as a way of making money. The book consists for the most part of charming stories about Lorenz's experiences as a raiser and observer of jackdaws, greylag geese, and other creatures. Along the way, it offers wise advice about choosing and keeping pets. It is also very humorous. The book ends, nonetheless, on a somber note. Allowing that he was quoting from an article he had published in a Viennese journal back in 1935, Lorenz wrote: "The day will come when two warring factions will be faced with the possibility of each wiping the other out completely. The day may come when the whole of mankind is divided into two such opposing camps. Shall we then behave like doves or like wolves? The fate of mankind will be settled by the answer to this question." To his 1935 statement Lorenz in 1949 added the comment: "We may well be apprehensive."[8]

Lorenz prepared the way for his "doves and wolves" question by suggesting that if humans had the option of behaving like doves or wolves, then behaving like wolves was much to be preferred. Wolves, he explained, had been equipped by evolution not only with powerful weapons—their sharp teeth and strong jaws—but also powerful, instinctive inhibitions against using these weapons on members of their own species. As he represented it, when two wolves fight, and one gets the better of the other, if the loser submissively exposes its neck to its adversary, the victor is instinctively inhibited from finishing the loser off. To Lorenz, this was a general law of nature: "When, in the course of its evolution, a species of animals develops a weapon which may destroy a fellow-member at one blow, then, in order to survive, it must develop, along with the weapon, a social inhibition to prevent a usage which could endanger the existence of the species."[9]

Doves were another story. Doves, Lorenz explained, lack powerful natural weapons and hence they have not had to develop inhibitions against injuring their own kind. In nature, when two doves fight, the bird that loses the contest can simply fly away. In a cage, however, fleeing is impossible, and the weaker bird may be pecked to death. This occurs because the winner has no innate inhibitions against continuing to the bloody end. Lorenz recounted how he had put a male turtledove and an African blond ringdove together in a cage, hoping to have them mate, only to return to find that the ringdove had nearly pecked the turtledove to death.

The problem for the human species, Lorenz proceeded to argue, is that it is more like the dove than the wolf when it comes to dealing with its own kind. Not having powerful natural weapons like wolves do, humans, in the course of their long biological evolution, have not needed to develop instinctive inhibitions against killing each other. Unfortunately, in the course of their more recent and rapid cultural evolution, they have developed artificial weapons of tremendous destructive potential but have not developed instinctive prohibitions against using them.

The image was a compelling one, made all the more compelling, as it turns out (although Lorenz did not acknowledge this), by the fact that Lorenz had improved on his story of 1935 by inserting doves in the place of *hares*. To Lorenz's mind, clearly, the two kinds of creatures were inter-

changeable for the sake of his argument—they both lacked powerful biological weapons and inhibitions against killing their own kind. The dove
was a more striking example, however, because it was the very "symbol of
peace." What Lorenz failed to address was that his account ran directly
contrary to the analysis of aggression and dove behavior spelled out explicitly years earlier by Wallace Craig.[10]

It is worth revisiting Craig's views here. Craig, the reader will recall,
was a specialist on ringdoves in particular and the pigeon family in general.
He grounded his 1921 paper "Why Do Animals Fight?" on his extensive
knowledge of doves and pigeons. But the image he offered of the fighting
that occurs in the pigeon family was entirely the opposite of what Lorenz
was to claim. The pigeon, Craig insisted, has no special appetite for fighting. In other words, it has no instinctive drive that makes it aggressive.

Craig described what happens in a fight between a stronger bird and a
weaker bird when the weaker bird does not fly away but instead simply
submits. Here, astonishingly enough, one finds Craig's victorious pigeon
acting not like the murderous dove of Lorenz's parable but instead like
the wolf. To cite again a passage cited in chapter 2: "If the enemy submits,
the agent ceases fighting. In pigeons this is witnessed again and again.
In the heat of battle the agent may rush upon his enemy, jump on his back,
peck him with all his might, and pull out his feathers. But if the reagent lies
down unresisting, the agent's blows quickly diminish into gentle taps, he
jumps off his prostrate foe, walks away, and does not again attack the enemy so long as he is quiet. This behavior is typical, and it proves that the pigeon is devoid of any tendency to destroy his rival."[11] Craig concluded that
"no distinctively 'biological' need for fighting" exists. Fights do indeed take
place among animals and among humans, but this, Craig insisted, arises
out of conflicts of interests, not out of any fundamental biological need to
behave aggressively.[12]

Did Lorenz know of Craig's paper? Yes: Craig sent it to him late in 1934
or early in 1935, together with other papers he had written, and Lorenz proceeded to cite it in his "Kumpan" monograph of 1935. Perhaps Lorenz had
it in mind in 1935 when he chose to contrast the fighting behavior of wolves
with that of hares. Be that as it may, the story of wolves and *doves* that Lorenz told in *King's Solomon Ring* ignored the fundamental claim of Craig's
discussion of why animals fight. And this was despite the fact that Craig
(as the reader will remember) was the person whom Lorenz once described
to Stresemann as "the best animal-understanding man" he knew, after
Heinroth.[13]

Lorenz in the 1960s developed his discussion of aggression further, in his book *On Aggression*. The thrust of his argument is clearer in the original title: *Das sogennante Böse: Zur Naturgeschichte der Aggression* (*The So-Called Evil: On the Natural History of Aggression*). His primary claim was that "aggression, far from being the diabolical, destructive principle that classical psychoanalysis makes it out to be, is really an essential part of the life-preserving organization of instincts." Not only was aggression essential to species survival, he maintained, but it also constituted phylogenetically "the rough and spiny shoot" which eventually in the higher forms of life and social organization bore "the blossoms of personal friendship and love." [14] His message was that the human race had to come to understand its instinctive aggressive drives in order to learn how to deal with them, and that essential to this understanding was a recognition of the positive as well as the negative aspects of aggression. Although man was faced with a predicament of the most urgent sort —"in his hand the atom bomb, the product of his intelligence, in his heart the aggression drive inherited from his anthropoid ancestors"—Lorenz was prepared to offer an "avowal of optimism." He believed the biologist could rescue humankind from its precarious state by teaching humans to change for the better. [15]

On Aggression is a book that ranges widely. Discussions of animal behavior appear there along with Lorenz's evolutionary reinterpretation of Kantian philosophy and his belief that Olympic games can serve as a means of discharging internationally dangerous aggressions. The book is not easily summarized. We will focus here on Lorenz's model of how instincts work and on the particular attention he gave to the form of aggression he termed "militant enthusiasm."

Lorenz presented the same view of instincts in *On Aggression* that he had been presenting for the previous quarter of a century. Instincts, he explained, are not just reactions to external stimuli. Rather, they are drives that build up spontaneously within the organism and that need ultimately to be discharged. He credited in particular Erich von Holst with having demonstrated through physiological experiments "that the central nervous system does not need to wait for stimuli, like an electric bell with a push-button, before it can respond, but that it can itself produce stimuli which give a natural, physiological explanation for the 'spontaneous' behavior of animals and humans." He credited Wallace Craig with having discovered through behavioral research essentially the same thing. Studying blond ringdoves, Lorenz said, Craig found that, over time and in the absence of the female bird, the threshold value of the stimuli needed to elicit the male bird's courtship behavior became less and less until finally the male simply directed his bowing and cooing toward the corner of his cage. [16]

Lorenz allowed that aggression is an instinct functioning in precisely these ways, that is, it is spontaneously generated internally, and it eventually requires release. Never mentioning that Craig had specifically denied that aggression is an instinct that works this way, Lorenz went on to observe that "present-day civilized man suffers from insufficient discharge of his aggressive drive." The job for "responsible morality" was one of "reestablishing a tolerable equilibrium between man's instincts and the requirements of a culturally evolved social order."[17]

Lorenz devoted special attention to the particular form of communal aggression he called "militant enthusiasm." He identified this as "a powerful, phylogenetically evolved behavior" with distinctive, subjective correlates: "Every man of normally strong emotions knows, from his own experience, the subjective phenomena that go hand in hand with the response of militant enthusiasm. A shiver runs down the back and, as more exact observation shows, along the outside of both arms. One soars elated, above all the ties of everyday life, one is ready to abandon all for the call of what, in the moment of this specific emotion, seems to be a sacred duty."[18] Although he did not say so in *On Aggression,* he was simply restating here much of what he had already said in his 1943 paper "On the Inborn Forms of Possible Experience." Missing from his 1960s account, however, was his belligerent comment of twenty years earlier: "I charge with emotional feebleness [*Gefühlschwache*] every young man who has not himself experienced this reaction in politically meaningful situations!"[19]

In *On Aggression* Lorenz explained that he was certain that "human militant enthusiasm evolved out of a communal defense response of our prehuman ancestors" and furthermore that "it must have been of high survival value." Calling militant enthusiasm "a true autonomous instinct," he explained: "It has its own appetitive behavior, its own releasing mechanisms, and, like the sexual urge or any other strong instinct, it engenders a specific feeling of intense satisfaction." The problem, however, as history had shown, was that "unbridled militant enthusiasm" was the greatest of dangers.[20] Strikingly, Lorenz selected a distant historical figure, rather than a recent one, to exemplify a leader capable of inspiring collective aggression. His choice was Napoleon. He did not mention Hitler.

What solutions were there to the problem of human aggression? Lorenz was convinced humankind could not be shielded from all the stimulus situations eliciting aggressive behavior. Nor would it be wise, he said, to try breeding aggression out of the race by means of "eugenic planning." This was not, apparently, because he had come to believe that eugenics itself was a bad idea. He still regarded "genetic decay" to be one of the major threats to the future of humankind. Rather, he was afraid that if aggression were to be bred out of the race, then "everything associated with ambition, ranking order, and countless other equally indispensable behavior patterns would probably also disappear from human life."[21] The better course, Lorenz allowed, would be to develop an applied science of human behavior that could promote the intelligent channeling of aggression in useful directions. Human aggressive instincts could still be aroused, as long as they were then enlisted in causes that were worthy. Lorenz offered the pursuit of science as a particularly attractive candidate. As he put it, "Scientific truth is one of the best causes for which a man can fight and although, being based on irreducible fact, it may seem less inspiring than the beauty of art or some of the older ideals possessing the glamour of myth and romance, it surpasses all others in being incontestable, and absolutely independent of cultural, national, and political allegiances."[22]

What proved immediately contestable to Lorenz's contemporaries was the soundness of the "scientific truths" of *On Aggression*. The book became a best seller, but a controversial one. Numerous readers commented on the manifest humanity of Lorenz himself, but not everyone was convinced by his arguments. Marston Bates, writing in the *New York Times Book Review,* was one of the positive readers. He wrote: "Mr. Lorenz, with his profound knowledge of animals and deep sympathy for the human condition, has been more successful than any author I know in explaining human actions in biological terms." In contrast, the American biologist John Paul Scott, writing in the *Nation,* was more critical. In an article entitled "That Old-Time Aggression," Scott accused Lorenz of being fifty years behind in his biology.[23]

The ethologist Robert Hinde, fully established by this time as a leader of the discipline, was rather more diplomatic than Scott when it came to reviewing Lorenz's book, but he was no less critical of Lorenz's key claims. Hinde agreed with Lorenz that human aggression must have some sort of biological basis, forged by natural selection in the course of the species' history. He challenged, however, Lorenz's basic claim about the *spontaneity* of aggression. The problem here, Hinde explained, was Lorenz's energy model of motivation. Hinde had already on several occasions pointed out the flaws in this model, but, as he put it, "Lorenz does not even attempt to answer the numerous criticisms of such models which have been made during the last two decades, and seems unaware of the dangers which they have for the theoretician." Hinde was likewise critical of the prescriptions Lorenz gave for how to deal with aggression. Rather than attempting to redirect or sublimate aggression (Lorenz's recommendation), Hinde wrote, "the solution lies rather in an endeavour to understand the roots of behavior, to tease out the aspects of experience which influence aggressiveness and to assess the nature of their effects. . . . Let us not merely plan a society to cope with man at his worst, but remember that society can influence the nature of its ingredients."[24]

What about Tinbergen? What sense did Lorenz's longtime colleague make of the phenomena of human aggression? The issue was one that Tinbergen certainly took seriously. In 1964, the year after the publication of the original German version of Lorenz's *On Aggression*, Tinbergen delivered at Oxford a lecture entitled "The Search for Animal Roots of Human Behaviour."[25] Four years later, he made the general subject of animal and human aggression the theme of his inaugural lecture as professor of animal behavior in the Department of Zoology at Oxford (the lecture that so displeased his friend Lorenz). His approach to the topic, "On War and Peace in Animals and Man," was, as we have seen, to emphasize "how much we do not know." He urged that ethology's methods, more than its results, were what needed to be brought to bear on the issue of human aggression. It was unwise, he observed further, to extrapolate from a few selected animal species to humans. The human species, he said, needed to be studied in its own right. He was prepared, nonetheless, to hazard at least one generalization about early man: "As a social hunting primate, man must originally have been organized on the principle of group territories."[26]

Eight years later, the subject of human aggression remained on Tinbergen's mind. Others had debated whether aggressive behavior was fueled by an internally generated drive or instead was essentially reactive, but this to Tinbergen seemed not so important as "the fact that we have inherited the potential . . . for using force in a number of diverse situations." Citing information about hunter-gatherer societies existing in the present, Tinbergen concluded that in the distant past the majority of clashes between human individuals or groups would not have been calamitous. Instead, they would have been "of the relatively harmless type of 'fight-flight' balance that characterizes similar intra-specific frictions in the vast majority of higher animals in which killing is relatively rare." Within the past ten thousand years or so, however, the situation had changed: "[Modern man] has taken the disastrous step to war by using his unique capacity for foresight and experience, and recognizing that under certain circumstances killing does pay, because a dead man will not return to fight again."[27]

Tinbergen's explanation was thus reminiscent of Lorenz's, but not identical to it. He did not endorse the idea that aggression is a drive fed by its own action-specific energy. Nor did he think that the main problem with the human species was that it was deficient in its instinctive inhibitions. The problem instead, as Tinbergen perceived it, was that the ancestors of *Homo sapiens,* millions of years ago, had made the "fateful evolutionary 'decision'" to leave their arboreal habitat, where they had been vegetarians, to enter "a new ecological niche, that of hunter-gatherers who sought their food on more open terrain." The "decision" to take up a new ecological niche inevitably had profound consequences, among the most far-reaching of which was that man, out and about in his hunting range, had used his reason to become a killer.

If Lorenz and Tinbergen each looked toward science for salvation, here too their perspectives were appreciably different. Lorenz was not disposed to see science as a social process. In *On Aggression* he wrote: "Scientific truth is universal, because it is only discovered by the human brain and not made by it, as art is."[28] Tinbergen, commenting some ten years later on the future of ethology, looked at the interrelations between science and society from another angle:

> We have to realise that all scientific endeavour is a function of attitudes in society as a whole. This is certainly true of "pure" or exploratory science, but to a lesser degree it also holds for the executive, applied branches, the various technologies. In particular exploratory science, the intellectual game of "science for the sake of science" flourishes only in rich countries and in affluent periods, i.e. when and where, apart from and beyond the resources needed for sheer survival (food, water, housing, clothing, communication, health, education, defense), there is a surplus of manpower and cash that can be allocated to the "luxury" occupation of being curious. This means that, when trying to forecast the future of our (as of any) science we shall have to make inspired guesses about the future of society as a whole.[29]

In discussing what he called the "two-way traffic between society and science," Tinbergen focused not on the ideological or political dimensions of science, but rather on the importance of making ethology into a more applied science. For this to work, he indicated, "a better knowledge of the interplay of reason and our deep-rooted, typically human motivations will be essential—knowledge, in other words, of our 'true nature,' which includes that of both our largely genetic programming and the genetically imposed potential and the limitations of our flexibility. There are signs that economic and social planners are beginning to see that they will need this better understanding of human behaviour; we shall have to try to help them as well as we can."[30]

Tinbergen's comments on the relations between science and society appeared in a book entitled *Growing Points in Ethology,* edited by P. P. G. Bateson and Robert Hinde and published in 1976. The book represented the proceedings of a conference held the previous year at Cambridge on the occasion of the twenty-fifth anniversary of the founding of the Sub-department of Animal Behaviour at Madingley. Although illness prevented Tinbergen from attending the conference, his influence on ethology was still manifest in the volume of conference papers—not only in his own contribution, but also in the very structure of the volume itself. The editors grouped the various conference papers largely according to Tinbergen's four "questions traditional to ethological research." They added a final section, "Human Social Relationships," wherein "questions of all four types" were raised once more.[31]

Tinbergen had missed the conference because he was in the midst of a prolonged period of severe depression. However, the views he expressed in his paper were not primarily a function of his psychological distress. They represented the frank assessment he had been developing for some time of the problems facing modern humankind. He believed it was urgently important to make ethology into a more applied science so that it might help deal with these problems.

Tinbergen's contribution was perhaps not quite what the editors of the volume were looking for from him. They noted of his contribution, "Certainly we are wholly in sympathy with his view. But he is of course not implying that ethologists should give up fundamental work." They warned that "an exclusive focus on problems of immediate practical importance could deprive ethology of the possibility of theoretical growth." They explained further: "The conceptual structure necessary for the practical problems of the 1990s will not necessarily arise from examination of the problems of today. Thus the urgent need for a concerted attack on applied problems should not weaken the vigour of research on basic theoretical and conceptual issues."[32]

The irony for the field of ethology, so soon after it had been graced by a Nobel Prize, was that the field's sense of its "basic theoretical and conceptual issues" was considerably less clear in 1975 than it had been twenty-five years earlier. Ethologists, Bateson and Hinde acknowledged, were in an "eclectic manner" straying farther and farther "from what could be called classical ethology." The field was in some danger of losing its identity because of this willingness of ethologists "to engage in dialogues all round." Bateson and Hinde chose to put a happy gloss on this diffusion of efforts and energies. In their words, "The nature of the specifically ethological contribution ceases to be an issue as disciplines join forces in coming to grips with common problems." Nonetheless, they stressed the importance of "asking the right questions" in ethology. They insisted it was crucial to keep an eye on Tinbergen's four questions and to maintain a balance in dealing with them.[33]

Fourteen years later Bateson and the American ethologist Peter Klopfer were more blunt in talking about the conceptual foundations of modern ethology. Devoting a volume of their journal, *Perspectives in Ethology*, to the question Whither ethology? they stated baldly: "We take the view that ethology as a coherent body of theory ceased to exist in the 1950s. . . . One by one the concepts and theories succumbed to critical analysis and, by the beginning of the 1960s, any vestiges of common belief in an ethological theory had disappeared." Gone were such ideas of "classical ethology" as action-specific energy, sign stimuli, fixed action patterns, and the sharp distinction between innate and learned behavior.

In what other ways had the field changed? One was that it had become increasingly mathematical. By the 1980s, comparing the work of "modern students of animal behaviour" with that of the classical ethologists, Felicity Huntingford could write:

> Perhaps the most obvious difference between the research of the early ethologists and that of modern students of animal behaviour is the enormous increase in the complexity of the exercise; techniques of data collection have become much more sophisticated, pencil and paper having largely given way to the multichannel event recorder, analysis of large blocks of multivariate data is commonplace and a highly complex body of theory has been developed, which itself requires precise quantitative results to test its predictions. These developments all feed on one another and whereas Lorenz boasted that he had never published a paper with graph in it (1975), ethology is now very much a quantitative subject.[34]

Huntingford's characterization of the change in the discipline was an apt one, even if her statement regarding Lorenz was not exactly right. Lorenz had indeed complained about the increasing amount of quantification in ethological publications. He had also allowed that he had never published a paper with a graph in it. But he had stopped just a whisker short of boasting about it. What he said in 1973 (the correct date of the paper in question) was the following: "Even I myself must confess to a slight feeling of inferiority when I realize that I have never in my life published a paper with a graph in it. I am impressed, against my better judgement, when any one of my co-workers deftly formulates a problem, for instance of motivation analysis, in such a way that he or she can feed it to a computer."[35]

Lorenz offered this statement in a polemical piece entitled "The Fashionable Fallacy of Dispensing with Description." The paper appeared in the German journal *Die Naturwissenschaften*—the same journal in which, thirty-six years earlier, he had moved beyond the ornithological circles that had first nurtured him and broadcast to the scientific community at large his claims as an instinct theorist. His basic thesis and complaint was that an infatuation with quantification was being allowed to crowd out other important cognitive processes. He objected to the "deplorable fallacy" that "only quantitative procedures are scientific and that the description of structure is superfluous." He also offered a sociocultural explanation for this fallacy. It came, he said, from the "'technomorphic' thought-habits acquired by our culture when dealing preponderantly with inorganic matter."[36]

Lorenz was not alone in 1973 in lamenting the field's increasing emphasis on quantification. That was the same year in which Bateson and Klopfer published the first volume of their new journal, *Perspectives in Ethology*. They expressed the concern that projects in ethology were being chosen on the basis of the ease with which they generated quantifiable data, not according to their promise for developing new theoretical approaches.[37]

If any single thing in the 1970s gave ethologists a new sense of direction and/or a renewed sense of what they were about, it was the challenge that the American naturalist E. O. Wilson presented to them in 1975 in his book entitled *Sociobiology: The New Synthesis*. Wilson's synthesis was an attempt to restructure the life sciences and ultimately the social sciences as well. Animal behavior studies were at the heart of the synthesis he was proclaiming. Ethology, however, was not.

By Wilson's later recounting, it was Konrad Lorenz who inspired Wilson himself to move into animal behavior studies. Wilson heard Lorenz lecture at Harvard in the fall of 1953. He heard Tinbergen lecture at about the same time, but it was Lorenz who made the greater impression on him. Lorenz issued a "call to arms" to recapture animal behavior from the comparative psychologists and return it to biology.[38] By 1975 Wilson had adopted an aggressive posture of his own with respect to the future of animal behavior studies. He believed, however, that ethology was past its prime. In his massive new book he wrote: "Although behavioral biology is traditionally spoken of as if it were a unified subject, it is now emerging as two distinct disciplines centered on neurophysiology and on sociobiology, respectively. The conventional wisdom also speaks of ethology, which is the naturalistic study of whole patterns of animal behavior, and its companion enterprise, comparative psychology, as the central, unifying fields of behavioral biology. They are not; both are destined to be cannibalized by neurophysiology and sensory physiology from one end and sociobiology and behavioral ecology from the other."[39]

The most noted response to Wilson's book was its sharp condemnation by the Boston-based Sociobiology Study Group, a group of scientists, teachers, doctors, and students on the political Left who saw the book as promoting "a new biological determinism." In a letter to the *New York Review of Books* in November 1975 the group characterized Wilson's theories as simply another version of the theorizing that had "provided an important basis for the enactment of sterilization laws and restrictive immigration laws by the United States between 1910 and 1930 and also for the eugenics policies which led to the establishment of gas chambers in Nazi Germany." The letter identified other thinkers who had promoted biological determinism in one form or another, among them Herbert Spencer, Konrad Lorenz, Robert Ardrey, Charles Davenport, Arthur Jensen, and William Schockley. Not in this published letter, but in a widely circulated manuscript entitled "Sociobiology: A New Biological Determinism," the Sociobiology Study Group made particular note of how Konrad Lorenz in Germany in 1940 had advocated the idea of race purification.[40]

The debate over the scientific merits and political implications of Wilson's *Sociobiology* proceeded with charges and countercharges between Wilson and his critics, various twists and turns, and each side complaining that its own views were not being properly represented. Each side, as the sociologist of science Ullica Segerstråle has aptly put it, regarded itself as the "defender of the truth."[41] Much less visible to the science journalists of the day, and hence to the broader public, was the response of ethologists and comparative psychologists to Wilson's book. They were less inclined than the Sociobiology Study Group was to cast the issue in political and moral terms. On the other hand, no one is pleased to hear that others are about to cannibalize his or her field.

Rebuttals of the cannibalization scenario were sounded by several of the fourteen scientists who contributed to the multiple review of Wilson's

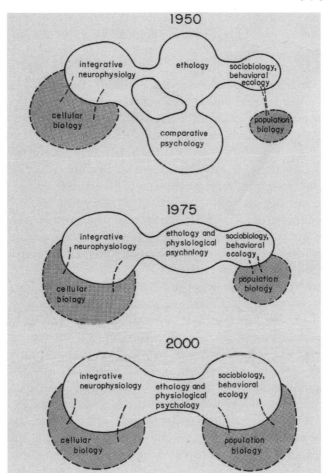

10.1 E. O. Wilson's diagram of the cannibalization of ethology. (From Wilson, *Sociobiology: The New Synthesis* [Cambridge, MA: Harvard University Press, 1975].)

Sociobiology in 1976 in the journal *Animal Behaviour*.[42] They were inclined to see Wilson's prediction more as a public relations ploy than anything else—"sales talk," as Robert Hinde put it—and they allowed that it might not really matter anyway what name was used for the evolving field that studied animal and human social behavior. They were confident that ethologists would have a major role in new developments. Several of the reviewers noted that Wilson himself was behind the times in drawing a sharp distinction between instinct and learning, for most psychologists and ethologists had given up this distinction some years earlier. On the other hand, reviewers acknowledged that ethology had been remiss in the past in not forging connections with population genetics. Wilson's major contribution in *Sociobiology: The New Synthesis* was to show that there was a new field waiting to be developed. Following conceptual leads provided by W. D. Hamilton, Robert Trivers, and others, Wilson called for a more fruitful union between animal behavior studies and evolutionary theory than had existed previously in the camp of the ethologists.

Very quickly, animal behavior studies were transformed. In the 1960s Tinbergen had regarded the study of behavioral function as woefully under-represented in the work of ethologists. Studies of causation were still dominating the discipline. Tinbergen thought of his own attempts to promote behavioral-ecological studies as his particular "hobby horse."[43] Then, in the 1970s, studies of function became the rage. There was a research boom in behavioral ecology and sociobiology. Questions of behavioral causation, development, and evolutionary history took a backseat to the new interest in function and adaptation.

How happy was Tinbergen to see these developments? His view of Wilson's *Sociobiology,* expressed in a review of the book in 1975, was that Wilson had effectively offered "*a* new synthesis," but not "*the* synthesis" that students of social behavior still needed. He did not elaborate on the implication that Wilson's "synthesis" was incomplete, but one can imagine that he found Wilson's effort to be lopsided. Wilson focused on "ultimate" causes to the exclusion of "proximate" causes. The causal and developmental questions so important in Tinbergen's vision of ethology were not part of Wilson's enterprise. Tinbergen in addition felt that Wilson's chapter on altruism suffered "from a lack of the very kind of analytical information on behavioural ecology which Wilson so urgently demands." As for Wilson's general account of human evolutionary history, Tinbergen found this to be generally solid, though Wilson's "rather pronounced 'scientism'" bothered him a bit. What surprised him most about Wilson's chapter on man was Wilson's astonishing optimism about man's future. Anyone familiar with the "human predicament," Tinbergen thought, had no reason to write so cheerily about humankind's achieving "an ecological steady state" some time around the end of the twenty-first century."[44]

With respect to specific developments in behavioral ecology after 1975, Tinbergen made no pretense of feeling any longer like a critical authority. His extended post–Nobel Prize depression had sapped much of his confidence in his analytical skills. He wanted to devote what energies he had to his studies of autism. Nonetheless, he took some satisfaction in recognizing that his own field studies had been a contributing factor in the growth and efflorescence of behavioral ecology.[45] In the fall of 1978 he wrote to his friend Ernst Mayr saying that although he could not "follow much of what is now going on even in etho-ecology," he could still "enjoy seeing a new generation such as Kruuk, Krebs, Dawkins a.o. [et al.] do things that I 'feel in my bones' are very worthwhile."[46]

Lorenz in the late 1970s was much less happy with the way ethology had developed. As early as the 1950s he had been angered by the way his concept of the "innate" had been gradually picked apart. Nor was he pleased when his ideas about action-specific energy, innate releasing mechanisms, and the like fared poorly when subjected to experimental and analytical scrutiny. He was also distressed by the general loss of interest in comparative studies. He was further irritated in the late 1970s to find his understanding of Darwinism impugned by Richard Dawkins, who took him to task for being "a 'good of the species' man" who "misunderstood how evolution works."[47]

When in 1978 Lorenz finally published a textbook on ethology—forty years after he had first planned such a volume—he ended up restating many of the ideas that had been constitutive of ethology's initial conceptual construction. He did so even though many of these ideas were no longer providing the field with its sense of identity and purpose. Thus, he offered a revised version of his old "psycho-hydraulic" model of instinctive action of 1950, despite the fact that in the minds of most other ethologists there was no hope that a few adjustments to the old model could save the day.[48] Three years later, in *The Foundations of Ethology* (the English translation of this text), he added a preface wherein he sharply voiced his dissatisfaction with modern trends in ethology. He compared the recent development of his field to the growth of a coral colony. "The more [a science] thrives and the faster it grows," he said, "the quicker its first beginnings—the vestiges of the founders and the contributions of the early discoverers—become overgrown and obscured by their own progeny." The same was true, he explained, of coral: "The polyps at the end of its branches have a much better chance of further development than those situated near the foundation. The ends go on growing faster and faster without considering the necessity for strengthening, in proportion, the base that must carry the weight of the whole structure. Unlike an oak tree, the coral colony does nothing to solidify its support. Consequently, there is a lot of coral rubble detached from points of departure, and this is either dead or, if still partly alive, growing in indeterminate directions and getting nowhere." Contemporary ethology, Lorenz complained, was like a coral. Its branches were "losing contact with their foundation" and "producing quite a lot of

rubble." Younger ethologists were in the process of forgetting "really important discoveries" the field's founders had made. This was partly a matter of fashion, Lorenz felt, and partly a matter of ideology.[49]

Curiously enough, a decade or so later in 1992, the German ethologist
Hanna-Maria Zippelius was inspired to write about the same situation, but
from a wholly different angle. Modern ethologists, she indicated, no longer
embraced the major ideas of classical ethology, but this was because they
had demonstrated that these ideas were untenable. She quoted Wolfgang
Wickler, Lorenz's successor at the Max Planck Institut in Seewiesen, saying
"action-specific energy proves to be a modern Phlogiston, and the psycho-
hydraulic model, despite refined alterations, proves unfit to depict adequately the preparedness and the changes in condition of animals." What
bothered Zippelius was that general biology textbooks, despite these modern developments, continued to set forth the ideas of classical ethology as
if they were still the coin of the realm. In her book *Die vermessene Theorie*
(*The Presumptuous Theory*), she spelled out the ways in which the ideas
of classical ethology had been refuted. She brought under her scrutiny
Lorenz's concepts of fixed action patterns, action-specific energy, innate
releasing mechanisms, the principle of dual quantification, appetitive behavior and consummatory action; Tinbergen's model of the hierarchical
organization of behavior; and so on. Then, in a short final chapter on classical ethology and modern behavioral ecology, she asked whether Lorenzian ethology had led into behavioral ecology or was instead just the opposite of it. She concluded that Lorenz's strongly typological thinking and his
notion of "the good of the species" were incompatible with the new ideas
of kin selection, evolutionarily stable strategies, and the like. She did not
discuss Tinbergen's functional studies. Her overall judgment regarding Lorenz's ideas was a firm one: they had had great meaning for the beginnings
of animal behavior study, but they were no longer of any use.[50]

The way that the interests of ethologists shifted in the 1970s and 1980s
to questions involving evolutionary theory is well illustrated in the introductory text entitled *Unravelling Animal Behaviour* published by the Oxford ethologist Marian Stamp Dawkins in 1986. Her chapter titles show
how the balance of interest had shifted over the previous decade and a half.
Five of ten chapters were devoted to evolutionary matters: "Adaptation"
(chapter 1), "Optimality" (chapter 2), "Inclusive Fitness" (chapter 3), "Evolutionarily Stable Strategies" (chapter 9), and "Sexual Selection" (chapter
10). Of the five remaining chapters, four had titles that were conventional
enough: "Genes and Behavior" (chapter 4), "Innate Behaviour" (chapter 5),
"The Machinery of Behaviour" (chapter 7), and "Communication" (chap-

ter 8). Chapter 6, however, had an awkward title signaling just how far ethology had come since 1950. Dawkins called it "Some Obstinate Remnants: Instinct, Displacement Activities, and Fixed Action Patterns." What had once been at the very heart of the conceptual structure of ethological theory had been demoted to "remnants"—"obstinate," to be sure, but "remnants" nonetheless. The awkwardness of this chapter was solved in a later edition of Dawkins's book. The chapter was eliminated altogether.[51]

One should not suppose, however, that Marian Dawkins herself was pleased with how ethology had developed in the 1970s and 1980s. Contributing to the 1989 Bateson and Klopfer volume, "Whither Ethology?" she allowed that the four legs on which ethology had stood traditionally had been growing very unequally. While ethology might "still be a four-legged animal," it was "an animal hopping around on one big leg, with the other three dangling somewhat ineffectively." Questions about function had come to dominate the field, virtually to the exclusion of questions of mechanism, development, and phylogeny. As she eloquently put it:

> Our understanding of evolution and fitness is greatly advanced over what it was only a few years ago. What we need to do now is to balance this with an equal understanding of the mechanisms that are responsible for behavior. We have a much clearer idea of gene selection down the generations. But genes operate through making bodies do things. These bodies have to develop and they need machinery (sense organs, decision centers, and means of executing action) to be able to pass the genes on to the next generation. To understand this process fully, we need a science that is not only aware of the evolutionary ebb and flow of genotypes over evolutionary time, but can also look at the bridge between generations, at the bodies that grow and move and court and find food and pass their genetic cargo on through time with the frailest and most marvelous of flesh-and-blood machinery. That science, with its eyes on both the long term and the short term, on evolution and mechanism, is ethology. It has been so since its birth and, with a bit of readjustment of its weight onto all four of its legs, it can still be so in the future.[52]

The problem, as the organizers of the "Whither Ethology?" volume framed it, was one of surviving and doing first-rate, creative science in a "harsh competitive world" where other fields were making their own claims on the resources of funding agencies and academic institutions. In their words, "The dilemma is how to be competitive with other fields in order to be in a position to do research, while at the same time cooperating with them in order to maintain the vitality of the study of behavior." Their view was that what ethology had to offer to animal behavior studies, and what animal behavior studies needed if they were genuinely to flourish, was "not a theory but an approach." The approach they recommended was Tinbergen's. It was the approach of addressing the four questions of ethology—questions about the control, development, evolution, and function of behavior—in such a way that they mutually informed one another.[53]

The idea of this approach, as we have seen, was not new. Tinbergen had spelled it out a full quarter of a century earlier. The specific circumstances under which he had come to see the importance of this approach were not identical to the circumstances facing his disciplinary descendants in the late 1980s. The field's intellectual, institutional, and even sociopolitical ecologies had been transformed in a variety of ways. But the same sorts of questions about ethology's identity, its centrality or marginality in the life sciences, its relations to neighboring disciplines, and so forth, continue to apply, and with force. It remains to be seen how successful ethologists will be in using Tinbergen's guiding vision—along with many other possible sorts of resources—simultaneously to further their enterprise within the life sciences and to achieve ever richer and deeper understandings of animal behavior.

CONCLUSION: HISTORY MATTERS

Konrad Lorenz once asked Wallace Craig to tell him more about Craig's teacher, Charles Otis Whitman. Lorenz explained: "I always find it important to know the *man,* to be able to appreciate fully his work and his philosophical opinions."[54] But Lorenz, it is clear, had no thought of approaching this topic by reading the whole of Whitman's scientific corpus. His emphasis here on "the *man*" (the italics were his) reflected his belief that if he grasped another person's character he could then judge the scientific claims of that person accordingly.

Lorenz's reading habits were certainly not those of the pioneering American student of insect behavior George Peckham, of whom William Morton Wheeler once remembered: "Every year he most conscientiously read, as a devout priest might read his breviary, Darwin's *Origin* and *Animals and Plants under Domestication.*"[55] Although the founders of ethology thought of themselves as Darwinians, they certainly did not read Darwin religiously. Indeed, they seem to have scarcely read him at all. As for those figures whom Lorenz happily identified as his immediate precursors—especially Whitman, Craig, and Heinroth—he came to their works almost haphazardly. It seems that someone else always had to tell him about them. As he embarked on his own studies of bird behavior, he had not much grasp of what other investigators of animal behavior had done before him. He then borrowed selectively from them when it suited him. In a number of instances where it seems, retrospectively, that something *should* have suited him, he overlooked it at first.

We have encountered numerous examples of the latter phenomenon. We have seen how Lorenz identified Whitman's comparison of instincts and structures as the "Archimedean point" of comparative behavior studies. Yet Lorenz entirely missed Whitman's existence (and Whitman's point) when Lorenz borrowed from Ziegler's work on instinct to write his own first paper on instinct. And this was despite the fact that Ziegler, in identifying Whitman as belonging to the "new direction" in psychology, went directly to what Lorenz later treated as Whitman's most important statement: "Instinct and structure are to be studied from the common standpoint of phyletic descent." Lorenz did not know of Whitman's work until Craig highlighted its importance to him. As for Craig, Lorenz had the opportunity to recognize *his* importance when Lorenz first read E. C. Tolman's *Purposive Behavior in Animals and Men,* a book which made much of Craig's distinction between the appetitive and consummatory parts of instinctive behavior patterns. But Lorenz did not grasp Craig's significance upon reading Tolman. He had to be alerted to it by Margaret Morse Nice. And it was more than three years after Lorenz first came in contact with Heinroth that Heinroth apparently felt he should gently mention to Lorenz his own most important scientific paper, his 1911 paper on the ethology and psychology of the ducks and geese. Even before that, what occasioned Lorenz's discovery of his affinities with Heinroth was Bernhard Hellmann's birthday gift to Lorenz of the first volume of *Die Vögel Mitteleuropas.*

These examples are not mentioned primarily as a criticism of Lorenz's rather spotty attention to the scientific literature of his time. After all, it is doubtful that bibliographic thoroughness and scientific creativity go hand in hand. Furthermore, to suggest that Lorenz at first "missed" some elements that he later made central to his thinking is not intended to imply that these very elements would inevitably have been incorporated in a science of animal behavior in just the way the ethologists ultimately used them. It is simply not the case that in 1900, say, ethology already existed in some ideal space, preformed and just waiting to be unveiled and activated. The field was yet to be conceived and constructed. What was essential to it and what was incidental, what was central and what was marginal, were issues that had to be worked out in the first place and then periodically if not continually renegotiated.

The point about Lorenz's reading habits is basically a reminder that what constitutes an investigator's own, immediate, effective context at any given time is not necessarily identical to what is "available" in the scientific literature. This should give us some pause as we think about how the work done in the period before Lorenz and Tinbergen relates to their establishment of ethology as a robust scientific enterprise. Whitman, Craig, and the various British field naturalists discussed here cannot be best understood in the role of precursors. Each had his own particular concerns, and it is just as instructive to see how they differed in their aims, methods, and ideas from the ethologists as to see how they were similar to them.

One difference between Whitman and Lorenz, for example, was Whitman's abiding interest in the mechanisms of evolution. Lorenz exhibited very little concern with ongoing discussions about how evolution works. Not surprisingly, then, he took no notice of Whitman's studies of directed variation and support of the idea of orthogenesis. Instead, he promoted that particular part of Whitman's work that was most useful to him, namely, the idea that instincts could be used like organs in reconstructing phylogenies. He also cited Whitman's observation that under the conditions of domestication, instinctive behavior patterns in animals tend to break down. Whitman argued that this favors greater freedom of action, which in turn allows for the development of intelligence. When Lorenz, in contrast, first wrote about the breakdown of instinctive behavior patterns under domestication, he did so to argue for the sociopolitical utility of animal behavior studies. In 1938, as a new member of the Third Reich, he argued that "domestication phenomena" in humans, just as in animals, causes the breakdown of instinctive inhibitions against mating with a type other than one's own. Only later (though interestingly enough, still during the Third Reich), did he stress, while citing Whitman again, the more liberating as opposed to the more degenerative side of domestication.

There are innumerable other examples of concerns, practices, and achievements that were winnowed out, used for different purposes, or simply passed by as the modern ethologists constructed their own identity and heritage. Edmund Selous was keenly interested as a field naturalist in verifying the reality of sexual selection and, in particular, the occurrence of "female choice." Although his *practice* of watching animals in nature became a prominent feature of the ethological enterprise, the *issue* of sexual selection was to drop out of the purview of students of animal behavior for an

entire generation. Eliot Howard, the other pathbreaking amateur field ornithologist of Selous's day, became famous for his theory of territory. Tinbergen took a strong interest in territorial behavior, but neither he nor any
of the other ethologists thought seriously of following Howard in his attempt, in *A Waterhen's Worlds*, to capture the bird's "viewpoint." Huxley, it
is true, nurtured an interest in the animal mind and animal subjective experience, but Tinbergen managed for years to keep such concerns distant
from the mainstream of ethology.

This sort of selection applied to aspects of the work of the founders of
ethology themselves. Tinbergen himself, in the years immediately after the
war, promoted ethology as the work of "Lorenz's school." However, he did
not equate ethology with the whole of Lorenz's thinking. He represented
Lorenz's ideas on releasers, innate releasing mechanisms, action-specific
energy, and the like, as the core of Lorenzian ethology. He left out Lorenz's
wartime ideas on race hygiene. And he left out other parts of Lorenz's
thinking as well, for example, Lorenz's efforts to unite animal behavior theory with philosophy.

Likewise Tinbergen's successors, in thinking about *Tinbergen's* later
work, did much the same thing. They distinguished between Tinbergen's
mainstream ethological studies and the other interests he developed. They
would have preferred to see him devote his Nobel Prize address of 1973
to his field studies on behavioral function rather than to his work on autism or his endorsement of the stress-reducing benefits of the Alexander
technique.

There is nothing surprising about this selective appropriation of the
ideas, practices, and interests of one's predecessors (or of one's colleagues).
We all recognize that the same thinker can have good ideas and bad ones.
We also appreciate that the ideas or practices that are relevant to a particular enterprise may or may not be helpful in another arena. And we recognize that ideas cherished at one time may be rejected later. Action-specific
energy and innate motor patterns were key concepts of the field of ethology
from the late 1930s to the early 1950s, but they were then undermined and
to a greater or lesser degree discarded—or at least marginalized (though it
depends on whom you talk to).

We do need to remember that individuals looking at the very same
event from different perspectives can understand it in different ways, and
especially so when political undercurrents are involved. Tinbergen at the
Cambridge physiological mechanisms meeting in 1949 had to explain to
Lorenz and Koehler why the very Germanic tone of Holst's criticisms of
Gray fell so gratingly on British ears so soon after the war. He also felt the

need to remind Koehler why Jewish scientists could be resentful of Lorenz's writings. Baerends compared Lorenz's lecture at the Paris instinct conference to the roaring of an angry lion. Koehler to the contrary thought that Lorenz's behavior in Paris was admirably restrained. In the years following Daniel Lehrman's critique of Lorenzian theory, representatives of the two opposing sides of the debate interpreted quite differently which side was making the more telling points in the argument. They also held contrasting views of which side was bringing ideological biases to the discussion. Differences of opinion on this score continue to this day.

Scientists of the present might imagine a variety of motives to take an interest in the writings and work of their predecessors. One is to gather materials for constructing an account that lends authority to the work of the present. Even a simple chronological alignment of "precursors" can serve this function. This is not a bad rhetorical strategy when one wants to represent one's own activities as the continuation or culmination of a line of efforts heading in the right direction. But these sequences are like the familiar cartoon sequences of hominids, beginning with an early, ape-like creature and proceeding through various stages to a modern human. They leave one with little sense of the contingent and erratic character that the path in question is likely to have had; or of the complexity of the intentions, achievements, ideas, and concerns of the investigators thus named; or of *other* actors and paths that also belong in the historical record.

This can be seen in Lorenz's recital of the sequence of contributions critical for the development of his own instinct concept. First came Whitman's and Heinroth's recognition that instincts could be used like organs in reconstructing phylogenies. This was followed by Wallace Craig's distinction between appetitive behavior and consummatory actions. To this Lorenz added his own view of the functioning of innate releasing mechanisms, and then incorporated "Holst's discovery of automatic stimulus production in the central nervous system."[56] This is an attractive and fair enough accounting of Lorenz's assembly of the primary elements of his instinct theory. But it leaves out twists, turns, and other options. The same is true of Lorenz's cheerful representation of ethology as "the discipline which applies to the behavior of animals and humans all those questions asked and those methodologies used as a matter of course in all the other branches of biology since Charles Darwin's time." One imagines Lorenz would have certainly been brought up short had he read the title of Robert Boakes's excellent history of animal psychology, *From Darwin to Behaviourism.*[57] It would have stuck in the Austrian ethologist's craw to find behaviorism—anathema to an ethologist—portrayed as one of Darwin's proper historical outcomes. Nonetheless, as Darwin clearly established in his *Origin of Species,* "descent with modification" involves the production of diversity, not just the evolution of a single line.

A different reason one might care to explore the history or prehistory of a field might be to find inspiration in the efforts of kindred spirits. Had Tinbergen taken an interest in Whitman's writings, for example, he might have found as much reason to feel a real affinity with the American biologist as Lorenz did. In Tinbergen's case, however, this sense of common purpose might not have been so immediately tied to Whitman's reconstructing of phylogenies as to his identification of his approach as "experimental natural history," and to his observation that "we need to get more deeply saturated with the meaning of the word 'biological,' and to keep renewing our faith in it as a governing conception." [58] Both of these could very well have served as mottos for Tinbergen's efforts half a century after Whitman's death.

If inspiration is one thing that modern scientists might draw from a more careful study of what motivated their predecessors, another possibility is that scientists might find in earlier works potentially productive ideas or observations that somehow never caught hold in the first place or otherwise dropped out of sight. Even if we do not endorse Lorenz's likening of ethology to a coral losing contact with its original base, it is not inconceivable that a new generation of students of animal behavior might find, in the writings of the founders, ideas or observations worthy of renewed consideration.

There is a final and perhaps most important way that an appreciation of the history of science could prove of genuine importance to modern scientists and to nonscientists as well. A deeper sense of the historical situatedness of the work of one's predecessors can become a means of reflecting critically on one's own activities and ideas in the present.

I have sought in this book to portray ethology as a complex, contingent, emerging, and evolving set of practices and concepts, forged by alliances of researchers working in specific institutional, social, and cultural settings. My aim has been to make historical sense of the ideas, practices, individual career paths, and institutions that collectively and creatively came to constitute the new field of ethology. The heterogeneity of the efforts that contributed to the making of a science of the biology of behavior in the twentieth century, combined with the historical, cultural, social, and political situatedness of these efforts, are what I have signaled with the phrase "ethology's ecologies."

The cases examined here illuminate the considerable importance of places and practices in the history of biology. Whitman and Craig in the United States; Selous, Howard, and Huxley in Britain; Lorenz in Austria and Germany; Tinbergen in Holland and then Britain—these individuals were all animal lovers (as Lorenz insisted), but their ideas and actions were not simply a function of a direct engagement with their animal subjects. The backgrounds they brought to their researches, the settings in which they did their work, and the methods they employed intersected in distinctive and reinforcing ways.

We found some of the ramifications of this by following up on Lorenz's astute characterization of his own practice as that of the "farmer" and Tinbergen's practice as that of the "hunter." This distinction illuminates differences not only in the methods of the two men but also in their thinking. In brief, Lorenz, the "farmer," was a raiser and breeder of animals. He was also keenly concerned with the health of his charges (in this respect his training as a physician came into play). Living in proximity to his animals gave him the opportunity to judge which behaviors were "normal" and which were "pathological." The observation of behaviors that were exhibited in effect in the "wrong" contexts—the starling that "performed the entire fly-catching behavioral sequence without a fly"—enabled him effectively to distinguish between the physiological performance of instinctive acts and the biological purpose of these acts. At the same time, he saw a happy combination between his practices as an animal raiser and his aspirations as a comparative anatomist. He championed the comparative insights one could get from observing species living side by side in a zoo or aviary, preferring these to the ecological insights one might get from studying animals in their natural habitats. In addition to raising animals he conducted breeding experiments with them. Manipulating the interbreeding of different bird species (an achievement facilitated by imprinting the young of one species on adults of the other) gave him a further entry into the study of phylogenetic relationships. This work, together with his observations on the behavior of domesticated animals, provided the basis of his claims to authority as he expostulated on the dangers of interbreeding and genetic decay in humans.

Tinbergen, the hunter, preferred to stalk or spy on animals in their native settings. In doing so he was better situated than Lorenz when it came to appreciating just how well animals are adapted to the ecological demands of the environments in which they live. Out in the field, he was better able to recognize how the behavioral machinery of an animal species is the product of ecologically significant "decisions" in the species' past. But this recognition was not solely a function of the habits he formed as a

10.2 The scientist as farmer: Lorenz in the mid-1960s. (Courtesy of the Konrad Lorenz Institute for Evolution and Cognition Research.)

young field naturalist in the Netherlands. At least as he later saw it, he would never have developed the "ecological-evolutionary outlook" if he had not moved to Britain and imbibed there the intellectual tradition represented by Darwin, Julian Huxley, Charles Elton, and David Lack.[59] His own change of environment, as he saw it, was intimately related to the research program he subsequently developed on the survival value of behavior patterns in their natural settings. Similarly, the view of human aggression he developed was also decidedly ecological in character. He believed that aggression is closely related to the settings in which it is exhibited and

10.3 The scientist as field naturalist: Niko Tinbergen in the spring of 1972 on Skomer Island, off the coast of Wales, where he had gone to film cliff-nesting birds for his Time/Life-BBC series *Behaviour and Survival*. (Photo by Lary Shaffer. Courtesy of Lary Shaffer.)

reflects the early history of the species, when the ancestors of *Homo sapiens* moved from an arboreal habitat to become hunter-gatherers who defended territories. Thus, whereas Lorenz emphasized breeding, innate character, and drives welling up from within, Tinbergen emphasized how the behavioral machinery of the organism was a product of the environmental demands placed upon it.

The most troublesome questions of practice and place in the history of ethology arise when we seek to understand Lorenz's behavior in the Third Reich. Lorenz's relations with the Nazi regime and the differences this part of Lorenz's career made for the subsequent history of ethology are both complicated topics on which observers are bound to disagree. These topics will continue to stimulate differences of opinion for some time to come. Simply condemning Lorenz as a Nazi or, alternatively, ignoring or excusing this part of his career does nothing to enhance our understanding of science in the Third Reich, or Lorenz's part in it. Painting Lorenz in black or white provides no opportunity to consider the shades of gray represented in the varying degrees of complicity or resistance—or complicity *and* resistance—that scientists under Hitler displayed. And dismissing the issue of Lorenz's wartime behavior as irrelevant to ethology's development begs the question of the extent to which postwar suspicions and resentments of each side for the other (or of *multiple* camps for one another) contributed to discussions of ostensibly scientific issues.

I have argued here that the seminal contributions Lorenz made to the study of animal behavior in the mid-1930s preceded the biopolitical ideas he promoted during the Third Reich, and that the latter were not a necessary outgrowth of the former, though Lorenz at the time was happy to construct a connection between them. To Tinbergen, Lorenz's attempt at a synthesis of animal behavior studies, race hygiene, and Kantian philosophy appeared not as a compelling synthesis but as a mosaic. Rebuilding ethology after the war, Tinbergen took Lorenz's early concepts, methods, and observations as the essential core of Lorenz's contribution to ethology and ignored the wartime political and philosophical accretions. By treating Lorenz's support of National Socialism as a mistake of the past to be put behind them, Tinbergen served as an important example for the field as a whole.

But at least in some quarters, and to some degree, Lorenz's political past continued to shadow the field. Among other things, it may have contributed to the ways in which individual scientists sought to distinguish their positions from Lorenz's, most notably in matters of human behavior. Interestingly enough, I have encountered no evidence that individual scientists made systematic efforts to find out more about Lorenz's wartime activities. There is no indication that either Thorpe or Huxley, for example, ever bothered to check when Lorenz told them that if they read his *Der Biologe* article of 1940, they "would actually find that I never had been a Nazi at all." Nor for that matter does it appear that Daniel Lehrman tried to track down all of Lorenz's wartime writings. He did not cite Lorenz's *Der Biologe* article in his famous *Quarterly Review of Biology* critique of Lorenz. Lorenz's friends and disciples seem to have been willing to believe the best of Lorenz or at least to put the issue aside, while his detractors were disposed to believe the worst and not forget.

The reader is referred back to chapter 5 for an account of Lorenz's career in the Third Reich. Different readers are likely to reach different interpretations of Lorenz's behavior based on the information in this and other accounts. Whatever one decides in this regard, I would suggest that ethologists, scientists in general, and indeed all of us would do well to remember Lorenz's wartime career as an instructive and cautionary example. Notwithstanding his immense talents as a creative scientist—indeed, while continuing to view himself as a highly talented, essentially apolitical, and independently minded scientist—Lorenz offered his intellectual services to a regime that ultimately proved murderous. If he never considered his own science to be genuinely entangled with politics, this itself was a posture—cultivated perhaps to reassure himself as much as anyone else—and he used it, among other postures, to advance his work and career in a time and place fraught with dangers and opportunities.

After the war, Lorenz and his fellow Austrian and German scientists were not disposed to elaborate on the sorts of day-to-day politics and self-positioning that were constitutive of the practice of science in the Third Reich. Instead, some two decades later, in his book *On Aggression,* Lorenz offered the following pious pronouncement about science and politics in general: "Scientific truth is wrested from reality existing outside and independent of the human brain. Since this reality is the same for all human beings, all correct scientific results will always agree with each other, in whatever national or political surroundings they may be gained. Should a scientist, in the conscious or even unconscious wish to make his results agree with his political doctrine, falsify or color the results of his work, be it ever so slightly, reality will put in an insuperable veto: these particular results will simply fail on practical application." [60]

Lorenz cited the case of genetics in the Soviet Union to support this assertion. Russian geneticists, he said, claimed to have demonstrated the inheritance of acquired characters, but their results could not be confirmed elsewhere. But Lorenz did not mention the case much closer to home of how, under the Third Reich, the "science" of "race hygiene" was a site for the interpenetration of science and politics. He did not address whether with a topic such as race hygiene one could ever expect "reality" to "put in an insuperable veto."

The theme of scientists cultivating the support of powerful patrons has been a familiar one since the origins of modern science. The temptations—or opportunities—to bring one's scientific work into alignment with a patron's interests have been many. It is instructive that Francis Bacon, perhaps the first great champion of the idea that "knowledge is power," already in the seventeenth century expressed concerns about the uses to which science might be put. In his utopian sketch *New Atlantis,* written around 1624 and then published after his death, Bacon specifically expressed reservations about putting scientific knowledge in the hands of the state. He described his version of an ideal scientific academy, which he called Salomon's House. When the scientists of Salomon's house made a discovery or invention, they consulted among themselves regarding whether it should be made public or kept secret. Of those they kept secret instead of publishing, some they revealed to the state, but others they did not.[61]

The question of the uses of knowledge becomes ever more acute as science becomes increasingly powerful. In the new "post-Genomic" era, when the knowledge generated by the Human Genome Project is simultaneously transforming the life sciences and the possibilities for shaping life, scientists have a responsibility to acknowledge and articulate what they know and what they do not know. As the French Nobel Prize–winning geneticist François Jacob has put it, "The danger for the scientist is to not test the limits of his science, and thus of his knowledge. It's to mix what he believes and what he knows. And especially, it's the certainty of being right."

Today, Lorenz himself is sometimes cited as a prime example of what went wrong in the relations between science and politics in the Third Reich. François Jacob is someone who has mentioned him in just this way. Writing on the "imperceptible progression" linking the eugenic theorizing of Francis Galton and the murders committed by Joseph Mengele in Auschwitz, Jacob names Lorenz and the German geneticist Eugen Fischer. Of Lorenz he states: "Konrad Lorenz compared the elimination of individuals who are asocial by reason of their deficient constitutions with the elimination of a malignant tumor, an operation that seemed to him easier and less risky in the first case than in the second."[62] And that is indeed the *only* thing Jacob says about Lorenz in the book from which this quote is taken.

Jacob's straight and simple line from Galton to the Holocaust, with Lorenz in between, is not very illuminating history. It does not tell us about the diversity of eugenic visions in the early twentieth century. It represents the Holocaust too straightforwardly as the consequence of earlier eugenic thinking. More subtle analysts insist instead on the contingency, incommensurability, and perhaps ultimate incomprehensibility of what finally happened. Yet straight-line narratives hold an attraction that makes them easy to grasp, even if they do not always enrich our understanding of history.

Significantly, and not surprisingly, this is *not* how Niko Tinbergen thought Lorenz should be remembered. Tinbergen felt he understood how Lorenz in the late 1930s had "derailed" and "fallen for Hitler." He had read the worst of Lorenz's statements of the 1940s about eliminating cancerous elements from the social body. And the brutal experience of the Nazi occupation of Holland was a thing Tinbergen could not ever entirely forgive or forget. Nonetheless, in the fall of 1973, when the announcement of the Nobel Prize awards reawakened attacks on Lorenz because of his political past, Tinbergen stood beside his colleague. To Ernst Mayr, Tinbergen wrote: "Of course it would have been good if Konrad had been able to bring himself once to recant for what he wrote in 1943, but for God's sake, we know the circumstances, we know the man, we know how he recoiled in horror when he saw what the Germans did in Poland. And he has paid dearly in Russian camps." As Tinbergen prepared to go to the awards ceremony in Stockholm, he recognized he might have "to pour oil on troubled waters."[63]

But how, then, did Tinbergen feel Lorenz should be remembered? The clearest answer to this is to be found in an obituary notice he wrote of Lorenz. That such a document exists may come as a surprise to those who know that Tinbergen died before Lorenz did. Tinbergen died on 21 December 1988. Lorenz died on 27 February 1989. The obituary notice in question, however, was written many years earlier. The *Times* of London made a practice of having on file "obituaries" of fellows or foreign members of the Royal Society of London. When Lorenz was named a foreign member of the Royal Society in 1964, Tinbergen was asked to prepare one of these obituaries-in-waiting.[64] He completed a draft the following year. Indeed, he gave Lorenz a look at it. Tinbergen wrote:

In a time when zoologists practically ignored the phenomena of animal behaviour; when animal psychology concentrated on extremely limited laboratory experiments mainly on domesticated rats; when human psychology was partly the domain of philosophers, partly an infant branch of medicine; and when neurophysiologists were concentrating on the study of nervous conduction and simple reflexes, Lorenz applied inductive scientific thinking and analysis to the immense range of phenomena of animal behavior. Not an experimenter himself, he had an unparalleled gift for creating order in the seemingly chaotic phenomena. His penetrating theoretic analyses, leading to new and fertile formulation of problems and hypotheses, coupled with unrivalled first-hand knowledge of behaviour patterns, opened up many new avenues of approach. His method— keeping tame individuals of wild species under free-ranging conditions —is now recognized as one of the basic methods of behaviour research. While he was the first to acknowledge the influence which earlier workers have had on this thinking—notably O. Heinroth, Ch. O. Whitman, and W. Craig—he must be considered the principal designer of a coherent biological science of behaviour. He proposed many new concepts which, in spite of subsequent modifications, have remained extremely fertile: Appetitive behavior and Fixed motor pattern ("Erbkoordination"), Imprinting, the "Innate Releasing Mechanism," Social Releasers or Signals, Sign Stimuli, the Innate Disposition to Learn. The influence of Lorenz's approach has been felt not only in various branches of zoology and neurophysiology but also in human psychology, psychiatry and psychoanalysis.[65]

Repeating the above statement in an updated draft of 1970, Tinbergen went on to add (though perhaps without running it by his friend):

> Lorenz was undoubtedly a great pioneer of scientific thought who carried the behavioral sciences a significant step further towards a more unified and biologically oriented science. He was by nature more a visionary, and an intuitive interpreter than an experimental verifier. While he himself had a great admiration for the art of the experimenter, his critics objected increasingly to what they considered his missionary zeal, and failure to lend a sympathetic ear to proposed modifications of his views. In their turn, many critics, having turned to more penetrating but less broad studies, have shown an unwarranted lack of appreciation for the unparalleled first-hand knowledge of animal behaviour that Lorenz possessed.

Tinbergen concluded the 1970 version of the obituary by observing: "Like all pioneers, Lorenz will be irreplaceable, and the presence of his imposing, deeply committed, and exuberant personality at international gatherings will be sadly missed."[66]

This, for Tinbergen, was the Lorenz who deserved to be remembered and respected for all he had done for the study of animal behavior. The obituary sketch captured in broad strokes the great strengths of the Austrian naturalist. Some light shading in the portrait suggested that Lorenz was not a man free of flaws. Nonetheless, Tinbergen's respect for Lorenz as a colleague and friend shone through the piece. At the same time, Tinbergen acknowledged privately to his contact person at the *Times*, "I cannot help agreeing in part with what some of [Lorenz's] critics say, in particular since the publication of his book "On Aggression"; and being a personal friend of long standing, I feel too near for comfort."[67]

Tinbergen did not spell out in the obituary the downsides of his friendship with Lorenz. He did not say that being Lorenz's friend was at times hard work because Lorenz could be just as exasperating as he was engaging and inspiring. He did not mention that although he had forgiven Lorenz for having supported the Nazis, he had not failed to notice that Lorenz had never publicly disavowed the pro-Nazi statements he made in the late 1930s and early 1940s. Nor did he elaborate on how poorly, as it seemed to him, Lorenz responded to criticism. There were a number of issues on which he wanted to press Lorenz—the most prominent of these being the whole question of what could be called "innate"—but Lorenz was very defensive about his own position, and this made it difficult to argue with him. As Tinbergen saw it, Lorenz was too wedded to the notion of innateness and too limited in his conception of development. Beyond this he was too indisposed to appreciate the importance of ecological context, and furthermore too inclined simply to speculate when his ideas should instead have been subjected to careful analysis and experimentation.

When Lorenz published in the 1970s his slender tract entitled *Civilized Man's Eight Deadly Sins,* Tinbergen found the book seriously flawed. He drafted his own view of the predicaments facing modern humankind, but he never brought the book to completion. One of the reasons he failed to do so was his expectation that Lorenz would feel hurt if Tinbergen publicly contradicted him. Still and all, Tinbergen was steadfast in his defense of Lorenz as the primary founder of ethology, the man whose energies, passions, and seminal ideas had given the field its initial life.

Historians of science have an obvious self-interest in emphasizing the importance of history. It may be, however, that no historian of science summed up the case for history any better than did the British biologist J. B. S. Haldane. In 1938, the same year in which Konrad Lorenz published his first article on the degenerative effects of domestication and civilization (and years before Haldane's life became entangled with Lorenz's in a personal way), Haldane published a book entitled *Heredity and Politics.* Combining his commanding knowledge of genetics with his great antipathy for fascism, Haldane spelled out the fundamental flaws of Nazi views of genetics and race. He concluded his book with the following observation: "If we hope to be successful in any political or social endeavour there are two prerequisites besides good will. We must examine the system with which we have to deal, and we must examine ourselves. We must find out what we take for granted in the field of social science, and then ask ourselves why we take it for granted, a much more difficult question. We must remember that the investigator, whether a biologist, an economist, or a sociologist, is himself a part of history, and that if he ever forgets that he is a part of history he will deceive his audience and deceive himself." [68]

Lorenz's case demonstrates Haldane's point. Lorenz insisted on the significance of the *biological* history of the human species for understanding human behavior, but he was never particularly reflective on how his own, more immediate, cultural, political, and social history affected his scientific ideas and practices and the stories he told about animals and hu-

10.4 Through the eyes of the hunter: Niko Tinbergen's photograph of fox tracks in the dunes.
(Photo courtesy of the Tinbergen family. Copy provided by Hans Kruuk and Jaap Tinbergen.)

mans. Tinbergen, in contrast, was more disposed to appreciate the importance of context for the behavior of animals and scientists alike. He explained the behavior of kittiwakes and black-headed gulls in the same
terms he used to describe how he set up his program at Oxford, namely, as
compromises worked out in the face of the diverse selective pressures of
particular environments. He likewise appealed to context when he vari-
ously pointed out to Lorenz that the assertive style of Germanic academic
disclosure was not well suited to winning Anglo-American converts, and
that the postwar scientific climate in Germany was influenced by the Nazis'
previous misapplication of biology. More generally, he recognized that "all
scientific endeavour is a function of attitudes in society as a whole."[69]

One of Tinbergen's virtues was a hardheaded, critical attitude toward
what any series of observations, experiments, and arguments proved or did
not prove. He at least once chided his friend Lorenz for using the word "in-

dubitably" in precisely those places where Lorenz himself seemed to recognize that his argument was weakest.[70] Tinbergen was much more prone in his own lectures and writings to emphasize "how much we do not know." But along with this virtue was his ability to appreciate the value of the insights and intuitions of a thinker like Lorenz, even when Lorenz's scientific and personal styles were so very different from Tinbergen's own.

Indeed, from the standpoint of the overall history of ethology, it is not a final weighing of the different contributions of the two men that seems most important but instead an appreciation of what they accomplished together in their long and productive interaction with each other. Crucial for the construction of ethology as a scientific discipline was the way each man recognized and benefited from what was best about the other. In their later years, they wistfully recalled to one another their early days together before the war. "What happy times they were!" remembered Tinbergen, and Lorenz agreed: "The summer in Altenberg where we rolled greylag eggs and dug ponds together was probably the most beautiful in my life."[71] The scars of war notwithstanding, they continued to be friends—and critical resources for each other—for the rest of their lives. In the course of reflecting on "ethology's ecologies," it is important to remember the host of other actors whose efforts, together with those of Lorenz and Tinbergen, helped create the new science of ethology in the twentieth century. At the same time, it is necessary to underscore the special roles Lorenz and Tinbergen each played in the intellectual landscape of the other.

As for the issue of where animal behavior studies belong within the life sciences, this was not the kind of issue that the founders of ethology could resolve once and for all. Ethology and the life sciences are continuing to evolve. Scientists who study animal behavior today have new social, institutional, and political ecologies to confront. In the complex collaborations and competitions that will be part of this interaction, matters of place are certain to play an important major role. We can imagine that those charting directions for the future will use Tinbergen's identification of the "biological approach" as a touchstone. We can also predict that philosophical, political, social, and ideological constructions of human nature will intersect, if not always in straightforward ways, with scientists' understandings of the behavior of animals. Might an appreciation of ethology's history have a role to play here as well? Haldane's words appear worth heeding. If ethologists, biologists, historians, and laypersons alike remember that we are a part of history, this may help us keep some perspective on how much we know about the behavior of animals and ourselves and how much we have yet to learn.

INTRODUCTION

1. R. A. Hinde and W. H. Thorpe, "Nobel recognition for ethology," *Nature* 245 (1973): 346.

2. E. B. Poulton, *The Colours of Animals, Their Meaning and Use, Especially Considered in the Case of Insects* (New York: Appleton, 1890), p. 286.

3. For an early critique of the experimentalists' view, see the special section in the spring 1981 issue of the *Journal of the History of Biology* introduced by Jane Maienschein, Ronald Rainger, and Keith R. Benson, "Introduction: were American morphologists in revolt?" *Journal of the History of Biology* 14 (1981): 83–87. See, more recently, Lynn K. Nyhart, "Natural history and the 'new' biology," in *Cultures of Natural History,* ed. N. Jardine, J. A. Secord, and E. C. Spary (Cambridge: Cambridge University Press, 1996), pp. 426–443.

4. See Richard W. Burkhardt, Jr., "Ethology, natural history, the life sciences, and the problem of place," *Journal of the History of Biology* 32 (1999): 489–508.

5. William Morton Wheeler, "'Natural history,' 'oecology' or 'ethology,'" *Science,* n.s., 15 (1902): 971–976.

6. Edmund Selous, *Bird Life Glimpses* (London: Allen, 1905), pp. 49–50.

7. Works relating to the history of ethology include P. H. Gray, "The early animal behaviorists: prolegomenon to ethology," *Isis* 59 (1968): 372–383; Julian Jaynes, "The historical origins of 'ethology' and 'comparative psychology,'" *Animal Behaviour* 17 (1969): 601–606; W. H. Thorpe, *The Origins and Rise of Ethology* (London: Heinemann, 1979); John R. Durant, "Innate character in animals and man: a perspective on the origins of ethology," in *Biology, Medicine and Society, 1840–1940,* ed. Charles Webster (Cambridge: Cambridge University Press, 1981), pp. 157–192; Kerstin Berminge, *Två Etologer: En vetenskapsteoretisk analys av Konrad Lorenz' forskarutveckling jämförd med Niko Tinbergens* (Göteborg: Institutionen för Vetenskapsteori, Göteborgs Universitet, 1988); Philippe Chavot, "Histoire de l'éthologie: recherches sur le développement des sciences du comportement en Allemagne, Grande-Bretagne et France, de 1930 à nos jours" (thèse de doctorat en histoire des sciences, Université Louis Pasteur—Strasbourg I, 1994); and D. R. Röell, *De wereld van instinct: Niko Tinbergen en het ontstaan van de ethologie in Nederland (1920–1950)* (Rotterdam: Erasmus Publishing, 1996), translated as *The World of Instinct: Niko Tinbergen and the Rise of Ethology in the Netherlands (1920–1950)* (Assen: Van Gorcum: 2002).

See too the articles by the present author: R. W. Burkhardt, Jr., "On the emergence of ethology as a scientific discipline," *Conspectus of History* 1 (1981): 62–81, "The development of an evolutionary ethology," in *Evolution from Molecules to Men,* ed. D. S. Bendall (Cambridge: Cambridge University Press, 1983), pp. 429–444, and "Charles Otis

Whitman, Wallace Craig, and the biological study of behavior in America, 1898–1924," in *The American Development of Biology,* ed. Ronald Rainger, Keith R. Benson, and Jane Maienschein (Philadelphia: University of Pennsylvania Press, 1988), pp. 185–218; Gregg Mitman and Richard W. Burkhardt, Jr., "Struggling for identity: the study of animal behavior in America, 1930–1945," in *The Expansion of American Biology,* ed. Keith Benson, Ronald Rainger, and Jane Maienschein (New Brunswick, NJ: Rutgers University Press), pp. 164–194; and Richard W. Burkhardt, Jr., "Le comportement animal et l'idéologie de domestication chez Buffon et les éthologistes modernes," in *Buffon 88: Actes du Colloque international pour le bicentenaire de la mort de Buffon,* ed. J.-C. Beaune et al. (Paris: J. Vrin, 1992), pp. 569–582, "Julian Huxley and the rise of ethology," in *Julian Huxley: Biologist and Statesman of Science,* ed. C. Kenneth Waters and Albert Van Helden (Houston: Rice University Press, 1992), pp. 127–149, "Le comportement animal et la biologie française (animal behavior and French biology), 1920–1950," in *Les Sciences biologiques et médicales en France, 1920–1950,* ed. Claude Debru, Jean Gayon, and Jean-François Picard (Paris: CNRS Éditions, 1994), pp. 99–111, and "The founders of ethology and the problem of animal subjective experience," in *Animal Consciousness and Animal Ethics: Perspectives from the Netherlands,* ed. Marcel Dol et al. (Assen: Van Gorcum, 1997), pp. 1–13.

8. Letter from Niko Tinbergen to Richard W. Burkhardt, Jr., 28 March 1979.

9. On Frisch's career see especially Karl von Frisch, *A Biologist Remembers* (Oxford: Oxford University Press, 1967). See also Bert Holldobler and Martin Lindauer, eds., *Experimental Behavioral Ecology and Sociobiology: In Memoriam Karl von Frisch, 1886–1982* (Sunderland, MA, 1985); and Richard W. Burkhardt, Jr., "Karl von Frisch," *Dictionary of Scientific Biography* 17, supp. 2 (1990): 312–320.

10. C. O. Whitman, "A biological farm: for the experimental investigation of heredity, variation and evolution and for the study of life-histories, habits, instincts and intelligence," *Biological Bulletin* 3 (1902): 214–224, quote on p. 222.

11. Georges Bohn, *La nouvelle psychologie animale* (Paris: Félix Alcan, 1911), pp. 1–2.

12. Daniel S. Lehrman, "A critique of Konrad Lorenz's theory of instinctive behavior," *Quarterly Review of Biology* 28 (1953): 337–363, quote on p. 337. Lehrman did not note in his review the existence of different traditions of animal behavior study within Europe. Ethology as developed by Lorenz and Tinbergen was distinct from the subjectivist animal psychology of J. A. Bierens de Haan and others and from the French research that had focused primarily on insect sociality.

13. Tinbergen to Peter and Jean Medawar, 18 January 1974, Nikolaas Tinbergen Papers, Bodleian Library, Oxford University.

14. Konrad Z. Lorenz, *Vergleichende Verhaltensforschung: Grundlagen der Ethologie* (Vienna: Springer-Verlag, 1978), quote on p. 1, translated as *The Foundations of Ethology* (New York: Springer-Verlag, 1981), quote on p. 1

15. For treatments of scientific knowledge from an ecological perspective, see Charles Rosenberg, "Toward an ecology of knowledge: on discipline, context, and history," in *The Organization of Knowledge in Modern America, 1860–1920,* ed. Alexander Oleson and John Voss (Baltimore: Johns Hopkins University Press 1979), pp. 440–455; and Susan Leigh Star and James R. Griesemer, "Institutional ecology, 'translations' and boundary objects: amateurs and professionals in Berkeley's Museum of Vertebrate Zoology, 1907–1939," *Social Studies of Science* 19 (1989): 387–420.

16. See, for example, Bruno Latour, *Science in Action* (Cambridge, MA: Harvard University Press, 1987); and Andrew Pickering, *The Mangle of Practice: Time, Agency, and Science* (Chicago: University of Chicago Press, 1995).

17. Peter Medawar, "Is the scientific paper fraudulent?" *Saturday Review,* 1 August 1964, pp. 42–43. See too the discussion of Medawar's point in Gerald L. Geison, *The Private Science of Louis Pasteur* (Princeton, NJ: Princeton University Press, 1995), pp. 13–15.

18. Niko Tinbergen to R. W. Burkhardt, 16 June 1982, followed by an abbreviated version of the same letter, 19 June 1982. Copies of both letters are in the Tinbergen Papers Oxford University. I have the original of the second.

CHAPTER ONE

1. C. O. Whitman, "Animal behavior," in *Biological Lectures from the Marine Biological Laboratory of Wood's Holl, Mass., 1898* (Boston: Ginn and Company, 1899), pp. 285–338. Whitman's lecture is reprinted in *Defining Biology: Lectures from the 1890s,* ed. Jane Maienschein (Cambridge, MA: Harvard University Press, 1986). Lorenz's statement comes from Konrad Lorenz, *Vergleichende Verhaltensforschung: Grundlagen der Ethologie* (Vienna: Springer-Verlag, 1978), p. 3.

2. Whitman, "Animal behavior," pp. 310n, 328.

3. C. O. Whitman, "Some of the functions and features of a biological station," in *Biological Lectures Delivered at the Marine Biological Laboratory of Wood's Holl,, 1896–1897* (Boston: Ginn and Company, 1998), pp. 231–242, quote on p. 241, and "A biological farm, for the experimental investigation of heredity, variation and evolution and for the study of life-histories, habits, instincts and intelligence," *Biological Bulletin* 3 (1902): 214.

4. Whitman, "Some of the functions and features of a biological station," pp. 240–241. On the differences between Whitman and Loeb, see Philip J. Pauly, *Controlling Life: Jacques Loeb and the Engineering Ideal in Biology* (New York: Oxford University Press, 1987), pp. 80–81.

5. On Whitman, see Frank R. Lillie, "Charles Otis Whitman," *Journal of Morphology* 22 (1911): xv–lxxvii; Charles B. Davenport, "The personality, heredity and work of Charles Otis Whitman, 1843–1910," *American Naturalist* 51 (1917): 5–30; Edward S. Morse, "Charles Otis Whitman," *Biographical Memoirs, National Academy of Sciences* 70 (1912): 269–288; Ernst Mayr, "Whitman, Charles Otis," *Dictionary of Scientific Biography* 14 (1976): 313–315; Jane Maienschein, "Whitman at Chicago: establishing a Chicago style of biology?" in *The American Development of Biology,* ed. Ronald Rainger, Keith R. Benson, and Jane Maienschein (Philadelphia: University of Pennsylvania Press, 1988), pp. 151–182; and Philip J. Pauly, "From adventism to biology: the development of Charles Otis Whitman," *Perspectives in Biology and Medicine* 37 (1994): 395–408.

6. On the childhood interest in animals on the part of ethologists see Konrad Lorenz, "Introduction: the study of behavior," in *Bird Life,* by Jürgen Nicolai (New York: G. P. Putnam's Sons; London: Thames and Hudson, 1974), p. 15. On Whitman's youth, see Davenport, "Whitman," p. 20; and Lillie, "Whitman," p. xvi. The quote is not directly from Whitman but is instead third hand. It is Lillie's report of Craig's account of what Whitman told him.

7. Pauly, "From adventism to biology."

8. Lillie, "Whitman," p. xviii.

9. C. O. Whitman, "The embryology of Clepsine," *Quarterly Journal of Microscopical Science* 18 (1878): 215–315. The quote, on p. 263, is cited here from Maienschein, "Whitman at Chicago," p. 169.

10. Though Whitman appears to have had an interest in behavior throughout his career, his serious studies of behavior, according to Harvey Carr, were concentrated in two periods, 1895–1898 and 1903–1907. The first period resulted in his publication of two papers, "Animal behavior" and "Myths in animal psychology," *Monist* 9 (1899): 524–537. In the second period, Whitman focused primarily on the reproductive behavior of his pigeons. This second period did not result in any publications on behavior in Whitman's lifetime. Selections from Whitman's behavioral manuscripts from this period are incorporated in *Posthumous Works of Charles Otis Whitman,* vol. 3, *The Behavior of Pigeons,* ed. Harvey A. Carr (Washington, DC: Carnegie Institution of Washington, 1919). On the periodization of Whitman's behavior studies, see Carr, "Editorial statement," in *The Behaviour of Pigeons,* p. v. Whitman's definition of biology is from "Myths in animal psychology," p. 524. Mary Alice Evans and Howard Ensign Evans, *William Morton Wheeler, Biologist* (Cambridge, MA: Harvard University Press, 1970), p. 9, report the anecdote of how Whitman's students greeted one another.

11. Lillie, "Whitman," p. lvii.

12. Whitman, "Animal behavior," pp. 288, 296.

13. Whitman, "Animal behavior," p. 302 (Whitman's italics). Four years later in his "biological farm" proposal Whitman made it clear that he regarded freedom from harassment to be as important for the scientist as for his animal subjects. Only under special circumstances could the investigator "have the unbroken quiet required in delicate observation, or expect natural behavior from the forms occupying his attention." Zoological and botanical parks—so familiar to *and so frequented by* the public—were not adequate for the purpose. See "A biological farm," p. 221.

14. Whitman, "Animal behavior," pp. 311, 299.

15. Whitman, "Animal behavior," pp. 309, 332.

16. Whitman, "Animal behavior," pp. 310n, 328. On Whitman's use of behavioral characters as well as color patterns to reconstruct the evolutionary history of the pigeons, see Lillie, "Whitman," p. lxiv. On the retrospective appreciation of Whitman's insight on using behavior patterns for phylogenetic purposes, see, for example, Konrad Lorenz, "Part and parcel in animal and human societies," in *Studies in Animal and Human Behaviour,* trans. Robert Martin, 2 vols. (Cambridge, MA: Harvard University Press, 1970–1971), 2:130–131.

17. Whitman, "Animal behavior," pp. 322, 323. See also Whitman's comments on the incubation instinct in birds, p. 328.

18. Whitman, "Animal behavior," pp. 306, 308–309.

19. Whitman, "Animal behavior," p. 321.

20. Whitman, "Animal behavior," p. 329. Whitman displayed a view typical of biologists and psychologists of the time when he wrote: "In the human race instinctive actions characterize the life of the savage, while they fall more and more into the background in the more intellectual races" (p. 311). Whitman also discussed the relations of instinct and intelligence in a letter of 25 September 1906 to the *Chicago Tribune* (in response to an article by E. Ray Lankester on the origins of intelligence).

21. Whitman, "Animal behavior," pp. 335–336.

22. Whitman, "Animal behavior," pp. 336, 338.

23. Whitman, "Myths in animal psychology," pp. 537, 538.

24. Whitman, "Myths in animal psychology," pp. 524–525.

25. In the last decade of his life, Whitman published only eight papers: three brief notices regarding his research on the heredity and evolution of the color patterns of pigeons, two articles on the Marine Biological Laboratory, his "biological farm" article, the address he delivered at the St. Louis exposition of 1904, "The problem of the origin of species," and another, shorter discussion of the problem of the origin of species. See Lillie, "Whitman," pp. lxxvi–lxxvii, for the complete references.

26. Lillie, "Whitman," p. lxxi.

27. Emily Whitman to R. S. Woodward, 27 September 1911, Charles Otis Whitman file, Carnegie Institution of Washington Archives.

28. Whitman, *Posthumous Works*, 3:28.

29. Konrad Lorenz was the first to underscore the importance of imprinting as a phenomenon for scientific study, but he was not the first to discover the phenomenon. See below, chapter 3.

30. On Whitman's use of this technique in his hybridization experiments, see Whitman, *Posthumous Works*, 3:28.

31. Charles Otis Whitman, "The problem of the origin of species," in *Congress of Arts and Science, Universal Exposition, St. Louis, 1904*, ed. Howard J. Rogers (Boston: Houghton, Mifflin and Company, 1906), 5:41–58, quotes from pp. 51, 50.

32. Whitman, "The problem of the origin of species," p. 57.

33. Whitman, "The problem of the origin of species," p. 44.

34. In a February 1908 address to the Wisconsin Natural History Society Whitman called Mendel's work "brilliant," and he stated further, "a finer model of experimental work and careful analysis has not been seen." Nonetheless, ten years of hybridization experiments of his own on wild and tame species of pigeons disposed Whitman to believe, as he put it, that "Mendel's experiments with different varieties of peas, important as they are, do not reveal any fundamental and universal law of heredity." See *Posthumous Works of Charles Otis Whitman*, ed. Oscar Riddle, vol. 2, *Inheritance, Fertility, and the Dominance of Sex and Color in Hybrids of Wild Species of Pigeons* (Washington, DC: Carnegie Institution of Washington, 1919), pp. 180, 182. See too C. O. Whitman, "The origin of species," *Bulletin of the Wisconsin Natural History Society* 5 (1907): 6–14.

35. Lillie, "Whitman," pp. xxxix, lxxi.

36. Emily Whitman to Woodward, 27 September 1911. In a later letter to Woodward (undated, but apparently from late February 1912), Emily Whitman stated: "In the purchase of the expensive birds we were aided by my brother Mr. L. L. Nunn—who helped us with these as with other enterprises—as the Morphological Journal and the Woods Holl Laboratory." Whitman had kept herons and flickers as well as pigeons, but his wife disposed of the herons and flickers after his death because they were too expensive and too difficult to keep. Whitman file, Carnegie Institution of Washington Archives.

37. Riddle to Woodward, 26 September 1911; Woodward to Davenport, n.d.; Davenport to Woodward, 4 October 1911 (replying to a letter from Woodward, 30 September 1911), Whitman file, Carnegie Institution of Washington Archives.

38. Mathews to Woodward, 27 November 1911, Whitman file, Carnegie Institution of Washington Archives.

39. The letter from Woodward to Riddle, 19 February 1912, and the articles of agreement are in the Whitman file, Carnegie Institution of Washington Archives. Riddle edited volumes 1 (*Orthogenetic Evolution in Pigeons*) and 2 of Whitman's *Posthumous Works*, while Carr edited volume 3.

40. Mayr, "Whitman, Charles Otis"; Stephen Jay Gould, *The Structure of Evolutionary Theory* (Cambridge, MA: Harvard University Press, 2002), pp. 383–395.

41. Konrad Lorenz, "Inductive and teleological psychology," in Lorenz, *Studies in Animal and Human Behaviour*, 1:361.

42. A brief obituary of Craig, written by A. W. Schorger, appeared in *Auk* 71 (1954): 496. Additional information on Craig's life up to 1918 appears in *The Alumni Record of the University of Illinois*, ed. Franklin W. Scott ([Urbana]: University of Illinois, 1918), p. 109. See also Richard W. Burkhardt, Jr., "Charles Otis Whitman, Wallace Craig, and the biological study of behavior in America, 1898–1924," in Rainger, Benson, and Maienschein, *The American Development of Biology*, pp. 185–218; and Theodora J. Kalikow and John A. Mills, "Wallace Craig (1876–1954), ethologist and animal psychologist," *Journal of Comparative Psychology* 103 (1989): 282–288. Further information on Craig's activities is to be found in Craig's manuscript correspondence. The present study has employed Craig letters from the American Philosophical Society (the C. B. Davenport and Leonard Carmichael Papers) and the archives of the University of Illinois at Urbana-Champaign (the Papers of the Illinois Natural History Survey), Western Michigan University (the C. C. Adams Papers, hereafter cited as Adams Papers, UWM), and Yale University (the Robert M. Yerkes Papers, hereafter RMYP/Yale). I am grateful to Robert A. Croker for providing me information about the Craig correspondence in the Adams Papers and to Robert Kohler for calling my attention to the Craig letters in the Papers of the Illinois Natural History Survey (see Natural History Survey Chief's Office, Director's Letterbooks, 1877–1911, University Archives, University of Illinois at Urbana-Champaign, henceforth UIUC Archives).

43. Wallace Craig, "On the early stages of the development of the urogenital system of the pig" (bachelor of science thesis, College of Science, University of Illinois, 1898), UIUC Archives, quote from p. 1.

44. According to a letter from Craig to Forbes dated 6 February 1908, Craig worked for three months as an apprentice without pay before he began earning a modest salary. Natural History Survey Chief's Office, Director's Letterbooks, 1877–1911, UIUC Archives.

45. Stephen Forbes, *Biennial Report of the Director for 1897–1898, Illinois State Laboratory of Natural History* (Urbana, IL, 1899), p. 4. See also Forbes's letter to Craig, dated 19 August 1898, spelling out Craig's duties as assistant naturalist. Natural History Survey Chief's Office, Director's Letterbooks, 1877–1911, UIUC Archives.

46. George W. Bennett, "A century of biological research: aquatic biology," *Illinois Natural History Survey Bulletin* 27 (1958): 163–178, see p. 165; Stephen Alfred Forbes and Robert Earl Richardson, *The Fishes of Illinois* (Urbana: Illinois State Laboratory of Natural History, 1908), p. xi; Craig to Adams, 12 December 1898, Adams Papers, UWM.

47. Craig to Forbes, 19 March 1899, Natural History Survey Chief's Office, Director's Letterbooks, 1877–1911, UIUC Archives.

48. Wallace Craig, "On the fishes of the Illinois River System at Havana, Ill." The title page of this manuscript identifies it as "Thesis for the Degree of Master of Science in the University of Illinois" and is dated 1898. The date would appear to be an error, given that Craig only began his fish studies in 1898 and did not submit the final version of his thesis until 1901, the year his M.S. degree was awarded. He had hopes that his report on his fish research would be published, but this never happened. Craig's correspondence with Forbes on this matter is in the University of Illinois Archives. The manuscript of the thesis is located at the Forbes Biological Station of the Illinois Natural History Survey at Havana, Illinois. I thank Katie Roat for providing me with a copy of the manuscript.

49. Craig to Adams, 24 April 1901. See also Craig to Adams, 2 May 1901, and 11 May 1901, Adams Papers, UWM; and Craig to Forbes, 24 June 1901, UIUC Archives.

50. Wallace Craig, "The voice of pigeons," pt. 1, "The voice of the ring dove (*Turtur risorius*)," *Biological Bulletin of the Marine Biological Laboratory* 6 (1903–1904): 323–324. This report appears along with reports by other researchers who made informal presentations to the research seminar of the Marine Biological Laboratory in summer 1903.

51. Wallace Craig, "The expressions of emotion in the pigeons," pt. 1, "The blond ring dove (*Turtur risorius*)," *Journal of Comparative Neurology and Psychology* 19 (1909): 29–80, quote on p. 31.

52. Craig, "The expressions of emotion in the pigeons," pt. 1, pp. 30–31.

53. Wallace Craig, "North Dakota life: plant, animal and human," *Bulletin of the American Geographical Society* 40 (1908): 321–332, 401–415.

54. For Shelford's response to Craig's paper, which was delivered in abbreviated form in December 1907 at a meeting of the American Association of Geographers, see Gregg Mitman, *The State of Nature: Ecology, Community and American Social Thought, 1900–1950* (Chicago: University of Chicago Press, 1992), pp. 43–47.

55. Wallace Craig, "The voices of pigeons regarded as a means of social control," *American Journal of Sociology* 14 (1908): 86–100. Here, as in his dissertation, Craig acknowledged the "constant, generous, and invaluable" aid Whitman had given him (p. 86n).

56. Craig, "The voices of pigeons regarded as a means of social control," pp. 86, 87.

57. Craig, "The voices of pigeons regarded as a means of social control," p. 88.

58. Craig, "The voices of pigeons regarded as a means of social control," pp. 89–90. Whereas Lorenz tended to theorize about imprinting in the context of distinguishing between innate and learned behavior, Craig treated the phenomenon as illustrative of his general point about the interactions through which the bird becomes a social entity.

59. Craig, "The voices of pigeons regarded as a means of social control," p. 93.

60. Craig, "The voices of pigeons regarded as a means of social control," p. 95–96.

61. Craig, "The voices of pigeons regarded as a means of social control," p. 96.

62. Craig, "The voices of pigeons regarded as a means of social control," p. 100.

63. Craig, "The voices of pigeons regarded as a means of social control," pp. 100,

86. Craig's thoughts on writing a book for Yerkes' animal behavior series are expressed in letters from Craig to Yerkes, 10 January, 13 February, and 13 September 1907, RMYP/Yale.

64. On Craig's feelings about teaching at Maine, see Craig to Adams, 24 May 1913, Adams Papers, UWM. On the courses he offered in his years at Maine, see the *Catalog of the University of Maine, 1908–1909* (Waterville, Me.: Sentinel Publishing Company, 1908), and subsequently.

65. Wallace Craig, "The expressions of emotion in the pigeons," pt. 2, "The mourning dove (*Zenaidura macroura* Linn.)," *Auk* 28 (1911): 398–407, and "The expressions of emotion in the pigeons," pt. 3, "The passenger pigeon (*Ectopistes migratorius* Linn.)," *Auk* 28 (1911): 408–426.

66. Wallace Craig, "Oviposition induced by the male in pigeons," *Journal of Morphology* 22 (1911): 299–305, "Observations on doves learning to drink," *Journal of Animal Behavior* 2 (1912): 273–279, "Behavior of the young bird in breaking out of the egg," *Journal of Animal Behavior* 2 (1912): 296–298, "The stimulation and the inhibition of ovulation in birds and mammals," *Journal of Animal Behavior* 3 (1913): 215–221, and "Male doves reared in isolation," *Journal of Animal Behavior* 4 (1914): 121–133.

67. Craig, "Oviposition induced by the male in pigeons." See also Craig, "The stimulation and inhibition of ovulation in birds and mammals."

68. On the extinction of the passenger pigeon, see especially A. W. Schorger, *The Passenger Pigeon: Its Natural History and Extinction* (Madison: University of Wisconsin Press, 1955).

69. Craig, "The expressions of emotion in the pigeons," pt. 3, p. 409.

70. Craig, "The expressions of emotion in the pigeons," pt. 3, pp. 419–420.

71. Craig, "The expressions of emotion in the pigeons," pt. 3, p. 420.

72. Craig, "The expressions of emotion in the pigeons," pt. 3, p. 417.

73. Craig, "The expressions of emotion in the pigeons," pt. 3, p. 426.

74. Craig to Adams, 24 May 1913, Adams Papers, UWM.

75. Craig to Adams, 24 May 1913, Adams Papers, UWM.

76. Craig to Adams, 24 May 1913, Adams Papers, UWM.

77. Craig to Adams, 18 December 1913, Adams Papers, UWM.

78. Craig to Adams, 18 December 1913, Adams Papers, UWM.

79. Craig to Adams, 14 February 1914, Adams Papers, UWM.

80. Wallace Craig, "Attitudes of appetition and of aversion in doves," *Psychological Bulletin* 11 (1914): 56–57.

81. Wallace Craig, "Appetites and aversions as constituents of instincts," *Biological Bulletin of the Marine Biological Laboratory*, 34 (1918): 91–107. See William McDougall, *An Outline of Psychology*, 6th ed. (London: Methuen, 1933); and Edward Chace Tolman, *Purposive Behavior in Animals and Men* (New York: Century Co., 1932), pp. 236, 245, 272–273, 276, quote from p. 272.

82. Craig, "Appetites and aversions," p. 91.

83. Craig, "Appetites and aversions," p. 92.

84. Craig, "Appetites and aversions," pp. 106–107.

85. University of Maine catalogs, 1914–1920.

86. For Craig's plans with regard to publishing see, for example, Craig to Yerkes,

22 February 1909, RMYP/Yale; Craig to Adams, 10 September 1911, and 4 February 1914, Adams Papers, UWM.

87. Craig to C. B. Davenport, 8 August 1920, Davenport Papers, American Philosophical Society.

88. C. B. Davenport to Craig, 16 August 1920, Davenport Papers, American Philosophical Society. Craig's more theoretical project was perhaps his study of what he called "The Space System of the Perceiving Self," the project for which he received support from the American Philosophical Society in the 1940s. On this subject, see the correspondence from and relating to Craig in the Carmichael Papers, American Philosophical Society.

89. Wallace Craig, "Why do animals fight?" *International Journal of Ethics* 31 (1921): 264–278, and "A note on Darwin's work on the expression of the emotions in man and animals," *Journal of Abnormal and Social Psychology* 16 (1921–1922): 356–366. Together with Oscar Riddle, Craig also prepared the chapter "Voice and instinct in pigeon hybridization and phylogeny" for volume 3 of the *Posthumous Works of Charles Otis Whitman* (above, n. 10). Craig's later work included "The song of the wood pewee *Myiochanes virens* Linnaeus: a study of bird music," *New York State Museum Bulletin*, no. 334 (1943): 1–186.

90. For a more extended analysis of American biologists' denial of biological justifications of militarism, see Gregg Mitman, *The State of Nature*, pp. 58–64, and "Evolution as gospel: William Patten, the language of democracy, and the Great War," *Isis* 81 (1990): 446–463. Mitman does not include Craig's paper in his discussion.

91. Craig, "Why do animals fight?" p. 264.

92. Craig, "Why do animals fight?" pp. 264–265.

93. Craig, "Why do animals fight?" p. 265.

94. Craig, "Why do animals fight?" p. 266.

95. Craig, "Why do animals fight?" p. 267.

96. Craig, "Why do animals fight?" p. 268.

97. Craig, "Why do animals fight?" p. 274. For Lorenz's writings on aggressive behavior, see below, chapter 10.

98. Craig, "Why do animals fight?" p. 276.

99. Craig, "Why do animals fight?" pp. 276–277.

100. Craig, "A note on Darwin's work." A reader today of Darwin's *Expression of the Emotions in Man and Animals* is also likely to be struck by Darwin's reluctance to suppose that animal emotional expression might have evolved in ways serving communicative functions. Interestingly enough, Konrad Lorenz said nothing about this in the preface he wrote for the 1965 University of Chicago Press edition of Darwin's book. In fact, what he does say provides no indication that he ever read the text he was introducing. The present writer, in a paper first delivered in 1982, suggested that Darwin's stance on the noncommunicative character of animal expression was tied to Darwin's desire to refute Charles Bell's claim that God, at the Creation, had endowed humans with special features for purposes of expression. I find that not only did Craig express this idea years earlier, but he also did so in the course of a discussion that was more wide ranging and nuanced than what I offered. See Richard W. Burkhardt, Jr., "Darwin on animal behav-

ior and evolution," in *The Darwinian Heritage,* ed. David Kohn (Princeton, NJ: Princeton University Press, 1985), pp. 435–481.

101. Craig, "A note on Darwin's work," p. 361.

102. Craig, "A note on Darwin's work," p. 366.

103. Craig to Pearl, 13 May 1923; and Craig to Yerkes, 15 May 1923. Pearl wrote to Yerkes on 16 May 1923 stating: "It seems to me that in this case specifically we owe a certain duty to pure science to see that the activity of a valuable brain is not lost to science. . . . Craig does not need a great deal of money, indeed never has had much. His requirements in every way to keep him at useful work are very modest." All three letters are in RMYP/Yale.

104. Craig to Yerkes, 30 July 1923, RMYP/Yale.

105. Craig to Carmichael, 3 February 1936, from Sheffield, England, in the Carmichael Papers, American Philosophical Society.

106. Grace Andrus de Laguna, *Speech: Its Function and Development* (New Haven, CT: Yale University Press, 1927). See especially chapter 2. I gladly thank Georgina Hoptroff for calling de Laguna's use of Craig's ideas to my attention. In an as yet unpublished paper on the work of Clarence Ray Carpenter, Hoptroff describes the influence that de Laguna's work had on Carpenter's thinking about animal calls.

107. On Nice's role in animal behavior studies, see Gregg Mitman and Richard W. Burkhardt, Jr., "Struggling for identity: the study of behavior in America, 1930–1945," in *The Expansion of American Biology,* ed. Keith R. Benson, Jane Maienschein, and Ronald Rainger (New Brunswick, NJ: Rutgers University Press, 1991), pp. 164–194.

108. Margaret Morse Nice, *Research Is a Passion with Me* (Toronto: Consolidated Amethyst Communications, 1979), pp. 140–141; Konrad Lorenz, "Der Kumpan in der Umwelt des Vogels: der Artegenosse als auslösendes Moment sozialer Verhaltungsweisen," *Journal für Ornithologie* 83 (1935): 137–215, 289–413, "Über die Bildung des Instinktbegriffes," *Die Naturwissenschaften* 25 (1937): 289–300, 307–318, 324–331, and *Studies in Animal and Human Behaviour,* 1:xix. A letter from Craig to Lorenz written from Edinburgh and dated 27 April 1937 indicates that Lorenz was not in touch with Craig in 1936. I thank Ingo Brigandt for calling my attention to this and two other letters from Craig to Lorenz in the Lorenz family papers.

109. Lorenz acknowledged that it was with the help of Craig and Erich von Holst that he "at last arrived at the clear conceptualization of appetitive behaviour, innate releasing mechanism and innate motor pattern." See Lorenz, *Studies in Animal and Human Behaviour,* 1:xx. Lorenz defined instinctive and appetitive behavior more narrowly than Craig had originally done, though he allowed in 1942 in his paper "Inductive and teleological psychology" that Craig had come to agree "down to the last detail with my more narrow and precise definition of the two concepts." *Studies in Animal and Human Behaviour,* 1:361. For more on Lorenz's debts to Craig, see chapter 3.

110. Griffin to Carmichael, 5 February 1952, Carmichael Papers, American Philosophical Society.

111. W. H. Thorpe, *The Origins and Rise of Ethology* (London: Heinemann, 1979), pp. 48–49.

112. Items in the Carmichael Papers, American Philosophical Society, including a copy of a letter from C. C. Adams to Gordon Allport of the Harvard Psychology De-

partment, dated 22 June 1954, indicate that Adams passed a copy of Craig's manuscript on to Allport. Kalikow and Mills, "Wallace Craig," report that in their search for Craig papers they were unable to find a copy of this manuscript. On the subject of Craig's death, see Adams to Carmichael, 18 May 1954, Carmichael Papers, American Philosophical Society.

113. In addition to belonging to the APA, Craig belonged to the American Ornithologists' Union, Sigma Xi, and the American Philosophical Society. Of these, only the AOU was specifically oriented toward biological concerns. For Craig's thoughts on psychological influences on physiological processes, see his "Oviposition induced by the male in pigeons."

114. Jennings to Yerkes, 21 January 1906; and Watson to Yerkes, 29 October 1909, RMYP/Yale; Robert M. Yerkes, "Psychology in its relations to biology," *Journal of Philosophy* 7 (1910): 113–124. On the significance of Yerkes' experiences at Harvard for his thinking about the needs of psychology as a discipline, see John M. O'Donnell, *The Origins of Behaviorism: American Psychology, 1870–1920* (New York: New York University Press, 1985), pp. 191–200.

115. On the history of this journal see Richard W. Burkhardt, Jr., "The *Journal of Animal Behavior* and the early history of animal behavior studies in America," *Journal of Comparative Psychology* 101 (1987): 223–230.

116. *Carnegie Institution of Washington Yearbook* 1 (1902): 167, 274–283, and subsequent *Yearbooks.* Eventually, in 1921, Davenport's two primary enterprises at Cold Spring Harbor, the Department of Experimental Evolution and the Eugenics Records Office (the latter having come under the Carnegie wing in 1918), were united and named the Department of Genetics of the Carnegie Institution.

117. On the Tortugas Laboratory, see Alfred Goldsborough Mayer, "Marine Biological Laboratory at Tortugas, Florida," *Carnegie Institution of Washington Yearbook* 3 (1904): 50–54, and subsequent reports in the *Yearbook.* Information on Watson's early applications to the Carnegie Institution is to be found in the Carnegie Institution of Washington Archives. For Watson's early plans to go to the Tortugas, see Watson to Yerkes, 7 February 1907, and 29 March 1907, RMYP/Yale. See also John B. Watson, "The behavior of noddy and sooty terns," *Papers from the Tortugas Laboratory of the Carnegie Institution of Washington* 2 (1908): 187–255; and John B. Watson and K. S. Lashley, "Homing and related activities of birds," *Papers from the Tortugas Laboratory of the Carnegie Institution of Washington* 8 (1915): 1–104.

118. In the first few years of the Carnegie Institution's existence, Jennings had received a grant, an assistantship, and the use of a research table at the Marine Laboratory at Naples. The Carnegie Institution also published Jennings's first monograph on invertebrate behavior. Shortly afterward, the institution's policies with respect to "minor grants" became very strict. As of 1910 the only money being given to zoology by the Carnegie Institution in its "minor grants" category was that given to W. E. Castle for his work on genetics and that given to the Marine Laboratory at Naples for the support of two tables.

119. See Craig to Adams, 4 February 1914, Adams Papers, UWM; and Craig to Pearl (copy), 13 May 1923, RMYP/Yale.

120. On the decline of the study of tropisms, see Philip J. Pauly, "The Loeb-Jennings

debate and the science of animal behavior," *Journal of the History of the Behavioral Sciences* 17 (1981): 504–515.

121. Jennings to Yerkes, 30 October 1907, RMYP/Yale.

122. Jennings to Yerkes, 10 June 1916, RMYP/Yale. The failure of animal behavior studies to take deep root in American zoology departments is illustrated by the fate of behavior courses in several major departments in the first half of the century, notably Columbia, Johns Hopkins, and Berkeley (as seen in the course catalogs of these universities). At Columbia, T. H. Morgan, the professor of experimental zoology, introduced in 1906 a course on tropisms (in addition to his other courses on experimental zoology, experimental embryology, and regeneration), and by 1910 he was also offering a course entitled "The Experimental Study of Instincts." The Columbia catalog continued to list Morgan's courses on tropisms and instincts up into the mid-1920s. How often he actually taught these courses is not clear, however, and the courses were dropped either just after he left Columbia for the California Institute of Technology or shortly before. Morgan, needless to say, had identified other, more rewarding research problems. Samuel J. Holmes, upon his arrival at Berkeley in 1912, took over the course on animal behavior that had been introduced by Harry B. Torrey the previous year and continued to teach it until his retirement in the 1930s. When Holmes retired, however, the course was dropped from the books, and as of 1950 no behavior course had found its way back into the biological curriculum at Berkeley. At Johns Hopkins, H. S. Jennings turned his behavior course over to S. O. Mast around 1910. Mast continued to teach animal behavior until he retired in the 1930s, but after Mast's retirement behavior ceased to be listed among the topics offered by the Hopkins biologists.

123. On the fate of Yerkes' students, see O'Donnell, *The Origins of Behaviorism,* pp. 195–197. On the research interests of APA members, see the American Psychological Association's *Yearbook,* 1921. See also Burkhardt, "The *Journal of Animal Behavior.*"

124. C. C. Adams to Yerkes, 16 February 1932, RMYP/Yale.

125. See especially Gregg Mitman, *The State of Nature;* Nicholas E. Collias, "The role of American zoologists and behavioural ecologists in the development of animal sociology, 1934–1964," *Animal Behavior* 41 (1991): 613–631; and Mitman and Burkhardt, "Struggling for identity."

126. See Allee to Yerkes, 21 June 1943, RMYP/Yale. For Allee's revised comments on Whitman, see W. C. Allee et al., *Principles of Ecology* (Philadelphia: W. B. Saunders Company, 1949), p. 24.

CHAPTER TWO

1. Edmund Selous, *The Bird Watcher in the Shetlands: With Some Notes on Seals— and Digressions* (London: J. M. Dent and Co., 1905), p. 323.

2. Frederick B. Kirkman to H. Eliot Howard, 30 October 1910, H. Eliot Howard Papers, Edward Grey Institute of Field Ornithology, Oxford University.

3. John Ray, *The Wisdom of God Manifested in the Works of Creation* (London: Samuel Smith, 1691), pp. 93–96. I do not mean to suggest here that Ray's work was unprecedented. For an excellent discussion of Ray that puts his work into historical context, see Neal C. Gillespie, "Natural history, natural theology, and social order: John Ray and the 'Newtonian ideology,'" *Journal of the History of Biology* 20 (1987): 1–49.

4. Ray, *Wisdom of God,* pp. 117–118, 124.

5. Gilbert White, *The Natural History and Antiquities of Selborne* (London: Bensely, 1789), pp. 115, 144.

6. On the social function of the activities of the clergyman-naturalist see David Elliston Allen, *The Naturalist in Britain: A Social History* (London: Penguin, 1978), p. 23.

7. White, *The Natural History and Antiquities of Selborne,* p. 94.

8. *The Autobiography of Charles Darwin, 1809–1882, with the Original Omissions Restored,* ed. Nora Barlow (London: Collins, 1958), p. 45; Francis Darwin quoted by James R. Moore, "Darwin of Down: the evolutionist as squarson-naturalist," in *The Darwinian Heritage,* ed. David Kohn (Princeton, NJ: Princeton University Press, 1985), p. 460.

9. Darwin's "C" notebook, p. 79, published in *Charles Darwin's Notebooks, 1836– 1844,* ed. Paul H. Barrett, Peter J. Gautrey, Sandra Herbert, David Kohn, and Sydney Smith (Ithaca, NY: Cornell University Press, 1987), p. 264.

10. Charles Darwin, *On the Origin of Species by Means of Natural Selection* (London: J. Murray, 1859), pp. 243–244. Complex instincts appeared to Darwin to be a serious test for his theory of evolution by natural selection. Ultimately, he found that the instincts of neuter castes of insects constituted a case where his theory of natural selection clearly triumphed over the Lamarckian idea of the inheritance of acquired characters. For further discussion of this point, see especially Robert Richards, *Darwin and the Emergence of Evolutionary Theories of Mind and Behavior* (Chicago: University of Chicago Press, 1987).

11. Charles Darwin, *The Descent of Man, and Selection in Relation to Sex* (London: John Murray, 1871), and *The Expression of the Emotions in Man and Animals* (London: John Murray, 1872). For more on Darwin's thoughts on behavior, see Richards, *Darwin and the Emergence of Evolutionary Theories of Mind and Behavior;* Richard W. Burkhardt, Jr., "The development of an evolutionary ethology," in *Evolution from Molecules to Men,* ed. D. S. Bendall (Cambridge: Cambridge University Press, 1983), pp. 429–444, and "Darwin on animal behavior and evolution," in *The Darwinian Heritage,* ed. David Kohn (Princeton, NJ: Princeton University Press, 1985), pp. 435–481; John R. Durant, "The ascent of nature in Darwin's *Descent of Man,*" in Kohn, *The Darwinian Heritage,* pp. 283–306; and Janet Browne, "Darwin and the expression of the emotions," in Kohn, *The Darwinian Heritage,* pp. 307–326.

12. Darwin, *The Descent of Man* (1871), pp. 48–49.

13. Charles Darwin, *The Descent of Man, and Selection in Relation to Sex,* 2d ed. (New York: D. Appleton and Company, 1896), p. 79.

14. Darwin, *The Descent of Man* (1871), 1:253–320, quotes from pp. 296, 257–258, respectively.

15. Darwin, *The Descent of Man* (1871), 2:122–123.

16. Darwin, *The Descent of Man* (1871), 1:64, 2:123.

17. On the positions Darwin and Wallace held regarding sexual selection, see especially Malcolm Kottler, "Darwin, Wallace, and the origin of sexual dimorphism," *Proceedings of the American Philosophical Society* 124 (1980): 203–226; and Helena Cronin, *The Ant and the Peacock* (Cambridge: Cambridge University Press, 1991).

18. George W. Peckham and Elizabeth G. Peckham, "Observations on sexual selection in spiders of the family Attidae," *Wisconsin Natural History Society Occasional Papers* 1 (1889–1990): 3–60, and "Additional observations on sexual selection in spiders of

the family Attidae, with some remarks on Mr. Wallace's theory of sexual ornamentation," *Wisconsin Natural History Society Occasional Papers* 1 (1889–1990): 117–151.

19. The words are F. B. Kirkman's. See below.

20. In particular by Richards, *Darwin and the Emergence of Evolutionary Theories of Mind and Behavior.*

21. Darwin, *The Descent of Man* (1871), 1:238, 2:326–329.

22. Edmund Selous, *Realities of Bird Life: Being Extracts from the Diaries of a Life-Loving Naturalist* (London: Constable, 1927), p .v.

23. "Men of the day: no. 585: Mr. Frederic Courtney [*sic*] Selous," *Vanity Fair Album*, vol. 26 (1894).

24. Stephen Taylor, *The Mighty Nimrod: A Life of Frederick Courteney Selous, African Hunter and Adventurer, 1851–1917* (London: Collins, 1989), p. 7.

25. E. B. Poulton, *The Colours of Animals, Their Meaning and Use, Especially Considered in the Case of Insects* (New York: Appleton, 1890), p. 287.

26. For biographical details on Selous see Jacques Delamain, "Edmund Selous," *Alauda* 6 (1934): 388–393; Margaret Morse Nice, "Edmund Selous: an appreciation," *Bird-Banding* 6 (1935): 90–96; K. E. L. Simmons, "Edmund Selous (1857–1934): fragments for a biography," *Ibis* 126 (1984): 595–596; and Richard W. Burkhardt, Jr., "Edmund Selous," *Dictionary of Scientific Biography* 18, supp. 2 (1990): 801–803.

27. W. L. Distant was a fellow of the Entomological Society of London, a member of the Anthropological Institute, a longtime friend of the famous naturalist traveler Henry Walter Bates, and a man who himself had collected zoological and anthropological specimens in the Transvaal. He had held forth at the Anthropological Society on the mental differences between men and women, and he had things to say about the habits of the Boers and the natives he encountered in South Africa. His standing and his experiences, colonial and otherwise, made him feel confident he could speak of mental states as they manifested themselves across race, class, gender, and zoological borders. See W. L. Distant, *A Naturalist in the Transvaal* (London: R. H. Porter, 1892).

28. "The shrine of the naturalist," *Saturday Review*, 30 December 1899, p. 826.

29. Edmund Selous, "An observational diary of the habits of night jars (*Caprimulgus europaeus*), mostly of a sitting pair: notes taken at time and on spot," *Zoologist* 3 (1899): 398.

30. "The unknown bird world," *Saturday Review* 91 (1901): 638.

31. On Fowler, see Percy Ewing Matheson, "William Warde Fowler (1847–1921)," in *Dictionary of National Biography, 1912–1921* (London: Oxford University Press, 1927), pp. 194–195; and Julian Huxley, "Obituary: W. Warde-Fowler," *British Birds* 15 (1921): 143–144. In addition to *A Year with the Birds* (1886), Fowler wrote *Kingham Old and New* (Oxford Blackwell, 1913). Like Selous, Fowler also wrote animal books for children.

32. W. Warde Fowler, "In the ways of birds," *Saturday Review* 92 (1901): 104–105.

33. "South African fauna," *Saturday Review* 90 (1900): 397.

34. "New fields for an old hunter," *Saturday Review* 91 (1901): 82. In addition to the review of Edmund Selous's *Bird Watching*, which appeared in July, the *Saturday Review* in November began publishing the first of several years' worth of essays by Edmund Selous on animal watching. See Edmund Selous, "Rabbits and hares," *Saturday Review* 92 (1901): 618–619, 677–679.

35. Frederick Courteney Selous, *A Hunter's Wanderings in Africa* (London: Richard Bentley and Son, 1881).

36. Edmund Selous, *Bird Watching* (London: J. M. Dent and Co., 1901), p. 335.

37. Edmund Selous, "The old zoo and the new," *Saturday Review* 91 (1901): 330–332, 365–366, 397–398, 433–434. Though the series was unsigned, it was subsequently identified as the work of Selous. Selous did not declare himself opposed to zoos per se. In the concluding installments of his article he offered suggestions regarding how an ideal zoo might be constituted. He made it quite clear, however, that the London zoo, as it then existed, was in great need of reform. The Humanitarian League later published this series of articles as a separate pamphlet.

38. Among books Selous aimed at a young rather than an adult audience were his *Beautiful Birds* (London: J. M. Dent and Co., 1901) and a series entitled "Tommy Smith's Animals."

39. On W. H. Hudson, Charles Dixon, and the Society for the Protection of Birds, see Allen, *The Naturalist in Britain.*

40. Edmund Selous, *Bird Life Glimpses* (London: Allen, 1905), pp. v–vi.

41. Edmund Selous, "An observational diary of the habits—mostly domestic—of the great crested grebe (*Podicipes cristatus*)," *Zoologist* 5 (1901): 161–183, quote on p. 173. On how he elaborated on his notes after returning home, see his "An observational diary of the habits of the great plover (*Oedicnemus crepitans*) during September and October," *Zoologist* 4 (1900): 173. Note to the attentive reader: page 173 is indeed the correct citation in the two successive volumes

42. Edmund Selous, "Observations tending to throw light on the question of sexual selection in birds, including a day-to-day diary on the breeding habits of the ruff (*Machetes pugnax*)," *Zoologist* 10 (1906): 201 (henceforth "Sexual selection in birds").

43. "Lion hunting in Somaliland," anonymous review of *Lion Hunting in Somaliland*, by Captain C. J. Melliss, *Saturday Review* 79 (1895): 794; "Natural history and sport," anonymous review of *Letters to Young Shooters*, by Sir Ralph Payne-Gallwey, *Saturday Review* 84 (1897): 199.

44. Selous, "Sexual selection in birds," pp. 215–216.

45. Selous, "Sexual selection in birds," pp. 286, 293.

46. Edmund Selous, "An observational diary of the nuptial habits of the blackcock (*Tetrao tetrix*) in Scandinavia and England," *Zoologist* 13 (1909): 401–413; 14 (1910): 23–29, 51–56, 176–182, 248–265; quote from 14 (1910): 259 (henceforth "Blackcock").

47. Selous, "An observational diary of the habits—mostly domestic—of the great crested grebe (*Podicipes cristatus*)," *Zoologist* 5 (1901): 161–183, and "An observational diary of the habits—mostly domestic—of the great crested grebe (*Podicipes cristatus*) and of the peewit (*Vanellus vulgaris*), with some general remarks," *Zoologist* 5 (1901): 339–350, 454–462; 6 (1902): 133–144.

48. Julian Huxley, "Introduction," in Selous, *Realities of Bird Life*, p. xii.

49. Selous, *The Bird Watcher in the Shetlands*, p. 44.

50. Edmund Selous, "Variations in colouring of *Stercorarius crepidatus*," *Zoologist* 6 (1902): 368–373, and *Bird Watching*, pp. 169, 170, 176.

51. Selous, *The Bird Watcher in the Shetlands*, pp. 261–283.

52. Selous, "Sexual selection in birds," p. 216.

53. Selous, "Sexual selection in birds," pp. 175–176.

54. Selous, "Sexual selection in birds," p. 163.

55. Selous, "Sexual selection in birds," pp. 374–375.

56. Selous, "Sexual selection in birds," p. 381.

57. Selous, "Blackcock," *Zoologist* 14 (1910): 260–261, 263.

58. Selous, "Blackcock," *Zoologist* 14 (1910): 55.

59. Selous, "Blackcock," *Zoologist* 14 (1910): 257–258.

60. Selous, "Blackcock," *Zoologist* 14 (1910): 261.

61. Selous, "Blackcock," *Zoologist* 14 (1910): 265.

62. Edmund Selous, "The finches" (pp. 83–156), "The buntings" (pp. 169–198), "The wagtails" (pp. 239–259), and "The pipits" (pp. 260–277), in *The British Bird Book*, ed. F. B. Kirkman, vol. 1 (London: T. C. and E. C. Jack, 1910).

63. See Edmund Selous Papers, Edward Grey Institute of Field Ornithology, Oxford University: copies of letters from Selous to Kirkman, dated 29 July 1910 in Selous notebook entitled "Notebook 4, July 27th 1910 North Roe."

64. For example, Selous, "The buntings," p. 189 n. 3.

65. Kirkman to Selous, 3 Jan 1910, Selous Papers, Oxford University, "The finches," p. 88, and "The buntings," pp. 183, 187, 198; Kirkman to Howard, 30 October 1910, Howard Papers, Oxford University. Selous had been scheduled to write for the work's later volumes the chapters on the nightjar, the auks, the skuas, the rails, the heron, the bittern, and the petrels.

66. Edmund Selous, "Origin of the social antics and courting displays of birds," *Zoologist* 16 (1912): 197–199, apparently responding to W. Farren, "The ringed-plovers" and "The lapwing," in Kirkman, *The British Bird Book*, vol. 3 (1912), pp. 340–356 and 370–386; Selous, "The nuptial habits of the blackcock," *Naturalist* (1913), pp. 96–98, responding to W. P. Pycraft, "The grouse subfamily," in Kirkman, *The British Bird Book*, vol. 4 (1913), pp. 12–36.

67. Selous, "The nuptial habits of the blackcock," *Naturalist* (1913), p. 96.

68. Selous, *The Bird Watcher in the Shetlands*, p. 114.

69. Selous, *The Bird Watcher in the Shetlands*, p. 266.

70. Selous, *The Bird Watcher in the Shetlands*, p. 254. To Lubbock's (Lord Avebury's) additional claim that travel was "one great source of pleasure which civilised people enjoy, but which savages do not," Selous responded: "He should have restricted the proposition to civilised women. No word more terrible in the ears of a husband than 'Paris' on the lips of a wife. What worry, what anxiety, fear of adventurers, horror of waiters, hatred of hotels—what misery, in short, of almost every degree and kind, do not men go through who have to travel with their families! How they would all stay home if they only could, and how glad they are . . . when they get back!" (p. 255).

71. See Edmund Selous, *Thought-Transference (or What?) in Birds* (London: Constable, 1931), and *Evolution of Habit in Birds* (London: Constable, 1933).

72. Nice, "Edmund Selous," 90–96; Delamain, "Edmund Selous," pp. 388–393.

73. Nice, "Edmund Selous," p. 95.

74. Selous, *Bird Life Glimpses*, pp. 49–50.

75. This discussion of Howard's daily activities and immediate surroundings is based on Percy R. Lowe, "Henry Eliot Howard: an appreciation," *British Birds* 34 (1941):

195–197. See also David Lack, "Some British pioneers in ornithological research, 1859–1939," *Ibis* 101 (1959): 71–81. Howard's discussion of the reed warbler indicates that his bird-watching was not wholly restricted to the vicinity of his home, for he mentions seeing these birds in Hungary in 1905. See H. Eliot Howard, "Reed warbler," in *The British Warblers: A History with Problems of Their Lives* (London: R. H. Porter, 1907–1914; pt. 5, 1910), p. 5.

76. H. E. Howard, "The grasshopper-warbler (*Locustella naevia*) in north Worcestershire," *Zoologist* 5 (1901): 60–63, "On Mr. Selous' theory of the origin of nests," *Zoologist* 6 (1902): 145–148, "On sexual selection and the aesthetic sense in birds," *Zoologist* 7 (1903): 407–417. Darwin, in *The Descent of Man*, chapter 13, cites the observations of his correspondent, Jenner Weir, on the behavior of captive native birds. He also cites Abraham Bartlett's observations on foreign birds in captivity at the London Zoological Gardens, and he cites observations from abroad.

77. Howard, "On sexual selection and the aesthetic sense in birds," pp. 411–413. The passage quoted represents Howard's rendering of this part of Wallace's argument, not Wallace's own words.

78. W. L. Distant, "Preface," *Zoologist* 7 (1903): iii–iv.

79. H. E. Howard, "Grasshopper warbler," in Howard, *The British Warblers*, pt. 1 (1907), especially pp. 14–18.

80. H. E. Howard, "Chiff-chaff," in Howard, *The British Warblers*, pt. 2 (1908), pp. 8–9.

81. See especially H. E. Howard, "Reed warbler," p. 38, and "General summary and concluding remarks," in Howard, *The British Warblers*, pt. 9 (1914), pp. 2–7.

82. Margaret Morse Nice, "The theory of territorialism and its development," in *Fifty Years' Progress of American Ornithology (1883–1933)*, by American Ornithologists' Union (Lancaster, PA: American Ornithologists' Union, 1933), pp. 89–100, quote on p. 90.

83. Howard, "Chiff-chaff," pp. 26n, 28.

84. The Howard Papers at the Edward Grey Institute, Oxford University, contain more than 120 letters and postcards from Morgan to Howard, dating from 1910 to 1935, the year before Morgan's death. Morgan recommended to Howard in a letter of 5 February 1915 that Howard publish "a little book embodying your leading conclusions." Morgan evidently did not save his letters from Howard with the care that Howard saved his letters from Morgan, for the Conwy Lloyd Morgan Papers in the Bristol University Archives contain only ten letters from Howard to Morgan. According to David Lack, there were those who felt that Morgan had too strong an influence on Howard. See Lack, "Some British pioneers in ornithological research," p. 73.

85. Morgan to Howard, 29 May 1912, Howard Papers, Oxford University.

86. Morgan to Howard, 15 December 1912, Howard Papers, Oxford University.

87. Howard, "General summary and concluding remarks," pp. 20–28.

88. Morgan to Howard, 29 November, 1913, Howard Papers, Oxford University. Morgan's ideas in this regard and the same basic ideas offered by Mark Baldwin and Henry Fairfield Osborn are discussed in Richards, *Darwin and the Emergence of Evolutionary Theories of Mind and Behavior*, pp. 398–404.

89. Morgan to Howard, 6 December, 1914, Howard Papers, Oxford University.

90. Eliot Howard, *Territory in Bird Life* (London: John Murray, 1920); quotations from Howard, *Territory in Bird Life* (London: Collins, 1948), pp. 94–95, 127.

91. On Altum, see Ernst Mayr, "Bernard Altum and the territory theory," *Proceedings of the Linnaean Society of New York,* nos. 45–46 (1933–1934): 24–38. For a review of the extensive literature on territory, see, among others, Margaret Morse Nice, "The role of territory in bird life," *American Midland Naturalist* 26 (1941): 441–487. On the delay in the appreciation of Howard's territory thesis see David Lack, "Some British pioneers in ornithological research," p. 73. See also E. M. Nicholson, *How Birds Live* (London: Williams and Norgate, 1927).

92. See especially Morgan to Howard, 5 April 1935, 9 May 1935, and 12 June 1935, Howard Papers, Oxford University.

93. Morgan to Howard, 18 February 1935, Howard Papers, Oxford University. Morgan's references are to E. H. Howard, *The Nature of a Bird's World* (Cambridge: Cambridge University Press, 1935), p. 6.

94. Eliot Howard, *A Waterhen's Worlds* (Cambridge: Cambridge University Press, 1940), p. vii.

95. Lack, "Some British pioneers in ornithological research," pp. 74–75.

96. Biographical information on Kirkman in *Who Was Who, 1941–1950.*

97. F. B. Kirkman, "Preface," in Kirkman, *The British Bird Book,* 1:iii.

98. Kirkman to Howard, 30 October 1910, Howard Papers, Oxford University.

99. Kirkman, "Preface," p. ix.

100. F. B. Kirkman, "The study of animal mind and natural history," *Wild Life* 3 (1914): 146–157, quote on p. 150.

101. F. B. Kirkman, "Editorial note," in Kirkman, *The British Bird Book,* 4:461–462.

102. F. B. Kirkman, "Study of bird behaviour," in Kirkman, *The British Bird Book,* 4:597.

103. Kirkman, "Study of bird behaviour," pp. 598.

104. Kirkman, "Study of bird behaviour," pp. 600–603.

105. Kirkman, "The study of animal mind and natural history," p. 148.

106. F. B. Kirkman, *Bird Behaviour: A Contribution Based Chiefly on a Study of the Black-Headed Gull* (London: T. Nelson and Sons and T. C. and E. C. Jack, 1937). On Kirkman's role in the ISAB see John R. Durant, "The making of ethology: the Association for the Study of Animal Behaviour," *Animal Behaviour* 34 (1986): 1601–1616.

107. On T. H. Huxley see Mario A. Di Gregorio, *T. H. Huxley's Place in Natural Science* (New Haven, CT: Yale University Press, 1984); and Adrian Desmond, *Huxley: The Devil's Disciple* (London: M. Joseph, 1994).

108. Juliette Huxley, *Leaves of the Tulip Tree* (London: John Murray, 1986), p. 70.

109. On Julian Huxley see especially C. Kenneth Waters and Albert van Helden, eds., *Julian Huxley: Biologist and Statesman of Science* (Houston: Rice University Press, 1992); and John R. Baker, "Julian Sorell Huxley," *Biographical Memoirs of Fellows of the Royal Society* 22 (1976): 207–238, and *Julian Huxley, Scientist and World Citizen, 1887–1975* (Paris: UNESCO, 1978). Huxley's autobiographical comments on his behavioral fieldwork are in Julian Huxley, *Memories,* 2 vols. (London: George Allen and Unwin, 1970, 1973), 1:79, 83.

110. Huxley told an audience of radio listeners in 1930 that a good bird-watcher would tend to become a good naturalist "almost by a natural momentum, provided he has enough time and energy to spend on his hobby." See Julian S. Huxley, *Bird-Watching and Bird Behaviour* (London: Chatto and Windus, 1930), pp. 13 and 17. Nonetheless, Huxley did on several occasions acknowledge the influence that Selous, Howard, and others had upon the study of bird behaviour. See especially his *Bird-Watching and Bird Behaviour*, p. vii, and his "Introduction," in *Realities of Bird Life: Being Extracts from the Diaries of a Life-Loving Naturalist*, by Edmund Selous (London: Constable, 1927). Huxley described his encounter with the green woodpecker in "Memorable incidents with birds," *Listener* 3 (1930): 841–842, and *Memories*, 1:36.

111. In *Bird Life Glimpses*, Selous wrote (p. 38): "With ourselves definite ideas have become greatly developed; but animals may live, rather, in a world of emotions, which would then be much more a cause of their actions, and, consequently, of the cries which accompanied them." Huxley's clearest exposition of this idea was in his popular essay "Ils n'ont que de l'âme: an essay on bird mind," *Cornhill Magazine* 54 (1923): 415–427, reprinted in *Essays of a Biologist* (London: Chatto and Windus, 1923), pp. 103–129. John R. Durant, "Innate character in animals and man: a perspective on the origins of ethology," in *Biology, Medicine and Society, 1840–1940*, ed. Charles Webster (Cambridge: Cambridge University Press, 1981), pp. 157–92, pays particular attention to this essay.

112. Huxley, "Obituary: W. Warde-Fowler," pp. 143–144.

113. Julian Huxley, "Habits of birds," Huxley notebook A, Julian S. Huxley field notebooks, Edward Grey Institute of Field Ornithology, Oxford University.

114. Julian Huxley, *The Individual in the Animal Kingdom* (Cambridge: Cambridge University Press, 1912), p. 154.

115. Julian Huxley, "A 'disharmony' in the reproductive habits of the wild duck (*Anas bochas*, L.)," *Biologisches Zentralblatt* 32 (1912): 621–623. Mary Bartley discusses further Huxley's interpretation of this case and the way he used bird examples to make moral points. See her "Courtship and continued progress: Julian Huxley's studies on bird behavior," *Journal of the History of Biology* 28 (1995): 91–108.

116. Huxley, *Memories*, 1:68.

117. Huxley to A. C. Hardy, 12 May 1921, Julian S. Huxley Papers, Rice University Archives. In his *Memories*, Huxley says more about the psychological difficulties he experienced from his guilty feelings regarding sex than about his anxieties over his prospects as a scientist.

118. For Huxley's work in experimental embryology see J. A. Witkowski, "Julian Huxley and the laboratory: embracing inquisitiveness and widespread curiosity," in Waters and Van Helden, *Julian Huxley: Biologist and Statesman of Science,* pp. 79–103.

119. Julian Huxley, "The courtship of birds," *Listener* (1930), p. 935, *Bird-Watching and Bird-Behaviour*, pp. 61–62; *Memories*, 1:79.

120. Julian Huxley, "A first account of the courtship of the redshank (*Totanus calidris* L.)," *Proceedings of the Zoological Society of London* (1912), p. 647.

121. Quote from Huxley, *Bird-Watching and Bird Behaviour*, p. 63.

122. Letter from Huxley dated 10 April 1911, in file box 162, Huxley Papers, Rice University Archives. In his published paper ("Redshank," p. 648) , Huxley reported of his

practice of note taking: "I made a number of notes on the spot, and usually within twenty-four hours embodied what I had seen the day before in a letter to an ornithological friend."

123. Huxley, "Redshank," p. 648. See Selous, "Sexual selection in birds."

124. Huxley, "Redshank," pp. 651, 654. The italics are Huxley's. In his autobiography (*Memories*, 1:79), Huxley wrote regarding his redshank paper: "I am not a little proud that I used the word 'formalized' for some of the male's actions, for we now know that much courtship behaviour is indeed stereotyped in a special formalism; and much prouder of having made field natural history scientifically respectable." In fact, Huxley did not use the word "formalized" in his redshank paper. He used the word "formal" there, but he did so in crediting *Selous* with the idea that the combats between the males were formal in character (Huxley, "Redshank," p. 652). In his *Bird-Watching and Bird Behaviour* of 1930 Huxley also credited Selous with having been the first to stress how hostility between males found an outlet in "formal posturings and mock combats instead of in genuine fighting" (p. 96). When it came to writing his autobiography, however, Huxley failed to mention that it was from Selous that he got the idea of the "formal" nature of the male's actions. Selous in his article did not claim that the combats between the redshank males were always merely formal. He described several male-to-male combats that seemed to him to be entirely genuine.

125. Julian Huxley, "The courtship-habits of the great crested grebe (*Podiceps cristatus*); with an addition to the theory of sexual selection," *Proceedings of the Zoological Society of London* (1914): 491–562, quote on p. 501. See Selous, "An observational diary of the habits—mostly domestic—of the great crested grebe (*Podicipes cristatus*) and of the peewit (*Vanellus vulgaris*)."

It seems that Selous's writings not only "dovetailed" with Huxley's observations but also influenced Huxley's published accounts of what he himself saw. Selous's article on the great crested grebe described how the great crested grebe in certain of its courtship actions resembles a penguin (*Zoologist* 5 [1901]: 344). Huxley's field notes do not mention that the bird looked like a penguin, but in his 1914 paper he proceeded to name two of the birds' displays the "ghostly penguin" and the "penguin dance."

John R. Baker notes that Selous used the verb "exhaust" in 1905 in describing how in the mutual ceremony of a pair of little grebes the birds broke the ceremony off "as though it exhausted the matter." Huxley subsequently wrote of the mutual ceremonies of the great crested grebe being "self-exhausting." In Baker's words, "That two independent workers should have chosen words derived from the verb 'exhaust' seems remarkable." Baker, "Huxley," p. 214.

126. Huxley, "The great crested grebe and the idea of secondary sexual characteristics," *Science*, n.s., 36 (1912): 601–602, quote on p. 602.

127. Huxley, "Grebe" (1912), p. 602. Darwin had struggled with the question of the transmission and development of secondary sexual characters. Huxley, in his "Redshank" paper, stated that it appeared to be "both more primitive and easier for hereditary characters to be transmitted equally to both sexes" (p. 655). E. B. Poulton, however, upon reading Huxley's assertion, told Huxley that he had found the reverse: "In butterflies I find a great tendency for colours & patterns to be associated with one sex, & often some resistance & apparent difficulty in transferring them to the other." E. B. Poulton

to Huxley, 28 September 1912, Huxley Papers, Rice University Archives. For Poulton's thoughts on epigamic characters, see his *Essays on Evolution* (Oxford: Clarendon Press, 1908), pp. 379–381, and *Charles Darwin and the Origin of Species* (London: Longmans, Green, and Co., 1909), pp. 132–143.

128. Huxley, "Grebe" (1914), p. 491n. John R. Baker's memoir of Huxley notes that Huxley's earlier (1912) paper on the great crested grebe attributed the bird's ruff and tufts to sexual, not natural, selection. Baker does not address the question of the intellectual influences that may have helped change Huxley's mind on this score. In addition to Howard's writings, W. P. Pycraft's *The Courtship of Animals* (London: Hutchinson, 1913) may have been important to Huxley at this time.

129. Huxley, "Grebe" (1914), p. 498.

130. Huxley, "Grebe" (1914), pp. 499–500.

131. Huxley, "Grebe" (1914), p. 509. See also pp. 514–515.

132. Huxley, "Grebe" (1914), p. 516.

133. Huxley, "Grebe" (1914), p. 556. The idea that behavioral displays were built upon actions arising from the bird's surplus energy was not new. Both Selous and Howard had used this explanation in their writings. As for the idea of mutual selection, Darwin had raised the possibility of "a double or mutual process of sexual selection" in *The Descent of Man* (1871), 1:277, and Selous had discussed the same, calling it "*intersexual* selection," in his *The Bird Watcher in the Shetlands,* chapter 30, pp. 261–283, quote on p. 280.

134. Huxley, "Grebe" (1914), pp. 516–517.

135. W. P. Pycraft (1868–1942) was not a field naturalist himself. He worked as a museum curator first in Leicester, then in Oxford, and finally in the zoological department of the British Museum of Natural History, where he was assistant in charge of the osteological collections. He promoted the study of animal life by writing numerous popular and semipopular science books, such as *A History of Birds* (1910), *The Infancy of Animals* (1912), and *The Courtship of Animals* (1913).

136. Huxley, "Grebe" (1914), p. 559. Huxley allowed that he also agreed with Pycraft's views on "the principle of Orthogenesis and its importance for the origin of sexual (and other) forms of ornamentation."

137. On Huxley's anthropomorphism, see John Durant, "The tension at the heart of Huxley's evolutionary ethology," in Waters and Van Helden, *Julian Huxley: Biologist and Statesman of Science,* pp. 150–160; and Eileen Crist, *Images of Animals: Anthropomorphism and Animal Mind* (Philadelphia: Temple University Press, 1999), pp. 172–180.

138. Huxley, "Grebe" (1914), p. 510.

139. John Durant notes that being a spectator at a play was a common trope of English naturalists. See Durant, "Innate character in animals and man."

140. Huxley, "Grebe" (1914), pp. 553–554.

141. Huxley's great crested grebe manuscript notes are preserved in the Huxley Papers in the Rice University Archives, box files 162:2. The quote comes from Huxley's field notes for 18 April 1912, p. 3.

142. Huxley, "Grebe" (1914), p. 498.

143. J. S. Huxley, "The absence of 'courtship' in the avocet," *British Birds* 19 (1925): 88–94, quote on p. 90.

144. Huxley, "Grebe" (1914), pp. 549–551. Grebe field notes, 18 April 1912, p. 3, box files 162:2, Huxley Papers, Rice University Archives.

145. For the use of film in American studies of animal behavior, see Gregg Mitman, "Cinematic nature: Hollywood technology, popular culture, and the American Museum of Natural History," *Isis* 84 (1993): 637–661.

146. Lack to Huxley, 15 March 1953; Lorenz to Huxley, 2 November 1954, and 26 November 1963, Huxley Papers, Rice University Archives.

147. Huxley, "Grebe" (1914), p. 492.

148. Huxley discusses the end of his engagement, but not his anxieties about his career, in *Memories*, 1:92.

149. Huxley to Hardy, 12 May 1921, Huxley Papers, Rice University Archives.

150. Huxley, "Bird-watching and biological science: some observations on the study of courtship in birds," *Auk* 33 (1916): 143–161, 256–270.

151. Juliette Huxley, *Leaves of the Tulip Tree*, pp. 75–76.

152. Huxley to Hardy, 12 May 1921, Huxley Papers, Rice University Archives.

153. Prior to his work on the red-throated diver Huxley published several short papers on bird behavior, including "Some points in the sexual habits of the little grebe, with a note on the occurrences of vocal duets in birds," *British Birds* 13 (1919): 155–158, "The accessory nature of many structures and habits associated with courtship," *Nature* 108 (1921): 565–566, and a review of *Territory in Bird Life*, by H. Eliot Howard, *Discovery* 2 (1921): 135–136. He also published in 1922 a short paper, "Preferential mating in birds with similar coloration in both sexes," *British Birds* 15 (1922): 99–101. Before his major paper on the red-throated diver appeared in 1923, a shorter version of his observations on the bird, together with those of G. J. van Oordt, appeared as J. S. Huxley and G. J. van Oordt, "Some observations on the habits of the red-throated diver in Spitsbergen," *British Birds* 16 (1922): 34–46.

154. Huxley, "Courtship activities in the red-throated diver (*Colymbus stellatus* Pontopp.); together with a discussion of the evolution of courtship in birds," *Journal of the Linnean Society of London, Zoology* 35 (1923): 253–292, quote on p. 269.

155. Huxley, "Red-throated diver" (1923), p. 283. Similarly, "It becomes increasingly clear that to interpret the behavior and evolution of a bird, even in apparently only one regard, it is necessary to take into account all the circumstances of its life" (p. 290).

156. Huxley, "Red-throated diver" (1923), pp. 274–275.

157. Huxley, "Red-throated diver" (1923), p. 276–277.

158. Huxley, "Red-throated diver" (1923), p. 278.

159. Huxley, "Red-throated diver" (1923), p. 286.

160. Huxley, "Red-throated diver" (1923), p. 288.

161. Huxley, "The absence of 'courtship' in the avocet"; J. S. Huxley, assisted by F. A. Montague, "Studies on the courtship and sexual life of birds," pt. 5, "The oyster-catcher," *Ibis*, 12th ser., 1 (1925): 868–897. Huxley's explanation of the absence of "courtship" in the avocet was that the bird possessed a "very placid 'temperament,'" and that as a result there was not sufficient emotional tension overflowing into action for courtship displays to develop. Though he did not do so in many other places, Huxley here credited Howard, Selous, "and various other writers" with the view "that 'courtship' dis-

plays arise immediately from the excited state of the unsatisfied male bird (or of both sexes in species with mutual displays" (p. 92).

162. Huxley, "Oyster-catcher," p. 895.

163. After spending a weekend of his spring vacation of 1922 with Howard, Huxley wrote: "I feel stimulated to take up systematic observation again: & I feel more than ever impelled to write a general book on the subject!" Huxley to Howard, 24 April 1922. See also a previous letter from Huxley to Howard dated 19 February 1922, as well as Huxley to Howard, 6 December 1923, and 3 March 1924. Howard Papers, Oxford University.

164. See the Huxley manuscript entitled "The courtship of birds: a biological study," unpublished ms., no. 59.9, Huxley Papers, Rice University Archives. On the cover page, in Huxley's hand, is the identification: "Notes by JSH 1925 for a book on Bird Courtship that never got written."

165. Julian Huxley, "Biology of bird courtship," in *Proceedings of the VIIth International Ornithological Congress at Amsterdam, 1930* (Amsterdam: Prof. Dr. L. F. de Beaufort, 1931), pp. 107–108, "Threat and warning coloration in birds, with a general discussion of the biological functions of colour," in *Proceedings of the Eighth International Ornithological Congress, Oxford, July 1934* (Oxford: Oxford University Press, 1938), pp. 430–455, "Darwin's theory of sexual selection and the data subsumed by it, in the light of recent research," *American Naturalist* 72 (1938): 416–433, "The present standing of the theory of sexual selection," in *Evolution: Essays on Aspects of Evolutionary Biology Presented to Professor E. S. Goodrich on His Seventieth Birthday*, ed. G. R. de Beer (Oxford: Clarendon Press, 1938), pp. 11–42.

Mary Jane West-Eberhard identifies the "forgotten era of sexual selection theory" as beginning in the 1930s as the result of population geneticists' redefining fitness in terms of change in gene frequencies and Huxley's suggesting that the term sexual selection be eliminated. See Mary Jane West-Eberhard, "Sexual selection, social competition, and speciation," *Quarterly Review of Biology* 58 (1983): 155–183, see p. 156. Huxley's treatment of sexual selection in his writings of the 1930s and 1940s deserves further historical examination.

166. Huxley, "The present standing of the theory of sexual selection," p. 23.

167. Durant, "The tension at the heart of Huxley's evolutionary ethology," p. 151.

168. Huxley, "Natural selection and evolutionary progress," in *British Association for the Advancement of Science: Report of the Annual Meeting, 1936* (London: Office of the British Association, 1936), pp. 81–100, quotes from p. 95.

169. Huxley, "Natural selection and evolutionary progress," p. 99.

170. *The Private Life of the Gannets*, which Huxley and Lockley filmed in 1934, received an Academy Award for 1937 in the "short subjects" category as the best one-reel documentary. The movie was filmed amid the enormous breeding colony of gannets on the tiny island of Grassholme, off the coast of Wales. The birds were filmed in June and again in August. More than a mile of film was shot (including several days' shooting from a herring boat, which produced some superb footage of gannets diving for fish). The title of the film reflected not simply the film's subject matter but also the wishes of the Hungarian-born British film mogul Alexander Korda, whose company, London Film, produced the picture. Korda had had considerable success with a series of "private lives" films, including *The Private Life of Henry VIII* (1933) and *The Private Life of Don*

Juan (1934). See Huxley, *Memories,* 1:210–122, and "Making and using nature films," *Listener* 13 (1935): 595–597, 629; and the letters from Ronald Lockley to Huxley in the Huxley Papers, Rice University Archives. See also Gregg Mitman, *Reel Nature: America's Romance with Wildlife on Film* (Cambridge, MA: Harvard University Press, 1999).

171. See John R. Durant, "The making of ethology."

172. At the Amsterdam zoo, an alternative approach to the study of animal behavior was developed by the zoo's charismatic director, A. F. J. Portielje. As early as 1925, Portielje began publishing papers on bird "ethology." He was more influenced by Oskar Heinroth than by Huxley. On Portielje see chapter 4.

173. On 17 April 1925, C. Lloyd Morgan wrote to Huxley: "Yes. We must spur the good E. H. to get on even if he has to leave some problems unsolved." Huxley Papers, Rice University Archives.

174. Huxley to Howard, 16 April 1925, Howard Papers, Oxford University.

175. Edmund Selous, *Bird Life Glimpses,* p. vi. Elsewhere Selous allowed that only those who had tried it knew "how very tantalising it is to watch animals for a long time, wanting them all the while to do something interesting, which they, all the while, persist in not doing." Edmund Selous, "Rabbits and Hares," p. 619.

176. Huxley to Howard, 19 February 1922, Howard Papers, Oxford University.

177. Huxley, "Some further observations on the courtship behaviour of the great crested grebe," *British Birds* 18 (1924): 129–134, and "Some points in the breeding behaviour of the common heron," *British Birds* 18 (1924): 155–163. On the Oxford Ornithological Society, see Allen, *The Naturalist in Britain,* pp. 252–262. The prime mover of the Oxford Ornithological Society, according to Allen, was Huxley's student Bernard Tucker. Baker, "Huxley," p. 210, indicates that Huxley upon returning to Oxford after the war gave one course of lectures on experimental zoology, another on genetics, and a third on animal behavior.

178. Tinbergen to Huxley, 18 February 1953, Huxley Papers, Rice University Archives.

179. Fowler to Huxley, 23 October 1916, Huxley Papers, Rice University Archives. Fowler said further that he especially liked the analogies Huxley drew between animal behavior and human behavior.

180. Huxley, "Bird-watching and biological science," p. 161. The failure to distinguish between different kinds of biological causation has proved to be a continuing problem in the life sciences. It is addressed by Ernst Mayr in his classic article "Cause and effect in biology," *Science* 134 (1961): 1501–1506, reprinted in *Evolution and the Diversity of Life: Selected Essays* (Cambridge, MA: Harvard University Press, 1976), pp. 359–371.

181. Huxley, "The outlook for biology," *Rice Institute Pamphlet* 11 (1924): 241–338, see esp. pp. 262–263, 272–273, and "The courtship of birds: a biological study."

182. Huxley, "The courtship of birds: a biological study," chapter 1, p. 14.

183. Huxley, *Memories,* 1:5.

184. Juliette Huxley, *Leaves of the Tulip Tree,* p. 165.

CHAPTER THREE

1. Oskar Heinroth, "Beiträge zur Biologie, namentlich Ethologie und Psychologie der Anatiden," in *Verhandlungen des 5. Internationalen Ornithologen-Kongresses in Berlin, 30. Mai bis 4. Juni 1910,* ed. Herman Schalow (Berlin: Deutsche Ornithologische Ge-

sellschaft, 1911), pp. 589–702, quote on p. 702. An English translation of extensive excerpts from this paper appears in Gordon M. Burghardt, ed., *Foundations of Comparative Ethology* (New York: Van Nostrand Reinhold Company, 1985), pp. 246–301.

2. Letter from Erwin Stresemann to Konrad Lorenz, 7 March 1934, in the Erwin Stresemann Papers, Preußischer Kulturbesitz, Staatsbibliothek zu Berlin, Nachlaß 150 (Stresemann), Ordnung 40 (Lorenz). This collection will hereafter be abbreviated as ESP/Berlin. Lorenz's correspondence with Erwin Stresemann together with his correspondence with Oskar Heinroth constitute the two most valuable sources for information on Lorenz's career in the 1930s. The bulk of Lorenz's correspondence with Heinroth appears in Oskar Heinroth and Konrad Lorenz, *Wozu aber hat das Vieh diesen Schnabel? Briefe aus der frühen Verhaltensforschung, 1930–1940*, ed. Otto Koenig (Munich: Piper, 1988) (hereafter cited as Heinroth and Lorenz, *Wozu*). The Lorenz-Stresemann correspondence remains for the most part unpublished. I first saw it in the summer of 1996, had a microfilm copy of it made in 1997, and cited the letters in my drafts of this chapter and chapter 5. Passages from some of the more striking of Lorenz's letters to Stresemann have been quoted by other scholars, including Jürgen Haffer, Erich Rutschke, and Klaus Wunderlich, "Erwin Stresemann (1889–1972): Leben und Werk eines Pioniers der wissenschaftlichen Ornithologie," *Acta Historica Leopoldina* 34 (2000): 1–465; Veronika Hofer, "Konrad Lorenz als Schüler von Karl Bühler: Diskussion der neu entdeckten Quellen zu den persönlichen und inhaltlichen Positionen zwischen Karl Bühler, Konrad Lorenz und Egon Brunswick," *Zeitgeschichte* 28 (2001): 135–159; Benedikt Föger und Klaus Taschwer, *Die andere Seite des Spiegels: Konrad Lorenz und der Nationalsozialismus* (Vienna: Czernin Verlag, 2001); and Klaus Taschwer and Benedikt Föger, *Konrad Lorenz: Biographie* (Vienna: Paul Zolnay Verlag, 2003). In citing the Lorenz-Stresemann correspondence here, I give the date of the letter in question and, where the same text has been already published elsewhere, an appropriate citation insofar as possible (see below). In cases where a passage has not previously been published, I provide the original German text in the footnote. English translations in the text are my own unless otherwise noted. The passage of the letter from Stresemann to Lorenz quoted above appears in Haffer et al., "Erwin Stresemann," pp. 278–279; and Föger and Taschwer, *Die andere Seite des Spiegels*, p. 50.

Klaus Taschwer and Benedikt Föger's 2003 biography of Lorenz reached me (courtesy of Klaus Taschwer) just as I was coming to the deadline for sending my revised manuscript off to the press and thus too late for me to make full use of it in my text and notes. This rich, well-documented biography offers an excellent account of the whole of Lorenz's life and career. The present book differs by concentrating more specifically on the development Lorenz's ideas, practices, and career specifically as they related to the science of ethology.

3. Psychologists now like to tell the story of how "Clever Hans," the horse who was alleged to be able to answer complicated mathematical questions, was, upon closer analysis, found to be taking cues from the nearly imperceptible movements of the person posing the questions to him. This case was famously debunked by Oskar Pfungst in his book *Clever Hans (The Horse of Mr. Von Osten): A Contribution to Experimental Animal and Human Psychology* (New York: Henry Holt, 1911), a work that appeared originally in German in 1907. However, contemporaries did not take Pfungst's account as definitive

with respect to the subject of "thinking animals." The Gesellschaft für Tierpsychologie was founded in September 1912 and began publication of its *Mitteilungen* in 1913. Heinrich Ernst Ziegler, professor of zoology at the Stuttgart Technical College and author of *Der Begriff des Instinktes einst und jetzt* (1904; 2nd ed., Jena: Gustav Fischer, 1910), was the society's first president. The society appears, however, to have been short-lived.

4. Lorenz describes his early childhood experience with animals in his autobiographical statements, "Konrad Lorenz," in *Les Prix Nobel en 1973* (Stockholm: Imprimerie Royale P. A. Norstedt & Söner, 1974), pp. 177–184, and "My family and other animals," in *Studying Animal Behavior: Autobiographies of the Founders*, ed. Donald A. Dewsbury (Chicago: University of Chicago Press, 1989; first published in 1985 by Associated University Presses as *Leaders in the Study of Animal Behavior: Autobiographical Perspectives*), pp. 259–287. For more biographical information on Lorenz, see especially Alec Nisbett, *Konrad Lorenz* (New York: Harcourt Brace Jovanovich, 1976); and, above all, Taschwer and Föger, *Konrad Lorenz: Biographie*. See too Föger and Taschwer, *Die andere Seite des Spiegels;* Franz M. Wuketits, *Konrad Lorenz: Leben und Werk eines großen Naturforschers* (Munich: Piper, 1990); Norbert Bischof, *Gescheiter als Alle die Laffen: Ein Psychogramm von Konrad Lorenz* (Hamburg: Rasch und Röhring, 1991; and, more briefly, Richard W. Burkhardt, Jr., "Konrad Zacharias Lorenz (1903–1989)," in *Darwin & Co.: Eine Geschichte der Biologie in Porträts*, ed. Ilse Jahn and Michael Schmitt (Munich: C. H. Beck, 2001), 2:422–441. On Lorenz's development as a theorist, see Robert J. Richards, "The innate and the learned: the evolution of Konrad Lorenz's theory of instinct," *Philosophy of the Social Sciences* 4 (1974): 111–133; and three papers by Theo J. Kalikow, "History of Konrad Lorenz's ethological theory, 1927–1939: the role of meta-theory, theory, anomaly and new discoveries in a scientific 'evolution,'" *Studies in History and Philosophy of Science* 6 (1975): 331–341, "Konrad Lorenz's ethological theory, 1939–1943: 'explanations' of human thinking, feeling and behaviour," *Philosophy of the Social Sciences* 6 (1976): 15–34, and "Konrad Lorenz's ethological theory: explanation and ideology, 1938–1943," *Journal of the History of Biology* 16 (1983): 39–73. See also Hofer, "Konrad Lorenz als Schüler von Karl Bühler"; and Ingo Brigandt, "The instinct concept of the early Konrad Lorenz" (unpublished manuscript).

5. See Adolf Lorenz's autobiography, *My Life and Work: The Search for a Missing Glove* (New York: Charles Scribner's Sons, 1936).

6. Adolf Lorenz, *My Life and Work*, pp. 217–218; Nisbett, *Konrad Lorenz*, pp. 14–16.

7. Lorenz, "Konrad Lorenz," and "My family and other animals." See also Selma Lagerlöf, *Nils Holgerssons underbara resa genom Sverige*, 2 vols. (Stockholm: Albert Bonniers Förlag, 1907–1910). Intended to be a primer for children at the elementary school level, Lagerlöf's book proved immensely popular in Sweden and was quickly translated into Danish, German, and English. For the English translation of volume 1 of the work, see *The Wonderful Adventures of Nils*, trans. Velma Swanston Howard (New York: Grosset and Dunlap, 1907). Already recognized as one of Sweden's most distinguished authors, Lagerlöf received the Nobel Prize for Literature in 1909.

8. Konrad Lorenz, "Foreword," in *The Herring Gull's World*, by Niko Tinbergen (London: Collins, 1953), p. xii. Lorenz made the same claim elsewhere, as, for example, in "The comparative method in studying innate behaviour patterns," *Symposia of the Society for Experimental Biology* 4 (1950): 235, and "My family and other animals," p. 263.

9. Konrad Z. Lorenz, *King Solomon's Ring* (London: Methuen, 1952).

10. On Lorenz's excellence as a student, see Wuketits, *Konrad Lorenz*, p. 35. Lorenz mentions Heberdey in "Konrad Lorenz," p. 177. Nisbett, *Konrad Lorenz*, p. 24, indicates that when Lorenz first began going to the Schottengymnasium at the age of eleven he took only chemistry, physics, and natural history, but by his fourth year he was studying "Shakespeare, Homer, and the humanities" as well as science. On the *Bildung* ideal, see Fritz K. Ringer, *The Decline of the German Mandarins: The German Academic Community, 1890–1933* (Cambridge, MA: Harvard University Press, 1969); and Jonathan Harwood, *Styles of Scientific Thought: The German Genetics Community, 1900–1933* (Chicago: University of Chicago Press, 1993).

11. On Lorenz's reverence for Hochstetter, see Konrad Lorenz, *Behind the Mirror: A Search for a Natural History of Human Knowledge* (London: Methuen, 1977), pp. 203–204. Lorenz was named assistant in Hochstetter's institute in 1928. His 1936 paper "Über eine eigentümliche Verbindung branchialer Hirnnerven bei *Cypselus apus*," *Gegenbaurs morphologisches Jahrbuch* 77 (1936): 305–325, is identified as "Aus dem II. Anatomischen Institut der Universität Wien."

12. Konrad Lorenz, *Vergleichende Verhaltensforschung: Grundlagen der Ethologie* (Vienna: Springer-Verlag, 1978), p. 3. The phrase "Archimedean point" appears also in Lorenz, *Evolution and Modification of Behavior* (Chicago: University of Chicago Press, 1965), p. 82.

13. Lorenz, *King Solomon's Ring*, p. 130.

14. Konrad Z. Lorenz, "Beobachtungen an Dohlen," *Journal für Ornithologie* 75 (1927): 511–519. In addition to Lorenz's observations on his first jackdaw, the paper included his brief comments on three adult jackdaws that he received in the spring of 1927 from the zoological garden at Schönbrunn. The Heinroth-Lorenz correspondence preserved at the Staatsbibliothek in Berlin unfortunately does not include the first exchanges between the two men. The first letters in the collection begin in 1930. See Heinroth and Lorenz, *Wozu*.

15. On Stresemann see Rolf Nöhring, "Erwin Stresemann," *Journal für Ornithologie* 114 (1973): 455–471; and Ernst Mayr, "Erwin Stresemann," *Dictionary of Scientific Biography* 18, supp. 2 (1990): 888–890.

16. In a letter to Stresemann dated 31 January 1933, Lorenz thanked Stresemann for all of Stresemann's previous help to him, including asking the questions that inspired Lorenz to raise a substantial number of jackdaws. ESP/Berlin.

17. A letter from Lorenz to Stresemann, 30 August 1928, describes some of Lorenz's early trials with the jackdaw colony. ESP/Berlin.

18. On zoology in Vienna see Wilhelm Marinelli, "Die Geschichte der Zoologie in Wien," *Verhandlungen der Deutschen Zoologischen Gesellschaft* 56 (1962): 50–55; and Rupert Riedl, "Die Wiener Schule," *Verhandlungen der Deutschen Zoologischen Gesellschaft* 78 (1985): 5–9.

19. On the relations between Lorenz and Bühler, see Hofer, "Konrad Lorenz als Schüler von Karl Bühler."

20. See Heinroth and Lorenz, *Wozu*, pp. 27–32. The paper by Mathilde Hertz, a gifted young experimentalist who had worked with Wolfgang Koehler, was "Beobachtungen an gefangenen Rabenvögeln," *Psychologische Forschung* 8 (1926): 336–397.

21. The principle source of biographical information on Heinroth is Katharina Heinroth, *Oskar Heinroth, Vater der Verhaltensforschung* (Stuttgart: Wissenschaftliche Verlagsgesellschaft, 1971). See too Erwin Stresemann, *Ornithology from Aristotle to the Present* (Cambridge, MA: Harvard University Press, 1975), pp. 345–348; and Richard W. Burkhardt, Jr., "Oskar August Heinroth," *Dictionary of Scientific Biography* 17, supp. 2 (1990): 392–394.

22. On the problem of finding zoological positions in Germany at the turn of the century see Lynn Nyhart, *Biology Takes Form: Animal Morphology and the German Universities, 1800–1900* (Chicago: University of Chicago Press, 1995), and "Natural history and the 'new' biology," in *Cultures of Natural History*, ed. N. Jardine, J. A. Secord, and E. C. Spary (Cambridge: Cambridge University Press, 1996), pp. 426–443.

23. The word "ethology" was not new, but Heinroth was certainly among the first to use it in the specific sense of studying the instinctive behavior of animals. On earlier uses of the word, see especially Julian Jaynes, "The historical origins of 'ethology' and 'comparative psychology,'" *Animal Behaviour* 17 (1969): 601–606; and John R. Durant, "Innate character in animals and man: a perspective on the origins of ethology," in *Biology, Medicine and Society, 1840–1940*, ed. Charles Webster (Cambridge: Cambridge University Press, 1981), pp. 157–92.

24. Heinroth, "Beiträge zur Biologie," p. 619, translated in Burghardt, *Foundations of Comparative Ethology*, p. 262. Burghardt, p. 219, notes that although Huxley is usually given credit for the idea of ritualization, Heinroth anticipated him in this regard.

25. The name "Kaspar Hauser" comes from a famous, early nineteenth-century case of a foundling child in Germany who was apparently raised under conditions of extreme sensory deprivation. Hence an experiment depriving an animal of learning experiences otherwise normal for its development, such as contact with other members of its species.

26. Oskar Heinroth and Magdalena Heinroth, *Die Vögel Mitteleuropas in allen Lebens- und Entwicklungsstufen photographisch aufgenommen und in ihrem Seelenleben bei der Aufzucht vom Ei ab beobachtet* (Berlin: H. Bermühler, 1924–1934).

27. Heinroth and Lorenz, *Wozu*, pp. 75, 91, 103.

28. Lorenz to Heinroth, 22 February 1931, in Heinroth and Lorenz, *Wozu*, pp. 40–41. Lorenz reported to Heinroth that Bernhard Hellmann had given him the first volume of *Die Vögel Mitteleuropas* as a birthday present. Hellmann allowed that he would have believed Lorenz had written the volume himself had there been no indication otherwise. Likewise, the Austrian zoologist W. Marinelli, upon reading Lorenz's jackdaw paper, lauded it for having been written in the style of Heinroth.

29. Quoted in Margaret Morse Nice, *Research Is a Passion with Me* (Toronto: Consolidated Amethyst Communications, 1979), p. 172.

30. Lorenz to Heinroth, 22 February 1931, in Heinroth and Lorenz, *Wozu*, p. 42.

31. Konrad Lorenz, "Beiträge zur Ethologie sozialer Corviden," *Journal für Ornithologie* 79 (1931): 67–127, translated as "Contributions to the study of the ethology of social Corvidae," in *Studies in Animal and Human Behaviour*, by Konrad Lorenz, trans. Robert Martin, 2 vols. (Cambridge, MA: Harvard University Press, 1970–1971), 1:1–56.

32. Of Lorenz's original colony of fifteen jackdaws in 1927, only three birds remained by the following June. In 1928, he introduced nineteen more birds to the colony,

sixteen of which he hand raised himself. In 1929 two of the birds remaining from the original 1927 colony mated and reared two nestlings, and Lorenz also added more hand-raised birds to the colony. By the winter of 1929–1930, the colony numbered twenty individuals, but eighteen of these disappeared the following spring when Lorenz was temporarily hospitalized, and another died shortly thereafter.

33. Lorenz letters to Heinroth, September 1930 (?), 30 March 1931, and 11 May 1931, in Heinroth and Lorenz, *Wozu*, pp. 27–30, 47–48, 51–54. On the goose named Martina, see Heinroth and Lorenz, *Wozu*, p. 200; and Konrad Lorenz, *Here Am I—Where Are You? The Behavior of the Greylag Goose* (New York: Harcourt Brace Jovanovich, 1991).

34. Konrad Lorenz, "Der Kumpan in der Umwelt des Vogels: der Artegenosse als auslösendes Moment sozialer Verhaltungsweisen," *Journal für Ornithologie* 83 (1935): 137–215, 289–413, see pp. 150–151. Translated as "Companions as factors in the bird's environment: the conspecific as the eliciting factor for social behaviour patterns," in Lorenz, *Studies in Animal and Human Behaviour*, 1:101–258, see p. 113.

35. This undated proposal for the station is preserved in ESP/Berlin. Stresemann's official response to the plan, in his capacity as president of the German Ornithological Society, is dated 22 April 1931. The paragraph of the proposal from which the quotes are taken reads: "Die Leitung sieht ihre Aufgabe hauptsächlich in der Erforschung der feineren Biologie unserer einheimischen Vogelwelt, von der ja erst so wenig bekannt ist, insbesondere aber darin, jene verwickelten Reflexverkettungen, die wir als Instinkte zu bezeichnen gewohnt sind und die bisher fast ausschliesslich vom psycholgischen Standpunkte betrachtet wurden, physiologisch zu bearbeiten."

36. Heinroth and Lorenz, *Wozu*, p. 50.

37. Lorenz to Stresemann, 8 May 1931. ESP/Berlin. "Sie können sich denken, Herr Professor, mit welcher Freude ich dem Entstehen der Station entgegensehe, wenn Sie sich vor Augen halten, wie viele Versuchsmöglichkeiten mir dort zu Gebote stehen, die mir als Privatmann verschlossen sind! Bei den weitaus meisten Vogelarten ist ja die genauere Trieblehre ja ein so unbeackertes Gebiet, dass eigentlich jeder Vogel, den man überhaupt freifliegend halten kann, Ergebnisse verspricht!!"

38. Konrad Lorenz, "Betrachtungen über das Erkennen der arteigenen Triebhandlungen der Vögel," *Journal für Ornithologie* 80 (1932): 50–98, translated as "A consideration of methods of identification of species-specific instinctive behaviour patterns in birds," in Lorenz, *Studies in Animal and Human Behaviour*, 1:57–100. After the fact, Lorenz explained to Heinroth that he did not dare at first to send him the original draft because he was afraid Heinroth would be unhappy with the unreliability of Lorenz's ideas and the abundance of unsubstantiated assertions. See Lorenz to Heinroth, 9 September 1931, in Heinroth and Lorenz, *Wozu*, p. 71.

39. Ziegler, *Der Begriff der Instinktes einst und jetz*, p. 46.

40. Lorenz, "Betrachtungen über das Erkennen der arteigenen Triebhandlungen der Vögel," p. 66, quoted from "A consideration of methods of identification of species-specific instinctive behaviour patterns in birds," p. 72. On Lorenz's use of Ziegler see Ingo Brigandt, "The instinct concept of the early Konrad Lorenz." Brigandt observes that although Ziegler quoted directly from Whitman's "animal behavior" paper of 1898, Lorenz failed to perceive Whitman's importance at this time.

41. See Friedrich Alverdes, *Tiersoziologie* (Leipzig: C. L. Hirschfeld, 1925), p. 5. See

also Alverdes, *Die Tierpsychologie in ihren Beziehungen zur Psychologie des Menschen* (Leipzig: C. L. Hirschfeld, 1932). Alverdes distinguished between instinctive (constant) and intelligent (variable) components in every action.

42. Lorenz, "Betrachtungen über das Erkennen der arteigenen Triebhandlungen der Vögel," pp. 67–70, "A consideration of methods of identification of species-specific instinctive behaviour patterns in birds," 1:pp. 72–75.

43. Paraphrased from Lorenz, "Betrachtungen über das Erkennen der arteigenen Triebhandlungen der Vögel," pp. 96–97, "A consideration of methods of identification of species-specific instinctive behaviour patterns in birds," p. 99.

44. Lorenz, "Betrachtungen über das Erkennen der arteigenen Triebhandlungen der Vögel," p. 98, "A consideration of methods of identification of species-specific instinctive behaviour patterns in birds," p. 100.

45. This account is based on G. Kramer, "50. Jahresversammlung der Deutschen Ornithologischen Gesellschaft in Wien vom 1.–4. Oktober 1932," *Journal für Ornithologie* 81 (1933): 360–375. On Ferdinand, see the anonymous article "König Ferdinand von Bulgarien zum 75. Geburtstage am 26. Februar 1936," *Journal für Ornithologie* 84 (1936): 1–2, most likely written by Stresemann.

46. Adolf Lorenz, *My Life and Work*, pp. 319–321.

47. Lorenz to Heinroth, 20 October 1932, in Heinroth and Lorenz, *Wozu*, p. 109.

48. Lorenz, "The comparative method in studying innate behaviour patterns," p. 234.

49. Lorenz to Stresemann, letters of 12 October 1932, December 1932? (undated), January 1933, 22 April 1933, 28 May 1933, and 17 July 1933, ESP/Berlin.

50. Konrad Lorenz, "Beobachtetes über das Fliegen der Vögel und über die Beziehungen der Flügel- und Steuerform zur Art des Fluges," *Journal für Ornithologie* 81 (1933): 107–236.

51. See Lorenz to Stresemann, letters of 8 January 1933, 31 January 1933, 22 April 1933, 17 July 1933, ESP/Berlin; and Lorenz to Heinroth, 17 July 1933, in Heinroth and Lorenz, *Wozu*, pp. 150–151.

52. Lorenz, "Konrad Lorenz," p. 178. See also Lorenz, "My family and other animals," p. 265.

53. This quote comes from p. 3 in a series of extracts of comments that Lorenz apparently sent to Craig. The first page of the set is missing. This quote is dated 23 February 1935. The manuscript is among the Margaret Morse Nice Papers, 1917–1968, collection number 2993, courtesy of the Division of Rare and Manuscript Collections, Cornell University Library.

54. Edward Chace Tolman's *Purposive Behavior in Animals and Men* was first published in 1932 (New York: Century Co.). For the similarities between Lorenz's views and McDougall's, see J. B. S. Haldane, "The sources of some ethological notions," *British Journal of Animal Behaviour* 4 (1956): 162–164; and the sections from William McDougall's *An Outline of Psychology* (London: Methuen, 1923) reprinted in Burghardt, *Foundations of Comparative Ethology*, pp. 106–121. In his autobiographical comments for *Les Prix Nobel en 1973* Lorenz did acknowledge that he was positively impressed by Karl Bühler and Bühler's student Egon Brunswick, who led him to see the value of bringing epistemological questions to bear on questions of animal behavior. In "My family and

other animals," p. 266, Lorenz again noted his debt to Brunswick, mentioning especially Brunswick's book *Wahrnehmung und Gegenstandswelt, Psychologie vom Gegenstand her.* This book was published in Leipzig and Vienna in 1934. For more on Lorenz's debts to Bühler and Brunswick, see Hofer, "Konrad Lorenz als Schüler von Karl Bühler."

55. Heinroth and Lorenz, *Wozu*, p. 161.

56. Heinroth and Lorenz, *Wozu*, p. 169. See G. H. Brückner, "Untersuchungen zur Tiersoziologie, insbesondere zur Auflösung der Familie," *Zeitschrift für Psychologie* 128 (1933): 1–110.

57. Lorenz to Heinroth, 27 January 1934, in Heinroth and Lorenz, *Wozu*, p. 171. See J. A. Bierens de Haan, "Der Stieglitz als Schöpfer," *Journal für Ornithologie* 81 (1933): 1–22.

58. Lorenz to Heinroth, 28 March to 9 April 1933, in Heinroth and Lorenz, *Wozu*, p. 140.

59. Jakob von Uexküll and Georg Kriszat, *Streifzüge durch die Umwelten von Tieren und Menschen* (Berlin: J. Springer, 1934), translated into English as "A stroll through the worlds of animals and men," in *Instinctive Behavior: The Development of a Modern Concept*, ed. Claire H. Schiller (New York: International Universities Press, 1957), pp. 5–80. In the foreword to the original German edition of the book (which was not included in the English translation), Uexküll indicates that he wrote the text of the book and his coauthor, Kriszat, was responsible for the pictures.

60. On Uexküll, see Friedrich Brock, "Jakob Johann Baron von Uexküll," *Sudhoffs Archiv für Geschichte der Medizin und der Naturwissenschaften* 27 (1934): 193–203, and "Verzeichnis der Schriften Jakob Johann v. Uexkülls und der aus dem Institut für Umweltforschung zu Hamburg hervorgegangenen Arbeiten," *Sudhoffs Archiv für Geschichte der Medizin und der Naturwissenschaften* 27 (1934): 204–212. See too the excellent historical evaluation of Uexküll by Anne Harrington, *Reenchanted Science: Holism in German Culture from Wilhelm II to Hitler* (Princeton, NJ: Princeton University Press, 1996).

61. Th. Beer, A. Bethe, and J. von Uexküll, "Vorschläge zu einer objektivierenden Nomenklature in der Physiologie des Nervensystems," *Biologisches Zentralblatt* 19 (1899): 517–521.

62. Jakob von Uexküll, "Umweltforschung," *Zeitschrift für Tierpsychologie* 1 (1937): 33–34.

63. Jakob von Uexküll, *Umwelt und Innenwelt der Thiere* (Berlin: J. Springer, 1909; 2d 2nd ed., 1921).

64. Although Uexküll states that "the effector cue or meaning extinguishes the receptor cue or meaning," in his account of the tick's behavior it is the tactile cue of falling on the hairs of the mammal, not the tick's letting go from its resting place, that extinguishes the previous cue of the butyric acid; likewise, it is the new sensation of heat, not the running about itself, that replaces the previous cue provided by the mammal's hair. See Uexküll, "A stroll through the worlds of animals and men," p. 11.

65. Uexküll, "A stroll through the worlds of animals and men," pp. 11–12.

66. In the foreword to Uexküll and Kriszat, *Streifzüge durch die Umwelten von Tieren und Menschen*, dated December 1933, Uexküll expressed his special indebtedness to Lorenz for enhancing the book by sending Uexküll illustrations of some of Lorenz's particularly significant experiences with jackdaws and starlings.

67. Lorenz to Stresemann, 4 March 1934, ESP/Berlin. The original reads: "Nun

fragt es sich: darf in den heutigen Zeiten ein Mann mit zwei Kindern eine staatliche sichere Stellung aufgeben, um seinem Dämon zu folgen? Es bleibt dabei zu bedenken, daß wir a) vorläufig auch ohne die Anatomie (300S = 150RM) zur fressen haben, wenn wir den Viehbestand etwas ändern (weniger Frischfresser, allgemein billig fressende Versuchstiere) b) daß ich einen Schwall von wirklich genauen tierpsychologischen Arbeiten produzieren würde. c) Als Docent für Tierpsychologie immerhin Mitglied der Universität bliebe. d) Solange ich an der Anatomie bin, keinerlei Chancen habe, eine psychologischen Universitätsstelle zu bekommen. Die ich mir vom Joche befreit ja doch wohl früher oder später erringen würde. Last not least, daß ich es als Anatom, als ausgesprochener Anti-Lieblingsschüler meines Chefs ohnehin nicht weit bringen würde, die gerühmte Sicherheit der anatomischen Stellung also nicht so weit her ist. Ich glaube, ich werde den Sprung ins Hungerleiderdasein eines Privatdocenten tun. Es wird sich schon irgendwann irgendwo eine psychologische Stelle auftun. Then I will start at the bottom and work my way still farther down! Der gegenwärtige Zustand ist jedenfalls wirklich unhaltbar und ich habe als Wärter in einem Zoo eine besser Möglichkeit was zu arbeiten und es zu etwas zu bringen, als als Assistant dieses ständig leicht gereizten schweren Neurasthenikers. . . . Hols der Teufel, ich glaube, ich *soll* diesen Schritt tun, solange ich noch jung bin und das Geld noch reicht, mir eine academische Stellung als Psychologe gehaltlos zu ersitzen." The English-language sections in this passage were Lorenz's.

68. Lorenz to Stresemann, 11 March 1934, ESP/Berlin: "er wäre so begeistert, wenn ein psychologischer Biologe eine Benehmenslehre lesen würde, er würde dem Betreffenden sofort so viele Studenten schicken, daß im Hörsall kein Platz wäre." See too Lorenz to Heinroth, 21 March 1934, in Heinroth and Lorenz, *Wozu,* p. 180. In his letter to Heinroth, Lorenz described with pride Bühler's intention of making Lorenz's animal psychological work part of Bühler's own students' education. He additionally indicated his intention of shifting over from the Anatomical Institute to Bühler's Psychological Institute as of 1 October 1934, though ultimately he did not do so until October 1935. Lorenz's letter to Stresemann is also cited by Hofer, "Konrad Lorenz als Schüler von Karl Bühler"; and Föger and Taschwer, *Die andere Seite des Spiegels,* though the latter date it incorrectly as 2 March 1934.

69. Lorenz to Stresemann, 14 May 1934, ESP/Berlin.

70. Back from his positive experience with the Anglo-American ornithologists at the Oxford meeting, Lorenz complained to Stresemann about Viennese philosophers who cared nothing for his ornithological knowledge but were prepared to notice any "purely philosophical mistakes" he made. Lorenz to Stresemann, 7 July 1934. On 21 November 1934 Lorenz wrote Stresemann saying that Brunswick had reviewed the whole manuscript to see that its psychological side was satisfactory. ESP/Berlin.

71. Lorenz to Stresemann, 2 March 1935 (see also Lorenz to Stresemann, no date, but evidently between 29 January 1935 and 2 March 1935), ESP/Berlin.

72. *Proceedings of the Eighth International Ornithological Congress, Oxford, July 1934* (Oxford: Oxford University Press, 1938).

73. Konrad Lorenz, "A contribution to the comparative sociology of colonial-nesting birds," in *Proceedings of the Eighth International Ornithological Congress,* pp. 207–218.

74. Lorenz to Stresemann, undated letter written between October 1933 and March 1944, ESP/Berlin.

75. Margaret Morse Nice to G. K. Noble, 25 January 1935, Nice Papers, Cornell University Library.

76. On Margaret Morse Nice's career and contributions to the study of behavior, see Gregg Mitman and Richard W. Burkhardt, Jr., "Struggling for identity: the study of animal behavior in America, 1930–1945," in *The Expansion of American Biology*, ed. Keith R. Benson, Jane Maienschein, and Ronald Rainger (New Brunswick, NJ: Rutgers University Press, 1991), pp. 164–194. See also Nice, *Research Is a Passion with Me*, pp. 128–135.

77. The monograph was long in production. Lorenz referred to his "Kumpan-work" in a letter to Stresemann as early as 15 October 1933. He was still sending final corrections of his manuscript to Stresemann as late as 16 April 1935. See ESP/Berlin.

78. Lorenz, "Kumpan" and "Companions as factors." The bird list appears on p. 113 of the translation.

79. Lorenz, "Kumpan," pp. 139–140, "Companions as factors," p. 103.

80. Lorenz, "Kumpan," p. 145, "Companions as factors," p. 108.

81. Konrad Lorenz, "The companion in the bird's world," *Auk* 54 (1937): 245–273, quote on p. 260.

82. Lorenz, "Kumpan," pp. 146–148, "Companions as factors," pp. 109–110.

83. Lorenz, "Kumpan," p. 152, "Companions as factors," p. 114.

84. Lorenz, "Kumpan," p. 174, "Companions as factors," p. 134.

85. Lorenz, "Kumpan," p. 142, "Companions as factors," p. 105. Lorenz noted that although a mother duck initially responds independently to a "thing to be brooded" and a "thing to be defended," she has the ability to acquire a knowledge of her own ducklings, and in time this allows her to unite the identity of a "thing to be brooded" and a "thing to be defended" into a "child companion." See Lorenz, "The companion in the bird's world," p. 262.

86. More than once in his writings, Lorenz cited Oskar Heinroth to the effect that "next to the wings of the Argus pheasant, the hectic life of Western civilized man is the most stupid product of intra-specific selection!" See, for example, Konrad Lorenz, *On Aggression* (New York: Harcourt Brace and World, 1966), p. 41.

87. Lorenz did not mention Darwin's *The Expression of the Emotions in Man and Animals* until he wrote his abbreviated, English version of his "Kumpan" paper. It was probably from Craig that Lorenz learned that Darwin had generally denied that forms of emotional expression evolved according to their communicative value. See Wallace Craig, "A note on Darwin's work on the expression of the emotions in man and animals," *Journal of Abnormal and Social Psychology* 16 (1921–1922): 356–366; and Lorenz, "The companion in the bird's world," p. 250. By the 1930s sexual selection had lost some of its appeal as a topic as it became blurred in classificatory enterprises such as Huxley's account of the functions of color in the animal kingdom. See above, chapter 2. Lorenz, like Huxley, felt that intraspecific competition was detrimental to the general good of the species.

88. Lorenz, "Kumpan," pp. 314–315, "Companions as factors," pp. 189–190.

89. Lorenz, "Kumpan," p. 319, "Companions as factors, p. 193.

90. Lorenz was conscious of his boldness in this regard, as evidenced by his comments in a letter to Stresemann, 29 December 1935, ESP/Berlin.

91. Lorenz readily acknowledged that, in comparing types of pair formation in birds with types of pair formation in reptiles and fish, he was talking about convergent phenomena rather than phylogenetic relationships.

92. Lorenz allowed that he could identify only a few species of birds that seemed to approximate the "lizard-type" of pair formation, namely, the Muscovy duck and some related forms. In contrast, the "labyrinth fish-type" of pair formation was, by his account, extremely common. Most bird species, he said, exhibit this particular type of pair formation, albeit with countless variations. In these species, he said, "each individual possesses not only the complete complement of species-specific, sex-determined instinctive behaviour patterns of its own sex; it also possesses (though normally in latent form) the sexual instinctive behaviour patterns of the other sex." The natural tendency in such birds, he suggested, is to perform the male behavior patterns. Female behavior patterns are exhibited only in the presence of a dominant conspecific. The birds that Lorenz characterized as displaying the "Cichlid fish-type" of pair formation were the birds that Huxley and other English bird-watchers had identified as performing "mutual display." These included the herons, cormorants, petrels, and grebes. In these birds, Lorenz allowed, there are no observable dominance relationships. In support of his claim Lorenz cited examples from his own observations on jackdaws, A. A. Allen's study of the American ruffed grouse (*Bonasa umbellus*), and Wallace Craig's observations on sexual reversibility in pigeons. See Lorenz, "Kumpan," pp. 321–337, "Companions as factors," pp. 195–208. See also A. A. Allen, "Sex rhythm in the ruffed grouse (*Bonasa umbellus* Linn.) and other birds," *Auk* 51 (1934): 180–199; and Wallace Craig, "The voices of pigeons regarded as a means of social control," *American Journal of Sociology* 14 (1908): 86–100.

93. Lorenz, "Kumpan," p. 382, "Companions as factors," p. 249.

94. Lorenz, "Kumpan," pp. 391–392, "Companions as factors," p. 258.

95. See especially Lorenz to Stresemann, 5 March 1933, ESP/Berlin.

96. Margaret Morse Nice, review of "The *Kumpan* in the bird's world: the fellow-member of the species as releasing factor of social behavior," by Konrad Lorenz, *Bird-Banding* 6 (1935): 113–114, 146–147.

97. Uexküll's inscription is reported by Lorenz in a letter from Lorenz to Stresemann, 21 November 1934, ESP/Berlin.

98. Lorenz reported on Howard's letter in a letter to Stresemann, 4 October 1935. For Howard's earlier statement on not being able to read a word of German, see Lorenz to Stresemann, 7 December 1934. Both letters in ESP/Berlin.

99. Craig to Lorenz, 27 April, 1937, Konrad Lorenz family papers. I thank Ingo Brigandt for calling this letter to my attention and Agnes von Cranach for allowing me to cite it.

100. Lorenz to Stresemann, 17 July 1933, ESP/Berlin.

101. Lorenz to Stresemann, 7 December 1934, ESP/Berlin.

102. Lorenz reported the package of reprints from Nice in a letter to Stresemann dated 7 December 1934, ESP/Berlin.

103. Lorenz to Stresemann, letter dated 2 March 1935 (but filed after Lorenz's letter of 25 March 1935 in the Stresemann Papers), ESP/Berlin.

104. Nice, *Research Is a Passion with Me*, pp. 140–141, mentions the seven-page letter from Craig to Lorenz. Lorenz's reaction to Craig's letter is described by Lorenz in a letter to Stresemann, 13 March 1935, ESP/Berlin. There Lorenz speaks of Craig's letter inspiring him to work furiously ("in pansenlosen Judische Hast') to broaden his "English manuscript" to take McDougall into account as well as Lloyd Morgan. Lorenz may also at this point have added a discussion of McDougall to the "Kumpan's" concluding remarks, which he had yet to send to Stresemann.

105. Lorenz, "Kumpan," pp. 383–384. For a slightly different translation see Lorenz, "Companions as factors," p. 250.

106. Lorenz entitled chapter 19 of his "Russian" manuscript "The discovery of 'appetitive behavior' by Wallace Craig." See Konrad Lorenz, *The Natural Science of the Human Species: An Introduction to Comparative Behavioral Research: The "Russian Manuscript" (1944–1948)* (Cambridge, MA: MIT Press, 1996), pp. 259–270. For more on the "Russian manuscript," see chapter 5. Lorenz likewise claimed in 1950 that "Craig's great discovery" was that "it is the discharge of consummatory actions and not [their] survival value which is the goal of appetitive behavior." See Lorenz, "The comparative method in studying innate behaviour patterns," p. 236.

107. Lorenz to Stresemann, 8 August 1935, ESP/Berlin. "Craig ist einer der schärfsten und präzisesten Denker und *Vielwisser* auf diesen Gebiete, Howard ist ein Dichter der zu ¾ intuitiv die richtigsten Ansichten von instinktiven Verhalten hat, der einzige der gleich mir den Einfluß der Erfahrung auf die Instinkthandlung *leugnet!*" An alternative to writing the English book and having a German version appear simultaneously, Lorenz said, would be to do only the English book, while publishing in German a short piece entitled something like "A critique of the modern instinct concept," designed for the journal *Die Naturwissenschaften*. Then, in addition, he would begin writing, in German, a "textbook of animal psychology."

108. Lorenz to Nice, 19 September 1935, Nice Papers, Cornell University Library.

109. Lorenz to Stresemann, 4 October 1935, ESP/Berlin.

110. On the career of Antonius, see Veronika Hofer, "Bühne—Wohnung—Territorium: Der Schönbrunner Tiergarten unter der Leitung von Otto Antonius (1924–1945)," in *Menagerie des Kaisers: Zoo der Wiener: 250 Jahre Tiergarten Schönbrunn*, ed. Mitchell Ash and Lothar Dittrich (Vienna: Pichler-Verlag, 2002), pp. 181–215.

111. Lorenz to Heinroth, 5 April 1934 (?), in Heinroth and Lorenz, *Wozu*, p. 181. The uncertainty of the date is signaled by the editor, Otto Koenig.

112. Lorenz to Stresemann, 27 November 1934, ESP/Berlin.

113. Lorenz to Stresemann, 10 January 1935, ESP/Berlin. Cited in Föger and Taschwer, *Die andere Seite des Spiegels*, p. 61. On 22 September 1935, with Antonius's status at Schönbrunn still unsettled, Lorenz wrote to Stresemann saying, "I really hope that he will be appointed again, but please keep all fingers crossed for me!" (Ich hoffe wirklich, dass er wieder eingesetzt wird, aber halten Sie mir bitte auf älle Fälle Daumen!). ESP/Berlin. Föger and Taschwer, *Die andere Seite des Spiegels*, pp. 62–63, indicate that a letter on Lorenz's behalf was sent in August 1936, albeit to no avail, from the office of Aus-

trian Chancellor Kurt Schuschnigg to the Ministry of Commerce and Trade, which had charge of the position.

114. Lorenz to Stresemann, a March 1935 (see also Lorenz to Stresemann, no date, but evidently written between 29 January 1935 and 2 March 1935), ESP/Berlin.

115. Lorenz to Nice, undated (but prior to 19 September 1935), Nice Papers, Cornell University Library; emphasis in original.

116. Klaus Taschwer was the first to call attention to Lorenz's popular lectures and writings in this period. See Taschwer, "'Rendezvous mit Tier und Mensch': wissenschaftssoziologische Anmerkungen zur Geschichte der Verlhaltensforschung in Österreich," in "Soziologische und historische Analysen der Sozialwissenschaften," ed. Christian Fleck, special issue, Österreichische Zeitschrift für Soziologie, Sonderband 5 (2000): 309–341. See also Föger and Taschwer, Die andere Seite des Spiegels, pp. 57–59.

117. The work appeared as Lorenz, "Über eine eigentümliche Verbindung branchialer Hirnnerven bei Cypselus apus."

118. See Lorenz to Heinroth, 18 December 1935, in Heinroth and Lorenz, Wozu, p. 203.

119. Lorenz to Stresemann, 4 October 1935, ESP/Berlin.

120. Lorenz to Heinroth, 18 December 1935, in Heinroth and Lorenz, Wozu, p. 203. Kramer's role in setting up Lorenz's Harnackhaus lecture is mentioned in Lorenz, Vergleichende Verhaltensforschung, p. 4, though here Lorenz incorrectly gives the date of his lecture as 1935 instead of 1936.

121. Heinroth to Lorenz, 1 and 14 January 1936, in Heinroth and Lorenz, Wozu, pp. 204–205, 211. Information on the building, functions, and sizes of rooms in Harnackhaus comes from Jahrbuch der Max-Planck-Gesellschaft zur Förderung der Wissenschaften E. V. Teil II (Göttingen: Max-Planck-Gesellschaft, 1961), pp. 863–865. Unfortunately, although the Journal für Ornithologie published reports on the Hauptsitzungen (general meetings) and Fachsitzungen (meetings typically devoted to reviews of recent literature) that the DOG held later in the year, it did not report on the February meeting. I have not found any source that gives the actual attendance at Lorenz's February lecture or confirms that he showed his two films along with his lecture.

122. See Konrad Lorenz, "Über die Bildung des Instinktbegriffes," Die Naturwissenschaften 25 (1937): 289–300, 307–318, 324–331. Lorenz acknowledged his debt to Craig in a footnote on p. 289. The paper appears in English translation as "The establishment of the instinct concept," in Lorenz, Studies in Animal and Human Behaviour, 1:259–315.

123. Lorenz, "Über die Bildung des Instinktbegriffes," p. 295.

124. Lorenz, "Über die Bildung des Instinktbegriffes," p. 329.

125. Lorenz, "Konrad Lorenz," p. 179; Lorenz, Vergleichende Verhaltensforschung, p. 5.

126. See the following chapter.

127. Lorenz, "Über die Bildung des Instinktbegriffes," p. 296, "The establishment of the instinct concept," p. 273.

128. Lorenz, "Über die Bildung des Instinktbegriffes," p. 297, "The establishment of the instinct concept," p. 274.

129. Lorenz, "Betrachtungen über das Erkennen der arteigenen Triebhandlungen

der Vögel," p. 90, A consideration of methods of identification of species-specific instinctive behaviour patterns in birds, p. 93.

130. Lorenz, "Über die Bildung des Instinktbegriffes," p. 327, quotations from "The establishment of the instinct concept," pp. 307–308. As indicated above, McDougall, in discussing instinctive actions, had already used the analogy of a gas under pressure. See n. 54, this chapter.

131. Lorenz to Nice, 12 March 1936, Nice Papers, Cornell University Library.

132. As reported by Lorenz to Stresemann, 2 July 1936, and 2 September 1936, ESP/ Berlin; Süffert expressed his enthusiasm to Lorenz in a letter dated 20 August 1936, Lorenz family papers.

133. Lorenz to Stresemann, 13 October 1936, ESP/Berlin.

134. On the founding of the DGT, see "Gründungsbericht der Deutschen Gesellschaft für Tierpsychologie," *Zeitschrift für Angewandte Psychologie und Charakterkunde* 51 (1936): 255–256; and J. Effertz, "Bericht über die Gründung der Deutschen Gesellschaft für Tierpsychologie," *Zeitschrift für Tierpsychologie* 1 (1937): 1–8. On Kronacher, see the obituary notice by V. Stang, O. Koehler, and J. Effertz, "Prof. emer. Dr. Dr. h. C. Dr. agr. h. c. Carl Kronacher," *Zeitschrift für Tierpsychologie* 2 (1938–1939). i–iv.

135. Heinroth to Lorenz, 4 November 1936, in Heinroth and Lorenz, *Wozu*, p. 220.

136. Lorenz to Heinroth, 16 November 1936, in Heinroth and Lorenz, *Wozu*, p. 221. Lorenz said much the same three days earlier in a letter to Stresemann. Lorenz to Stresemann, 13 November 1936, ESP/Berlin.

137. Lorenz to Stresemann, 13 November 1936, notes how the lecture would come from the introductory chapter of the book manuscript. ESP/Berlin.

138. Konrad Lorenz, "Biologische Fragestellung in der Tierpsychologie," *Zeitschrift für Tierpsychologie* 1 (1937): 24–32, quote on p. 25. This paper lacks a bibliography. Elsewhere, Lorenz indicates that the paper by Koehler that especially impressed him was Otto Koehler, "Das Ganzheits Problem in der Biologie," *Schriften der Königsberger Gelehrten Gesellschaft* 9 (1932): 139–204.

139. Lorenz, "Biologische Fragestellung," p. 28.

140. Lorenz to Heinroth, 5 January 1937, in Heinroth and Lorenz, *Wozu*, pp. 222–223.

141. Lorenz was not unhappy with the result. As he explained in a letter to Stresemann, it was a joy to see Antonius happily back in his old post, and this also meant, contrary to what Lorenz had come to fear, that the job would not be given to a certain "well-documented total idiot" with whom Lorenz would have found it impossible to work. Lorenz to Stresemann, 23 January 1937, ESP/Berlin. Cited by Föger and Taschwer, *Die andere Seite des Spiegels*, p. 63.

142. Judging from the pages of the *Zeitschrift für Tierpsychologie*, the influence of the three founders of the DGT on the subsequent intellectual life of the society was rather minimal. Klein never published in the journal, Effertz's only paper was his account of the society's founding, and Kronacher published only a paragraph-long note indicating that the Deutsche Gesellschaft für Tierpsychologie was *not* the continuation of the Gesellschaft für Tierpsychologie founded in Stuttgart, which had occupied itself exclusively with the achievements of so-called clever animals. See Carl Kronacher, "Erklärung zur Frage der 'zahlensprechenden' Tiere," *Zeitschrift für Tierpsychologie* 1 (1937): 91. Nonetheless, the interest in having the society pursue both pure and applied research

was manifested in at least the first two annual meetings of the society, where in each case the first day was devoted to "pure science" papers and the second day was devoted to applied science papers. Lorenz and Effertz were to give a joint presentation entitled "Examples of applied animal psychology and their utilization in animal breeding and animal keeping" at the third annual meeting of the society, scheduled for Leipzig from 21 to 23 September, 1939. Wartime events intervened and the meeting was first postponed and then ultimately never held.

CHAPTER FOUR

1. Letter from Niko Tinbergen to Richard W. Burkhardt, Jr., 6 June 1979.

2. Lorenz to Stresemann, 26 March 1937: "[Tinbergen] . . . macht wunderschöne Experimente über angeborene Schematen. Er hat ein besonderes und mir abgehendes (ich lerne massenhaft von Ihm!) Talent für präzise Versuchsanordnungen." See also Lorenz to Stresemann, 13 May 1937, Erwin Stresemann Papers, Preußischer Kulturbesitz, Staatsbibliothek zu Berlin (hereafter ESP/Berlin).

3. A letter from Lorenz to Stresemann dated 13 November 1936 (ESP/Berlin) describes Lorenz's plans to earn some money by picking up a Hudson automobile in Antwerp and driving it to Berlin for his brother-in-law, an automobile dealer. In this letter, Lorenz tells Stresemann that Tinbergen, having heard of Lorenz's travel plans, has arranged for Lorenz to come give lectures in Leiden. Tinbergen first suggested this in a letter dated 5 November 1936, mentioning to Lorenz the idea of a symposium in Leiden in which Lorenz and the Dutch animal psychologist J. A. Bierens de Haan could present their contrasting views on instinct and to which other prominent students of animal behavior would be invited. Van der Klaauw wrote to Lorenz on 18 November 1936 to invite him officially to the event. Both of these letters and two other preconference letters from Tinbergen to Lorenz are in the Lorenz family papers. D. R. Röell describes the symposium and cites correspondence between van der Klaauw and Bierens de Haan regarding it. See D. R. Röell, *The World of Instinct: Niko Tinbergen and the Rise of Ethology in the Netherlands (1920–1950)* (Assen: Van Gorcum, 2002), p. 111. Originally published as *De wereld van instinct: Niko Tinbergen en het ontstaan van de ethologie in Nederland (1920–1950)* (Rotterdam: Erasmus Publishing, 1996).

4. Niko Tinbergen interview with Richard W. Burkhardt, Jr., 30 April 1979. Tinbergen did not save his early letters from Lorenz, but several of Tinbergen's early letters to Lorenz remain among the Lorenz family papers. I am grateful to Ingo Brigandt for first making these letters and other letters in this collection known to me.

5. For biographical information on Tinbergen, see his own autobiographical comments in "Nikolaas Tinbergen," in *Les Prix Nobel en 1973* (Stockholm: Imprimerie Royale P. A. Norstedt & Söner, 1974), pp. 197–200, and especially his "Watching and wondering," in *Studying Animal Behavior: Autobiographies of the Founders*, ed. Donald A. Dewsbury (Chicago: University of Chicago Press, 1989), pp. 431–463. The most extensive study of Tinbergen's life is now Hans Kruuk, *Niko's Nature: A Life of Niko Tinbergen and His Science of Animal Behaviour* (Oxford: Oxford University Press, 2003). Kruuk's book appeared after the revised manuscript of the present book had been returned to the University of Chicago Press and thus is noted here only briefly. See also R. A. Hinde,

"Nikolaas Tinbergen," *Biographical Memoirs of the Royal Society* 36 (1990): 547–565; Gerard Baerends, Colin Beer, and Aubrey Manning, "Introduction," in *Function and Evolution in Behaviour: Essays in Honour of Professor Niko Tinbergen, F.R.S.,* ed. Gerard Baerends, Colin Beer, and Aubrey Manning (Oxford: Clarendon Press, 1975), pp. xi–xxii; Gerard P. Baerends, "Early ethology: growing from Dutch roots," in *The Tinbergen Legacy,* ed. M. S. Dawkins, T. R. Halliday, and R. Dawkins (London: Chapman and Hall, 1991), pp. 1–17; and Röell, *The World of Instinct.*

6. Tinbergen, "Watching and wondering," pp. 432–433.

7. Tinbergen letter to Richard W. Burkhardt, Jr., 5 May 1979.

8. Charles E. Raven, *In Praise of Birds: Pictures of Bird Life* (London: Martin Hopkinson and Co., 1925), pp. 137–138.

9. On the NJN, see Gerard P. Baerends, "Two pillars of wisdom," in Dewsbury, *Studying Animal Behavior,* p. 14; Baerends, Beer, and Manning, "Introduction," p. xii; Hinde, "Nikolaas Tinbergen," pp. 549–550; and Röell, *The World of Instinct,* pp. 44–48. Baerends, Beer, and Manning report that it was a lecture that Tinbergen gave to promote the NJN that led Baerends to join the society in 1928.

10. Tinbergen interview with Richard W. Burkhardt, Jr., 30 April 1979.

11. Tinbergen letter to Richard W. Burkhardt, Jr., 5 May 1979 and interview with Richard W. Burkhardt, Jr., 30 April 1979.

12. Jan Verwey, "Die Paarungsbiologie des Fischreihers (*Ardea cinerea* L.)," in *Verhandlungen des VI. Internationalen Ornithologen-Kongresses in Kopenhagen 1926,* ed. F. Steinbacher (Berlin, 1929), pp. 390–413, "Die Paarungsbiologie des Fischreihers," *Zoologische Jahrbücher* 48 (1930): 1–120. An undated letter from Tinbergen to W. H. Thorpe, probably written in the 1970s when Thorpe was writing his history of ethology, says that Tinbergen's earliest studies of the social behavior of the herring gull were influenced by Heinroth and Huxley but "above all" by Verwey's gray heron study. William Homan Thorpe Papers Cambridge University Library, MS.Add.8784, M17.

13. Tinbergen letter to Richard W. Burkhardt, Jr., 5 May 1979; Baerends, Beer, and Manning, "Introduction," p. xiii.

14. Tinbergen letter to Richard W. Burkhardt, Jr., 5 May 1979; Tinbergen, "Watching and wondering"; Hinde, "Nikolaas Tinbergen"; and Röell, *The World of Instinct.* Tinbergen also found inspiration in the writings of the British ornithologists Edmund Selous, H. Eliot Howard, and Julian Huxley and, once he became interested in organisms other than birds, in the entomological studies of the French naturalist J.-H. Fabre and the Austrian comparative physiologist Karl von Frisch.

15. Others who deserve mention as promoters of Dutch field ornithology include G. J. van Oordt, who collaborated with Huxley in Huxley's study of the red-throated diver in Spitsbergen, and E. D. van Oort, who pioneered bird banding in Holland and supported the establishment of bird migration study centers at the Ringstation Wasenaar and the Vogeltrekstation Texel. On Holland's rich system of bird sanctuaries, see G. J. van Oordt, "The ornithological reservations of the Netherlands in 1930," *Ardea* 19 (1930): 1–12.

16. A. F. J. Portielje, "Zur Ethologie bzw. Psychologie der *Rhea americana* L.," *Ardea* 14 (1925): 1–14, "Zur Ethologie bzw. Psychologie von *Botaurus stellaris* (L.)," *Ardea* 15

(1926): 1–15, "Zur Ethologie bzw. Psychologie von *Phalacrocorax carbo subcormoranus* (Brehm)," *Ardea* 16 (1927): 107–123, "Zur Ethologie bzw. Psychologie der Silbermowe, *Larus argentatus argentatus* Pont.," *Ardea* 17 (1928): 112–149.

17. J. A. Bierens de Haan, "Die Balz des Argusfasans," *Biologisches Zentralblatt* 46 (1926): 428–435.

18. Tinbergen interview with Richard W. Burkhardt, Jr., 30 April 1979. Röell, *The World of Instinct,* pp. 54–58, describes Portielje as a theorist and discusses Tinbergen's friendly relations with him.

19. Portielje, "Zur Ethologie bzw. Psychologie von *Botaurus stellaris.*" Tinbergen (interview with Richard W. Burkhardt, Jr., 30 April 1979) described how he preferred Huxley's work, while Verwey preferred Heinroth's.

20. Julian Huxley, "Biology of bird courtship," in *Proceedings of the VIIth International Ornithological Congress at Amsterdam, 1930* (Amsterdam: Prof. Dr. L. F. de Beaufort, 1931), pp. 107–108; N. Tinbergen, "Zur Paarungsbiologie der Flussseeschwalbe (*Sterna hirundo hirundo* L.)," *Ardea* 20 (1931): 1–18. The claim by Huxley to which Tinbergen objected does not appear in the published abstract of Huxley's paper. Tinbergen in a 1964 paper credited Professor Boschma with having "generously allowed one of his undergraduates [i.e., Tinbergen] to spend an entire spring away from the laboratory, observing the love rituals of Terns. He even accepted the rather incoherent account this young man wrote of his observations as part of the work to be submitted for his 'doctoraal' examination." See N. Tinbergen, "On adaptive radiation in gulls (tribe *Larini*)," *Zoologische Mededeelingen* 39 (1964): 209–223, quote on p. 209.

21. N. Tinbergen, "Over het voedsel van de Blauwe Reiger (*Ardea cinerea cinerea* L.)," *Ardea* 19 (1930): 89–93, "Beobachtungen am Baumfalken (*Falco subbuteo subbuteo* L.)," *Journal für Ornithologie* 80 (1932): 40–50, "Ueber die Ernährung einer Waldohreulenbrut (*Asio otus otus* L.)," *Beiträge zur Fortpflanzungsbiologie der Vögel* 8 (1932): 54–55, "Over het voedsel van de Sperwer (*Accipiter nisus nisus* L.) in de Nederlandsche duinstreek," *Ardea* 21 (1932): 77–89.

22. As reported by Tinbergen in a letter to M. M. Nice, 5 December 1945, Margaret Morse Nice Papers, 1917–1968, collection number 2993, courtesy of the Division of Rare and Manuscript Collections, Cornell University Library (hereafter MMNP/Cornell). Tinbergen described how Verwey told him that Verwey too had at first been interested primarily in birds but had subsequently come to think that all animals were interesting.

23. N. Tinbergen, "Über die Orientierung des Bienenwolfes (*Philanthus triangulum* Fabr.)," *Zeitschrift für vergleichende Physiologie* 16 (1932): 304–334. Part of this article is reproduced in English translation as "On the orientation of the digger wasp *Philanthus triangulum* Fabr. I," in *The Animal in Its World: Explorations of an Ethologist, 1932–1972,* vol. 1 (London: George Allen and Unwin, 1972), pp. 103–127. Tinbergen gives a more popular account of his *Philanthus* work in *Curious Naturalists* (London: Country Life, 1958), pp. 15–30.

24. Hinde, "Nikolaas Tinbergen"; Tinbergen, *Curious Naturalists,* and "Watching and wondering."

25. Tinbergen, *Curious Naturalists,* p. 31.

26. Tinbergen, "The behavior of the snow bunting in spring," *Transactions of the Linnaean Society of New York* 5 (1939): 1–94 (see p. 16 on "substitute activities"), "Field

observations of east Greenland birds," pt. 1, "The behaviour of the red-necked phalarope (*Phalaropus lobatus* L.) in spring," *Ardea* 24 (1935): 1–42, "Watching and wondering," Tinbergen letter to R. W. Burkhardt, 6 June 1979.

27. The information in this paragraph is based primarily on a letter from Tinbergen to W. H. Thorpe, 8 September 1976, Nikolaas Tinbergen Papers, Bodleian Library, Oxford University (hereafter NTP/Oxford). Tinbergen evidently provided this information to Thorpe when Thorpe was writing his history of ethology. Thorpe left these details on Tinbergen's teaching program at Leiden out of his account. See W. H. Thorpe, *The Origins and Rise of Ethology* (London: Heinemann, 1979).

28. On van der Klaauw see H. Boschma, "A concise review of the scientific activities of C. J. van der Klaauw," in "Volume jubilaire dédié à C. J. van der Klaauw," *Archives néerlandaises de zoologie* 13, supp. (1958): 5–9. Tinbergen's feelings about the block practicals are specified in his letter to Thorpe, 8 September 1976, NTP/Oxford.

29. See N. Tinbergen, "Waarnemingen en proeven over de sociologie van een zilvermeeuwenkolonie," *De Levende Natuur* 40 (1935): 263–280, 304–308, "Zur Soziologie der Silbermöwe, *Larus a. argentatus* Pont.," *Beiträge zur Fortpflanzungsbiologie der Vögel* 12 (1936): 89–96. See also H. L. Booy and N. Tinbergen, "Nieuwe feiten over de sociologie van de zilvermeeuwen," *De Levende Natuur* 41 (1937): 325–334. The date of 1934 as the beginning of stickleback work by Tinbergen and ter Pelwijk is given by J. J. A. Van Iersel, "An analysis of the parental behaviour of the male three-spined stickleback (*Gasterosteus aculeatus* L.)," *Behaviour*, supp. 3 (1953): 1.

30. In the field, G. van Deusekom began following up on Tinbergen's studies of the orientation of *Philanthus*. In the lab, P. J. van Eck undertook a dissertation entitled "Color vision and cone function in the thrush (*Turdus ericetorum ericetorum* Turton)," while P. Meyknecht's was "Color vision and brightness differentiation in the little owl (*Athene noctua vidalii* A. E. Brehm)."

31. Tinbergen, "Watching and wondering," p. 444.

32. Tinbergen letter to his parents, 11 December 1934, NTP/Oxford. The position in the East Indies is described in an undated letter Tinbergen also wrote to his parents.

33. N. Tinbergen, "Über die Orientierung des Beinenwolfes (*Philanthus triangulum* Fabr.)," pt. 2, "Die Bienenjagd," *Zeitschrift für vergleichende Physiologie* 21 (1935): 699–716; N. Tinbergen and W. Kruyt, "Über die Orientierung des Beinenwolfes (*Philanthus triangulum* Fabr.)," pt. 3, "Die Bevorzugung bestimmter Wegmarken," *Zeitschrift für vergleichende Physiologie* 25 (1938): 292–334; N. Tinbergen and R. J. van der Linde, "Über die Orientierung des Beinenwolfes (*Philanthus triangulum* Fabr.)," pt. 4, "Heimflug aus unbekanntem Gebiet," *Biologisches Zentralblatt* 58 (1938): 425–435; G. Schuyl, L. Tinbergen, and N. Tinbergen, "Ethologische Beobachtungen am Baumfalken (*Falco s. subbuteo* L.)," *Journal für Ornithologie* 84 (1936): 387–433; N. Tinbergen, B. J. D. Meeuse, L. K. Boerema and W. Varossieau, "Die Balz des Samtfalters, *Eumenis semele* (L.)," *Zeitschrift für Tierpsychologie* 5 (1942): 182–226.

34. J. J. ter Pelwijk and N. Tinbergen, "Roodkaakjes," *De Levende Natuur* 41 (1936): 129–137.

35. Tinbergen letter to W. H. Thorpe, 8 September 1976, NTP/Oxford.

36. The papers published as the result of the symposium were J. A. Bierens de Haan, "Über den Begriff des Instinktes in der Tierpsychologie," *Folia Biotheoretica, Se-*

ries B 2 (1937): 1–16; Konrad Lorenz, "Über den Begriff der Instinkthandlung," *Folia Bio-theoretica, Series B* 2 (1937): 17–50; L. Verlaine, "Qu'est-ce que l'instinct?" *Folia Biotheo-retica, Series B* 2 (1937): 51–64; and E. S. Russell, "Instinctive behaviour and bodily de-velopment," *Folia Biotheoretica, Series B* 2 (1937): 67–76. The presence at the symposium of Portielje and Pontus Palmgren as well as Verlaine and Bierens de Haan is mentioned in Tinbergen, "Watching and wondering," p. 445.

37. In the introduction to his published paper Lorenz wrote, "As pleased as I was at the time to accept Professor C. J. van der Klaauw's request to speak at the symposium on animal psychology in Leiden on the problem of instinctive actions, I am now equally pleased to accept his invitation to extract the purely conceptual material from my recent paper on this topic in the 'Naturwissenschaften' and to compile it in a manner appro-priate for the 'Folia Biotheoretica.'" Quoted from the English translation given in *Foun-dations of Comparative Ethology,* ed. Gordon M. Burghardt (New York: Van Nostrand Reinhold Company, 1985), p. 406. The paper also identified itself as an "elaboration" of the lecture Lorenz gave at the symposium. However, despite numerous entreaties from van der Klaauw for the final version of the paper, Lorenz did not send it until late May 1937. One reason for the delay was apparently that in correspondence with Erich von Holst in April 1937 Lorenz was led to a significantly revised view of instinct. On the im-portance of Holst's work for Lorenz's thinking, see below. Van der Klaauw's and Holst's letters to Lorenz are in the Lorenz family papers. Röell, *The World of Instinct,* p. 111, also cites a letter from van der Klaauw to Bierens de Haan complaining of Lorenz's slow de-livery of the final version. Although Lorenz did not confront Bierens de Haan's views di-rectly in his Leiden symposium paper, he did so in his later paper "Induktive und teleo-logische Psychologie," *Die Naturwissenschaften* 30 (1942): 133–143.

38. Here I use the published version of Lorenz's paper as evidence of what he said at the Leiden symposium, though, as indicated in the previous footnote, the presenta-tion he gave and the paper he later published were not identical.

39. Lorenz letter to Oskar Heinroth, 5 January 1937, in *Wozu aber hat das Vieh die-sen Schnabel? Briefe aus der frühen Verhaltensforschung, 1930–1940,* by Oskar Heinroth and Konrad Lorenz, ed. Otto Koenig (Munich: Piper, 1988), pp. 222–223 (hereafter cited as Heinroth and Lorenz, *Wozu*).

40. Tinbergen, "Watching and wondering," p. 445, interview with Richard W. Burkhardt, Jr., 30 April 1979. The second quote in the text is what Tinbergen reported to me. Tinbergen was quite explicit in his statement to me that Lorenz spoke of "what *I* need" rather than "what *we* need." Tinbergen recounted this quite good-naturedly, ex-plaining: "That was Konrad, you see."

41. The word "clicked" is Tinbergen's. See Tinbergen, "Nikolaas Tinbergen," p. 198. See also Tinbergen, "Aus der Kinderstube der Ethologie," in Heinroth and Lorenz, *Wozu,* pp. 309–314.

42. Tinbergen to Huxley, 18 February 1953, Julian S. Huxley Papers, Rice University Archives.

43. Konrad Lorenz, "The present state of ethology" (unpublished and undated manuscript of a talk delivered two years before Lorenz's retirement), p. 3. I am grateful to Ernst Mayr for providing me with a copy of this manuscript.

44. Letter from Niko Tinbergen to Richard W. Burkhardt, Jr., 6 June 1979. Cer-

tainly Tinbergen was impressed by Lorenz's focus on behavioral *causation*. In a letter of 17 January 1937 he told Lorenz, "I now believe that your conception of the instinctive action [*Instinkhandlung*] is the sole useful thing (for causal analytical research)." What Bierens de Haan took as the *plasticity* of instincts, Tinbergen suggested, was in fact a function of multiple causal factors operating together at the same time. The letter is in the Lorenz family papers.

45. Lorenz, *The Foundations of Ethology*, pp. 6–7. Lorenz went on to say: "Its [the IRM's] further elaboration and refinement, and the exploration of its physiological characteristics, especially its functional limitations, are all due to Niko Tinbergen's experiments." The same basic account occurs also in Konrad Lorenz, "My family and other animals," in Dewsbury, *Studying Animal Behavior*, p. 269.

46. Konrad Lorenz, "Der Kumpan in der Umwelt des Vogels: der Artegenosse als auslösendes Moment sozialer Verhaltungsweisen," *Journal für Ornithologie* 83 (1935): 140, 141, and further.

47. Lorenz, "Konrad Lorenz," p. 180.

48. Tinbergen arrived in Altenberg on 2 March 1937, according to a letter from Lorenz to Stresemann, 26 March 1937, ESP/Berlin. In the Tinbergen manuscripts at Oxford, the earliest letter Tinbergen wrote from Altenberg to his parents is dated 10 March 1937; the last date in his Altenberg bird diary is 12 June 1937.

49. Tinbergen to Lorenz, 18 February 1937, Lorenz family papers.

50. Lorenz announced the arrival of Tinbergen and the ducks in a letter to Heinroth, 16 March 1937, in Heinroth and Lorenz, *Wozu*, p. 231.

51. The mention of Tinbergen's lecturing in Vienna is in a letter from Tinbergen to Mayr, 25 March 1938, Ernst Mayr Papers, Harvard University Archives (henceforth EMP/Harvard), HUG (FP), 14.7, box 2, folder 68. All items from the Ernst Mayr Papers are cited courtesy of the Harvard University Archives.

52. Tinbergen interview with Richard W. Burkhardt, Jr., 30 April 1979. Lorenz to Stresemann, 26 March 1937, ESP/Berlin. See also Lorenz to Stresemann, 13 May 1937. Lorenz's and Tinbergen's jointly published paper on the egg-rolling behavior of the greylag goose contains a footnote stating that the theoretical part of the paper was largely the work of Lorenz while the devising and the carrying out of the experiments was largely the work of Tinbergen. See K. Lorenz and N. Tinbergen, "Taxis und Instinkthandlung in der Eirollbewegung der Graugans, ," pt. 1, *Zeitschrift für Tierpsychologie* 2 (1938): 1n. The idea of spelling out this division of labor was Tinbergen's (in letters to Lorenz dated 1 and 13 October 1937, Lorenz family papers). This paper also identified itself as a sequel to the two papers Lorenz published in 1937 based upon his Berlin and Leiden lectures of 1936. The egg-rolling paper bore the roman numeral I to signify that it would be followed by additional work. However, no follow-up study was ever reported. The original paper appears in English translation (but without any indication that it was coauthored by Tinbergen) as "Taxis and instinctive behaviour pattern in egg-rolling by the greylag goose," in *Studies in Animal and Human Behaviour*, by Konrad Lorenz, trans. Robert Martin, 2 vols. (Cambridge, MA: Harvard University Press, 1970–1971), 1:316–350.

53. As indicated above, sorting out whose idea this was initially is very difficult to do. Tinbergen had already embarked on work of this sort with D. J. Kuenen, as Röell indicates in *The World of Instinct*, pp. 120–121. Tinbergen was eager to point out to Lorenz,

however, that even before Tinbergen came to Altenberg in March 1937 Lorenz had mentioned the egg-rolling reaction as a "simultaneous Instinkthandlung-Taxis intercalation." Tinbergen to Lorenz, 1 October 1937, Lorenz family papers.

54. Lorenz and Tinbergen, "Taxis und Instinkthandlung," p. 10; Lorenz, "Taxis and instinctive behaviour pattern," pp. 327–328.

55. On Bethe, see Erich von Holst, "Albrecht Bethe," *Die Naturwissenschaften* 42 (1955): 99–101.

56. Albrecht Bethe, "Dürfen wir den Ameisen und Bienen psychische Qualitäten zuschreiben?" *Archiv für die gesamte Physiologie des Menschen und der Tiere* 70 (1898): 15–99.

57. See, for example, Erich von Holst, "Versuche zur Theorie der relativen Koordination," *Pflügers Archiv* 237 (1936): 93–121.

58. Holst to Lorenz, 1 April 1937, Lorenz family papers; Lorenz to Stresemann, 26 April 1937, ESP/Berlin. Ingo Brigandt has informed me that in a letter of 9 January 1937 Otto Koehler urged Lorenz to read Holst's recent papers because Holst's findings corresponded well to Lorenz's theorizing. Letter in the Lorenz family papers.

59. C. J. van der Klaauw to Lorenz, 25 May 1937, Lorenz family papers.

60. Lorenz, "Über den Begriff der Instinkthandlung," pp. 35, 36.

61. Tinbergen to Lorenz, 1 October 1937, Lorenz family papers.

62. Lorenz and Tinbergen, "Taxis und Instinkthandlung," p. 17; Lorenz, "Taxis and instinctive behaviour pattern," p. 336.

63. These experiments were never published in detail, though both Tinbergen and Lorenz referred to them in published writings. See Konrad Lorenz, "Vergleichende Verhaltensforschung," *Verhandlungen der Deutschen Zoologischen Gesellschaft, Zoologischer Anzeiger,* supp. 12 (1939): 69–102, notably pp. 92–94; N. Tinbergen, "Social releasers and the experimental method required for their study," *Wilson Bulletin* 60 (1948): 6–51, notably p. 7. Tinbergen's field notes of the experiments, beginning with observations made on 16 March and concluding with observations made on 11 June, remain among his papers at Oxford.

64. Later work by Wolfgang Schleidt demonstrated that the young turkeys were not responding to the shapes of the dummies but instead to how they moved. See W. M. Schleidt, "Reaktionen von Truthühnern auf fliegende Raubvögel und Versuche zur Analyse ihres AAM's," *Zeitschrift für Tierpsychologie* 18 (1961): 534–560.

65. Tinbergen letter from Altenberg to his father and mother, 3/29/37, NTP/Oxford.

66. Undated letter from Tinbergen to Lorenz in the Lorenz family papers.

67. Tinbergen, "Nikolaas Tinbergen," p. 199; repeated in Tinbergen, "Watching and wondering," p. 447.

68. J. J. ter Pelkwijk and N. Tinbergen, "Eine reizbiologische Analyse einiger Verhaltensweisen von *Gasterosteus aculeatus* L.," *Zeitschrift für Tierpsychologie* 1 (1937): 193–200. Aside from three articles describing ecological and experimental studies of the three-spined stickleback, the only other references cited by ter Pelkwijk and Tinbergen were Oskar Heinroth's 1910 "Beiträge zur Biologie, namentlich Ethologie und Psychologie der Anatiden," in *Verhandlungen des 5. Internationalen Ornithologen-Kongresses in, Berlin, 30. Mai bis 4. Juni 1910,* ed. Herman Schalow (Berlin: Deutsche Ornithologische

Gesellschaft, 1911), pp. 589–702; " and Lorenz's 1935 "Kumpan." Tinbergen knew of Lorenz's "Kumpan" paper as early as August 1935, for he referred to it in a footnote though not in the bibliography of his paper on the red-necked phalarope, a paper that appeared in the August 1935 issue of *Ardea*. See Tinbergen, "Field observations of east Greenland birds," pt. 1, p. 31.

69. See Röell, *The World of Instinct*, especially pp. 108–110.

70. Tinbergen to Mayr, 25 March 1938 (in German), EMP/Harvard, HUG (FP) 14.7, box 2, folder 68. The letter to Nice, referred to in Tinbergen's letter to Yerkes of 3 April 1938 is not in the Nice correspondence at Cornell. See Robert M. Yerkes Papers, Yale University (henceforth RMYP/Yale).

71. Tinbergen to Yerkes, 3 April 1938, RMYP/Yale.

72. Tinbergen to Yerkes, 24 October 1938, RMYP/Yale. See also Yerkes to G. K. Noble, 8 November 1938: "Doubtless you had opportunity to become acquainted with Doctor Tinbergen. We very much enjoyed having him in our laboratories and we are sorry that he could not stay longer and undertake some work on his own account." RMYP/Yale.

73. Tinbergen, "Watching and wondering," p. 447.

74. Tinbergen to Mayr, 5 February 1977, copy courtesy of Ernst Mayr. See below, chapter 7, n. 23, for an explanation of how letters from the Mayr correspondence are cited in this book.

75. See Tinbergen, "Watching and wondering," p. 447; Tinbergen to Nice, 28 July 1938, MMNP/Cornell; and Tinbergen interview with Richard W. Burkhardt, Jr., 30 April 1979. According to a letter from Libbie Hyman to W. C. Allee, 13 December 1939 (W. C. Allee Papers, University of Chicago), Lehrman and Tinbergen at that time were corresponding regularly (I am indebted to Gregg Mitman for providing me with a copy of this letter). William Vogt, "Preliminary notes on the behavior and ecology of the eastern willet," *Proceedings of the Linnaean Society of New York* 49 (1938): 8–42, thanks Tinbergen along with Mayr, Nice, and Lehrman for their criticism and suggestions concerning his paper. Vogt's paper was awarded the 1938 Linnaean Prize for Ornithological Research (a prize that was designated for work either by an amateur or by a professional working in his spare time). Among the others whom Tinbergen met at the Linnaean Society of New York (and to whom Tinbergen at later dates asked Mayr to pass on greetings) were Robert P. Allen and Frederick P. Mangels, authors of "Studies of the nesting behavior of the black-crowned night heron," *Proceedings of the Linnaean Society of New York* 50–51 (1940): 1–28.

76. Tinbergen gave a lecture to the Linnaean Society of New York on 11 October 1938 entitled "A year in Greenland." In addition, he wrote "Why do birds behave as they do?" *Bird-Lore* 40 (1938): 389–395, 41 (1939): 23–30. He also published an article entitled "In the life of a herring gull" in *Natural History* 43 (1939): 222–229.

77. N. Tinbergen, "On the analysis of social organization among vertebrates, with special reference to birds," *American Midland Naturalist* 21 (1939): 210–234. Tinbergen made clear at the beginning that he was not going to talk about different types of animal communities or the functions of social behavior. His topic instead was the ethological analysis of the causes of the innate behavior patterns essential to social integration. While acknowledging the value the work of such American authors as W. C. Allee, C. R.

Carpenter, and G. K. Noble, he indicated that he would employ the theoretical framework offered by Lorenz in his "Kumpan" monograph of 1935.

78. F. B. Kirkman, *Bird Behaviour: A Contribution Based Chiefly on a Study of the Black-Headed Gull* (London: T. Nelson and Sons and T. C. and E. C. Jack, 1937), p. 78, had previously used the phrase "substitute reaction" to refer to "a reaction substituted for one that is prevented from taking place." Tinbergen later decided that the word "substitute" was flawed because it conveyed a sense of intention. I thank Colin Beer for this observation.

79. Tinbergen, "On the analysis of social organization among vertebrates," p. 225.

80. Tinbergen, "On the analysis of social organization among vertebrates," p. 228.

81. See Francis Herrick, "The individual vs. the species in behavior studies," *Auk* 56 (1939): 244–249; Frederick C. Lincoln, "The individual versus the species in migration studies," *Auk* 56 (1939): 250–254; Margaret M. Nice, "The social Kumpan and the song sparrow;" *Auk* 56 (1939): 255–262; and G. K. Noble, "The role of dominance in the social life of birds," *Auk* 56 (1939): 263–273. Tinbergen's paper for the symposium was not published. Herrick's paper offered a genial call to study more individuals in a species before generalizing about entire populations, Lincoln's paper simply stressed the value bird banding could have for studying the migratory movements of individual birds, and Noble attempted unsuccessfully to amalgamate a wide range of studies in the service of a problematical distinction between "social dominance" and "sexual dominance."

82. Nice dropped the word "sure" from the published version of her paper in response to criticism from Noble, but she in turn criticized Noble's ideas of dominance and continued to press him regarding his unwillingness to accept Lorenz's idea of releasers. Noble to Nice, 24 October 1938; and Nice to Noble, 28 October 1938, MMNP/Cornell. For more on Noble and Nice, see Gregg Mitman and Richard W. Burkhardt, Jr., "Struggling for identity: the study of animal behavior in America, 1930–1945," in *The Expansion of American Biology*, ed. Keith R. Benson, Jane Maienschein, and Ronald Rainger (New Brunswick, NJ: Rutgers University Press, 1991), pp. 164–194, esp. p. 184. See Lorenz to Nice, 14 February 1939, MMNP/Cornell.

83. Tinbergen to Mayr, 11 December 1938 (in German), EMP/Harvard, HUG (FP) 14.7, box 2, folder 68.

84. See the letters from Noble to Tinbergen, 20 February 1939; and Tinbergen to Noble, 6 March 1939, 23 April 1939, G. K. Noble Papers, American Museum of Natural History. See too Mitman and Burkhardt, "Struggling for identity," p. 178.

85. Tinbergen to Mayr, 9 May 1939 (in German), EMP/Harvard, HUG (FP) 14.7, box 2, folder 68.

86. Tinbergen, "Practicum ethologie derdejaars—Zoologisch Laboratorium Leiden," undated, NTP/Oxford. The document is further titled "Inleiding tot de ethologie zoals die op het practicum bedreven wordt." The original passage cited in the text reads: "De vraagstelling dezer analytische ethologie zoals zij op dit practicum bedreven wordt is dus principieel dezelfde als die der physiologie, wanneer deze opgevat wordt als de studie der verrichtingen der dieren, en er is niets tegen deze ethologie als een tak der physiologie te beschouwen." I thank Tom Zuidema for translating this passage for me. The full translation reads: "The scientific approach of this analytic ethology, as it is used in this practicum, is thus in principle the same as that of physiology, when this is viewed

as the study of the actions of the animals, and there is no objection to considering this ethology a branch of physiology." I have identified this manuscript as dating from 1939 on the basis of items cited in the bibliography.

87. Tinbergen to Mayr, 9 June 1939 (in German), EMP/Harvard, HUG (FP) 14.7, box 2, folder 68.

88. Tinbergen to Mayr, 9 June 1939, EMP/Harvard, HUG (FP) 14.7, box 2, folder 68. N. Tinbergen, "Die Ethologie als Hilfswissenschaft der Oekologie," *Journal für Ornithologie* 88 (1940): 171–177.

89. Tinbergen's countrymen Frans Makkink and Adriaan Kortlandt had also taken notice of these apparently "irrelevant" actions that would later come to be called "substitute" or "displacement" activities. See G. F. Makkink, "An attempt at an ethogram of the European avocet (*Recurvirostra avosetta* L.), with ethological and psychological remarks," *Ardea* 25 (1936): 1–62; and A. Kortlandt, "Wechselwirkung zwischen Instinkten," *Archives néerlandaises de zoologie* 4 (1940): 443–520. For a detailed discussion of the Dutch origins of the concept of "displacement activities," see Röell, *The World of Instinct,* pp. 121–130. Although the German Society for Animal Psychology canceled the meeting scheduled for Leipzig, the society continued to publish its journal, which is where Tinbergen first discussed at length his ideas on displacement behavior. See Tinbergen, "Die Übersprungbewegung," *Zeitschrift für Tierpsychologie* 4 (1940): 1–40.

90. "Mitteilung der Deutschen Gesellschaft für Tierpsychologie," *Zeitschrift für Tierpsychologie* 3 (1939): 248.

91. Tinbergen to Mayr, 6 September 1939, EMP/Harvard, HUG (FP) 14.7, box 2, folder 68. Tinbergen's paper was published as "The behavior of the snow bunting in spring."

92. Tinbergen to Mayr, 7 October 1939, EMP/Harvard, HUG (FP) 14.7, box 2, folder 68.

93. Tinbergen to Mayr, 5 January 1940, EMP/Harvard, HUG (FP) 14.7, box 2, folder 88.

94. Letters from Noble to Tinbergen 12 December 1939; and Tinbergen to Noble, 24 January 1940 and 14 February 1940, G. K. Noble Papers, American Museum of Natural History.

95. Tinbergen to Mayr, 9 June 1940, EMP/Harvard, HUG (FP) 14.7, box 2, folder 88.

96. Tinbergen to Mayr, 9 June 1940, EMP/Harvard, HUG (FP) 14.7, box 2, folder 88.

97. G. P. Baerends, "Fortpflanzungsverhalten und Orientierung der Grabwespe *Ammophila campestris* Jur.," *Tijdschrift voor Entomologie* 84 (1941): 68–275.

98. For biographical information on Baerends see his autobiographical statement, "Two pillars of wisdom," pp. 13–40.

99. Baerends, "Fortpflanzungsverhalten und Orientierung der Grabwespe," p. 203.

100. Baerends, "Fortpflanzungsverhalten und Orientierung der Grabwespe," pp. 268.

101. Baerends, "Fortpflanzungsverhalten und Orientierung der Grabwespe," pp. 268–269.

102. In a letter to Lorenz dated 29 October 1941, Tinbergen mentions his independent discovery of the hierarchy of instinctive moods. This letter is in the Lorenz family papers.

103. On Kortlandt's observation tower (which replaced an earlier, eight-meter tower that he built) and on his cormorant studies see Adriaan Kortlandt, "Patterns of pair-formation and nest-building in the European cormorant *Phalacrocorax carbo sinensis*," *Ardea* 83 (1995): 11–25.

104. Adriaan Kortlandt letter to Richard W. Burkhardt, Jr., 6 January 2004. For a more extended discussion of Kortlandt's development of hierarchical models, see Röell, *The World of Instinct*, pp. 127–135; and Kortlandt, "Wechselwirkung zwischen Instinkten."

105. Tinbergen to Lack, 26 February 1940, David Lack Papers, Edward Grey Institute of Field Ornithology, Oxford University.

106. Tinbergen's aborted work on cichlid fish is mentioned in a letter to Joseph Hickey dated 6 September 1945, MMNP/Cornell.

107. The only known letter remaining from the Tinbergen-Lorenz correspondence in this period is dated 29 October 1941. It is in the Lorenz family papers. In the letter Tinbergen mentioned that he did not know the whereabouts of Bernhard Hellmann, but he knew of someone who knew Hellmann's address. Hellmann, Lorenz's friend, had fled from Austria to Holland and was subsequently put to death by the Nazis in a gas chamber.

108. Tinbergen, Meeuse, Boerema, and Varossieau, "Die Balz des Samtfalters, *Eumenis semele* (L.)." Tinbergen's misgivings about submitting the paper to the German journal are mentioned in a letter from Otto Koehler to Otto Antonius, 10 October 1942, Archive of the Schönbrunn Zoo, Vienna, Austria. I thank Veronika Hofer for this information. See below regarding the publication of the paper.

109. N. Tinbergen, "An objectivistic study of the innate behaviour of animals," *Bibliotheca Biotheoretica* 1 (1942): 39–98.

110. Tinbergen, "An objectivistic study," pp. 39, 40.

111. Tinbergen, "An objectivistic study," pp. 41–42, 77–78, 94.

112. Tinbergen's comments are in a letter from Tinbergen to Yerkes, 18 October 1946, RMYP/Yale. The events at Leiden are described in the same letter, as well as in Tinbergen to Nice, 23 June 1945, MMNP/Cornell, and in Werner Warmbrunn, *The Dutch under German Occupation, 1940–1945* (Stanford, CA: Stanford University Press, 1963), pp. 146–153.

113. Tinbergen to Nice, 23 June 1945, MMNP/Cornell.

114. The letters from Otto Koehler to Otto Antonius, 10 October 1942 and Antonius to Koehler 16 October 1942, are cited in Benedikt Föger and Klaus Taschwer, *Die andere Seite des Spiegels: Konrad Lorenz und der Nationalsozialismus* (Vienna: Czernin Verlag, 2001), pp. 156–158. As the authors explain, one of Koehler's concerns, as editor of the *Zeitschrift für Tierpsychologie* (with Antonius and Lorenz), was whether the journal could publish, without Tinbergen's final permission, Tinbergen's paper "Die Balz des Samtfalters." With Tinbergen in prison, it was impossible to contact him to secure his final permission, and the article, as Koehler told Antonius, had in fact already gone to press. I thank Veronika Hofer for copies of the letters from Koehler to Otto Antonius, 10 October 1942, and from Antonius to Koehler, 16 October 1942, and an additional letter from Koehler to Antonius, 19 October 1942, all in the Archive of the Schönbrunn

Zoo, Vienna. After the war, Tinbergen thanked Koehler for his efforts but also explained why it was impossible to accept any German favors. See below, chapter 6.

115. In the internment camp, Tinbergen continued to think about ethology. He completed a manuscript on social behavior in animals. He also wrote children's stories about herring gulls and sticklebacks, composed at the rate of one page per week to go with the one letter per week he was allowed to send his family. After the war, the stories were published as *Kleew: The Story of a Gull* (New York: Oxford University Press, 1947), in the Dutch edition *Klieuw* (['s-Gravenhage]: Boucher, 1948), and *The Tale of John Stickle* (London: Methuen, 1952).

116. Tinbergen to Yerkes, 18 October 1946, RMYP/Yale. See also Tinbergen to Nice, 23 June 1945, MMNP/Cornell.

CHAPTER FIVE

1. For entries into the now enormous literature on science in the Third Reich see Jonathan Harwood, "German science and technology under National Socialism," *Perspectives on Science* 5 (1997): 128–151; Arleen Marcia Tuchman, "Institutions and disciplines: recent work in the history of German science," *Journal of Modern History* 69 (1997): 298–319; Alan Beyerchen, "What we now know about Nazism and science," *Social Research* 59 (1992): 615–641; Monika Renneberg and Mark Walker, eds., *Science, Technology and National Socialism* (Cambridge: Cambridge University Press, 1994); Paul Weindling, *Health, Race and German Politics between National Unification and Nazism, 1870–1945* (Cambridge: Cambridge University Press, 1989); Kristie Macrakis, *Surviving the Swastika: Scientific Research in Nazi Germany* (New York: Oxford University Press, 1993); Ute Deichmann, *Biologen unter Hitler: Vertreibung, Karrieren, Forschungsförderung* (Frankfurt: Campus, 1992), translated by Thomas Dunlap as *Biologists under Hitler* (Cambridge, MA: Harvard University Press, 1996). See too Benno Müller-Hill, *Murderous Science: Elimination by Scientific Selection of Jews, Gypsies, and Others, Germany, 1933–1945* (Oxford: Oxford University Press, 1988). For the careers of individual scientists, see J. L. Heilbron, *The Dilemmas of an Upright Man: Max Planck as Spokesman of German Science* (Berkeley: University of California Press, 1986); David Charles Cassidy, *Uncertainty: The Life and Science of Werner Heisenberg* (New York: W. H. Freeman, 1992); and Anne Harrington *Reenchanted Science: Holism in German Culture from Wilhelm II to Hitler* (Princeton, NJ: Princeton University Press, 1996).

2. This is the case, for example, of W. H. Thorpe's *The Origins and Rise of Ethology* (London: Heinemann, 1979).

3. Theodora Kalikow was the first historian to tackle systematically the subject of Lorenz's career in the Third Reich. See Theodora J. Kalikow, "Konrad Lorenz's ethological theory: explanation and ideology, 1938–1943," *Journal of the History of Biology* 16 (1983): 39–73. See also Kalikow's earlier version of this paper, "Die ethologische Theorie von Konrad Lorenz: Erklärung und Ideologie, 1938 bis 1943," in *Naturwissenschaft, Technik und NS-Ideologie*, ed. S. Richter and H. Mehrtens (Frankfurt: Suhrkamp, 1980), pp. 189–214. More recently, Deichmann, *Biologists under Hitler*, discussed Lorenz at length, describing him as a scientist who not only joined the Nazi Party but who also advocated in various speeches and publications the "weeding out" (*Ausmerzung*) of the el-

ements of decay in the social body. The work that now offers the most detail on Lorenz's politics is Benedikt Föger and Klaus Taschwer, *Die andere Seite des Spiegels: Konrad Lorenz und der Nationalsozialismus* (Vienna: Czernin Verlag, 2001).

4. Peter Klopfer, "Konrad Lorenz and the National Socialists: on the politics of ethology," *International Journal of Comparative Psychology* 7 (1994): 202–208. I believe Klopfer is wrong in his suggestion that Nazi sympathies had anything to do with the inspiration of the primary concepts of Lorenzian ethology. Klopfer offers no evidence to support his claim. It appears to the contrary that Lorenz developed his basic concepts of releasers, innate releasing mechanisms, and action-specific energy well before he began to display any interest in ways that his work might be seen to intersect with the interests of the Nazis. Klopfer nonetheless provided a valuable service when he noted that Otto Koenig's published version of the correspondence between Lorenz and Oskar Heinroth left out certain politically embarrassing material. This led me to study the whole of the Lorenz-Heinroth correspondence at the Staatsbibliothek in Berlin. See Staatsbibliothek zu Berlin—Preußischer Kulturbesitz, Nachlass 137 (Oskar Heinroth), Ordner 27: Lorenz, Konrad (hereafter OHP/Berlin). Klopfer also comments on Lorenz's politics in Peter H. Klopfer, *Politics and People in Ethology* (Lewisburg, PA: Bucknell University Press; London: Associated University Presses, 1999), pp. 57–63. Klopfer's claims there have been criticized in detail by Wolfgang M. Schleidt, "Politik gegen und mit Konrad Lorenz," in *Konrad Lorenz und seine verhaltensbiologischen Konzepte aus heutiger Sicht,* ed. K. Kotrschal, G. Müller, and H. Winkler (Fürth: Filander Verlag, 2001), pp. 73–92.

5. Paul Leyhausen, review of *Konrad Lorenz,* by Franz Wuketits, *Ethology* 89 (1991): 344–346, and "The cat who walks by himself," in *Studying Animal Behavior: Autobiographies of the Founders,* ed. Donald A. Dewsbury (Chicago: University of Chicago Press, 1989), pp. 225–256, quote on 227. In the first of these pieces Leyhausen objects to Franz Wuketits's biography of Lorenz, *Konrad Lorenz: Leben und Werk eines großen Naturforschers (Konrad Lorenz: Life and Work of a Great Scientist)* (Munich: Piper, 1990). In that book Wuketits, an Austrian philosopher and admirer of Lorenz's overall intellectual achievement, acknowledges that Lorenz was naive and made unfortunate mistakes of judgment during the Third Reich. He suggests that many of Lorenz's pronouncements from this period need to be understood in the context of the language of the time, but he also says that such relativizing can go only so far, and that specific statements Lorenz made in several of his publications can undeniably be interpreted as a warrant for "an inhumane policy." In a more recent volume (1995) on the history of animal behavior studies, Wuketits finesses the issue of Lorenz's wartime politics by simply writing: "Lorenz was not among those scientists who in the Third Reich distanced themselves from the Nazis." Wuketits, *Die Entdeckung des Verhaltens: Eine Geschichte der Verhaltensforschung* (Darmstadt: Wissenschaftliche Buchgesellschaft, 1995), p. 84. Leyhausen for his part insists that Lorenz did not have the mind-set of a Nazi. Lorenz's ideas about race purity, Leyhausen maintains, were ideas that were common to biologists of the day, when the heterozygosity of individuals in wild populations was not yet generally appreciated. If Lorenz joined the Nazi Party, Leyhausen says, this was because Lorenz felt— naively, to be sure, but he was not alone in misreading the situation—that the only hope of correcting the Nazis' biological misconceptions was to attempt to reform these misconceptions from the inside.

6. See, for example, Erich von Holst, review of "Das Tier als Individualität in der Großhirnforschung" (1936), by Emil Diebschlag, *Berichte über die wissenschaftliche Biologie* 41 (1937): 366. There Holst observed of the article in question: "Important results of recent years in the field of animal behavioral science (e.g., the works of Lorenz) are unfortunately ignored, giving the impression that this branch of science is still stuck at its most primitive beginnings." See also Otto Koehler, "'Instinkt oder Verstand?' Wertung zweier neuer Bücher und ihrer Stellung in der heutigen Tierpsychologie," *Naturwissenschaften* 27 (1939): 179–184, where Koehler reviewed *The Behaviour of Animals*, by E. S. Russell, and *Psyche und Leistung der Tiere*, by Werner Fischel. See too Koehler's review of *Die tierischen Instinkte und ihr Umbau durch Erfahrung*, by J. A. Bierens de Haan, *Zeitschrift für Tierpsychologie* 4 (1940): 280–286.

7. See Lorenz to Heinroth, 19 February 1937 and 16 March 1937, in Oskar Heinroth and Konrad Lorenz, *Wozu aber hat das Vieh diesen Schnabel? Briefe aus der frühen Verhaltensforschung, 1930–1940*, ed. Otto Koenig (Munich: Piper, 1988), pp. 228–232 (hereafter cited as Heinroth and Lorenz, *Wozu*). It is not easy to judge the accuracy of Lorenz's claim to Tinbergen that scientific appointments in the Third Reich as of 1937 were less political than they had been when the Nazis first came to power. Historians of German psychology have noted such a shift for the discipline of psychology, but Ute Deichmann's monumental study of German biology shows no clear trend for biologists in this regard. The percentage of party membership in the groups of biologists appointed to full professorship for the periods 1933–1936, 1937–1940, and 1941–1945 was roughly the same (70%–75%). Deichmann observes that "political activity did not necessarily stand in opposition to professional quality." See Deichmann, *Biologists under Hitler*, pp. 66–70. For the psychologists, See Ulfried Geuter, *Die Professionalisierung der deutschen Psychologie im Nationalsozialismus* (Frankfurt: Suhrkamp, 1984), translated as *The Professionalization of Psychology in Nazi Germany* (Cambridge: Cambridge University Press, 1992).

8. Lorenz to Greite, 25 January 1937, and 13 February 1937; and Lorenz, "Antrag an die deutsche Forschungsgemeinschaft," 13 February 1937, in Bundesarchiv Koblenz file, "Deutsche Forschungsgemeinschaft," R73/12781.

9. Lorenz to Greite, 13 March 1937, Bundesarchiv Koblenz, R73/12781.

10. Lorenz to DFG, 14 May 1937, Bundesarchiv Koblenz, R73/12781.

11. Lorenz, "An die Deutsche Forschungsgemeinschft," four-page ms., undated by Lorenz but stamped 18 May 1937 by the DFG, Bundesarchiv Koblenz, R73/12781.

12. See Lorenz to Stresemann, 13 May 1937, Stresemann Letters, Erwin Stresemann Papers, Preußischer Kulturbesitz, Staatsbibliothek zu Berlin (hereafter ESP/Berlin). Deichmann, *Biologists under Hitler*, p. 183, cites Stresemann's comments to the DFG.

13. Letter from the Österreichische-Deutsche Wissenschaftshilfe to Lorenz, 2 June 1937, Bundesarchiv Koblenz, R73/12781.

14. See Lorenz to Heinroth 16 June 1937, OHP/Berlin, and Lorenz to Stresemann, also 16 June 1937, ESP/Berlin. Koenig, in editing Heinroth and Lorenz, *Wozu*, suppressed Lorenz's comments in this regard. Lorenz briefly told Heinroth that this "enemy" was trying "to mess things up" for him with the Kaiser Wilhelm Gesellschaft as well as with the DFG, and that the only grounds for this were that Lorenz had become a docent without involving the man in the process. To Stresemann, Lorenz described in much greater detail what he thought had happened. (The letter is cited in Föger and

Taschwer, *Die andere Seite des Spiegels,* pp. 72–73, but misdated 16 April 1937.) Judging from a letter from Lorenz to Stresemann dated 30 November 1937, Lorenz seems to have lost his initial certitude that Abel was the one who had caused him the trouble. There he wrote, "If I only knew who the primary denigrator was, I would yet strangle him today!" (wenn ich wüsste, wer der primäre Anschwärzer war, ich derwurgert ihn heut noch!). ESP/Berlin.

15. Kalikow cites extracts from the five letter writers who between 5 July and 17 October 1937 responded to von Wettstein's inquiries about Lorenz. See Kalikow, "Konrad Lorenz's ethological theory: explanation and ideology, 1938–1943," pp. 51–53. Deichmann, *Biologists under Hitler,* pp. 183–184; and Föger and Taschwer, *Die andere Seite des Spiegels,* pp. 74–77, make use of the same files. I am grateful to Theodora Kalikow for sharing her notes from the DFG files with me, and I thank also the Bundesarchiv in Koblenz for providing me photocopies of materials in the Lorenz file.

16. Antonius and Pichler as cited in Deichmann, *Biologists under Hitler,* p. 184.

17. Knoll to Wettstein, 17 October 1937, Bundesarchiv Koblenz.

18. Lorenz's exchange with Hartmann is reported in Klaus Taschwer and Benedikt Föger, *Konrad Lorenz: Biographie* (Vienna: Paul Zolnay Verlag, 2003), pp. 75–76.

19. Wettstein to DFG, 14 December 1937, Bundesarchiv Koblenz.

20. Lorenz filled out the forms entitled "Personalfragebogen zu dem Gesuch um ein Forschungsstipendium" and "Zum Nachweis der arischen Abstammung," dating the latter 11 January 1938.

21. Lorenz to Nice, 13 September 1937, Margaret Morse Nice Papers, 1917–1968, collection number 2993, courtesy of the Division of Rare and Manuscript Collections, Cornell University Library.

22. Lorenz to Stresemann, 30 November 1937, ESP/Berlin.

23. F. L. Carsten, *The First Austrian Republic, 1918–1938* (Aldershot: Gower Publishing Company, 1985), pp. 277–279; Gerhard Botz, *Nationalsozialismus in Wien* (Buchloe: Druck und Verlag Obermayer, 1988), pp. 51–76.

24. Radomír Luza, *Austro-German Relations in the Anschluss Era* (Princeton, NJ: Princeton University Press, 1975), p. 57.

25. Lorenz to Heinroth, 22 March 1938, OHP/Berlin. This paragraph was omitted in the version of the letter published by Otto Koenig in Heinroth and Lorenz, *Wozu,* but appears in Föger and Taschwer, *Die andere Seite des Spiegels,* p. 82.

26. Lorenz to Stresemann, 26 March 1938, ESP/Berlin. The original German reads: "Sie können sich keine blasse Vorstellung davon machen, welche Begeisterung hier herrschte und selbst jetzt noch herrscht, in welcher Ausnahms- und Festesstimmung selbst so unpolitische Menschen wie wir sind! Man muss 5 Jahre lang unter der Regierung der schwarzen Schweinehunde gestanden haben, um ein 'Deutschland Erwache' in seinem Inneren mit der vollen Intensitat [*sic*] zu erleben. Ich glaube, wir Oesterreicher sind die aufrichtigsten und überzeugtesten Nationalisozialisten überhaupt! Mann muss im Grunde genommen den Herren Schuschnigg und Konsorten dankbar sein, denn ohne ihr unbeabsichtigte Hilfe wären die Faulen unter ihren Nationalcharackter nach besonders meckerbereiten Oesterreicher lange nicht so schnell, gründlich und nachhaltig zu Hitler bekehrt worden. Und das sind sie jetz wirklich und zweifellos!" Föger und Taschwer, *Die andere Seite des Spiegels,* cite this letter but not this particular paragraph.

27. Lorenz to Stresemann, 26 March 1938, ESP/Berlin. The original text reads: "Hier in Altenberg ist es prächtig! Gänse brüten, Enten balzen, Kinder sind gesund und wachsen, die Dohlen sind heuer in nie dagewesene Anzahl eingetroffen und wieder viel zahmer, seit ich den Versuch aufgegeben habe, sie durch zeitweises Einsperren gewaltsam zahm zu machen. Diese Vorgehen hatten den gegenteiligen Erfolg, ganz wie Schuschnigg an uns selbst erfahren musste!"

28. Lorenz to DFG, 8 April 1938, Bundesarchiv Koblenz.

29. Lorenz to Stresemann, 11 April 1938, ESP/Berlin. Cited in Föger and Taschwer, *Die andere Seite des Spiegels*, p. 83.

30. Elsewhere, however, in a letter to Hochstetter dated 16 April 1938, Lorenz did express concern for Bühler. See Taschwer and Föger, *Konrad Lorenz*, pp. 82–83.

31. Lorenz to Stresemann, 11 April 1938, ESP/Berlin. Cited in Föger and Taschwer, *Die andere Seite des Spiegels*, pp. 84–85.

32. Lorenz to Stresemann, 19 April 1938, ESP/Berlin. The original sentences read: "Wenn mich irgend etwas dazu bestimmen kann, mich wirklich um eine Lehrkanzel zu bewerben, so ist es der Umstand, dass ich es als eine soziale und nationale Lebensaufgabe betrachte, das völlig 'in Geist aufgelöste' Studium der menschlichen Psychologie wieder zu einem Gebiet induktiver Naturforschung zu machen!" And *"mir die Altenberger Forschungsstation wesentlich wichtiger ist, als die schönste Lehrkanzel !!!!"* In the original of this last sentence, Lorenz emphasized these words not by underlining them but by spacing out the letters.

33. Bühler's chair in Vienna remained unoccupied until 1943. See Geuter, *The Professionalization of Psychology in Nazi Germany*, p. 66.

34. See Lorenz to Stresemann, 16 February 1939, ESP/Berlin.

35. Föger and Taschwer, *Die andere Seite des Spiegels*, p. 79. The authors report that Lorenz's membership in the Nazi Party was then backdated to 1 May 1937.

36. Lorenz to Heinroth, 22 March 1938, in Heinroth and Lorenz, *Wozu*, pp. 242–243.

37. Lorenz to Heinroth, 28 March 1938, in Heinroth and Lorenz, *Wozu*, pp. 243–244.

38. See Heinroth to Lorenz, 6 January 1937, in Heinroth and Lorenz, *Wozu*, p. 225; E. R. Jaensch, "Einladung zur XVI. Tagung der Deutschen Gesellschaft für Psychologie 2.–4. Juli 1938 in Bayreuth," *Zeitschrift für Tierpsychologie* 2 (1938–1939): 99.

39. Lorenz to Heinroth, 16 June 1938, in Heinroth and Lorenz, *Wozu*, p. 247.

40. See, for example, Konrad Lorenz, "Companions as factors in the bird's environment: the conspecific as the eliciting factor for social behaviour patterns," in *Studies in Animal and Human Behaviour*, trans. Robert Martin, 2 vols. (Cambridge, MA: Harvard University Press, 1970–1971), 1:138.

41. Konrad Lorenz, "Der Kumpan in der Umwelt des Vogels: der Artgenosse als auslösendes Moment sozialer Verhaltungsweisen," *Journal für Ornithologie* 83 (1935): 312. On the prevalence in Germany after the turn of the century of the idea of cultural and genetic decline, see George L. Mosse, *The Crisis of German Ideology* (New York: Grossett and Dunlap, 1964).

42. Lorenz, "Kumpan," p. 312, and "Companions as factors," p. 187: "We Europeans are . . . much more distinct from one another than are the racially less diverse Negroes or Chinese, and these in their turn differ from one another far more than members of non-domesticated animal species."

43. K. Lorenz, "Über Ausfallserscheinungen im Instinktverhalten von Haustieren und ihre sozialpsychologische Bedeutung," in *Charakter und Erziehung: Bericht über den 16. Kongress der deutschen Gesellschaft für Psychologie in Bayreuth,* ed. Otto Klemm (Leipzig: Johann Ambrosius Barth, 1938), pp. 139–147, quote on p. 146. The prevalence of cancer imagery is discussed in Robert N. Proctor, *The Nazi War on Cancer* (Princeton, NJ: Princeton University Press, 1999).

44. E. R. Jaensch, "Der Hühnerhof als Forschungs- und Aufklärungsmittel in menschlichen Rassenfragen," *Zeitschrift für Tierpsychologie* 2 (1938–1939): 252, fn. 2.

45. Jaensch, "Der Hühnerhof."

46. "Deutsche Ornithologische Gesellschaft: 56. Jahresversammlung (1938) in Berlin," *Journal für Ornithologie* 87 (1939): 165–188. See pp. 172–175 for the discussion of Lorenz's films. A more detailed description of Lorenz's greylag film appeared in the official report of the July 1937 DOG meeting in Dresden. See "55. Jahresversammlung (1937) in Dresden," *Journal für Ornithologie* 85 (1937): 684–685.

47. "Deutsche Ornithologische Gesellschaft: 56. Jahresversammlung (1938) in Berlin," p. 174.

48. See Rudolph Virchow, "Die Freiheit der Wissenschaft im modernen Staat," in *Amtlicher Bericht der 50. Versammlung Deutscher Naturforscher und Ärzte in München* (Munich: F. Straub, 1877), pp. 65–77; Ernst Haeckel, *Freie Wissenschaft und freie Lehre* (Stuttgart: E. Schwizerbart, 1878); Weindling, *Health, Race and German Politics between National Unification and Nazism.*

49. See Sheila Faith Weiss, *Race Hygiene and National Efficiency: The Eugenics of Wilhelm Schallmayer* (Berkeley: University of California Press, 1987); and Weindling, *Health, Race and German Politics.*

50. Heinrich Ernst Ziegler, *Die Naturwissenschaft und die socialdemokratische Theorie, ihr Verhältniss dargelegt auf Grund der Werke von Darwin und Bebel; zugleich ein Beitrag zur wissenschaftlichen Kritik der Theorien der derzeitigen Socialdemokratie* (Stuttgart: F. Enke, 1894).

51. See Fritz Lenz, "Psychological differences between the leading races of mankind," in *Human Heredity,* by Erwin Baur, Eugen Fischer, and Fritz Lenz (New York: Macmillan, 1931), pp. 674–675.

52. On Uexküll see Anne Harrington, *Reenchanted Science: Holism in German Culture from Wilhelm II to Hitler.* Harrington describes Uexküll in 1933 as having "social connections and bio-political publications" that "made it rather easy for him to be perceived by various National Socialists as a natural ally and intellectual resource—a perception that Uexküll seems to have permitted and even to have partly cultivated." As she goes on to explain, however, "it was also the case that Uexküll's biology and politics did not map so seamlessly onto the agenda of National Socialism as some would have liked." Yet even where he attempted to take a stand of his own, Harrington says, Uexküll "still deliberately avoided giving any impression that he was opposed to the goals of National Socialism in general" (pp. 68, 69, 71). Lorenz's stance toward National Socialism could be described in much the same terms.

53. On Tandler see Manfred D. Laubichler, "Biology and the state: the biopolitics of Julius Tandler and Jacob von Uexküll" (unpublished paper).

54. H. J. Muller, review of *Human Heredity,* by Erwin Baur, Eugen Fischer, and Fritz Lenz, *Birth Control Review* 17 (January 1933): 19–21.

55. Karl von Frisch, *Du and das Leben* (Berlin: Verlag Ullstein, 1936). See pp. 344–346. Postwar editions of the book omitted the section from which the quotes are taken.

56. See Konrad Lorenz, "Die angeborenen Formen möglicher Erfahrung," *Zeitschrift für Tierpsychologie* 5 (1943): 380, where he notes his father's saying that from a "race-biological" point of view, the whole art of medicine was a misfortune. Charles Darwin had noted this problem in 1871 in *The Descent of Man.* At the same time, however, Darwin opted for heeding one's higher moral sympathies and helping one's fellow man.

57. See K. Lorenz, "Durch Domestikation verursachte Störungen arteigenen Verhaltens," *Zeitschrift für angewandte Psychologie und Charakterkunde* 59 (1940): 2–81, "Psychologie und Stammesgeschichte," in *Die Evolution der Organismen,* ed. Gerhard Heberer (Jena: Gustav Fischer, 1943), pp. 105–127, "Nochmals: Systematik und Entwicklungsgedanke im Unterricht," *Der Biologe* 9 (1940): 24–36, and "Die angeborenen Formen möglicher Erfahrung." Kalikow, "Konrad Lorenz's ethological theory: explanation and ideology, 1938–1943," surveys each of Lorenz's writings in this period, assesses their political content, and identifies the recurring themes of Lorenz's political writings.

58. "Mitteilung der Deutschen Gesellschaft für Tierpsychologie," *Zeitschrift für Tierpsychologie* 3 (1939): 248.

59. On Lorenz's invitation to speak at the German Club, see Lorenz to Heinroth, 9 [11] January 1939, in Lorenz and Heinroth, *Wozu,* p. 257. On this lecture and Lorenz's other popular presentations of his biopolitical views in this period, see Föger and Taschwer, *Die andere Seite des Spiegels,* pp. 112–115, 133–135.

60. See the letters exchanged between Lorenz and Heinroth, 22 December 1938–16 February 1939, in Heinroth and Lorenz, *Wozu,* pp. 249–270.

61. Lorenz to Heinroth, 9 January 1939, in Heinroth and Lorenz, *Wozu,* p. 255.

62. Heinroth to Lorenz, 16 January 1939, in Heinroth and Lorenz, *Wozu,* p. 258.

63. Lorenz to Heinroth, 18 January 1939, in Heinroth and Lorenz, *Wozu,* pp. 259–263.

64. Lorenz, "Durch Domestikation verursachte Störungen arteigenen Verhaltens," pp. 13–52.

65. Lorenz, "Durch Domestikation verursachte Störungen arteigenen Verhaltens," pp. 28–29; Oskar Heinroth, "Beiträge zur Biologie, namentlich Ethologie und Psychologie der Anatiden," in *Verhandlungen des 5. Internationalen Ornithologen-Kongresses in Berlin, 30. Mai bis 4. Juni 1910,* ed. Herman Schalow (Berlin: Deutsche Ornithologische Gesellschaft, 1911), pp. 589–702, quote on p. 702.

66. Lorenz, "Durch Domestikation verursachte Störungen arteigenen Verhaltens," pp. 69–70. See Ute Deichmann, *Biologists under Hitler,* pp. 190–193, for a discussion of Lorenz's tendency to twist quotations to suit his own purposes.

67. Lorenz, "Durch Domestikation verursachte Störungen arteigenen Verhaltens," p. 79.

68. Lorenz, "Durch Domestikation verursachte Störungen arteigenen Verhaltens," p. 61.

69. Lorenz, "Durch Domestikation verursachte Störungen arteigenen Verhaltens,"

pp. 71–72. When Lorenz's paper was published, Jaensch was no longer alive to appreciate how Lorenz's work had developed. He died unexpectedly on 12 January 1939 following an operation.

70. Lorenz, "Durch Domestikation verursachte Störungen arteigenen Verhaltens," p. 76.

71. Lorenz to Heinroth, 21 November 1939, OHP/Berlin. This letter was omitted entirely from the published Lorenz-Heinroth correspondence, *Wozu aber hat das Vieh diesen Schnabel?*

72. The present account of Lehmann's founding of *Der Biologe* and his leadership of the DVB is based on Anne Bäumer-Schleinkofer, *NS-Biologie und Schule* (Frankfurt am Main: Peter Lang, 1992). Lehmann's attempts to establish himself as a leader in the area of political biology were thwarted by another claimant to that role, Robert Wetzel. On this score see Deichmann, *Biologists under Hitler,* pp. 84–87.

73. Compare Rossner's "Systematik und Entwicklungsgedanke im Unterricht," *Der Biologe* 8 (1939): 366–372, especially pp. 367–368, and his "Lebenskunde ist Tatsachenforschung!" *Der Biologe* 8 (1939): 73.

74. Rossner, "Systematik und Entwicklungsgedanke," p. 370.

75. Gerhard Heberer, "Die gegenwärtigen Vorstellungen über den Stammbaum der Tiere und die 'Systematische Phylogenie' E. Haeckels," *Der Biologe* 8 (1939): 264–273, quote on p. 272.

76. On Heberer see Deichmann, *Biologists under Hitler,* pp. 271–276; and Uwe Hoßfeld, *Gerhard Heberer (1901–1973): Sein Beitrag zur Biologie im 20. Jahrhundert* (Berlin: Verlag für Wissenschaft und Bildung, 1997).

77. Lorenz, "Nochmals: Systematik und Entwicklungsgedanke im Unterricht," p. 24: "Dass es im Schulwesen des Nationalsozialistischen Grossdeutschland Männer gibt, die tatsächlich noch immer Entwicklungsgedanken und Abstammungslehre als solche ablehnen."

78. Lorenz, "Nochmals: Systematik und Entwicklungsgedanke im Unterricht," pp. 24–25.

79. Lorenz, "Nochmals: Systematik und Entwicklungsgedanke im Unterricht," p. 30–31.

80. Lorenz, "Nochmals: Systematik und Entwicklungsgedanke im Unterricht," p. 32.

81. Lorenz to Thorpe, 17 August 1950, William Homan Thorpe Papers, Cambridge University Library, MS.Add.8784.

82. Lorenz to Julian Huxley, 19 August 1950, Julian S. Huxley Papers, Rice University Archives.

83. Geuter, *The Professionalization of Psychology in Nazi Germany,* p. xvii.

84. Rossner, "Systematik und Entwicklungsgedanke," p. 368; Walter Greite, "Zu Ernst Rüdins 65. Geburtstag," *Der Biologe* 8 (1939): 153–154.

85. Lorenz to Heinroth, 21 November 1939, OHP/Berlin; and Lorenz to Stresemann, 5 February 1940, ESP/Berlin.

86. "Aus der Pressestelle des Deutschen Biologen-Verbandes," *Der Biologe* 3 (1934): 191–192.

87. Georg Schwidetzky, *Sprechen Sie Schimpansisch? Einführung in die Tier- und Ursprachenlehre* (Leipzig: Deutsche Gesellschaft für Tier- und Ursprachenforschung, 1931). This was the first of the Schriften der Deutschen Gesellschaft für Tier- und Ursprachenforschung. The tenth of the series was Georg Schwidetzky, *Unsere Ausstellung: Rasse und Sprache: Eine Führung für die Abwesenden* (Leipzig: Deutsche Gesellschaft für Tier- und Ursprachenforschung, 1938).

88. See Lorenz to Heinroth, 18 Dec 1939, in Heinroth and Lorenz, *Wozu*, p. 290.

89. Heinroth to Lorenz, 27 Jan 1940, in Heinroth and Lorenz, *Wozu*, p. 293.

90. Lorenz to Heinroth, 29 January 1940, OHP/Berlin.

91. Heinroth to Lorenz, 29 March 1940, in Heinroth and Lorenz, *Wozu*, p. 296.

92. Lorenz to Stresemann, 5 February 1940, ESP/Berlin. Koehler's review of Fritsche's book appeared in *Der Biologe* under the heading "Eine biologische Veröffentlichung, derer wir nicht bedürfen" [A biological publication that we do not need]. See Otto Koehler, review of *Tierseele und Schöpfungsgeheimnis*, by Herbert Fritsche, *Der Biologe* 9 (1940): 48–50. See also Koehler, review of *Die Tierseele auf der Grundlage der grundwissenschaftlichen Philosophie von Johannes Rehmke*, by Bernhard Hecke, *Zeitschrift für Tierpsychologie* 3 (1939): 374–384

93. See Taschwer and Föger, *Konrad Lorenz*, p. 99.

94. Konrad Lorenz, "My family and other animals," in Dewsbury, *Studying Animal Behavior*, p. 270.

95. Geuter, *The Professionalization of Psychology in Nazi Germany*, p. 73. See too Deichmann, *Biologists under Hitler*, p. 401.

96. Kalikow, "Lorenz's ethological theory: explanation and ideology, 1938–1943," p. 5n. For the most extensive analysis of this to date, see Taschwer and Föger, *Konrad Lorenz*, pp. 100–105.

97. Konrad Lorenz, "Kants Lehre vom Apriorischen im Lichte gegenwärtiger Biologie," *Blätter für deutsche Philosophie* 15 (1941): 94–125, translated as "Kant's doctrine of the *a priori* in the light of contemporary biology," in *Konrad Lorenz: The Man and His Ideas*, by Richard I. Evans (New York: Harcourt Brace Jovanovich, 1975), pp. 181–217, see pp. 182–183. With respect to Kant's view of the fundamental inaccessibility of the thing-in-itself, Lorenz's view was that "the categorical forms of intuition and categories have proved themselves as working hypotheses in the coping of our species with the absolute reality of the environment (in spite of their validity being only approximate and relative)" (p. 192).

98. On the *Bildung* ideal, see Fritz K. Ringer, *The Decline of the German Mandarins: The German Academic Community, 1890–1933* (Cambridge, MA: Harvard University Press, 1969); and Jonathan Harwood, *Styles of Scientific Thought: The German Genetics Community, 1900–1933* (Chicago: University of Chicago Press, 1993).

99. Lorenz, "Kant's doctrine of the *a priori* in the light of contemporary biology," p. 216.

100. Lorenz to Stresemann, 16 August 1940, ESP/Berlin.

101. Lorenz, "Kant's doctrine of the *a priori* in the light of contemporary biology," p. 185.

102. Konrad Lorenz, "Oskar Heinroth 70 Jahre!" *Der Biologe* 10 (1941): 45–47.

103. Konrad Lorenz, "Vergleichende Bewegungsstudien an Anatiden," in "Festschrift O. Heinroth," Ergänzungsband 3, *Journal für Ornithologie* 89 (1941): 194–293, translated as "Comparative studies of the motor patterns of Anatinae," in Lorenz, *Studies in Animal and Human Behaviour,* 2:14–114.

104. Lorenz, "Comparative studies of the motor patterns of Anatinae," p. 111.

105. On the relation between degrees of fertility and degrees of relationship, Lorenz cited Heinrich Poll, "Über Vogelmischlinge," in Schalow, *Verhandlungen des 5. Internationalen Ornithologen-Kongresses in Berlin, 30. Mai bis 4. Juni 1910,* pp. 399–468.

106. Lorenz, "Comparative studies of the motor patterns of Anatinae," pp. 18, 93.

107. For example, in Lorenz, "My family and other animals," p. 273.

108. See Geuter, *The Professionalization of Psychology in Nazi Germany,* pp. 233–242. See too Föger and Taschwer, *Die andere Seite des Spiegels,* p. 144.

109. See Martin Broszat, *Nationalsozialistische Polenpolitik, 1939–1945* (Stuttgart: Deutsche Verlags-Anstalt, 1961); and Michael Burleigh, *Germany Turns Eastwards: A Study of Ostforschung in the Third Reich* (Cambridge: Cambridge University Press, 1988). See also Deichmann, *Biologists under Hitler.*

110. Quoted in Polish Ministry of Information, *The German New Order in Poland* (London: Hutchinson and Co., n.d. [1942?]), p. 409.

111. Polish Ministry of Information, *The German New Order in Poland,* p. 410.

112. Polish Ministry of Information, *The German New Order in Poland,* p. 445.

113. Rudolf Hippius, I. G. Feldman, K. Jellinek, and K. Leider, *Volkstum, Gesinnung und Charakter: Bericht über psychologische Untersuchungen an Posener deutsch-polnischen Mischlingen und Polen, Sommer 1942* (Stuttgart: W. Kohlhämmer, 1943). Deichmann was the first historian to call attention to Lorenz's connection to Hippius's work, which she describes in *Biologists under Hitler,* pp. 195–197. I thank Klaus Taschwer for his observations on the extended purposes of Hippius's study.

114. Hippius et al., *Volkstum, Gesinnung und Charakter,* pp. 25, 137.

115. Alec Nisbett, *Konrad Lorenz* (New York: Harcourt Brace Jovanovich, 1976), p. 94.

116. Lorenz, "Die angeborenen Formen möglicher Erfahrung," p. 313.

117. Lorenz, "Die angeborenen Formen möglicher Erfahrung," p. 296. Lorenz was citing here E. Fischer, "Die Rassenmerkmale des Menschen als Domestikationserscheinungen," *Zeitschrift für Morphologie und Anthropologie* 18 (1914): 479–524; and Arthur Schopenhauer, "Metaphysik der Geschlechtsliebe," in *Die Welt als Wille und Vorstellung* (Leipzig, 1877).

118. Lorenz, "Die angeborenen Formen möglicher Erfahrung," pp. 363–364. Lorenz offered no specific citation of Bolk's work here, but see L. Bolk, *Das Problem der Menschwerdung* (Jena, 1926).

119. Lorenz, "Die angeborenen Formen möglicher Erfahrung," p. 365.

120. Lorenz, "Die angeborenen Formen möglicher Erfahrung," p. 369.

121. Lorenz, "Die angeborenen Formen möglicher Erfahrung," p. 370.

122. Lorenz, "Die angeborenen Formen möglicher Erfahrung," p. 389.

123. Lorenz, "Die angeborenen Formen möglicher Erfahrung," p. 390.

124. Lorenz, "Die angeborenen Formen möglicher Erfahrung," p. 407.

125. Lorenz, "Die angeborenen Formen möglicher Erfahrung," p. 395.

126. Lorenz, "Die angeborenen Formen möglicher Erfahrung," p. 407.

127. Lorenz, "Durch Domestikation verursachte Störungen arteigenen Verhaltens," p. 76.

128. Lorenz to Stresemann, 5 April 1948, ESP/Berlin. The original German reads: "Die Bombe, von der Erwin so liebenswürdig sagte, sie könne mir nach Veröffentlichung der 'Angeborenen Formen möglicher Erfahrung' ruhig auf den Kopf fallen, ist einige Meter danebengegangen (was er allerdings auch vorausgesagt hat, wie ich der Berechtigkeit halber erwähnen muss!)." In a letter that Margaret Morse Nice received from Stresemann on 9 January 1946, Stresemann remarked that Lorenz in his paper of 1943 had "almost surpassed himself." Reported by Nice in a typescript circular letter, Nice Papers, Cornell University Library.

129. Ringer, *The Decline of the German Mandarins*, p. 447.

130. The statement appears on the second page of a letter from Lorenz to Heinroth in the Heinroth correspondence, OHP/Berlin.. The first page of the letter is missing (and thus the letter cannot be precisely dated).

131. Konrad Lorenz, *Civilized Man's Eight Deadly Sins* (New York: Harcourt Brace Jovanovich, 1974), previously published as *Die acht Todsünden der zivilisierten Menschheit* (Munich: Piper, 1973).

132. Benedikt Föger and Klaus Taschwer have noted that Lorenz's time as a military psychiatrist in Poznan is the part of his career about which the least is known. His job was to treat soldiers suffering from shell shock and other serious, war-related psychological disorders. The authors report that in certain other sectors of the Wehrmacht, military psychiatrists identified some of their patients as having "lives not worth living" and sent them off to their deaths. By Föger and Taschwer's account, Lorenz's wartime experiences as a psychiatrist represented the part of his life that he later found most painful to think about. Nothing definitive can be said about Lorenz's own practice as a military psychiatrist, however, because the medical records from Lorenz's sector apparently did not survive the war. See Föger and Taschwer, *Die andere Seite des Spiegels*, pp. 158–160; and Taschwer and Föger, *Konrad Lorenz*, pp. 121–124.

133. Konrad Lorenz, *The Natural Science of the Human Species: An Introduction to Comparative Behavioral Research: The "Russian Manuscript" (1944–1948)* (Cambridge, MA: MIT Press, 1996).

134. Lorenz, *The Natural Science of the Human Species*, p. 76.

135. Interestingly, Lorenz continued to draw a parallel between cancerous growths in organisms on the one hand and domestication effects in human civilizations on the other. Here, though, he did not describe these domestication effects as a threat to the *race* but rather as a threat to "modern human cultural systems." See Lorenz, *The Natural Science of the Human Species*, p. 98.

136. See Föger and Taschwer, *Die andere Seite des Spiegels*, chapter 9.

137. Lorenz, *The Natural Science of the Human Species*, p. 138: "Karl Marx's saying that quantity gives rise to quality is nowhere more appropriate than when applied to the phylogenetic developmental processes of organic creation." His comment on dialectical materialism appears on p. xxxi. Another reference to dialectical materialism appears on p. 196.

CHAPTER SIX

1. Tinbergen's letter to Huxley was one of two letters published under the heading "Scientific affairs in Europe" in the 10 November 1945 issue of *Nature* 156 (1945): 576–578.

2. Lorenz to Thorpe, 10 July 1948, William Homan Thorpe Papers, Cambridge University (henceforth WHTP/Cambridge), MS.Add.8784.

3. Tinbergen to Nice, 23 June 1945; and also H. N. Kluyver to Nice, 13 February 1946, Margaret Morse Nice Papers, 1917–1968, collection number 2993, courtesy of the Division of Rare and Manuscript Collections, Cornell University Library (hereafter MMNP/Cornell).

4. As indicated above, Tinbergen's letter to Huxley was one of two letters published under the heading "Scientific affairs in Europe," in the 10 November 1945 issue of *Nature*. *Nature* did not cite the date of Tinbergen's letter, and the original letter is not among the Julian S. Huxley Papers in the Rice University Archives.

5. Tinbergen to Nice, 23 June 1945, MMNP/Cornell.

6. See Tinbergen to Mayr, 5 September 1945, Ernst Mayr Papers, Harvard University Archives (hereafter EMP/Harvard), HUG (FP) 14.7, box 3, folder 142; Tinbergen to Lack, 11 October 1945, 2 November 1945, and 19 November 1945, David Lack Papers, Edward Grey Institute of Field Ornithology, Oxford University; Thorpe to Tinbergen, 1 December 1945, Lack Papers, Oxford University.

7. See Tinbergen to Nice, 7 November 1945; and Stresemann to Nice, 21 December 1945, MMNP/Cornell; unsigned letter to Tinbergen, 1 December 1945, Lack Papers, Oxford University. Textual evidence indicates the author of this last letter was Thorpe. He encouraged Tinbergen to communicate with Alastair Worden, the secretary of the Institute for the Study of Animal Behaviour (ISAB) in London, to clear up questions concerning the relation of Tinbergen's journal to the ISAB's *Bulletin*, which had begun publication in 1938. The British part of this story has been related by John Durant, "The making of ethology: the Association for the Study of Animal Behaviour, 1936–1986," *Animal Behaviour* 34 (1986): 1601–1616.

8. H. Hediger, P. Palmgren, W. H. Thorpe, and N. Tinbergen, "Behaviour" undated two-page flyer, copy in the Robert M. Yerkes Papers, Yale University.

9. Tinbergen to Mayr, 5 September 1945, EMP/Harvard, HUG (FP) 14.7, box 3, folder 142.

10. Tinbergen to Mayr, 11 November 1945, EMP/Harvard, HUG (FP) 14.7, box 4, folder 160.

11. Tinbergen to Mayr 11 November 1945, EMP/Harvard, HUG (FP) 14.7, box 4, folder 160.

12. Tinbergen to Nice, 26 February 1946, MMNP/Cornell; Tinbergen to Mayr, 15 March 1946, and 20 May 1946, EMP/Harvard, HUG (FP) 14.7, box 4, folder 160.

13. Tinbergen to Yerkes, 18 October 1948, Yerkes Papers, Yale University.

14. Reported by Tinbergen in Tinbergen to Mayr, 15 March 1946, EMP/Harvard, HUG (FP) 14.7, box 4, folder 160.

15. Tinbergen to Nice, 23 June 1945, MMNP/Cornell.

16. Tinbergen to Hickey, 6 September 1945, MMNP/Cornell.

17. Tinbergen to Huxley, in "Scientific affairs in Europe," p. 577.

18. Stresemann's letter had first appeared in the September–October issue of the

Swiss journal *Der Ornithologische Beobachter*. See "Zwei Briefe aus dem Ausland," *Der Ornithologische Beobachter* 42 (September–October 1945): 149. Huxley on a trip to Switzerland obtained a copy of Stresemann's letter and sent it to *Nature* along with the letter from Tinbergen, where it appeared under the heading "Scientific affairs in Europe," *Nature* 156 (1945): 576–579.

19. Tinbergen to Nice, 5 December 1945, MMNP/Cornell. At the same time that he wrote this, Tinbergen continued to express strong anti-German sentiments.

20. Tinbergen to Nice, 4 January 1946, MMNP/Cornell.

21. Stresemann to Nice, 4 February 1946, MMNP/Cornell.

22. The phrase is in Stresemann to Nice, 6 June 1946. The Belgian ornithologist C. Dupond similarly called Nice "la bonne mère des ornithologistes européens," in Dupond to Nice, 24 September 1946. Both letters are in MMNP/Cornell. In her autobiography, *Research Is a Passion with Me* (Toronto: Consolidated Amethyst Communications, 1979), pp. 248–252, Nice describes, though only briefly, the postwar efforts to send Care packages and other parcels to European ornithologists. Her manuscript correspondence at Cornell provides a rich sense of her role in these efforts and the importance of her activities in general.

23. Tinbergen to Nice, 26 February 1946, MMNP/Cornell.

24. Tinbergen to Mayr, 25 February 1946, EMP/Harvard, HUG (FP) 14.7, box 4, folder 160.

25. Tinbergen to Nice, 27 May 1946, MMNP/Cornell. See also Tinbergen to Nice, 4 January 1946, where he wrote: "My connections with German ornithology will not be taken up again so long as nazis like SS-man Niethammer and Ministerialdirigent Ludwig Schuster play a part in it."

26. Tinbergen to Koehler, 25 May 1946, Otto Koehler Papers, Universitätsarchiv Freiburg im Breisgau (hereafter OKP/Freiburg). Here Tinbergen thanked Koehler for a letter dated 10 May 1946, now presumably lost. He also thanked Koehler for having tried, with Lorenz, to get Tinbergen released from the prison camp, explaining that it was "entirely impossible to accept intervention from the German side, however much we also appreciated your friendly intentions."

27. Tinbergen to Mayr, 20 May 1946, EMP/Harvard, HUG (FP) 14.7, box 4, folder 160; Tinbergen to Nice, 27 May 1946, MMNP/Cornell. Whether Tinbergen ever received such a straightforward apology from Lorenz is unclear.

28. Tinbergen to Mayr, 30 June 1946, EMP/Harvard, HUG (FP) 14.7, box 4, folder 160; Tinbergen to Nice, 1 September 1946; and Margarethe Lorenz to Nice, 21 August 1946, MMNP/Cornell.

29. Tinbergen to Koehler, 25 May 1946, OKP/Freiburg.

30. Tinbergen to Mayr, 20 May 1946, EMP/Harvard, HUG (FP) 14.7, box 4, folder 160.

31. Tinbergen to Huxley, in "Scientific affairs in Europe," p. 577.

32. Tinbergen to Lack, 31 March 1946, Lack Papers, Oxford University.

33. Tinbergen to Mayr, 5 September 1945, EMP/Harvard, HUG (FP) 14.7, box 3, folder 142; Tinbergen to Huxley, in "Scientific affairs in Europe," p. 577.

34. Tinbergen to Mayr, 15 June 1946 addendum to letter of 20 May 1946, EMP/Harvard, HUG (FP) 14.7, box 4, folder 160.

35. Tinbergen to Lack, 31 March 1946, and 21 April 1946, Lack Papers, Oxford University; Tinbergen to Nice, 27 May 1946, MMNP/Cornell.

36. Tinbergen to Lack, 21 April 1946, Lack Papers, Oxford University.

37. Tinbergen to Nice, 27 May 1946, MMNP/Cornell.

38. N. Tinbergen and J. J. A. Van Iersel, "'Displacement reactions' in the three-spined stickleback," *Behaviour* 1 (1947): 56–63.

39. See N. Tinbergen, *Curious Naturalists* (London: Country Life, 1958), pp. 118–136; L. de Ruiter, "Some experiments on the camouflage of stick caterpillars," *Behaviour* 4 (1952): 222–232, and "Countershading in caterpillars: an analysis of its adaptive significance," *Archives néerlandaises de zoologie* 11 (1956): 285–342.

40. Tinbergen to Nice, 27 May 1946, MMNP/Cornell.

41. Tinbergen to Nice, 1 September 1946, MMNP/Cornell. It was actually a five-year rather than a six-year lapse since Tinbergen and his students had done summer fieldwork at Hulshorst.

42. Mayr to Tinbergen, 7 March 1946, EMP/Harvard, HUG (FP) 14.7, box 4, folder 160. For previous efforts, see Mayr to Tinbergen, 25 January 1946, EMP/Harvard, HUG (FP) 14.7, box 3, folder 142.

43. Tinbergen to Mayr, 20 May 1946, EMP/Harvard, HUG (FP) 14.7, box 4, folder 160.

44. Tinbergen to Mayr, 20 May 1946, and 30 June 1946; and Mayr to Tinbergen, 21 June 1946, EMP/Harvard, HUG (FP) 14.7, box 4, folder 160. Frank Beach had suggested to Mayr that Tinbergen might use his *Acta Biotheoretica* paper to form the nucleus of the lectures.

45. Beach to Yerkes, 11 February 1947, Yerkes Papers, Yale University. C. R. Carpenter also became an editor. In a letter of 27 January 1948 Tinbergen was able to report to the other editors that Otto Koehler had agreed to join the editorial board. Copy in OKP/Freiburg.

46. I have not found the titles of Tinbergen's individual lectures. The title of the series is given in a letter from Tinbergen to Koehler dated 20 January 1947, OKP/Freiburg. Their focus can be deduced from Tinbergen's 1942 *Acta Biotheoretica* paper, from his book *The Study of Instinct* (Oxford: Clarendon Press, 1951), and from his statement to Mayr: "I see the importance of making as good a job as possible of my book; I am spending every spare minute at writing. I am extending it a little beyond the scope of the six lectures and hope to give more space to sociology, ontogeny, ecology and evolution of behavior, but I am not sure whether I will be able to manage it." Tinbergen to Mayr, 10 October 1947, EMP/Harvard, HUG (FP) 14.7, box 5, folder 218.

47. Tinbergen to Koehler, 20 January 1947, OKP/Freiburg.

48. Tinbergen to Nice, 27 February 1947, MMNP/Cornell.

49. Tinbergen to Mayr, 28 March 1947, EMP/Harvard, HUG (FP) 14.7, box 4, folder 197.

50. Tinbergen gave his inaugural address 25 April 1947. The address, entitled "De Natuur is sterker dan de leer, of de lof van het veldwerk," was published separately by the University of Leiden as a twenty-six-page pamphlet.

51. Tinbergen to Nice, 29 June 1947, MMNP/Cornell; Tinbergen to Mayr, 29 June 1947, EMP/Harvard, HUG (FP) 14.7, box 5, folder 218.

52. In 1946 he told Margaret Nice he was sorry to say that "the political attitude of our people as a whole is rather the same as before the war." Tinbergen to Nice, 27 May 1946, MMNP/Cornell. That Tinbergen's postwar unhappiness with life in Holland was related to his wartime experiences is specified in a letter from David Lack to Alister Hardy, 28 April 1958, Alister Hardy Papers, Bodleian Library, Oxford University.

53. Tinbergen to Mayr, 26 August 1947, EMP/Harvard, HUG (FP) 14.7, box 5, folder 218. Earlier Tinbergen had told Mayr that the Dutch people were "still relatively lazy." Tinbergen to Mayr, 28 March 1947, EMP/Harvard, HUG (FP) 14.7, box 4, folder 197.

54. Tinbergen to Mayr, 10 October 1947, EMP/Harvard, HUG (FP) 14.7, box 5, folder 218. The other two professors were van der Klaauw and H. P. Wolvekamp. Van der Klaauw taught mostly anatomy and his version of "theoretical biology." Wolvekamp taught physiology.

55. Tinbergen to Mayr, 10 October 1947, EMP/Harvard, HUG (FP) 14.7, box 5, folder 218. In addition, one student undertook an anatomical study of stickleback fins.

56. Mayr to Tinbergen, 29 October 1947, EMP/Harvard, HUG (FP) 14.7, box 5, folder 218.

57. Tinbergen to Mayr, 22 February 1948, EMP/Harvard, HUG (FP) 14.7, box 5, folder 255.

58. See below, chapter 7, for a discussion of Thorpe's interactions with Hardy in June 1948.

59. Tinbergen to Mayr, 5 May 1948, EMP/Harvard, HUG (FP) 14.7, box 5, folder 255.

60. Tinbergen to Mayr, 3 June 1948, EMP/Harvard, HUG (FP) 14.7, box 6, folder 277.

61. Tinbergen to Mayr, 25 August 1948, EMP/Harvard, HUG (FP) 14.7, box 6, folder 277.

62. Mayr to Tinbergen, 20 September 1948; and Tinbergen to Mayr, 22 September 1948, EMP/Harvard, HUG (FP) 14.7, box 6, folder 277. See also Mayr to Beach, 13 September 1948; and Beach to Mayr, 14 September 1948, EMP/Harvard, HUG (FP) 14.7, box 5, folder 260.

63. N. Tinbergen, "Physiologische Instinktforschung," *Experientia* 4 (1948): 121–133; Tinbergen and Van Iersel, "'Displacement reactions' in the three-spined stickleback"; N. Tinbergen, "Social releasers and the experimental method required for their study," *Wilson Bulletin* 60 (1948): 6–51. Tinbergen wrote to Mayr, 25 August 1948: "Do you think I have been too rude to Rand in my Wilson Bulletin paper? My reason for being sharp was primarily that he was very hard on our work and thus might cause a defeatist attitude in others." EMP/Harvard, HUG (FP) 14.7, box 6, folder 277. See A. L. Rand, "Lorenz's objective method of interpreting bird behavior," *Auk* 58 (1941): 289–291, and "Nest sanitation and an alleged releaser," *Auk* 59 (1942): 404–409.

64. Tinbergen to Mayr, 22 February 1948, EMP/Harvard, HUG (FP) 14.7, box 5, folder 255.

65. See Thorpe to Lorenz, 6 January 1949; and Lorenz to Thorpe, 15 January 1949, WHTP/Cambridge; Tinbergen to Mayr, n.d., but evidently early January 1949, and 9 March 1949, EMP/Harvard, HUG (FP) 14.7, box 6, folder 313.

66. Tinbergen to Hardy, 17 March 1962, Hardy Papers, Oxford University; and Gerard Baerends, Colin Beer, and Aubrey Manning, "Introduction," in *Function and Evolu-*

tion in Behaviour: Essays in Honour of Professor Niko Tinbergen, F.R.S., ed. Gerard Baerends, Colin Beer, and Aubrey Manning (Oxford: Clarendon Press, 1975), p. xvii.

67. Tinbergen to Mayr, 9 March 1949, EMP/Harvard, HUG (FP) 14.7, box 6, folder 313.

68. Margarethe Lorenz to Nice, 26 November 1947, and 29 January 1948, MMNP/ Cornell; Alec Nisbett, *Konrad Lorenz* (New York: Harcourt Brace Jovanovich, 1976), pp. 98–99; Konrad Lorenz, *The Natural Science of the Human Species: An Introduction to Comparative Behavioral Research: The "Russian Manuscript" (1944–1948)* (Cambridge, MA: MIT Press, 1996), p. xiii.

69. Lorenz to Stresemann, 5 April 1948, Erwin Stresemann Papers, Preußischer Kulturbesitz, Staatsbibliothek zu Berlin. Stresemann soon reported the news of Lorenz's return to Margaret Morse Nice: "The best news I had for a long time is that of Konrad Lorenz being back home, confirmed by a long and very charming letter which he wrote to Verte and me. Now we all may look forward confidently to seeing him." Stresemann to Nice, 25 April 1948, MMNP/Cornell.

70. Lorenz to Stresemann, 5 April 1948, Stresemann Papers, Staatsbibliothek zu Berlin. For information on the founding of Max Planck Institutes in this period, see Eckart Henning and Marion Kazemi, "Chronik der Max-Planck-Gesellschaft zur Förderung der Wissenschaften unter der Präsidentschaft Otto Hahns (1946–1960)," *Veröffentlichungen aus dem Archiv zur Geschichte der Max-Planck-Gesellschaft* 4 (1992): 1–160.

71. This account is based on a letter from Lorenz to Lack, 21 June 1948, Lack Papers, Oxford University. Thorpe received a copy of the letter (WHTP/Cambridge) and subsequently used it in his account of the station's origin. See W. H. Thorpe, *The Origins and Rise of Ethology* (London: Heinemann, 1979), pp. 108–109; and also Irenäus Eibl-Eibesfeldt, "'Fishy, fishy, fishy': autobiographical sketches," in *Studying Animal Behavior: Autobiographies of the Founders,* ed. Donald A. Dewsbury (Chicago: University of Chicago Press, 1989), pp. 69–91.

72. Eibl-Eibesfeldt, "Fishy, fishy, fishy."

73. Lorenz to Lack, 21 June 1948, Lack Papers, Oxford, Oxford University.

74. Lorenz to Lack, 21 June 1948, Lack Papers, Oxford, Oxford University.

75. Thorpe to Huxley, 21 July 1948. See also Thorpe to James Gray, 28 July 1948, where Thorpe indicates that David Lack suggested to him that the Koenigs might be the ideal people to run the ornithological field station at Cambridge. WHTP/Cambridge. Thorpe quotes Lack as saying: "I will do nothing about this until I hear from you, as we had agreed that the E. G. Institute should concentrate on ecology and behaviour in the field, not on observations on captive birds which were to be your sphere." WHTP/ Cambridge.

76. Thorpe to Lorenz, 30 July 1948, WHTP/Cambridge.

77. Lorenz to Thorpe, 10 July 1948, WHTP/Cambridge.

78. Lorenz to Thorpe, 10 July 1948, WHTP/Cambridge.

79. Huxley to Lorenz, 17 July 1948, Huxley Papers, Rice University Archives. Huxley's critique does not correspond exactly to the introduction to Lorenz's now-published *The Natural Science of the Human Species,* so it seems that Huxley was dealing with a different draft.

80. Lorenz to Huxley, 11 August 1948. Huxley Papers, Rice University Archives; copy also in WHTP/Cambridge.

81. Lorenz to Huxley, 11 August 1948. Huxley Papers, Rice University Archives; copy also in WHTP/Cambridge.

82. Thorpe to Meister, October (no day given) 1948, WHTP/Cambridge.

83. Lorenz to Thorpe, 19 October 1948, WHTP/Cambridge.

84. See especially E. M. Nicholson to Thorpe, 2 September 1948; Thorpe to Mrs. J. B. Priestley, 14 September 1948; Mrs. J. B. Priestley to Thorpe, 9 October 1948; Nicholson to Thorpe, 10 and 12 October, 1948; Lorenz to Thorpe, 19 October 1948, WHTP/Cambridge.

85. Thorpe to Huxley, 9 November 1948, WHTP/Cambridge.

86. Meister to Thorpe, 17 November 1948; Thorpe to Meister, 26 November 1948, WHTP/Cambridge.

87. Lorenz to Thorpe, 20 December 1948. Lorenz recognized the embarrassing possibility that Priestley's money, when distributed through the Austrian Academy of Sciences, might come only to him and not be shared with the Wilhelminenberg Station. He therefore urged Thorpe to hint to the Priestleys that a word from them to Meister might help Koenig get some funding too.

88. Lorenz to Thorpe, 23 January 1949, WHTP/Cambridge.

89. Lorenz to Thorpe, 2 February 1949, WHTP/Cambridge.

90. Lorenz to Thorpe, 28 February 1948, and 9 May 1948, WHTP/Cambridge. Lorenz described his reinstallation problems at the University of Vienna in much the same terms to Margaret Morse Nice in a letter dated 10 May 1949, MMNP/Cornell. For a different account of the whitewashing of Lorenz's political past and the circumstances of his official rehabilitation, see Benedikt Föger and Klaus Taschwer, *Die andere Seite des Spiegels: Konrad Lorenz und der Nationalsozialismus* (Vienna: Czernin Verlag, 2001), pp. 181–188.

91. Lorenz to Thorpe, 29 June 1949, WHTP/Cambridge.

92. Meister to Thorpe, 8 June 1949, WHTP/Cambridge.

93. On Lorenz's travel schedule in England, see Lorenz to Thorpe, 9 May 1949. See also Thorpe to Lack, 24 June 1949; and Lorenz to Thorpe, 29 June 1949. In this last letter Lorenz indicates he will speak to the Oxford Society for Experimental Psychology (having been invited there by Professor Zangwill) and will devote his talk to a critique of mechanistic and vitalistic schools of psychology. WHTP/Cambridge.

94. Thorpe, *The Origins and Rise of Ethology*, p. 81.

95. W. H. Thorpe, untitled ms., numbered WHT/M/114, WHTP/Cambridge.

96. W. H. Thorpe, "The modern concept of instinctive behaviour," *Bulletin of Animal Behaviour* 1, no. 7 (1948): 2–12, especially 9–10.

97. W. H. Thorpe, "Suggestion to the S.E.B. council for 1949," ms. no. WHT/M/112, WHTP/Cambridge. The steering committee Thorpe suggested for the symposium consisted of C. F. A. Pantin, J. Z. Young, G. C. Grindley, D. L. Gunn, R. J. Pumphrey, G. P. Wells, and Thorpe himself.

98. W. H. Thorpe, "Plan for a symposium on behaviour," ms. no. WHT/M/113, WHTP/Cambridge.

99. See Thorpe, *The Origins and Rise of Ethology*, p. 81. Not only had Lorenz not seen Tinbergen since 1938, but he had not seen Koehler since 1941.

100. Konrad Lorenz, "My family and other animals," in Dewsbury, *Studying Animal Behavior*, p. 277.

101. R. A. Hinde, "Nikolaas Tinbergen," *Biographical Memoirs of Fellows of the Royal Society* 36 (1990): 547–565, quote on p. 553.

102. Durant, "The making of ethology," pp. 1601–1616.

103. *Symposia of the Society for Experimental Biology*, vol. 4 (1950).

104. Lissmann had studied under Uexküll in Hamburg, receiving his doctorate in 1932 for research on the behavior of Siamese fighting fishes. He then went from Germany to Hungary on a postdoctoral grant, and was there when Hitler came to power in 1933. Lissmann was not Jewish, but he despised the Nazis, and he refused to return to Germany. In 1934 James Gray offered him a place in Gray's laboratory at Cambridge. Working in part with Gray and in part independently, Lissmann over the next several years established a solid reputation for his experiments on locomotor control in earthworms, dogfish, and gastropods. For more on Lissmann see R. McN. Alexander, "Hans Werner Lissmann, 30 April 1909–21 April 1995," *Biographical Memoirs of Fellows of the Royal Society* 42 (1996): 235–245.

105. Holst to Lorenz, 15 July 1949; Lorenz to Holst, 25 July 1949; and Koehler to Holst, 27 July 1949, Erich von Holst Papers, Archive of the Max-Planck-Gesellschaft, Berlin. See also G. P. Baerends, "Turning points in the history of ethology," in *L'Histoire de la connaissance du comportement animal*, ed. Liliane Bodson (Liège: Université de Liège, 1993), pp. 59–76. Baerends writes: "Victory fell to the Cambridge crew and the ethologists were highly disappointed that Von Holst's work had not been shown to full advantage. At first the German speaking ethologists present found it difficult to understand the unfortunate course the meeting had taken. But, after Niko Tinbergen had made clear to them why it went wrong, Professor Otto Koehler took it upon himself to consult Von Holst about measures to make the damage undone" (p. 65). For the published version of the symposium, Holst's contribution was not his jarring critique of Gray, read by Lissmann, but instead his paper entitled "Quantitative Messung von Stimmungen im Verhalten der Fische." See *Symposia of the Society for Experimental Biology* 4 (1950): 143–172.

106. Hans W. Lissmann interview with Richard W. Burkhardt, Jr., 13 June 1983.

107. Colin Beer, personal communication, observes that J. E. Smith's excellent paper on locomotion in starfish would have been more appropriately included in the section "Central and Peripheral Control."

108. See Konrad Lorenz, "The comparative method in studying innate behaviour patterns," *Symposia of the Society for Experimental Biology* 4 (1950): 221–268.

109. See Lorenz to Thorpe, 23 January 1949, 28 February 1949, and 29 June 1949, WHTP/Cambridge. After the event, Lorenz provided the Austrian Academy of Sciences with a ten-page report on the symposium. In the report he also indicated that on the day after the Cambridge symposium (23 July 1949) he gave an address to the Oxford Society of Experimental Psychology entitled "Taxis, Raum-Repräsentation und Einsicht" (Taxis, space representation and insight). This information comes from an undated manuscript in Lorenz's personnel file in the Archiv der Österreichischen Akademie der

Wissenschaften, entitled "Kurzer Bericht über das von der Society for Experimental Biology im Juli 1949 in Cambridge abgehaltene 'Symposium on Physiological Mechanisms in Animal Behaviour' . . . " I thank Klaus Taschwer for making this document known to me.

110. The discussion that follows is based on the published paper, leaving it to the reader to recognize that Lorenz may not have offered exactly the same comments in his oral address in Cambridge.

111. Lorenz, "The comparative method in studying innate behaviour patterns," pp. 232–233.

112. Lorenz, "The comparative method in studying innate behaviour patterns," p. 222.

113. Lorenz, "The comparative method in studying innate behaviour patterns," p. 238.

114. Lorenz, "The comparative method in studying innate behaviour patterns," p. 246.

115. Lorenz, "The comparative method in studying innate behaviour patterns," p. 248.

116. Lorenz, "The comparative method in studying innate behaviour patterns," p. 251.

117. Lorenz, "The comparative method in studying innate behaviour patterns," p. 253.

118. Lorenz, "The comparative method in studying innate behaviour patterns," p. 255.

119. Lorenz, "The comparative method in studying innate behaviour patterns," p. 257.

120. Lorenz, "The comparative method in studying innate behaviour patterns," pp. 266–267.

121. N. Tinbergen, "The hierarchical organization of nervous mechanisms underlying instinctive behaviour," *Symposia of the Society for Experimental Biology* 4 (1950): 305–312.

122. Tinbergen, "The hierarchical organization of nervous mechanisms underlying instinctive behaviour," p. 307.

123. Tinbergen, "The hierarchical organization of nervous mechanisms underlying instinctive behaviour," p. 308.

124. Tinbergen, "The hierarchical organization of nervous mechanisms underlying instinctive behaviour," p. 308.

125. Tinbergen, "The hierarchical organization of nervous mechanisms underlying instinctive behaviour," p. 310.

126. Tinbergen, "The hierarchical organization of nervous mechanisms underlying instinctive behaviour," pp. 310–311.

127. Tinbergen, "The hierarchical organization of nervous mechanisms underlying instinctive behaviour," p. 311. See P. Weiss, "Self-differentiation of the basic patterns of coordination," *Comparative Psychology Monographs* 17 (1941): 1–96.

128. Tinbergen, "The hierarchical organization of nervous mechanisms underlying instinctive behaviour, p. 312. Colin Beer has argued that Weiss's model does not fit as well

with Tinbergen's model as Tinbergen claimed. See Beer, "Darwin, instinct, and ethology," *Journal of the History of the Behavioral Sciences* 19 (1983): 68–80.

129. Tinbergen, "The hierarchical organization of nervous mechanisms underlying instinctive behaviour, p. 312.

130. Three main categories of terms were identified for consideration at the round-table sessions: terms relating to "elementary behaviour patterns," terms relating to "instinct," and terms relating to "learning." One session each was devoted to "elementary behaviour patterns" and "learning," while two sessions, chaired by Lorenz and Tinbergen, were devoted to terms subsumed under the heading of "instinct." Among the terms discussed at the instinct sessions were "instinct," "drive," "appetitive behaviour," "consummatory act," "fixed action pattern," "reaction specific energy," "displacement activity," and "releaser." Holst was among those interested in the precise definition of terms. See Holst to Lorenz, 15 July 1949, Holst Papers, Archive of the Max-Planck-Gesellschaft, Berlin.

131. Thorpe indicates that at the beginning of the sessions the phrase "reaction specific energy" was used, but, as he put it, "it was generally agreed that the use of the term energy in this sense was undesirable and that 'specific action potential' would be a better version." See W. H. Thorpe, "The definition of some terms used in animal behaviour studies," *Bulletin of Animal Behaviour*, no. 9 (1951): 34–40, quote on p. 37.

132. Lorenz to Holst, 25 July 1949; Koehler to Holst, 27 July 1949, Holst Papers, Archive of the Max-Planck-Gesellschaft, Berlin.

133. Tinbergen to Lack, 25 July 1949, Lack Papers, Oxford University.

134. Lorenz, "Kurzer Bericht" (see above, n. 109).

135. Lorenz to Holst, 25 July 1949, Holst Papers, Archive of the Max-Planck-Gesellschaft, Berlin.

136. Koehler to Holst, 27 July 1949, Holst Papers, Archive of the Max-Planck-Gesellschaft, Berlin.

137. Tinbergen to Mayr, 10 August 1949, EMP/Harvard, HUG (FP) 14.7, box 6, folder 313.

138. Lorenz's itinerary in England (at least as it was shaping up before the fact) is referred to in Thorpe to Lack, 24 June 1949, WHTP/Cambridge.

139. The inscribed copy of the book is still on the shelf in the library of Lorenz's home at Altenberg, now the site of the Konrad Lorenz Institute for Evolution and Cognition Research.

140. Lorenz to Craig, 2 September 1949, copy in MMNP/Cornell. See also Koehler to Nice, 8 September 1949, MMNP/Cornell, where Koehler reports that Lorenz was in England for two months, and that both Lorenz and Tinbergen made very good impressions at the congress.

141. Lorenz to Thorpe, 2 September 1949, WHTP/Cambridge.

142. Holst to Lorenz, 15 July 1949, Holst Papers, Archive of the Max-Planck-Gesellschaft, Berlin.

143. Lorenz to Thorpe, 2 September 1949, WHTP/Cambridge. In this letter Lorenz told Thorpe that Thorpe and Tinbergen were his "two closest personal friends." He also said that the two things he was thinking about most from the conference were Thorpe's "conception of the fundamentals, in which perception, innate releasing mechanism and

insight fuse into ONE, that is to say into the elementary perception of relation" and how Tinbergen's ideas on hierarchy bore on the concept of "activity-specific-energy."

144. For Lorenz's comparison of his own situation with those of others, see Lorenz to Thorpe, 28 February 1949 and (following Lorenz's interaction with Holst and Tinbergen at the 1950 meeting in Wilhelmshaven) 31 June 1950 and 17 August 1950, WHTP/Cambridge.

CHAPTER SEVEN

1. Tinbergen to Mayr, 4 September 1950, Ernst Mayr Papers, Harvard University Archives (hereafter EMP/Harvard) , HUG (FP) 14.7, box 8, folder 372.

2. Lorenz to Thorpe, 1 November 1950, William Homan Thorpe Papers, Cambridge University Library (hereafter WHTP/Cambridge), MS.Add.8784.

3. Tinbergen to Mayr, 1 April 1950, EMP/Harvard, HUG (FP) 14.7, box 7, folder 351. See also Tinbergen to Mayr, 19 October 1949, EMP/Harvard, HUG (FP) 14.7, box 7, folder 332, and Tinbergen's 1961 BBC talk, "On turning native," no. C.236, Nikolaas Tinbergen Papers, Bodleian Library, Oxford University (hereafter NTP/Oxford).

4. The case of Groningen will not be pursued in this book, which has grown beyond its original anticipated dimensions, but the Groningen program would certainly afford another valuable example of the way that a program of ethological studies took shape in a particular institutional context. For an introduction to the subject, see Gerard P. Baerends, "Two pillars of wisdom," in *Studying Animal Behavior: Autobiographies of the Founders*, ed. Donald A. Dewsbury (Chicago: University of Chicago Press, 1989), pp. 13–40; and his comments and those of his colleagues and students in "The research programme of the zoological laboratory of the University of Groningen," reported in *Archives néerlandaises de zoologie* 16 (1964): 149–171.

5. Hardy to Huxley, 16 May 1921, Julian S. Huxley Papers, Rice University Archives (hereafter JSHP/Rice).

6. Hardy to Huxley, 27 April 1921, and 16 May 1921, JSHP/Rice.

7. Draft of a letter from Alister Hardy to Julian Huxley, 26 July 1926, Alister C. Hardy Papers, Bodleian University, Oxford University (hereafter ACHP/Oxford). The final version of this letter is not in JSHP/Rice.

8. Alister C. Hardy, "Natural history—old and new," inaugural address delivered on 28 April 1942 at Marischal College, University of Aberdeen, Aberdeen, 1942.

9. Alister Clavering Hardy, "Application for the Linacre Professorship of Zoology and Comparative Anatomy in the University of Oxford," ms. dated 2 July 1945, ACHP/Oxford.

10. Alister Hardy interview with Desmond Morris and Richard Burkhardt, 8 June 1983, ACHP/Oxford. On the history of zoology at Oxford just prior to Hardy's appointment as the Linacre Professor and on the history of the BAP and the Edward Grey Institute, see Jack Morrell, *Science at Oxford, 1914–1939: Transforming an Arts University* (Oxford: Clarendon Press, 1997), pp. 268–304.

11. John R. Baker to Alister Hardy, 21 October 1945, ACHP/Oxford.

12. Baker to Hardy, 11 December 1945, ACHP/Oxford.

13. Baker to Hardy, 11 December 1945, ACHP/Oxford. Baker's satisfaction with Hardy's handling of the situation may have been enhanced by the fact that Hardy in the

meantime arranged Baker's promotion to the position of reader in cytology. See Baker to Hardy, 30 November 1945, ACHP/Oxford. In a letter to Hardy dated 9 March 1949, Peter Medawar apologized "for the really contemptible behaviour of the resident staff on your first arrival in Oxford. It was not until I had had a bit of responsibility of my own, here, that I realized with horror and regret just how badly we had behaved and how upset you must have been by it. . . . Incidentally, wherever I go now the Oxford zoology department is being spoken of in a way that is altogether quite novel—as the leading department in the country. To have achieved this stature in 3–4 years is something you ought to be jolly proud of. . . . particularly as you did [it], I'm ashamed to say, in spite of your staff rather than with its help!" Medawar began his letter by thanking Hardy for advocating Medawar's candidacy for the Royal Society.

14. A. C. Hardy, "Zoology outside the laboratory," *Advancement of Science* 6 (1949): 213–223.

15. Tinbergen's trip with Hardy into the field to film bird flocks was reported to me by Robert Hinde in an interview of 10 June 1983. That Tinbergen "greatly deprecated" Hardy's idea of studying ESP in jackdaws is reported in W. S. Verplanck's manuscript "Newsletters" written in England and Europe in 1953, copies of which he sent back to his mother and others in the United States. In 1992 he made a photocopy of the typed original and gave it to Donald Dewsbury, adding page numbers to the photocopy and entitling the collection "Newsletters: late winter and spring, 1953 from William S. Verplanck's 1953 APS grant travels to centers of ethological research in the UK and the Continent." I thank Donald Dewsbury for making a copy of these newsletters available to me. Copies of the Verplanck "Newsletters" may also be among the W. S. Verplanck Papers conserved at the Archives of the History of American Psychology at the University of Akron.

16. Tinbergen told Mayr in his letter of 1 April 1950, "I am slow in finding my methods of teaching and research and to adapt myself to the intricate organisation of this university." EMP/Harvard, HUG (FP) 14.7, box 7, folder 351. See also Tinbergen to Thorpe, 8 September 1976, A4, NTP/Oxford, for his characterization of the Oxford system as "entirely alien" because it involved "condensed three year courses and three terms per year, with very little flexibility." Richard Dawkins reports that his first contact with Tinbergen was hearing him lecture on molluscs in Oxford's "Animal Kingdom" course. See Richard Dawkins, "Introduction," in *The Tinbergen Legacy*, ed. M. S. Dawkins, T. R. Halliday, and R. Dawkins (London: Chapman and Hall, 1991), p. x. See also Tinbergen, "Report on the activities of N. Tinbergen as university lecturer in animal behaviour from 1. October 1954 till Christmas 1958," which indicates that over the last two years of the period in question Tinbergen's teaching also included "16 lectures in elementary German" and "a ten-days' practical in animal behaviour." A45, NTP/Oxford.

17. Tinbergen to Huxley, 18 February 1953, JSHP/Rice.

18. The students who came later in Tinbergen's career reported experiences similar to those of students who had come earlier. Of the Friday evening seminars Richard Dawkins recalled, "Niko refused to let sloppy language pass." John Krebs likewise cited Tinbergen's "insistence on great precision of thought." See Richard Dawkins, "Introduction," and John R. Krebs, "Animal communication: ideas derived from Tinbergen's activities," in Dawkins, Halliday, and Dawkins, *The Tinbergen Legacy*, pp. xi and 60. Suc-

cessive cohorts of Tinbergen students referred to him as "the maestro." See, for example, Desmond Morris, *Animal Days* (New York: William Morrow, 1980), p. 67.

19. Tinbergen to Mayr, 4 September 1950, EMP/Harvard, HUG (FP) 14.7, box 8, folder 372.

20. Tinbergen to Mayr, 4 September 1950, EMP/Harvard, HUG (FP) 14.7, box 8, folder 372.

21. N. Tinbergen, "The work of the Animal Behaviour Research Group in the Department of Zoology, University of Oxford," *Animal Behaviour* 11 (1963): 206–209, quotations from p. 206.

22. Tinbergen to Thorpe, 8 September 1976, NTP/Oxford.

23. Tinbergen letter to Mayr, 29 October 1956, copy courtesy of Ernst Mayr (in personal archive of Richard W. Burkhardt, Jr.). Tinbergen also noted here that Oxford would be providing him the following year with a junior lecturer and an assistant.

I take this occasion to note that in 1982 Ernst Mayr arranged to have copied for me his correspondence with Niko Tinbergen and Konrad Lorenz, plus selected items from his correspondence with other writers. I later supposed that the original copies of these letters had all since become part of the collection of Ernst Mayr papers at the Harvard University Archives. However, comparing just recently my own collection of photocopies with the Harvard University Archives' inventory of the Mayr papers, I discovered that I have a sizable collection of correspondence between Ernst Mayr and Niko Tinbergen, covering the period from 16 March 1953 to 26 January 1982, that seems to be currently lacking at Harvard. Ernst has indicated to me that he does not know where the originals are. I plan to recopy the copies in my possession, so that they may be added to the Harvard collection, but for now I will cite these letters simply as "courtesy of Ernst Mayr."

24. Tinbergen to Mayr, 23 March 1961, copy courtesy of Ernst Mayr.

25. Tinbergen, "The work of the Animal Behaviour Research Group," p. 206.

26. For biographical information on Thorpe and a bibliography of his writings, see R. A. Hinde, "William Homan Thorpe," *Biographical Memoirs of Fellows of the Royal Society* 33 (1987): 621–639. For the interplay of Thorpe's philosophy and his science, see Neal C. Gillespie, "The interface of natural theology and science in the ethology of W. H. Thorpe," *Journal of the History of Biology* 23 (1990): 1–38.

27. Hinde, "William Homan Thorpe," pp. 622–623.

28. Quoted in Arthur Koestler, *The Case of the Midwife Toad* (London: Hutchinson, 1971), p. 76.

29. W. H. Thorpe interview with R. W. Burkhardt, 28 February 1979.

30. Paul Kammerer, "Breeding experiments on the inheritance of acquired characters," *Nature* 111 (1923): 637–640, quote on p. 640.

31. Letter from the Hon. Mrs. Onslow to William Bateson, 1 May 1923, quoted in Koestler, *The Case of the Midwife Toad*, p. 76.

32. J. W. Heslop Harrison, "Experiments on the egg-laying instincts of the sawfly, *Pontania salicis* Christ., and their bearing on the inheritance of acquired characters; with some remarks on a new principle in evolution," *Proceedings of the Royal Society of London, Series B* 101 (1927): 115–125.

33. W. H. Thorpe, "Biological races in insects and allied groups," *Biological Reviews of the Cambridge Philosophical Society* 5 (1930): 177–212.

34. W. H. Thorpe interview with R. W. Burkhardt, 28 February 1979.

35. W. H. Thorpe and F. G. W. Jones, "Olfactory condition in a parasitic insect and its relation to the problem of host selection," *Proceedings of the Royal Society of London, Series B* 124 (1937): 56–81, quote on p. 78.

36. W. H. Thorpe, "Further experiments on olfactory condition in a parasitic insect: the nature of the conditioning process," *Proceedings of the Royal Society of London, Series B* 126 (1938): 370–397.

37. W. H. Thorpe, "Ecology and the future of systematics," in *The New Systematics,* ed. Julian Huxley (Oxford, 1940), pp. 341–364, quote on p. 354.

38. W. H. Thorpe, "Types of learning in insects and other arthropods," *British Journal of Psychology* 33 (1943): 220–234; 34 (1943): 20–31; 34 (1944): 66–76.

39. At this point Thorpe cited only three papers by Lorenz: the "Kumpan" paper of 1935; the 1937 shortened version of the "Kumpan" paper that appeared in the *Auk,* and a manuscript English translation, prepared by Daniel Lehrman, of Lorenz's 1939 paper "Vergleichende Verhaltensforschung."

40. Edward A. Armstrong cited Lorenz frequently in his book *Bird Display: An Introduction to the Study of Bird Psychology* (Cambridge: Cambridge University Press, 1942).

41. W. H. Thorpe, "The evolutionary significance of habitat selection," *Journal of Animal Ecology* 14 (1945): 67–70, quote on p. 69.

42. W. H. Thorpe, "Animal learning and evolution," *Nature* 156 (1945): 46–47.

43. W. H. Thorpe, *The Origins and Rise of Ethology* (London: Heinemann, 1979), p. 121.

44. This story is told in Philippe Chavot, "Histoire de l'éthologie: recherches sur le développement des sciences du comportement en Allemagne, Grande-Bretagne et France, de 1930 à nos jours" (thèse de doctorat en histoire des sciences, Université Louis Pasteur—Strasbourg I, 1994), pp. 196–198.

45. Thorpe to Hardy, 4 June 1948, ACHP/Oxford.

46. Gray's views of the work of the ethologists and the operation at Madingley were reported by Hans Lissmann in an interview with Richard W. Burkhardt, Jr., 13 June 1983.

47. Robert Hinde interview with Richard W. Burkhardt, Jr., 10 June 1983.

48. Hinde mentions Moreau and Lorenz as Thorpe's early choices for the post of curator, in R. A. Hinde, "Ethology in relation to other disciplines," in Dewsbury, *Studying Animal Behavior,* p. 195. On the Koenigs, see above, chapter 6.

49. Department of Zoology, Ornithological Field Station, report for 1950–1951, ms. document at the Madingley Sub-department of Animal Behaviour.

50. Ornithological Field Station, reports for 1950–1951, 1951–1952, and 1952–1953. Also "Report for the academic year 1974–5," containing Thorpe's speech on the occasion of the twenty-fifth anniversary of the field station, ms. document at the Madingley Sub-department of Animal Behaviour.

51. Thorpe himself published extensively on the results of his study with birdsong. See, for example, W. H. Thorpe, "The learning of song patterns by birds, with especial reference to the song of the chaffinch (*Fringilla coelebs*)," *Ibis* 100 (1958): 535–570, and

Bird Song: The Biology of Vocal Communication and Expression in Birds (Cambridge: Cambridge University Press, 1961).

52. Peter Marler, "Hark ye to the birds: autobiographical marginalia," in Dewsbury, *Studying Animal Behavior,* pp. 315–345, quote on p. 323.

53. Lorenz to Craig, 2 September 1949, copy in Margaret Morse Nice Papers, 1917–1968, collection number 2993, courtesy of the Division of Rare and Manuscript Collections, Cornell University Library (hereafter MMNP/Cornell).

54. Lorenz to Mayr, 20 November 1949 (in German), EMP/Harvard, HUG (FP) 14.7, box 7, folder 344; Lorenz to Thorpe, 20 November 1949, WHTP/Cambridge.

55. Lorenz to Mayr, 20 November 1949 (in German), EMP/Harvard, HUG (FP) 14.7, box 7, folder 344; Lorenz to Thorpe, 20 November 1949 (in German), WHTP/Cambridge.

56. Lorenz to Mayr, 20 November 1949 (in German), EMP/Harvard, HUG (FP) 14.7, box 7, folder 344.

57. Lorenz to Mayr, 20 November 1949 (in German), EMP/Harvard, HUG (FP) 14.7, box 7, folder 344; Lorenz to Thorpe, 20 November 1949, and 18 December 1949, WHTP/Cambridge. Lorenz had also been invited to Paris for a March 1950 conference on animal societies. According to a letter from Margarethe Lorenz to Margaret Nice, dated 1 February 1950, Lorenz planned to meet with Rockefeller people while he was in Paris. Alfred Seitz in a letter to Nice, dated 31 January 1950, indicated he would go to Slimbridge in April and work there with Lorenz filming ducks and geese. Both letters are in MMNP/Cornell. As things turned out, Lorenz did not go to Paris or Slimbridge that spring. Seitz did not go to Slimbridge either, for he received in the meantime a position at the Nuremberg zoo.

58. Jost Franz to Margaret Nice, 15 January 1950, MMNP/Cornell. Franz indicates that Lorenz lectured to students as well as participated in the evening debate. He identifies Lorenz's debate as having been with von Uexküll.

59. Koehler to Margaret Nice, 3 February 1950, MMNP/Cornell.

60. Koehler to Margaret Nice, 14 January 1948. Of his own politics in the past, Koehler told Nice: "I always had the same [Weltanschauung], with Kaiser Wilhelm before the first war, with the after war German republic, during the nazi regime, and now, and always it would not fit to the official reading. I never was member of any party, because I always thought German parties were so awfully egoistical ones, only seeking personal advantages and not taking enough care for the common sort of what once was the so called nation. . . . I was most surprised when I heard [in] 1933 that 'Nationalsozialismus ist angewandte Biologie,' and my position, which I helt like it always had been, now was most dangerous." Koehler refers to Portmann's anti-Darwinism (Koehler to Nice, 3 February 1950, and 4 March 1956) and to German Catholics' anti-Darwinism (Koehler to Nice, 14 October 1952, MMNP/Cornell). Koehler's portrayal of his own distance from the Nazis contrasts with Ute Deichmann's note on him in *Biologists under Hitler,* trans. Thomas Dunlap (Cambridge, MA: Harvard University Press, 1996), p. 401.

61. Excerpt from the diary of "E.F.D." dated 19 November 1950, series 705, folder 3363, Rockefeller Foundation Archives, Rockefeller Archive Center. I thank Thomas Rosenbaum of the Rockefeller Archive Center for the identification of "E.F.D." as Edward F. D'Arms.

62. Koehler to Margaret Nice, 3 February 1950, MMNP/Cornell.

63. Lorenz to Thorpe, 4 March 1950, WHTP/Cambridge. Lorenz's short lecture tour to Germany took place in late January, judging from letters from Margarethe Lorenz to M. M. Nice, 1 February 1950; and Koehler to Nice, 3 February 1950, MMNP/Cornell.

64. Margarethe Lorenz to Margaret Nice, 1 February 1950, MMNP/Cornell.

65. Holst to Lorenz, 13 February 1950, Erich von Holst Papers, Archive of the Max-Planck-Gesellschaft, Berlin (henceforth EHP/MPG-Berlin).

66. Frisch to Holst, 6 February 1950; Holst to Frisch (plus review of candidates, written the same day), 11 February 1950, EHP/MPG-Berlin.

67. Holst to Lorenz, 13 February 1950, EHP/MPG-Berlin.

68. Thorpe to Frisch, 14 February 1950, and Lorenz to Thorpe, 4 March 1950, WHTP/Cambridge.

69. Lorenz to Thorpe, 4 March 1950, WHTP/Cambridge.

70. Lorenz to Thorpe, 4 March 1950, WHTP/Cambridge. See also Huxley to Lorenz, 9 February 1950, JSHP/Rice.

71. Lorenz to Huxley, 29 March 1950, JSHP/Rice.

72. Koehler to Holst, 20 April 1950, EHP/MPG-Berlin; Lorenz to Thorpe, 28 April 1950, WHTP/Cambridge.

73. Lorenz to Thorpe, 28 April 1950, WHTP/Cambridge.

74. Lorenz to Thorpe 31 June 1950, WHTP/Cambridge. In his letter, Lorenz emphasized the word "never" by wide spacing rather than by underlining.

75. Thorpe to Gray, 18 July 1950; Thorpe to Lorenz, 27 July 1950, WHTP/Cambridge.

76. Thorpe to Scott, 28 July 1950, WHTP/Cambridge.

77. See Holst to Koehler, 10 June 1950; Koehler to Holst, 22 June 1950; and Holst to Koehler, 18 July 1950, EHP/MPG-Berlin. See also Agnes von Cranach, "Editor's foreword," in *The Natural Science of the Human Species: An Introduction to Comparative Behavioral Research: The "Russian Manuscript" (1944–1948),* by Konrad Lorenz (Cambridge, MA: MIT Press, 1996), pp. ix–x.

78. Somewhat earlier Lorenz seems to have told Thorpe he thought he was in disfavor with the ministry because he was a Protestant. This, at least, is what Thorpe reported to Gray in his letter to Gray of 18 July 1950 (WHTP/Cambridge). This claim does not appear, however, in any of his letters to Thorpe now preserved in WHTP/Cambridge. Possibly Lorenz suggested the anti-Protestant explanation to Thorpe in person at the Wilhelmshaven meeting, when his nomination was running into difficulty but a final decision had not been made.

79. Lorenz to Thorpe, 17 August 1950, WHTP/Cambridge.

80. Lorenz to Thorpe 17 August 1950, WHTP/Cambridge.

81. Lorenz to Huxley 19 August 1950, JSHP/Rice. Lorenz also wrote a brief letter to Mayr on the same day: Lorenz to Mayr, 19 August 1950, EMP/Harvard, HUG (FP) 14.7, box 8, folder 365. Deichmann, *Biologists under Hitler,* p. 198, cites a letter from Lorenz to Gustav Kramer of 29 August 1950 saying the same things about the ministry's not wanting his ideas to reach middle school teachers in Styria.

82. Lorenz to Huxley, 19 August 1950, JSHP/Rice.

83. Holst to Koehler, 25 August, 1950: "Es ist ja unser alter Traum, daß wir zu dreien,

Konrad, Kramer und ich irgendwo nahe an einer Universität—am liegsten an der Frei-
burger—ein Institut für Verhaltensphysiologie aufmachen wollten; aber Lieblings-
träume pflegen nicht in Erfüllung zu gehen, ihr Wert liegt wohl nur darin, daß sie in der
Ferne schweben, wo die Dürre des Realen ihnen nichts anhaben kann." EHP/MPG-
Berlin. The account of these three men's having had the dream of building an institute
together is mentioned also by Lorenz in a letter to Huxley: "Since we were students, it
was our dream to have a joint institute, and, as we are in many ways complementary,
different as we are, I am really convinced that we shall do good work together." Lorenz
to Huxley, 9 November 1950, JSHP/Rice.

84. Holst to Lorenz, 1 September 1950, EHP/MPG-Berlin. This letter crossed in the
mail with a long letter from Lorenz, which unfortunately is not preserved in EHP/MPG-
Berlin. Holst's next letter to Lorenz, dated 10 September 1950, refers to this letter from
Lorenz and again offers arguments against Lorenz's going to Oxford.

85. Holst to Lorenz, 10 September 1950, EHP/MPG-Berlin.

86. See Scott to Thorpe, 4 October 1950; Thorpe to Harris, 14 October 1950; Tinbergen to Thorpe, 16 October 1950; Thorpe to Huxley, 21 October 1950; Tinbergen to
Thorpe, 31 October 1950, and 3 November 1950; Haldane to Thorpe, 22 November 1950.
Margarethe Lorenz in an undated letter also expressed anxieties to Thorpe and Thorpe's
wife regarding the political situation in Austria. WHTP/Cambridge.

87. Lorenz to Thorpe, 1 November 1950 (with additional comments added 3 November 1950), WHTP/Cambridge. In the original letter, the phrase "build an institute
for me!" was not emphasized by underlining but instead by the German practice of
spacing out the letters. Lorenz did, however, underline the second and third phrases
underlined in the text above. Detailed information on the development of the Max-
Planck-Gesellschaft's position for Lorenz is to be found in the Archive of the Max-
Planck-Gesellschaft, Berlin (hereafter MPG Archive), in the file entitled "Verhaltensphysiologie, allgemein (1)." Included are Hahn's official notes from the meeting with
Lorenz and Holst on 30 October 1950, indicating that Lorenz had an appointment offer
from England, which he would accept only if he did not find suitable research opportunities in Germany.

88. The Thorpe Papers include a draft entitled "Research programme," with statements by Tinbergen and Harris, outlining what Lorenz would do at Slimbridge and what
the significance of the work would be. This was written on or before 9 November 1950,
judging from a letter from Tinbergen to Thorpe, 9 November 1950, WHTP/Cambridge.

89. See Huxley to Thorpe, 8 November 1950, JSHP/Rice.

90. Lorenz to Huxley, 9 November 1950, JSHP/Rice. Lorenz offered similar accounts of his plans for Slimbridge in his letters to Tinbergen of 2 December 1950, WHTP/
Cambridge.

91. Lorenz to Huxley, 9 November 1950, JSHP/Rice. For Austrian comments on
Austria's loss of Frisch and Lorenz to Germany, see Benedikt Föger and Klaus Taschwer,
Die andere Seite des Spiegels: Konrad Lorenz und der Nationalsozialismus (Vienna: Czer-
nin Verlag, 2001), pp. 192–194.

92. Lorenz to Tinbergen, 2 December 1950, WHTP/Cambridge.

93. See Hahn to Rajewsky, 6 December 1950, MPG Archive.

94. See Senatsprotokoll, 19 December 1950, in "Verhaltensphysiologie, allgemein

(1)," MPG Archive. See also Holst to Hahn, 11 December 1950, regarding the perception that he had gone over his colleagues' heads; and Koehler to Holst, 21 December 1950, regarding that and other objections to the Buldern plan. That there were no dissenting votes is stated by Lorenz in a letter to Huxley, 21 December 1950, JSHP/Rice. Lorenz's account of the chronology of the events that took place at this time, given in his autobiographical chapter in Dewsbury, *Studying Animal Behavior,* is inaccurate. As he recalled it, he had received an offer of a lectureship from Bristol and had already accepted the offer when Erich von Holst spoke to Otto Hahn and Hahn intervened on Lorenz's behalf. In fact, the offer from Bristol came in January 1951, by which time the Max-Planck-Gesellschaft had already voted on Lorenz's appointment. Indeed, from the time of his meeting with Otto Hahn at the end of October 1950, Lorenz was committed to Buldern. At that point Bristol University was only just applying to the Nature Conservancy for a research fellowship for Lorenz. See Konrad Lorenz, "My family and other animals," in Dewsbury, *Studying Animal Behavior,* p. 278. See also the Lorenz letter beginning "Dear Sir," 16 February 1951, in which Lorenz responded to the Bristol offer dated 8 January 1951, in WHTP/Cambridge. Lorenz respectfully declined the Bristol offer. At the same time, he tried to convince the Bristol authorities that it would be a good idea to fund in his place a research assistant who would then take on the sort of comparative studies Lorenz had planned to conduct at Slimbridge (an idea suggested to him by Peter Scott). He allowed that he himself still intended to conduct studies at Slimbridge on a seasonal basis. On the timing of the Bristol application to the Nature Conservancy, see Tinbergen to Thorpe, 9 November 1950, WHTP/Cambridge.

95. Koehler to Holst, 21 December 1950; Holst to Koehler, 6 January 1951, EHP/ MPG-Berlin; Lorenz to Huxley, 21 December 1950, JSHP/Rice.

96. Lorenz to Thorpe, 31 December 1950, WHTP/Cambridge. Lorenz noted in this letter how his wife appeared younger and happier, and Margarethe Lorenz in turn reported to the Thorpes in April 1951 that "the whole change in our life seems to have a very good influence on Konrad and so on the whole family." This undated letter seems to have been sent with a letter from Lorenz dated Slimbridge, 28 April 1951; WHTP/ Cambridge.

97. See Lorenz to Huxley, 10 January 1951, JSHP/Rice; and Lorenz to Thorpe, 16 February 1951, WHTP/Cambridge.

98. See Holst to Lorenz, 24 June 1952; Lorenz to Holst and Kramer, 20 July 1952; and Lorenz's report (undated) to the Max-Planck-Gesellschaft's Generalverwaltung; EHP/ MPG-Berlin. The Holst Papers and the "Verhaltensphysiologie, allgemein (1)" file at the MPG Archive abounds with information on the difficulties between the Lorenz group and the new Baron Romberg. Helpful sketches of the early setup at Buldern are to be found in Alec Nisbett, *Konrad Lorenz* (New York: Harcourt Brace Jovanovich, 1976); and Irenäus Eibl-Eibesfeldt, " 'Fishy, fishy, fishy': autobiographical sketches," in Dewsbury, *Studying Animal Behavior,* pp. 69–91.

99. Edward Chace Tolman, *Purposive Behavior in Animals and Men* (New York: Century Co., 1932), p. xii.

100. See T. C. Schneirla, "Contemporary American animal psychology in perspective," in *Twentieth Century Psychology,* ed. P. L. Harriman (New York: Philosophical Library, 1946), pp. 306–316. On the distinctiveness of the American comparative psychol-

ogists' research focus, as compared with that of students of animal behavior in other countries, see R. W. Burkhardt, "Le comportement animal et la biologie française (animal behavior and French biology), 1920–1950," in *Les Sciences biologiques et médicales en France, 1920–1950,* ed. Claude Debru, Jean Gayon, and Jean-François Picard (Paris: CNRS Éditions, 1994), pp. 99–111.

101. T. C. Schneirla, "Contemporary American animal psychology in perspective"; F. A. Beach, "The snark was a boojum," *American Psychologist* 5 (1950): 115–124. For an analysis of the kinds of research undertaken in what Schneirla and Beach viewed as the golden age of American comparative psychology, see Richard W. Burkhardt, Jr., "The *Journal of Animal Behavior* and the early history of animal behavior studies in America," *Journal of Comparative Psychology* 101 (1987): 223–230.

102. Frank A. Beach, "Frank A. Beach," *History of Psychology in Autobiography* 6 (1974): 33–58, quote on p. 48. For more on Beach, see Donald A. Dewsbury, "Frank Ambrose Beach," *Biographical Memoirs of the National Academy of Sciences* 73 (1998): 64–85.

103. Gregg Mitman and Richard W. Burkhardt, Jr., "Struggling for identity: the study of animal behavior in America, 1930–1945," in *The Expansion of American Biology,* ed. Keith R. Benson, Jane Maienschein, and Ronald Rainger (New Brunswick, NJ: Rutgers University Press, 1991), pp. 164–194.

104. On Schneirla, see Lester R. Aronson et al., eds., *Development and Evolution of Behavior: Essays in Memory of T. C. Schneirla* (San Francisco: W. H. Freeman, 1970); and *Selected Writings of T. C. Schneirla,* ed. Lester R. Aronson et al. (San Francisco: W. H. Freeman, 1972). The textbook in question, N. R. F. Maier and T. C. Schneirla, *Principles of Animal Psychology* (New York: McGraw-Hill, 1935), promoted a strongly comparative approach to the study of animal behavior. The text was updated in a Dover edition of 1964. This updated version included critical comments on modern ethology's handling of issues of ontogeny.

105. "Objectives of the research program of the Department of Animal Behavior," ms. typescript with the annotation by hand "*Final copy:* 11/8/43," T. C. Schneirla Papers, Archives of the History of American Psychology, University of Akron.

106. Frank Beach to C. R. Carpenter, 30 September 1946. A copy of this letter is in the Robert M. Yerkes Papers, Yale University.

107. T. C. Schneirla to R. M. Yerkes, 8 March 1944, Yerkes Papers, Yale University.

108. T. C. Schneirla, "Contemporary American psychology in perspective."

109. "Provisional sketch of research program on general subject: studies on learning capacity and adaptive behavior as factors in relation to problems of heredity and evolution," ms. document dated 21 April 1947, folder M579, "Animal Behavior Department," Schneirla Papers, University of Akron.

110. Schneirla to Yerkes, 21 September 1950, Yerkes Papers, Yale University.

111. The information about Lehrman's family life and his relations with Noble comes from a letter from the invertebrate zoologist Libbie Hyman to W. C. Allee, 13 December 1939, in the W. C. Allee Papers, University of Chicago. This letter describes Lehrman's relations with Noble and mentions also that Lehrman was corresponding regularly with Tinbergen. I am indebted to Gregg Mitman for providing me with a copy of this letter. On Libbie Hyman and her own experiences with Noble see Judith E. Win-

ston, "Libbie Hyman and the American Museum of Natural History," *American Museum Novitates,* no. 3277 (12 November 1999): 12–25.

112. On Rand's critique of Lorenz, see above, chapter 6.

113. Daniel S. Lehrman, "Comparative behavior studies," *Bird-Banding* 12 (1941): 86–87. The review was of Lorenz, "Vergleichende Verhaltensforschung," *Verhandlungen der Deutschen Zoologischen Gesellschaft, Zoologische Anzeiger,* supp. 12 (1939): 69–102.

114. On Schneirla's politics, see Ethel Tobach and Lester R. Aronson, "T. C. Schneirla: a biographical note," in Aronson et al., *Development and Evolution of Behavior,* pp. xi–xviii.

115. Tinbergen to Mayr, 17 April 1953, copy courtesy of Ernst Mayr. Tinbergen was responding to a letter from Mayr dated 30 March 1953, in which Mayr mentioned a personal reason why Schneirla might have been unhappy with the European ethologists: "Our friend S. feels very much hurt about the neglect of his publications, as he said amazingly frankly after a seminar given by Gustav Kramer. The results are all sorts of misunderstandings and misinterpretations which I believe could be terminated only if someone of the European school was actually here in this country." Copy courtesy of Ernst Mayr.

CHAPTER EIGHT

1. The journals that did most to define the field in the 1950s were *Behaviour* (beginning publication in 1948), the *Zeitschrift für Tierpsychologie* (resuming publication in 1948 after a four-and-a-half-year hiatus), and the *British Journal of Animal Behaviour* (replacing in 1953 the earlier *Bulletin of Animal Behaviour*).

2. N. Tinbergen, *The Study of Instinct* (Oxford: Clarendon Press, 1951). Lorenz had been planning to write an introduction to "comparative behavior study" ever since the 1930s, but he had still not brought a text to completion. Had he done so, his book would have differed appreciably in style from Tinbergen's, though its basic view of instinctive behavior would certainly have been much the same.

3. Tinbergen, *The Study of Instinct* (1951), p. xiii.

4. Tinbergen, *The Study of Instinct* (1951), p. 101. Tinbergen noted that additional external stimuli might play a role in raising levels of motivation.

5. Tinbergen, *The Study of Instinct* (1951), p. 107.

6. Tinbergen, *The Study of Instinct* (1951), pp. 101, 127.

7. Tinbergen, *The Study of Instinct* (1951), pp. 128, 184, 185.

8. W. H. Thorpe interview with Richard W. Burkhardt, Jr., 28 February 1979.

9. W. H. Thorpe, *The Origins and Rise of Ethology* (London: Heinemann, 1979), p. 83.

10. Holst's and Mittelstaedt's idea was well beyond the "half-laid" stage when Holst presented it. It appeared in print later the same year in the prestigious German scientific journal *Die Wissenschaften.* See E. von Holst and H. Mittelstaedt, "Das Reafferenzprincip," *Die Naturwissenschaften* 37 (1950): 464–476.

11. Inge von Keiser, "Symposium über Verhaltensphysiologie im Max-Planck-Institut für Meeresbiologie in Wilhelmshaven," *Experentia* 6 (1950): 400–402.

12. On animal behavior studies in France, see R. W. Burkhardt, "Le comportement animal et la biologie française (animal behavior and French biology), 1920–1950," in *Les Sciences biologiques et médicales en France, 1920–1950,* ed. Claude Debru, Jean Gayon,

and Jean-François Picard (Paris: CNRS Éditions, 1994), pp. 99–111; and Philippe Chavot, "Histoire de l'éthologie: recherches sur le développement des sciences du comportement en Allemagne, Grande-Bretagne et France, de 1930 à nos jours" (thèse de doctorat en histoire des sciences, Université Louis Pasteur—Strasbourg I, 1994).

13. More precisely, Hinde's studies of mobbing in the chaffinch addressed the waning of responsiveness in the chaffinch in terms of a decrement in "specific action potential." He thus followed the decision of the participants at the 1949 Cambridge conference to use that neutral phrase instead of Lorenz's "action specific energy" or "reaction specific energy." See R. A. Hinde, "Factors governing the changes in strength of a partially inborn response, as shown by the mobbing behaviour of the chaffinch," pt. 1, "The nature of the response, and an examination of its course," *Proceedings of the Royal Society of London, Series B* 142 (1954): 306–331, and "Factors governing the changes in strength of a partially inborn response, as shown by the mobbing behaviour of the chaffinch," pt. 2, "The waning of the response," *Proceedings of the Royal Society of London, Series B* 142 (1954): 331–358, see especially p. 354. A third installment of this study appeared in 1961.

14. Lorenz to Thorpe, 30 April 1952, William Homan Thorpe Papers, Cambridge University Library (hereafter WHTP/Cambridge), MS.Add.8784.

15. For some of Hinde's earliest critical comments on Lorenz's and Tinbergen's models, see R. A. Hinde, "Changes in responsiveness to a constant stimulus," *British Journal of Animal Behaviour* 2 (1954): 41–55, especially pp. 43–44.

16. Lorenz to Thorpe, 30 April 1952, WHTP/Cambridge.

17. W. S. Verplanck, "Newsletters: late winter and spring, 1953" (see above, chapter 7, n. 15), quotes from pp. 34–35. Made available to me by Donald Dewsbury.

18. Verplanck, "Newsletters," pp. 5–6. Verplanck additionally records that he got his first view of Lorenz at one of the Oxford colleges where Lorenz and C. S. Lewis debated the topic of "instinct and ethics." In the audience were J. B. S. Haldane and his wife, the latter whom Verplanck described as "a tense (and I mean *really* tense) character."

19. Verplanck, "Newsletters," p. 9.

20. Most of the seventeen participants from Oxford were from Tinbergen's research group.

21. M. Bastock, D. Morris, and M. Moynihan, "Some comments on conflict and thwarting in animals," *Behaviour* 6 (1953): 66–84.

22. Hinde, "Changes in responsiveness to a constant stimulus," p. 44. Marga Vicedo, in a forthcoming paper entitled "Ethology comes to America: facts, semantics, and politics in the study of animal behavior," cites a letter from Tinbergen to Hinde dated 1 November 1953 in which Tinbergen suggested to Hinde that the latter's critique of the ethologists' "provisional" models might be presented in a more constructive fashion than Hinde was doing.

23. Robert A. Hinde, "Ethology in relation to other disciplines," in *Studying Animal Behavior: Autobiographies of the Founders,* ed. Donald A. Dewsbury (Chicago: University of Chicago Press, 1989), p. 194.

24. R. A. Hinde, "The behaviour of the great tit *(Parus major)* and some other related species," *Behaviour,* supp. 2 (1952): 79. See also p. 74.

25. R. A. Hinde, "Ethological models and the concept of 'drive,'" *British Journal for the Philosophy of Science* 6 (1956): 321–331, quote on p. 330.

26. Letter from N. Tinbergen to Desmond Morris, undated. I am grateful to Desmond Morris for providing me a copy of this document.

27. This account is based on Desmond Morris, *Animal Days* (New York: William Morrow, 1980), pp. 78–80.

28. D. O. Hebb, "Heredity and environment in mammalian behaviour," *British Journal of Animal Behaviour* 1 (1953): 43–47, quotes from pp. 46 and 47.

29. J. S. Kennedy, "Is modern ethology objective?" *British Journal of Animal Behaviour* 2 (1954): 12–19.

30. Regarding Kennedy's critique, Lorenz simply allowed that if the distinction he had drawn between endogenous generation of stimuli and afferent control made him a dualist and hence a vitalist, this did not bother him because it placed him in the company of neurophysiologists like Adrian, Lashley, and von Holst. See Lorenz, "The objectivistic theory of instinct," in *L'Instinct dans le comportement des animaux et de l'homme*, by M. Autuori et al. (Paris: Masson, 1956), pp. 60–61.

31. Daniel S. Lehrman, "A critique of Konrad Lorenz's theory of instinctive behavior," *Quarterly Review of Biology* 298 (1953): 337–363 (henceforth cited as Lehrman, "Critique").

32. On Lehrman's early draft of his critique of Lorenz, and his subsequent toning down of that critique, see especially Jay S. Rosenblatt, "Daniel Sanford Lehrman, June 1, 1919–August 27, 1972," *Biographical Memoirs of the National Academy of Sciences* 66 (1995): 227–246. Rosenblatt identifies Karl Lashley, Hans-Lukas Teuber, and Donald O. Hebb as individuals who advised Lehrman to tone down the political side of his critique. Gerard P. Baerends identifies Frank Beach as someone else who functioned in this role. See Baerends, "Turning points in the history of ethology," in *L'Histoire de la connaissance du comportement animal*, ed. Liliane Bodson (Liège: Université de Liège, 1993), pp. 59–76, on p. 67. See also on this subject Rae Silver and Jay S. Rosenblatt, "The development of a developmentalist: Daniel S. Lehrman," *Developmental Psychobiology* 20 (1987): 563–570. Marga Vicedo reports that the editors of the *Quarterly Review of Biology* advised Lehrman that the conclusion of the paper was too emotional. See Vicedo, "Ethology comes to America." I have not succeeded in finding a copy of the original draft of Lehrman's paper.

33. Allee to Frisch, 7 July 1949, cited in Ute Deichmann, *Biologists under Hitler*, trans. Thomas Dunlap (Cambridge, MA: Harvard University Press, 1996), p. 199.

34. Z. Y. Kuo, "Ontogeny of embryonic behavior in Aves," *Journal of Experimental Zoology* 61 (1932): 395–430; *Journal of Experimental Zoology* 62 (1932): 453–489; *Journal of Comparative Pyschology* 13 (1932): 245–272; 14 (1932): 109–122.

35. Lehrman, "Critique," p. 340.

36. Lehrman, "Critique," p. 348.

37. Lehrman, "Critique," p. 347.

38. Lehrman, "Critique," p. 351.

39. Lehrman, "Critique," p. 359.

40. Lehrman, "Critique," p. 354.

41. A letter from Schneirla to Beach dated 16 February 1954 noted that Lehrman's critique of Lorenz was "out in the current number of the Quarterly Review of Biology."

T. C. Schneirla Papers, Archives of the History of American Psychology, University of Akron.

42. Alfred E. Emerson to T. C. Schneirla, 13 August 1954, M578, folder Em-Fl, Schneirla Papers, University of Akron.

43. The claim that Lorenz was furious is based on Lorenz's initial public response to Lehrman, which took place at the Paris instinct conference in June 1954 (on which see below). The information regarding Koehler and Tinbergen comes from a letter from Koehler to Tinbergen dated 26 August 1954 and a letter in reply from Tinbergen dated 29 August 1954, Otto Koehler Papers, Universitätsarchiv Freiburg im Breisgau (hereafter OKP/Freiburg).

44. Tinbergen to Koehler, 17 August 1954, OKP/Freiburg; Tinbergen to Mayr, 23 March 1954, copy courtesy of Ernst Mayr.

45. A letter from Beach to Schneirla, dated 11 February 1954, indicates that Beach had initially invited Baerends and Hinde, but Hinde declined because his wife was going to have a baby at about the time of the conference. Schneirla Papers, University of Akron.

46. Gerard P. Baerends, "Two pillars of wisdom," in Dewsbury, *Studying Animal Behavior*, p. 29.

47. Haldane eventually took Lorenz to task for the latter's politics. See J. B. S. Haldane, "The argument from animals to men: an examination of its validity for anthropology, *Journal of the Royal Anthropological Institute of Great Britain and Ireland* 86, pt. 2 (1956): 1–14. Haldane also criticized Lorenz for not giving due credit to intellectual precursors, among others William McDougall, in Haldane, "The sources of some ethological notions," *British Journal of Animal Behaviour* 4 (1956): 162–164. Paul Griffiths examines further Haldane's scientific criticisms of Lorenz in "Instinct in the 50s," *Biology and Philosophy*, vol. 19 (2004; in press). For Spurway's published comments on Lorenz's ideas, see Helen Spurway, "Behold, my child, the Nordic dog," review of *King Solomon's Ring*, by Konrad Z. Lorenz, *British Journal for the Philosophy of Science* 3 (1952): 265–272, and "The causes of domestication: an attempt to integrate some ideas of Konrad Lorenz with evolution theory," *Journal of Genetics* 53 (1955): 325–362. The Lorenz-Spurway affair is described in Klaus Taschwer and Benedikt Föger, *Konrad Lorenz: Biographie* (Vienna: Paul Zolnay Verlag, 2003), pp. 166–168.

48. Lorenz, "The objectivistic theory of instinct," p. 51.

49. Lorenz, "The objectivistic theory of instinct," p. 56.

50. Lorenz, "The objectivistic theory of instinct," p. 59.

51. Lorenz, "The objectivistic theory of instinct," p. 60.

52. T. C. Schneirla, "Discussion," in Autuori et al., *L'Instinct dans le comportement des animaux et de l'homme*, pp. 67–70.

53. D. Lehrman, "Discussion," in Autuori et al., *L'Instinct dans le comportement des animaux et de l'homme*, p. 72.

54. K. Lorenz, "Discussion," in Autuori et al., *L'Instinct dans le comportement des animaux et de l'homme*, p. 74.

55. Schneirla, "Discussion," p. 70.

56. Henri Piéron, "Conclusions," in Autuori et al., *L'Instinct dans le comportement des animaux et de l'homme*, pp. 785–789.

57. Tinbergen to Koehler, 17 August 1954 (in German), OKP/Freiburg.

58. Koehler to Tinbergen, 21 August 1954 (in German), OKP/Freiburg. See too Otto Koehler, review of "A critique of Konrad Lorenz's theory of instinctive behaviour," by Daniel S. Lehrman, *Zeitschrift für Tierpsychologie* 11 (1954): 330–334.

59. Koehler to Tinbergen, 21 August 1954 and 26 August 1954 (both in German), OKP/Freiburg.

60. Tinbergen to Koehler, 23 August 1954 (in German), OKP/Freiburg.

61. Tinbergen to Koehler, 23 August 1954 (in German), OKP/Freiburg.

62. Koehler to Tinbergen, 26 August 1954 (in German), OKP/Freiburg.

63. Tinbergen to Koehler, 29 August 1954 (in German), OKP/Freiburg.

64. Koehler to Tinbergen, 1 September 1954 (in German), OKP/Freiburg. Koehler's review of Daniel S. Lehrman, "A critique of Konrad Lorenz's theory of instinctive behavior," appeared in *Zeitschrift für Tierpsychologie* 11 (1954): 330–334. There he allowed that it was Lehrman, not Lorenz, who was dogmatic, and further that Lehrman could "only maintain his position by concealing or misunderstanding the whole of our factual material."

65. Bertram Schaffner, ed., *Group Processes: Transactions of the First Conference, September 26, 27, 28, 29, and 30, 1954, Ithaca, New York* (New York: Josiah Macy, Jr. Foundation, 1955), pp. 300, 310, 311.

66. N. Tinbergen, "Psychology and ethology as supplementary parts of a science of behavior," in Schaffner, *Group Processes,* pp. 75–167.

67. Tinbergen, "Psychology and ethology as supplementary parts of a science of behavior," p. 77.

68. Tinbergen, "Psychology and ethology as supplementary parts of a science of behavior," pp. 79–85.

69. Tinbergen, "Psychology and ethology as supplementary parts of a science of behavior," p. 86.

70. Tinbergen, "Psychology and ethology as supplementary parts of a science of behavior," p. 107.

71. Tinbergen, "Psychology and ethology as supplementary parts of a science of behavior," p. 115.

72. Lorenz to Holst, 24 October 1954 (in German), Erich von Holst Papers, Archive of the Max-Planck-Gesellschaft, Berlin.

73. Lorenz to Thorpe, 11 March 1955, WHTP/Cambridge.

74. Niko Tinbergen to R. W. Burkhardt, 16 June 1982, and 19 June 1982, Nikolaas Tinbergen Papers, Bodleian Library, Oxford University.

75. Although the Cambridge conference followed the Fourth International Congress, held in Freiburg, Thorpe chose to identify the 1959 Cambridge conference as the *Sixth* International Ethological Congress, thereby crediting either the 1950 Wilhelmshaven roundtable or, more likely, the 1949 Cambridge conference on physiological mechanisms in behavior as the *first* international ethological congress. Successive conferences after the 1959 Cambridge conference followed this renumbering. I thank W. M. S. Russell for first calling my attention to this renumbering.

76. Donald A. Dewsbury, "Americans in Europe: the role of travel in the spread of European ethology after World War II," *Animal Behaviour* 49 (1995): 1649–1663.

77. Tinbergen to Koehler, 28 December 1960 (in German), OKP/Freiburg.

78. The organizing committee for the Stockholm congress consisted of Lorenz, Tinbergen, Baerends, Koehler, van Iersel, and three Americans: Beach, Lehrman, and Roeder.

79. C. G. Beer, "Was Professor Lehrman an ethologist?" *Animal Behaviour* 23 (1975): 957–964. This paper was a memorial lecture given for Daniel Lehrman at the Thirteenth International Ethological Congress, held in Washington, DC, in 1973, the year after Lehrman's death.

80. Lorenz to Tinbergen, 20 November 1967, Konrad Z. Lorenz Papers, Konrad-Lorenz-Institut für Evolutions- und Kognitionsforschung, Altenberg, Austria. Notwithstanding the sentiments expressed at the 1967 congress, Lehrman ultimately concluded that his differences with Lorenz continued to run deep. See Daniel S. Lehrman, "Semantic and conceptual issues in the nature-nurture problem," in *Development and Evolution of Behavior: Essays in Memory of T. C. Schneirla*, ed. Lester Aronson et al. (San Francisco: W. H. Freeman, 1970), pp. 17–52.

81. N. Tinbergen, "5. Internationaler Ethologenkongreß," *Zeitschrift für Tierpsychologie* 14 (1957): 377–380.

82. K. D. Roeder, "VII. Internationale Ethologenkongreß," *Zeitschrift für Tierpsychologie* 18 (1961): 491–494. Roeder playfully identified the widely different types of investigators studying behavior. They consisted of "the ethologist who makes himself uncomfortable in his blind, the psychologist who makes the animal uncomfortable by putting it in the blind, the 'optomotor boys' who make both themselves and the animal uncomfortable by placing themselves in the feedback loop, and the physiologists who climb right inside and mess the whole business up."

83. K. Lorenz, "VII. Internationaler Ethologenkongreß," *Zeitschrift für Tierpsychologie* 18 (1961): 495–497.

CHAPTER NINE

1. N. Tinbergen, "An objectivistic study of the innate behaviour of animals," *Bibliotheca Biotheoretica* 1 (1942): 40, 44.

2. For the last of these studies, see N. Tinbergen and A. C. Perdeck, "On the stimulus situation releasing the begging response in the newly hatched herring gull chick (*Larus argentatus argentatus* Pont.)," *Behaviour* 3 (1950): 1–39. This study described the results of experimental work begun in Holland after the war and pursued for three successive seasons and that Perdeck then continued in 1949.

3. N. Tinbergen, "Recent advances in the study of bird behaviour," in *Proceedings of the Xth International Ornithological Congress, Uppsala, 1950* (1951), pp. 368–369.

4. See N. Tinbergen, "A note on the origin and evolution of threat display," *Ibis* 94 (1952): 160–162; N. Tinbergen and M. Moynihan, "Head flagging in the black-headed gull: its function and origin," *British Birds* 45 (1952): 19–22; J. J. A. Van Iersel, "An analysis of the parental behaviour of the male three-spined stickleback (*Gasterosteus aculeatus* L.)," *Behaviour*, supp. 3 (1953).

5. Letter from Tinbergen to R. W. Burkhardt, 20 May 1979. Tinbergen went on to say in this regard: "I have always been annoyed by those physiologists who thought that these higher levels of integration were less accessible to analysis than the activity of, say,

a single neurone; my concern in those early days was to show that one's method of analysis had to be the same at every level and that 'exactness' of such analysis did not depend at all on the smallness of the system studied, but on discipline of thinking."

6. N. Tinbergen, "'Derived' activities: their causation, biological significance, origin, and emancipation during evolution," *Quarterly Review of Biology* 27 (1952): 1–32, and *The Herring Gull's World* (London: Collins, 1953). On Tinbergen's intention of refocusing his activities, see Mayr to Tinbergen, 6 November 1950; and Tinbergen to Mayr, 25 November 1950, Ernst Mayr Papers, Harvard University Archives (hereafter EMP/Harvard), HUG (FP) 14.7, box 8, folder 372. Tinbergen was aware that it would take a few years to refocus his interests, and this proved to be the case. At the first international ethological congress, at Buldern in 1952, he gave a paper on a theme he had been working on for some time: "Fighting and threat and their relations to courtship behaviour." It was not until the Second International Ethological Congress, held in Oxford in 1953, that Tinbergen spoke on comparative studies of the behaviour of gulls. He made his comment about his efforts "to colonize" England in a letter to Koehler dated 24 October 1952.

7. For different entries into this literature, see, for example, Eckhard H. Hess, "Ethology: an approach toward the complete analysis of behavior," in *New Directions in Psychology I,* by Roger Brown, Eugen Galanter, Eckhard H. Hess, and George Mandler (New York: Holt Rinehart and Winston, 1961), pp. 157–266; Colin G. Beer, "Ethology—the zoologist's approach to behaviour," pts. 1 and 2, *Tuatara* 11 (1963): 170–177; 12 (1964): 16–39; Gordon M. Burghardt, "Instinct and behavior: toward an ethological psychology," in *The Study of Behavior: Learning, Motivation, Emotion, and Instinct,* ed. John A. Nevin (Glenview, Ill.: Scott, Foresman and Company, 1973), pp. 323–400; and Robert A. Hinde, *Ethology: Its Nature and Relations with Other Sciences* (New York: Oxford University Press, 1982).

8. For an excellent discussion of Lorenz's activities and success as a public figure see Klaus Taschwer and Benedikt Föger, *Konrad Lorenz: Biographie* (Vienna: Paul Zolnay Verlag, 2003).

9. Tinbergen to Mayr, 9 November 1949, EMP/Harvard, HUG (FP) 14.7, box 7, folder 332. Mayr, like Konrad Lorenz and Jan Verwey, had been urging Tinbergen to do comparative work.

10. See Tinbergen to Mayr, 25 November 1950, EMP/Harvard, HUG (FP) 14.7, box 8, folder 372.

11. Tinbergen to Mayr, 4 September 1950, EMP/Harvard, HUG (FP) 14.7, box 8, folder 372; see also Tinbergen to Thorpe, 8 September 1976, William Homan Thorpe Papers, Cambridge University Library, MS.Add.8784.

12. In addition, Tinbergen set one of his new graduate students, Margaret Bastock, to work on the behavior of the fruit fly *Drosophila melanogaster.*

13. See Desmond Morris, *Animal Days* (New York: William Morrow, 1980), pp. 68–73.

14. On the stickleback work, see especially Van Iersel, "An analysis of the parental behaviour of the male three-spined stickleback"; D. Morris, "Homosexuality in the ten-spined stickleback (*Pygosteus pungitius* L.)," *Behaviour* 4 (1952): 233–261, and "The reproductive behaviour of the ten-spined stickleback (*Pygosteus pungitius* L.)," *Behaviour,*

supp. 6 (1958); R. Hoogland, D. Morris, and N. Tinbergen, "The spines of sticklebacks (*Gasterosteus* and *Pygosteus*) as means of defence against predators (*Perca* and *Esox*)," *Behaviour* 10 (1957): 205–236. In contrast to his feeling that he was most himself while doing research in the field, Tinbergen felt that "only half of me" was displayed in the lab or in the seminars he held at his home. On this score see Tinbergen to Desmond Morris, 4 July 1979, Nikolaas Tinbergen Papers, Bodleian Library, Oxford University (hereafter NTP/Oxford).

15. W. S. Verplanck, manuscript "Newsletters: late winter and spring, 1953" (see above, chapter 7, n. 15), p. 28. Made available to me by Donald Dewsbury. By Verplanck's account, the other "sore points" among Tinbergen's students were Tinbergen's "devout belief in the unlearned nature of everything important, and his theory about hierarchies," and his habit of telling his students how little he had been paid and how long his work day had been, relative to theirs, when he was a student.

16. N. Tinbergen, *Curious Naturalists* (London: Country Life, 1958), p. 242.

17. Tinbergen, *Curious Naturalists*, p. 228.

18. See Tinbergen, *Curious Naturalists*, for a general account of this work.

19. Tinbergen to Mayr, 20 July 1952, EMP/Harvard, HUG (FP) 14.7, box 11, folder 479, and 17 April 1953, copy courtesy of Ernst Mayr. On 27 March 1953, Tinbergen wrote to Mayr about the need to look not only at widely aberrant species (Mayr's recommendation) but also at very closely related species: "The more we analyse the threat and courtship displays, the more we become convinced that they are all complexes, and that to homologise them as wholes (i.e. to say that they are derived historically from the same thing in a common ancestor) is impossible: in one species an attitude roughly similar to that in another species may consist of element[s] a, b, and c; in the other species it may be built up of elements b, c, and d." Copy courtesy of Ernst Mayr.

20. Tinbergen to Mayr, 20 July 1952, EMP/Harvard, HUG (FP) 14.7, box 11, folder 479.

21. Tinbergen to Huxley, 10 June 1953, Julian S. Huxley Papers, Rice University Archives (hereafter JSHP/Rice).

22. Verplanck, "Newsletters," pp. 18–19.

23. See Tinbergen, *Curious Naturalists;* and Esther Cullen, "Adaptations in the kittiwake to cliff-nesting," *Ibis* 99 (1957): 275–302.

24. Verplanck in his manuscript "Newsletters" wrote home from the Inner Farne in 1953, "One learns some amusing things—that for example, Nikko manipulates his field parties—Mike and Esther are posted together, because he hopes that they will emulate the birds, as a couple did last year." Verplanck, "Newsletters," p. 20.

25. Tinbergen to Huxley, 11 November 1955, JSHP/Rice.

26. Cullen, "Adaptations in the kittiwake to cliff-nesting," p. 299.

27. N. Tinbergen, "Behaviour, systematics, and natural selection," *Ibis* 101 (1959): 318–330, quote and table on p. 324. This paper was subsequently published as Tinbergen's contribution to the Darwin centenary held at the University of Chicago in 1959. See Tinbergen, "Behaviour, systematics, and natural selection," in *Evolution after Darwin*, ed. Sol Tax, vol. 1, *The Evolution of Life* (Chicago: University of Chicago Press, 1960), pp. 595–613.

28. Tinbergen, "Behaviour, systematics, and natural selection" (*Ibis*), p. 326.

29. Tinbergen, "Behaviour, systematics, and natural selection" (*Ibis*), p. 328.

30. N. Tinbergen, "Comparative studies of the behaviour of gulls (Laridae): a progress report," *Behaviour* 15 (1959): 1–70. Tinbergen was not the only biologist thinking about behavioral convergence in the 1950s. Peter Marler pointed out the circumstances under which the warning calls of small birds had come to display striking convergences. See P. Marler, "The characteristics of some animal calls," *Nature* 176 (1955): 6.

31. R. A. Hinde and N. Tinbergen, "The comparative study of species-specific behavior," in *Behavior and Evolution,* ed. A. Roe and G. G. Simpson (New Haven, CT: Yale University Press, 1958), pp. 251–268, quote on p. 265.

32. N. Tinbergen, "On the functions of territory in gulls," *Ibis* 98 (1956): 400–411.

33. N. Tinbergen, "The behavior of the snow bunting in spring," *Transactions of the Linnaean Society of New York* 5 (1939): 1–94. He also discussed "function" in his 1942 "objectivistic study" paper. There, however, his stress was on the importance of not confusing questions of function with questions of causation.

34. See Tinbergen, *Curious Naturalists,* chapter 7; and L. de Ruiter, "Some experiments on the camouflage of stick caterpillars," *Behaviour* 4 (1952): 222–232, and "Countershading in caterpillars: an analysis of its adaptive significance," *Archives néerlandaises de zoologie* 11 (1956): 285–341.

35. A. D. Blest, "The function of eyespot patterns in the Lepidoptera," *Behaviour* 11 (1957): 209–256.

36. Interestingly enough, although Kettlewell's studies of the different effects of predation on the two forms of the peppered moth, *Biston betularia,* came to be hailed as the revelation of a particularly dramatic example of evolution in action, what Tinbergen later remembered most fondly about this work with Kettlewell were the simple joys of being out in the field as a field naturalist. See Tinbergen, "Happy moments with Bernard Kettlewell," C.161, NTP/Oxford.

37. Hoogland, Morris, and Tinbergen, "The spines of sticklebacks (*Gasterosteus* and *Pygosteus*) as means of defense against predators (*Perca* and *Esox*)."

38. Tinbergen to Huxley, 13 June 1958, JSHP/Rice.

39. Tinbergen to Huxley, 27 June 1958, JSHP/Rice. The special funding for which Huxley gave Tinbergen reason to hope did not materialize. Tinbergen nonetheless was able to proceed with his plans, thanks to his existing funding from the Nuffield and Ford foundations and the American air force. On this score see Tinbergen to Huxley, 13 June 1958, JSHP/Rice.

40. Colin Beer, "Incubation and nest-building by the black-headed gull," pt. 1, "Incubation behaviour in the incubation period." *Behaviour* 18 (1961): 62–106.

41. Tinbergen to Mayr, 29 April, 1960, copy courtesy of Ernst Mayr.

42. Tinbergen to Huxley, 19 April 1960, JSHP/Rice.

43. N. Tinbergen, G. J. Broekuysen, F. Feekes, J. C. W. Houghton, H. Kruuk, and E. Szulc, "Egg shell removal by the black-headed gull, *Larus ridibundus* L.: a behaviour component of camouflage," *Behaviour* 19 (1962): 74–117, reprinted in *The Animal in Its World: Explorations of an Ethologist, 1932–1972,* by Niko Tinbergen, vol. 1, *Field Studies* (London: George Allen and Unwin, 1972), pp. 250–294, quote from p. 287.

44. Tinbergen et al., "Egg shell removal by the black-headed gull, *Larus ridibundus* L.: a behaviour component of camouflage," pp. 288–291.

45. Tinbergen et al., "Egg shell removal by the black-headed gull, *Larus ridibundus* L.: a behaviour component of camouflage," pp. 291–292.

46. Tinbergen to Huxley, 26 March 1961, JSHP/Rice.

47. Tinbergen to Mayr, 23 November 1962, copy courtesy of Ernst Mayr.

48. Tinbergen to Mayr, 17 April 1963, copy courtesy of Ernst Mayr.

49. Tinbergen to Huxley, 26 January 1963, JSHP/Rice; also Tinbergen to Mayr, 17 April 1963, copy courtesy of Ernst Mayr.

50. Tinbergen to Mayr, 17 April 1963, copy courtesy of Ernst Mayr.

51. See letters from Tinbergen to Koehler dated 1 October 1962, 6 February 1963, and 17 February 1963, Otto Koehler Papers, Universitätsarchiv Freiburg im Breisgau.

52. Niko Tinbergen, "On aims and methods of ethology," *Zeitschrift für Tierpsychologie* 20 (1963): 410–433.

53. N. Tinbergen, *The Study of Instinct* (Oxford: Clarendon Press, 1951), p. 2.

54. Jay S. Rosenblatt, "Daniel Sanford Lehrman, June 1, 1919–August 27, 1972," *Biographical Memoirs of the National Academy of Sciences* 66 (1995): 236.

55. Tinbergen, "On aims and methods of ethology," p. 424.

56. Konrad Lorenz, "Phylogenetische Anpassung und adaptive Modifikation des Verhaltens," *Zeitschrift für Tierpsychologie* 18 (1961): 139–187.

57. Konrad Lorenz, *Evolution and Modification of Behavior* (Chicago: University of Chicago Press, 1965).

58. Lorenz, *Evolution and Modification of Behavior*, p. 5.

59. Lorenz, "Phylogenetische Anpassung," p. 140, and *Evolution and Modification of Behavior*, pp. 7–9.

60. Lorenz, *Evolution and Modification of Behavior*, p. 13.

61. Lorenz, *Evolution and Modification of Behavior*, p. 18.

62. Lorenz, *Evolution and Modification of Behavior*, pp. 80–81.

63. Tinbergen to Lorenz, 31 January 1961, NTP/Oxford. Tinbergen began the letter with two and a half sentences in German, the second of which I have rendered into English in the above text. The original reads: "Und zum ersten Mal in unserem Leben sind wir wirklich verschiedener Meinung!"

64. Tinbergen to Mayr, 5 February 1977, copy courtesy of Ernst Mayr.

65. Tinbergen, "On aims and methods of ethology," pp. 410–411.

66. Tinbergen, "On aims and methods of ethology," pp. 413–416.

67. In 1965 Tinbergen expressed again his belief that ethology had a particular contribution to make to evolutionary theory. See N. Tinbergen, "Behavior and natural selection," in *Ideas in Modern Biology,* ed. John A. Moore (Garden City, NY: Natural History Press, 1965), pp. 519–542. There he criticized Wynne-Edwards's account of altruistic spacing out and concluded by saying (pp. 538–539): "Many ethologists are by tradition field observers, and they are thus in a good position to see the animal in its continuous struggle with the very complex environment that has molded them. I believe that the type of work I have been discussing might well lead to a better knowledge of the powers of natural selection in general, and to a higher regard for these powers with respect to all functions and structures as we find them in present-day animals."

68. Tinbergen, "On aims and methods of ethology," p. 430.

69. For a contemporary view of the state of ethology by an ethologist of the gener-

ation after Tinbergen, see the admirable analysis by Colin G. Beer, "Ethology—the zoologist's approach to behaviour." See also Beer, "Instinct," in *International Encyclopedia of the Social Sciences,* ed. David L. Sills (New York: Macmillan, 1968), 7:363–372.

70. A field of "cognitive ethology" has developed since the 1970s. Gordon Burghardt has recently suggested that to Tinbergen's four questions of ethology there should now be added a fifth, that of "private experience." See Gordon M. Burghardt, "Amending Tinbergen: a fifth aim for ethology," in *Anthropomorphism, Anecdotes, and Animals,* ed. Robert W. Mitchell, Nicholas S. Thompson, and H. Lyn Miles (Albany: State University of New York Press, 1997), pp. 254–276. In the same volume see also Colin Beer, "Expressions of mind in animal behavior," pp. 198–209.

71. Tinbergen to Koehler, 4 July 1952, Koehler Papers, Universitätsarchiv Freiburg im Breisgau.

72. Tinbergen to Huxley, 21 January 1959, JSHP/Rice.

73. Tinbergen's comments appeared in "Panel four: the evolution of the mind," in *Evolution after Darwin,* vol. 3, *Issues in Evolution,* ed. Sol Tax and Charles Callender (Chicago: University of Chicago Press, 1960), p. 185.

74. Huxley to Tinbergen, 27 March 1962, JSHP/Rice.

75. Tinbergen to Huxley, 20 March 1965, JSHP/Rice.

76. R. A. Hinde, "Nikolaas Tinbergen," *Biographical Memoirs of Fellows of the Royal Society* 36 (1990): 556–559.

77. Tinbergen to Huxley, 13 January 1965, JSHP/Rice.

78. Niko Tinbergen to Julian Huxley, 30 April 1965, JSHP/Rice.

79. Tinbergen to Mayr, 1 November 1965, copy courtesy of Ernst Mayr.

80. Tinbergen to Mayr, 17 February 1972, copy courtesy of Ernst Mayr.

81. John Krebs and Richard Dawkins, "Michael Cullen, 1927–2001: the unsung hero of a golden age in the development of understanding of animal behaviour," *Guardian,* 10 April 2001, reprinted in the *Animal Behavior Society Newsletter* 46, no. 3 (August 2001): 3–4.

82. Tinbergen to Mayr, 9 January 1968, copy courtesy of Ernst Mayr.

83. David Lack, "Preface by the president," in *Proceedings of the XIV International Ornithological Congress: Oxford, 24–30 July 1966,* ed. D. W. Snow (Oxford: Blackwell Scientific Publications, 1967), p. vii.

84. N. Tinbergen, "Adaptive features of the black-headed gull *Larus ridibundus* L.," in Snow, *Proceedings of the XIV International Ornithological Congress,* pp. 43–59.

85. N. Tinbergen, "Adaptive features of the black-headed gull," pp. 55–58.

86. Tinbergen to Mayr, 9 January 1968, copy courtesy of Ernst Mayr. Tinbergen wanted to know what Mayr thought about Wynne-Edwards's (and even Theodosius Dobzhansky's) inclination to dismiss certain selection pressures as "inconceivable" without ever analyzing and testing them. As Tinbergen put it, "When I then see at the same time how our field analyses reveal time and again the most unexpected pressures (e.g. those exerted by predators) I feel like shouting from the rooftops: 'For God's sake boys, analyse these pressures, and do apply the experimental method.'"

87. Mayr to Tinbergen, 23 January 1968, copy courtesy of Ernst Mayr.

88. Tinbergen to Mayr 25 January 1968, copy courtesy of Ernst Mayr.

89. N. Tinbergen, "On war and peace in animals and man: an ethologist's approach

to the biology of aggression," *Science* 160 (1968): 1411–1418. In his correspondence with Mayr the previous month, Tinbergen agreed with Mayr's assessment that Konrad Lorenz often wrote things with which one had to disagree, and that Lorenz furthermore was "always weakest when he talks about man." Mayr to Tinbergen, 23 January 1968; and Tinbergen to Mayr 25 January 1968, copies courtesy of Ernst Mayr.

90. N. Tinbergen, "On war and peace in animals and man," p. 1412.

91. N. Tinbergen, "On war and peace in animals and man," p. 1412.

92. Tinbergen, "On war and peace in animals and man," p. 1418.

93. Lorenz to Tinbergen, 16 February 1968 (in German), Konrad Z. Lorenz Papers, Konrad-Lorenz-Institut für Evolutions- und Kognitionsforschung, Altenberg, Austria (hereafter KZLP/KLI). In his personal copy of Hinde's book *Animal Behaviour: A Synthesis of Ethology and Comparative Psychology* (New York: McGraw-Hill, 1966), Lorenz annotated the title to read: "Animal Behaviour: A Castration of Ethology in an attempt to save 'Comparative' Psychology." The book remains in his library at Altenberg, now the site of the Konrad Lorenz Institute for Evolution and Cognition Research.

94. Tinbergen to Lorenz, 2 March 1968, KZLP/KLI.

95. N. Tinbergen, "Ethology," in *Scientific Thought, 1900–1960*, ed. R. Harré (Oxford: Clarendon Press, 1969), pp. 238–268.

96. See Tinbergen to Lorenz, letters of 23 May and 11 June 1968, the second of which refers to a letter from Lorenz to Tinbergen dated 28 May 1968. KZLP/KLI.

97. On Tinbergen's use of film, see Gregg Mitman, *Reel Nature: America's Romance with Wildlife on Film* (Cambridge, MA: Harvard University Press, 1999), pp. 80–83.

98. N. Tinbergen, "Introduction to 1969 reprint," in *The Study of Instinct: With a New Introduction* (Oxford: Clarendon Press, 1969), pp. v–xi. In a letter to Ernst Mayr of 5 September 1973, Tinbergen returned to the connection between studies of causation and concerns with control, noting that Americans seemed "rather power-oriented." This was exemplified, he felt, in American approaches to American work in animal psychology in the twentieth century, where there was a strong emphasis on molding development through conditioning. Copy courtesy of Ernst Mayr.

99. Tinbergen described this to Ernst Mayr as "a book mainly on 'approach,' with facts only as examples of methods." Tinbergen to Mayr, 26 October 1973, copy courtesy of Ernst Mayr.

100. See Karel H. Voous, "Report of the secretary-general," in *Proceedings of the XVth International Ornithological Congress, The Hague, The Netherlands, 30 August–5 September 1970* (Leiden: E. J. Brill, 1972), pp. 1–12.

101. Niko Tinbergen, "Functional ethology and the human sciences," *Proceedings of the Royal Society of London, Series B* 182 (1972): 385–410.

102. In Daniel S. Lehrman, "Semantic and conceptual issues in the nature-nurture problem," in *Development and Evolution of Behavior: Essays in Memory of T. C. Schneirla*, ed. Lester Aronson et al. (San Francisco: W. H. Freeman, 1970), pp. 17–52. This particular paper represented Lehrman's single most important "direct" contribution to the debate about innate versus acquired elements of behavior after his 1953 critique of Lorenz. In the article, he directed his attention to Lorenz's recent book, *Evolution and Modification of Behavior*.

103. Tinbergen to Mayr, 26 March 1972, copy courtesy of Ernst Mayr.

CHAPTER TEN

1. Konrad Lorenz, *On Aggression* (New York: Harcourt Brace and World, 1966), pp. 288–289.

2. John Bowlby, *Attachment and Loss*, vol. 1, *Attachment* (New York: Basic Books, 1969), pp. 6–7.

3. See Bowlby, *Attachment and Loss*, 1:xviii, 166–172. See too John Bowlby, "Ethological light on psychoanalytical problems," in *The Development and Integration of Behaviour: Essays in Honour of Robert Hinde*, ed. Patrick Bateson (Cambridge: Cambridge University Press, 1991), pp. 301–314; and Robert Hinde's comment in the same volume, "Commentary 4" (p. 411): "It was John Bowlby who helped me to set up the monkey colony at Madingley in order to study the effects of maternal deprivation experimentally— for me a stepping stone to developmental psychology."

4. See Robert C. Stauffer, "Ecology in the long manuscript version of Darwin's *Origin of Species* and Linnaeus' *Oeconomy of Nature*," *Proceedings of the American Philosophical Society* 104 (1960): 235–241; Thomas Robert Malthus, *An Essay on the Principle of Population, as It Affects the Future Improvement of Society* (London: J. Johnson, 1798); *Charles Darwin's Notebooks, 1836–1844*, ed. Paul H. Barrett, Peter J. Gautrey, Sandra Herbert, David Kohn, and Sydney Smith (Ithaca, NY: Cornell University Press, 1987), p. 539; and Charles Darwin, *The Descent of Man, and Selection in Relation to Sex*, 2 vols. (London: John Murray, 1871), 2:403.

5. On the views of many American biologists toward war, see especially Gregg Mitman, "Evolution as gospel: William Patten, the language of democracy, and the Great War," *Isis* 81 (1990): 446–463, "Dominance, leadership, and aggression: animal behavior studies during the Second World War," *Journal of the History of the Behavioral Sciences* 26 (1990): 3–16, and *The State of Nature: Ecology, Community, and American Social Thought, 1900–1950* (Chicago: University of Chicago Press, 1992). Mitman does not discuss Wallace Craig's paper "Why do animals fight," *International Journal of Ethics* 31 (1921): 264–278.

6. Spencer Weart, *Nuclear Fear: A History of Images* (Cambridge, MA: Harvard University Press, 1988), p. 256.

7. See also Konrad Lorenz, "Über das Töten von Artgenossen," *Jahrbuch der Max-Planck-Gesellschaft zur Förderung der Wissenschaften* (Göttingen, 1955), pp. 105–140.

8. Konrad Lorenz, *King Solomon's Ring* (London: Methuen, 1952), p. 199. Lorenz's original piece, entitled "Moral und Waffen der Tiere," appeared on 15 November 1935 in the *Neues Wiener Tagblatt*. I thank Klaus Taschwer for identifying the original source for me.

9. Lorenz, *King Solomon's Ring*, p. 197.

10. It may be remarked further that the bloody encounter Lorenz reported between his male turtledove and his female ringdove involved two different species, and thus was of questionable relevance to his broader discussion of whether animals kill members of their own species.

11. Craig, "Why do animals fight?" p. 274.

12. Craig, "Why do animals fight?" p. 277.

13. See above, chapter 3. Despite Lorenz's neglect in the 1950s of Craig's analysis of why animals fight, Craig's position was not completely overlooked in this period. Craig's

views on this score were endorsed by Peter Marler, "Studies of fighting in chaffinches," pt. 4, "Appetitive and consummatory behaviour," *British Journal of Animal Behaviour* 5 (1957): 29–37.

14. Lorenz, *On Aggression*, p. 48.

15. Lorenz, *On Aggression*, pp. 49, 275.

16. Lorenz, *On Aggression*, pp. 51–52.

17. Lorenz, *On Aggression*, pp. 243, 246.

18. Lorenz, *On Aggression*, pp. 268.

19. Konrad Lorenz, "Die angeborenen Formen möglicher Erfahrung," *Zeitschrift für Tierpsychologie* 5 (1943): 293.

20. Lorenz, *On Aggression*, pp. 270–271.

21. Lorenz, *On Aggression*, p. 278.

22. Lorenz, *On Aggression*, p. 291.

23. Marston Bates, review of *On Aggression*, by Konrad Lorenz, *New York Times Book Review*, 19 June 1966, p. 3; John Paul Scott, "That old-time aggression," *Nation*, 9 January 1967, p. 53. Scott's article is included in the collection of critical reviews of Lorenz's *On Aggression* and Robert Ardrey's *The Territorial Imperative* edited by Ashley Montagu and entitled *Man and Aggression* (New York: Oxford University Press, 1968).

24. Robert Hinde, "The nature of aggression," *New Society* 9 (1967): 302–304.

25. Tinbergen's "The search for animal roots of animal behaviour" was first published in 1973 in Niko Tinbergen, *The Animal in Its World: Explorations of an Ethologist, 1932–1972*, 2 vols. (London: George Allen and Unwin, 1972, 1973), 2:161–174.

26. N. Tinbergen, "On war and peace in animals and man: an ethologist's approach to the biology of aggression," *Science* 160 (1968): 1411–1418. See too Tinbergen's 1972 Croonian Lecture, "Functional ethology and the human sciences," *Proceedings of the Royal Society of London, Series B* 182 (1972): 385–410, reprinted in Tinbergen, *The Animal in Its World*, 2:200–231.

27. N. Tinbergen, "Ethology in a changing world," in *Growing Points in Ethology*, ed. P. P. G. Bateson and R. A. Hinde (Cambridge: Cambridge University Press, 1976), pp. 507–527, quote on p. 517.

28. Lorenz, *On Aggression*, pp. 288–289.

29. Tinbergen, "Ethology in a changing world," p. 508.

30. Tinbergen, "Ethology in a changing world," p. 524.

31. Bateson and Hinde, "Introduction," in *Growing Points in Ethology*, p. 2.

32. P. P. G. Bateson and R. A. Hinde, "Editorial: 8," in Bateson and Hinde, *Growing Points in Ethology*, p. 531.

33. Bateson and Hinde, "Conclusion: on asking the right questions," in *Growing Points in Ethology*, p. 536.

34. Felicity Huntingford, *The Study of Animal Behaviour* (London: Chapman and Hall, 1984), p. 9.

35. Konrad Lorenz, "The fashionable fallacy of dispensing with description," *Naturwissenchaften* 60 (1973): 1–9, quote on p. 4.

36. Lorenz, "The fashionable fallacy of dispensing with description," p. 1.

37. P. P. G. Bateson and Peter H. Klopfer, "Preface," in "Whither ethology?" ed. P. G. Bateson and Peter H. Klopfer, *Perspectives in Ethology* 8 (1989): vi.

38. Edward O. Wilson, *Naturalist* (New York: Warner Books, 1995), pp. 285–287.

39. Edward O. Wilson, *Sociobiology: The New Synthesis* (Cambridge, MA: Harvard University Press, 1975).

40. Elizabeth Allen et al., "Against 'sociobiology,'" *New York Review of Books*, 13 November 1975, pp. 43–44, quote on p. 43. See also L. Allen et al., "Sociobiology: a new biological determinism" (undated ms.). The late Ed Banks provided me with a copy of this manuscript.

41. See Ullica Segerstråle, *Defenders of the Truth: The Sociobiology Debate* (Oxford University Press, 2000).

42. The fourteen contributors were G. P. Baerends, George Barlow, N. G. Blurton Jones, J. H. Crook, Eberhard Curio, J. F. Eisenberg, Robert Hinde, Jerry Hirsch, John R. Krebs, Hans Kruuk, N. J. Mackintosh, Ethel Tobach, J. S. Rosenblatt, and Wolfgang Wickler. See "Multiple review of Wilson's *Sociobiology*," *Animal Behaviour* 24 (1976): 698–718.

43. The phrase "hobby horse" appears in a letter from Tinbergen to Mayr, 9 January 1968, copy courtesy of Ernst Mayr.

44. Niko Tinbergen, "Sociobiology," review of *Sociobiology: The New Synthesis*, by E. O. Wilson, *New Humanist*, October 1975. Tinbergen and especially Lorenz offered sharper criticism of Wilson in their correspondence with one another, for example, in letters they exchanged in October 1978. In Konrad Z. Lorenz Papers, Konrad-Lorenz-Institut für Evolutions- und Kognitionsforschung, Altenberg, Austria (hereafter KZLP/KLI).

45. Behavioral ecology, it must be noted, had not developed exclusively from the work of Tinbergen. When John R. Krebs and Nicholas B. Davies in 1978 published the first edition of their text, *Behavioural Ecology*, they represented behavioral ecology as having arisen from four approaches developed in the 1960s. One source was Tinbergen's group at Oxford: "the procedure of testing questions about the survival value of behaviour with simple field experiments or observations." The others were J. H. Crooks's and David Lack's "comparative approach," connecting the social organization of animals (birds and primates) with ecological factors; the work of W. D. Hamilton and John Maynard Smith, especially their development of the ideas of "kin selection" and "inclusive fitness"; and the work of R. H. MacArthur, who established "that hypotheses about evolutionary questions in ecology could be couched in precise mathematical terms." J. R. Krebs and N. B. Davies, eds., *Behavioural Ecology: An Evolutionary Approach* (Oxford: Blackwell Scientific Publications, 1978), pp.1–2.

46. Tinbergen to Mayr, 9 September 1978, copy courtesy of Ernst Mayr.

47. Richard Dawkins, *The Selfish Gene* (New York: Oxford University Press, 1976), pp. 2, 9, 72. Konrad Lorenz to Ernst Mayr, 21 February 1980, copy courtesy of Ernst Mayr.

48. Konrad Lorenz's *The Foundations of Ethology* (New York: Springer-Verlag, 1981) was a "revised and enlarged version" of his *Vergleichende Verhaltensforschung: Grundlagen der Ethologie*, published in 1978 by Springer-Verlag in Vienna and New York. The 1978 German language version lacked the preface cited here.

49. Lorenz, *The Foundations of Ethology*, p. xi.

50. Hanna-Maria Zippelius, *Die vermessene Theorie: Eine kritische Auseinanderset-*

zung mit der Instinkttheorie von Konrad Lorenz und verhaltenskundlicher Forschungspraxis (Braunschweig: Friedr. Vieweg & Sohn, 1992).

51. Marian Stamp Dawkins, *Unravelling Animal Behaviour* (Harlow: Longman Scientific and Technical, 1986; 2nd ed., 1995).

52. Marian Stamp Dawkins, "The future of ethology: how many legs are we standing on?" *Perspectives in Ethology* 8 (1989): 47–54.

53. P. P. G. Bateson and Peter H. Klopfer, "Preface," in "Whither ethology?" ed. P. P. G. Bateson and Peter H. Klopfer, *Perspectives in Ethology* 8 (1989): v–viii. See too in the same volume George W. Barlow, "Has sociobiology killed ethology or revitalized it," pp. 1–45.

54. Lorenz to Craig, 2 September 1949, copy in Margaret Morse Nice Papers, 1917–1968, collection number 2993, courtesy of the Division of Rare and Manuscript Collections, Cornell University Library.

55. Cited in Mary Alice Evans and Howard Ensign Evans, *William Morton Wheeler, Biologist* (Cambridge, MA: Harvard University Press, 1970), p. 36.

56. This narrative can be found, for example, in Konrad Lorenz, *The Natural Science of the Human Species: An Introduction to Comparative Behavioral Research. The "Russian Manuscript" (1944–1948)* (Cambridge, MA: MIT Press, 1996), chapters 18–21.

57. Robert Boakes, *From Darwin to Behaviourism: Psychology and the Minds of Animals* (Cambridge: Cambridge University Press, 1984).

58. See chapter 1.

59. Tinbergen to Mayr, 26 October 1973, copy courtesy of Ernst Mayr.

60. Lorenz, *On Aggression*, pp. 288–289.

61. Francis Bacon, *New Atlantis*, in *Famous Utopias of the Renaissance*, ed. Frederick R. White (Chicago: Packard and Company, 1946), p. 249.

62. François Jacob, *Of Flies, Mice, and Men* (Cambridge, MA: Harvard University Press, 1998), p. 119.

63. Tinbergen to Mayr, 26 October 1973, copy courtesy of Ernst Mayr.

64. Tinbergen described this assignment in a letter to Otto Koehler of 9 September 1964, Otto Koehler Papers, Universitätsarchiv Freiburg im Breisgau.

65. Tinbergen letter to Lorenz, 18 November 1965, accompanied by a two-page draft entitled "Biographical notes on Konrad Zacharius Lorenz, F.M.R.S." KZLP/KLI.

66. Niko Tinbergen, ms. entitled "Professor Konrad Lorenz," dated 7 September 1970, E.11, Nikolaas Tinbergen Papers, Bodleian Library, Oxford University (hereafter NTP/Oxford). The *Times* of London ultimately published its obituary of Lorenz on 1 March 1989, two days after the naturalist's death. Much of the obituary was taken word for word from what Tinbergen had written for the purpose some two decades earlier.

67. Niko Tinbergen to C. Watson, note attached to the manuscript of 7 September 1970, E.11, NTP/Oxford.

68. J. B. S. Haldane, *Heredity and Politics* (London: G. Allen and Unwin, 1938), p. 182.

69. Cited above, this chapter, n. 29.

70. Tinbergen letter to Lorenz, 31 January 1961, NTP/Oxford.

71. Tinbergen to Lorenz, 13 June 1979; and Lorenz to Tinbergen, 27 June 1979, KZLP/KLI.

{BIBLIOGRAPHY}

ARCHIVAL MATERIALS

Archives of the History of American Psychology, University of Akron (T. C. Schneirla Papers).

American Museum of Natural History (Gladwyn Kingsley Noble Papers).

American Philosophical Society, Philadelphia (C. B. Davenport Papers; Leonard Carmichael Papers).

Bristol University Archives (Conwy Lloyd Morgan Papers).

Bundesarchiv Koblenz (Deutsche Forschungsgemeinschaft Papers, Konrad Lorenz file).

Cambridge University Library, Department of Manuscripts and University Archives (William Homan Thorpe Papers).

Carnegie Institution of Washington Archives (Charles Otis Whitman file).

Cornell University Archives (Margaret Morse Nice Papers).

Universitätsarchiv Freiburg im Breisgau (Otto Koehler Papers).

Harvard University Archives (Ernst Mayr Papers).

University of Illinois at Urbana-Champaign Archives (Papers of the Illinois Natural History Survey).

Konrad-Lorenz-Institut für Evolutions- und Kognitionsforschung, Altenberg, Austria (Konrad Z. Lorenz Papers).

Archiv zur Geschichte der Max-Planck-Gesellschaft, Berlin (Erich von Holst Papers).

Oxford University, Bodleian Library, Department of Special Collections and Western Manuscripts (Alister Hardy Papers; Nikolaas Tinbergen Papers).

Oxford University, Edward Grey Institute of Field Ornithology (H. Eliot Howard Papers; Julian S. Huxley notebooks; David Lack Papers; Edmund Selous Papers).

Rice University Archives(Julian S. Huxley Papers).

Rockefeller Archive Center (Rockefeller Foundation Archives).

Staatsbibliothek zu Berlin (West), Preußischer Kulturbesitz (Oskar Heinroth Papers; Erwin Stresemann Papers).

Western Michigan University, University Archives and Regional History Collections (Charles C. Adams Papers).

Yale University Library, Manuscripts and Archives (Robert Mearns Yerkes Papers).

PUBLISHED WORKS

Allen, A. A. "Sex rhythm in the ruffed grouse (*Bonasa umbellus* Linn.) and other birds." *Auk* 51 (1934): 180–199.

Allen, David Elliston. *The Naturalist in Britain: A Social History.* London: Penguin, 1978.

Allen, Elizabeth, et al. "Against 'sociobiology.'" *New York Review of Books,* 13 November 1975, pp. 43–44.

Allen, Robert P., and Frederick P. Mangels. "Studies of the nesting behavior of the black-crowned night heron." *Proceedings of the Linnaean Society of New York* 50–51 (1940): 1–28.

Alverdes, Friedrich. *Tiersoziologie.* Leipzig: C. L. Hirschfeld, 1925.

———. *Die Tierpsychologie in ihren Beziehungen zur Psychologie des Menschen.* Leipzig: C. L. Hirschfeld, 1932.

Armstrong, Edward A. *Bird Display: An Introduction to the Study of Bird Psychology.* Cambridge: Cambridge University Press, 1942.

Aronson, Lester R., et al., eds. *Development and Evolution of Behavior: Essays in Memory of T. C. Schneirla.* San Francisco: W. H. Freeman, 1970.

Avebury, The Right Hon. Lord [John Lubbock]. *Essays and Addresses, 1900–1903.* London: Macmillan, 1903.

Baerends, Gerard P. "Fortpflanzungsverhalten und Orientierung der Grabwespe *Ammophila campestris* Jur." *Tijdschrift voor Entomologie* 84 (1941): 68–275.

———. "Two pillars of wisdom." In *Studying Animal Behavior: Autobiographies of the Founders,* ed. Donald A. Dewsbury, pp. 13–40. Chicago: University of Chicago Press, 1989.

———. "Early ethology: growing from Dutch roots." In *The Tinbergen Legacy,* ed. M. S. Dawkins, T. R. Halliday, and R. Dawkins, pp. 1–17. London: Chapman and Hall, 1991.

———. "Turning points in the history of ethology." In *L'Histoire de la connaissance du comportement animal,* ed. Liliane Bodson, pp. 59–76. Liège: Université de Liège, 1993.

Baerends, Gerard, Colin Beer, and Aubrey Manning. "Introduction." In *Function and Evolution in Behaviour: Essays in Honour of Professor Niko Tinbergen, F.R.S.,* ed. Gerard Baerends, Colin Beer, and Aubrey Manning, pp. xi–xxii. Oxford: Clarendon Press, 1975.

Baerends, G. P., et al. "Multiple review of Wilson's *Sociobiology.*" *Animal Behaviour* 24 (1976): 698–718.

Baker, John R. "Julian Sorell Huxley." *Biographical Memoirs of Fellows of the Royal Society* 22 (1976): 207–238.

———. *Julian Huxley, Scientist and World Citizen, 1887–1975.* Paris: UNESCO, 1978.

Barlow, George W. "Has sociobiology killed ethology or revitalized it?" In "Whither ethology," ed. P. P. G. Bateson and Peter H. Klopfer. *Perspectives in Ethology* 8 (1989): 1–45.

———. "Nature-nurture and the debates surrounding ethology and sociobiology." *American Zoologist* 31 (1991): 286–296.

Bartley, Mary. "Courtship and continued progress: Julian Huxley's studies on bird behavior." *Journal of the History of Biology* 28 (1995): 91–108.

Bastock, M., D. Morris, and M. Moynihan. "Some comments on conflict and thwarting in animals." *Behaviour* 6 (1953): 66–84.

Bateson, P. P. G., and R. A. Hinde. *Growing Points in Ethology.* Cambridge: Cambridge University Press, 1976.

Bateson, P. P. G., and Peter H. Klopfer, eds. "Whither ethology?" *Perspectives in Ethology,* vol. 8 (1989).

Bäumer-Schleinkofer, Anne. *NS-Biologie und Schule*. Frankfurt am Main: Peter Lang, 1992.

Baur, Erwin, Eugen Fischer, and Fritz Lenz. *Human Heredity*. New York: Macmillan, 1931.

Beach, Frank A. "The snark was a boojum." *American Psychologist* 5 (1950): 115–124.

——. "Frank A. Beach." In *A History of Psychology in Autobiography*, ed. Edwin G. Boring, 6:33–58. Worcester, MA: Clark University Press, 1974.

Beer, Colin G. "Incubation and nest-building by the black-headed gull." Pt. 1, "Incubation behaviour in the incubation period." *Behaviour* 18 (1961): 62–106.

——. "Ethology: the zoologist's approach to behaviour." Pts. 1 and 2. *Tuatara* 11 (1963): 170–177; 12 (1964): 16–39.

——. "Instinct." In *International Encyclopedia of the Social Sciences*, 7:363–372. New York: Macmillan, 1968.

——. "Was Professor Lehrman an ethologist?" *Animal Behaviour* 23 (1975): 957–964.

——. "Darwin, instinct, and ethology." *Journal of the History of the Behavioral Sciences* 19 (1983): 68–80.

——. "Expressions of mind in animal behavior." In *Anthropomorphism, Anecdotes, and Animals*, ed. Robert W. Mitchell, Nicholas S. Thompson, and H. Lyn Miles, pp. 198–209. Albany: State University of New York Press, 1997.

Beer, Th., A. Bethe, and J. von Uexküll. "Vorschläge zu einer objektivierenden Nomenklature in der Physiologie des Nervensystems." *Biologisches Zentralblatt* 19 (1899): 517–521.

Bennett, George W. "A century of biological research: aquatic biology." *Illinois Natural History Survey Bulletin* 27 (1958): 163–178.

Benson, Keith R., Jane Maienschein, and Ronald Rainger, eds. *The Expansion of American Biology*. New Brunswick, NJ: Rutgers University Press, 1991.

Berminge, Kerstin. *Två Etologer: En vetenskapsteoretisk analys av Konrad Lorenz' forskarutveckling jämförd med Niko Tinbergens*. Göteborg: Institutionen för Vetenskapsteori, Göteborgs Universitet, 1988.

Bethe, Albrecht. "Dürfen wir den Ameisen und Bienen psychische Qualitäten zuschreiben?" *Archiv für die gesamte Physiologie des Menschen und der Tiere* 70 (1898): 15–99.

Beyerchen, Alan. "What we now know about Nazism and science." *Social Research* 59 (1992): 615–641.

Bierens de Haan, J. A. "Die Balz des Argusfasans." *Biologisches Zentralblatt* 46 (1926): 428–435.

——. "Der Stieglitz als Schöpfer." *Journal für Ornithologie* 81 (1933): 1–22.

——. "Über den Begriff des Instinktes in der Tierpsychologie." *Folia Biotheoretica, Series B* 2 (1937): 1–16.

Blest, A. D. "The function of eyespot patterns in the Lepidoptera." *Behaviour* 11 (1957): 209–256.

Boakes, Robert. *From Darwin to Behaviourism: Psychology and the Minds of Animals*. Cambridge: Cambridge University Press, 1984.

Bohn, Georges. *La nouvelle psychologie animale*. Paris: Félix Alcan, 1911.

Bolk, L. *Das Problem der Menschwerdung*. Jena, 1926.

Booy, H. L., and N. Tinbergen. "Nieuwe feiten over de sociologie van de zilvermeeu-wen." *De Levende Natuur* 41 (1937): 325–334.

Boschma, H. "A concise review of the scientific activities of C. J. van der Klaauw." In "Volume jubilaire dédié à C. J. van der Klaauw." *Archives néerlandaises de zoologie* 13, supp. 1 (1958): 5–9.

Botz, Gerhard. *Nationalsozialismus in Wien.* Buchloe: Druck und Verlag Obermayer, 1988.

Bowlby, John. *Attachment and Loss.* Vol. 1, *Attachment.* New York: Basic Books, 1969.

———. "Ethological light on psychoanalytical problems." In *The Development and Integration of Behaviour: Essays in Honour of Robert Hinde,* ed. Patrick Bateson, pp. 301–314. Cambridge: Cambridge University Press, 1991.

Brock, Friedrich. "Jakob Johann Baron von Uexküll." *Sudhoffs Archiv für Geschichte der Medizin und der Naturwissenschaften* 27 (1934): 193–203.

———. "Verzeichnis der Schriften Jakob Johann v. Uexkülls und der aus dem Institut für Umweltforschung zu Hamburg hervorgegangenen Arbeiten." *Sudhoffs Archiv für Geschichte der Medizin und der Naturwissenschaften* 27 (1934): 204–212.

Broszat, Martin. *Nationalsozialistische Polenpolitik, 1939–1945.* Stuttgart: Deutsche Verlags-Anstalt, 1961.

Browne, Janet. "Darwin and the expression of the emotions." In *The Darwinian Heritage,* ed. David Kohn, pp. 307–326. Princeton, NJ: Princeton University Press, 1985.

Brückner, G. H. "Untersuchungen zur Tiersoziologie, insbesondere zur Auflösung der Familie." *Zeitschrift für Psychologie* 128 (1933): 1–110.

Burghardt, Gordon M. "Instinct and behavior: toward an ethological psychology." In *The Study of Behavior: Learning, Motivation, Emotion, and Instinct,* ed. John A. Nevin, pp. 323–400. Glenview, IL: Scott, Foresman and Company, 1973.

———, ed. *Foundations of Comparative Ethology.* New York: Van Nostrand Reinhold Company, 1985.

———. "Amending Tinbergen: a fifth aim for ethology." In *Anthropomorphism, Anecdotes, and Animals,* ed. Robert W. Mitchell, Nicholas S. Thompson, and H. Lyn Miles, pp. 254–276. Albany: State University of New York Press, 1997.

Burkhardt, Richard W., Jr. "On the emergence of ethology as a scientific discipline." *Conspectus of History* 1 (1981): 62–81.

———. "The development of an evolutionary ethology." In *Evolution from Molecules to Men,* ed. D. S. Bendall, pp. 429–444. Cambridge: Cambridge University Press, 1983.

———. "Darwin on animal behavior and evolution." In *The Darwinian Heritage,* ed. David Kohn, pp. 435–481. Princeton, NJ: Princeton University Press, 1985.

———. "The *Journal of Animal Behavior* and the early history of animal behavior studies in America." *Journal of Comparative Psychology* 101 (1987): 223–230.

———. "Charles Otis Whitman, Wallace Craig, and the biological study of behavior in America, 1898–1924." In *The American Development of Biology,* ed. Ronald Rainger, Keith Benson, and Jane Maienschein, pp. 185–218. Philadelphia: University of Pennsylvania Press, 1988.

———. "Karl von Frisch." *Dictionary of Scientific Biography* 17, supp. 2 (1990): 312–320.

———. "Oskar August Heinroth." *Dictionary of Scientific Biography* 17, supp. 2 (1990): 392–394.

————. "Edmund Selous." *Dictionary of Scientific Biography* 18, supp. 2 (1990): 801–803.

————. "Le comportement animal et l'idéologie de domestication chez Buffon et les éthologistes modernes." In *Buffon 88: Actes du Colloque international pour le bicentenaire de la mort de Buffon,* ed. J.-C. Beaune et al., pp. 569–582. Paris: J. Vrin, 1992.

————. "Julian Huxley and the rise of ethology." In *Julian Huxley: Biologist and Statesman of Science,* ed. C. Kenneth Waters and Albert Van Helden, pp. 127–149. Houston: Rice University Press, 1992.

————. "Le comportement animal et la biologie française (animal behavior and French biology), 1920–1950." In *Les Sciences biologiques et médicales en France 1920–1950,* ed. Claude Debru, Jean Gayon, and Jean-François Picard, pp. 99–111. Paris: CNRS Éditions, 1994.

————. "The founders of ethology and the problem of animal subjective experience." In *Animal Consciousness and Animal Ethics: Perspectives from the Netherlands,* ed. Marcel Dol et al., pp. 1–13. Assen: Van Gorcum, 1997.

————. "Ethology, natural history, the life sciences, and the problem of place." *Journal of the History of Biology* 32 (1999): 489–508.

————. "Konrad Zacharias Lorenz (1903–1989)." In *Darwin & Co.: Eine Geschichte der Biologie in Porträts,* ed. Ilse Jahn and Michael Schmitt, 2:422–441. Munich: C. H. Beck, 2001.

————. "The founders of ethology and the problem of human aggression: a study in ethology's ecologies." In *The Animal/Human Boundary,* ed. Angela N. H. Creager and William Chester Jordans, pp. 265–304. Rochester, NY: Rochester University Press, 2002.

Burleigh, Michael. *Germany Turns Eastwards: A Study of Ostforschung in the Third Reich.* Cambridge: Cambridge University Press, 1988.

Carsten, F. L. *The First Austrian Republic, 1918–1938.* Aldershot: Gower Publishing Company, 1985.

Chavot, Philippe. "Histoire de l'éthologie: recherches sur le développement des sciences du comportement en Allemagne, Grande-Bretagne et France, de 1930 à nos jours." Thèse de doctorat en histoire des sciences, Université Louis Pasteur—Strasbourg I, 1994.

Cloud, Wallace. "Winners and sinners." *Sciences,* December 1973, pp. 16–21.

Collias, Nicholas E. "The role of American zoologists and behavioural ecologists in the development of animal sociology, 1934–1964." *Animal Behaviour* 41 (1991): 613–631.

Craig, Wallace. "The voice of pigeons." Pt. 1, "The voice of the ring dove (*Turtur risorius*)." *Biological Bulletin of the Marine Biological Laboratory* 6 (1903–1904): 323–324.

————. "North Dakota life: plant, animal and human." *Bulletin of the American Geographical Society* 40 (1908): 321–332, 401–415.

————. "The voices of pigeons regarded as a means of social control." *American Journal of Sociology* 14 (1908): 86–100.

————. "The expressions of emotion in the pigeons." Pt. 1, "The blond ring dove (*Turtur risorius*)." *Journal of Comparative Neurology and Psychology* 19 (1909): 29–80.

————. "The expressions of emotion in the pigeons." Pt. 2, "The mourning dove (*Zenaidura macroura* Linn.)." *Auk* 28 (1911): 398–407.

———. "The expressions of emotion in the pigeons." Pt. 3, "The passenger pigeon (*Ectopistes migratorius* Linn.)." *Auk* 28 (1911): 408–426.

———. "Oviposition induced by the male in pigeons." *Journal of Morphology* 22 (1911): 299–305.

———. "Observations on doves learning to drink." *Journal of Animal Behavior* 2 (1912): 273–279.

———. "Behavior of the young bird in breaking out of the egg." *Journal of Animal Behavior* 2 (1912): 296–298.

———. "The stimulation and the inhibition of ovulation in birds and mammals." *Journal of Animal Behavior* 3 (1913): 215–221.

———. "Attitudes of appetition and of aversion in doves." *Psychological Bulletin* 11 (1914): 56–57.

———. "Male doves reared in isolation." *Journal of Animal Behavior* 4 (1914): 121–133.

———. "Appetites and aversions as constituents of instincts." *Biological Bulletin of the Marine Biological Laboratory* 34 (1918): 91–107.

———. "Why do animals fight?" *International Journal of Ethics* 31 (1921): 264–278.

———. "A note on Darwin's work on the expression of the emotions in man and animals." *Journal of Abnormal and Social Psychology* 16 (1921–1922): 356–366.

———. "The twilight song of the wood peewee: a preliminary statement." *Auk* 43 (1926): 150–152.

———. "The music of the wood pewee's song and one of its laws." *Auk* 50 (1933): 174–178.

———. "The song of the wood pewee *Myiochanes virens* Linnaeus: a study of bird music." *Bulletin, New York State Museum*, no. 334 (1943): 1–186.

Cranach, Agnes von. "Editor's foreword." In *The Natural Science of the Human Species: An Introduction to Comparative Behavioral Research: The "Russian Manuscript" (1944–1948)*, by Konrad Lorenz, pp. ix–xv. Cambridge, MA: MIT Press, 1996.

Crist, Eileen. *Images of Animals: Anthropomorphism and Animal Mind*. Philadelphia: Temple University Press, 1999.

Cronin, Helena. *The Ant and the Peacock*. Cambridge: Cambridge University Press, 1991.

Cullen, Esther. "Adaptations in the kittiwake to cliff-nesting." *Ibis* 99 (1957): 275–302.

Darwin, Charles. *On the Origin of Species by Means of Natural Selection*. London: J. Murray, 1859.

———. *The Descent of Man, and Selection in Relation to Sex*. 2 vols. London: John Murray, 1871.

———. *The Expression of the Emotions in Man and Animals*. London: John Murray, 1872.

———. *The Autobiography of Charles Darwin, 1809–1882, with the Original Omissions Restored*. Edited by Nora Barlow. London: Collins, 1958.

———. *Charles Darwin's Notebooks, 1836–1844*. Edited by Paul H. Barrett, Peter J. Gautrey, Sandra Herbert, David Kohn, and Sydney Smith. Ithaca, NY: Cornell University Press, 1987.

Davenport, Charles B. "The personality, heredity and work of Charles Otis Whitman, 1843–1910." *American Naturalist* 51 (1917): 5–30.

Dawkins, Marian Stamp. *Unravelling Animal Behaviour*. Harlow: Longman Scientific and Technical, 1986; 2nd ed., 1995.

———. "The future of ethology: how many legs are we standing on?" *Perspectives in Ethology* 8 (1989): 47–54.

Dawkins, Marian Stamp, T. R. Halliday, and R. Dawkins, eds. *The Tinbergen Legacy*. London: Chapman and Hall, 1991.

Dawkins, Richard. *The Selfish Gene*. New York: Oxford University Press, 1976.

———. "Introduction." In *The Tinbergen Legacy*, ed. M. S. Dawkins, T. R. Halliday, and R. Dawkins, pp. ix–xii. London: Chapman and Hall, 1991.

Deichmann, Ute. *Biologists under Hitler*. Translated by Thomas Dunlap. Cambridge, MA: Harvard University Press, 1996. Translation of Ute Deichmann, *Biologen unter Hitler: Vertreibung, Karrieren, Forschungsförderung* (Frankfurt: Campus, 1992).

de Laguna, Grace Andrus. *Speech: Its Function and Development*. New Haven, CT: Yale University Press, 1927.

Delamain, Jacques. "Edmund Selous." *Alauda* 6 (1934): 388–393.

Desmond, Adrian. *Huxley: The Devil's Disciple*. London: M. Joseph, 1994.

Desmond, Adrian, and James Moore. *Darwin*. London: Michael Joseph, 1991.

Dewsbury, Donald A., ed. *Studying Animal Behavior: Autobiographies of the Founders*. Chicago: University of Chicago Press, 1989. Originally published as *Leaders in the Study of Animal Behavior: Autobiographical Perspectives* (London: Associated University Presses, 1985).

———. "Americans in Europe: the role of travel in the spread of European ethology after World War II." *Animal Behaviour* 49 (1995): 1649–1663.

———. "Frank Ambrose Beach." *Biographical Memoirs of the National Academy of Sciences* 73 (1998): 64–85.

Di Gregorio, Mario A. *T. H. Huxley's Place in Natural Science*. New Haven, CT: Yale University Press, 1984.

Distant, W. L. "On the mental differences between the sexes." *Journal of the Anthropological Institute of Great Britain and Ireland* 4 (1875): 78–87.

———. *A Naturalist in the Transvaal*. London: R. H. Porter, 1892.

———. "Preface." *Zoologist* 7 (1903): iii–iv.

Durant, John R. "Innate character in animals and man: a perspective on the origins of ethology." In *Biology, Medicine and Society, 1840–1940*, ed. Charles Webster, pp. 157–192. Cambridge: Cambridge University Press, 1981.

———. "The ascent of nature in Darwin's *Descent of Man*." In *The Darwinian Heritage*, ed. David Kohn, pp. 283–306. Princeton, NJ: Princeton University Press, 1985.

———. "The making of ethology: the Association for the Study of Animal Behaviour." *Animal Behaviour* 34 (1986): 1601–1616.

———. "The tension at the heart of Huxley's evolutionary ethology." In *Julian Huxley: Biologist and Statesman of Science*, ed. C. Kenneth Waters and Albert Van Helden, pp. 150–160. Houston: Rice University Press, 1992.

Effertz, J. "Bericht über die Gründung der Deutschen Gesellschaft für Tierpsychologie." *Zeitschrift für Tierpsychologie* 1 (1937): 1–8.

Eibl-Eibesfeldt, Irenäus. "'Fishy, fishy, fishy': autobiographical sketches." In *Studying*

Animal Behavior: Autobiographies of the Founders, ed. Donald A. Dewsbury, pp. 69–91. Chicago: University of Chicago Press, 1989.

Eisenberg, Leon. "The human nature of human nature." *Science* 176 (1972): 123–128.

Evans, Mary Alice, and Howard Ensign Evans. *William Morton Wheeler, Biologist.* Cambridge, MA: Harvard University Press, 1970.

Evans, Richard I. *Konrad Lorenz: The Man and His Ideas.* New York: Harcourt Brace Jovanovich, 1975.

Fischer, Eugen. "Die Rassenmerkmale des Menschen als Domestikationserscheinungen." *Zeitschrift für Morphologie und Anthropologie* 18 (1914): 479–524.

Föger, Benedikt, and Klaus Taschwer. *Die andere Seite des Speigels: Konrad Lorenz und der Nationalsozialismus.* Vienna: Czernin Verlag, 2001.

Forbes, Stephen. *Biennial Report of the Director for 1897–98, Illinois State Laboratory of Natural History.* Urbana, IL, 1989.

Forbes, Stephen, and Robert Earl Richardson. *The Fishes of Illinois.* Urbana: Illinois State Laboratory of Natural History, 1908.

Fowler, W. Warde. "In the ways of birds." *Saturday Review* 92 (1901): 104–105.

———. *Kingham Old and New.* Oxford: Blackwell, 1913.

Frisch, Karl von. *Du and das Leben.* Berlin: Deutsche Verlag, 1936.

———. *A Biologist Remembers.* Oxford: Oxford University Press, 1967.

Geison, Gerald L. *The Private Science of Louis Pasteur.* Princeton, NJ: Princeton University Press, 1995.

Geuter, Ulfried . *Die Professionalisierung der deutschen Psychologie im Nationalsozialismus.* Frankfurt: Suhrkamp, 1984. Translated as *The Professionalization of Psychology in Nazi Germany* (Cambridge: Cambridge University Press, 1992).

Gillespie, Neal C. "Natural history, natural theology, and social order: John Ray and the 'Newtonian ideology.'" *Journal of the History of Biology* 20 (1987): 1–49.

———. "The interface of natural theology and science in the ethology of W. H. Thorpe." *Journal of the History of Biology* 23 (1990): 1–38.

Gould, Stephen Jay. *The Structure of Evolutionary Theory.* Cambridge, MA: Harvard University Press, 2002.

Gray, P. H. "The early animal behaviorists: prolegomenon to ethology." *Isis* 59 (1968): 372–383.

Greite, Walter. "Zu Ernst Rüdins 65. Geburtstag." *Der Biologe* 8 (1939): 153–154.

Griffiths, Paul. "Instinct in the 50s." *Biology and Philosophy,* vol. 19 (2004; in press).

Haeckel, Ernst. *Freie Wissenschaft und freie Lehre.* Stuttgart: E. Schwizerbart, 1878.

Haffer, Jürgen, Erich Rutschke, and Klaus Wunderlich. "Erwin Stresemann (1889–1972): Leben und Werk eines Pioniers der wissenschaftlichen Ornithologie." *Acta Historica Leopoldina* 34 (2000): 1–465.

Haldane, J. B. S. *Heredity and Politics.* London: G. Allen and Unwin, 1938.

———. "The sources of some ethological notions." *British Journal of Animal Behaviour* 4 (1956): 162–164.

———. "The argument from animals to men: an examination of its validity for anthropology, *Journal of the Royal Anthropological Institute of Great Britain and Ireland* 86, pt. 2 (1956): 1–14.

Hardy, A. C. "Natural history—old and new." Inaugural address delivered on 28 April 1942 at Marischal College, University of Aberdeen, Aberdeen, 1942.

————. "Zoology outside the laboratory." *Advancement of Science* 6 (1949): 213–223.

————. "German science and technology under National Socialism." *Perspectives on Science* 5 (1997): 128–151.

Harrington, Anne. *Reenchanted Science: Holism in German Culture from Wilhelm II to Hitler.* Princeton, NJ: Princeton University Press, 1996.

Harrison, J. W. Heslop. "Experiments on the egg-laying instincts of the sawfly, *Pontania salicis* Christ., and their bearing on the inheritance of acquired characters; with some remarks on a new principle in evolution." *Proceedings of the Royal Society of London, Series B* 101 (1927): 115–125.

Harwood, Jonathan. *Styles of Scientific Thought: The German Genetics Community, 1900–1933.* Chicago: University of Chicago Press, 1993.

Hebb, D. O. "Heredity and environment in mammalian behaviour." *British Journal of Animal Behaviour* 1 (1953): 43–47.

Heberer, Gerhard. "Die gegenwärtigen Vorstellungen über den Stammbaum der Tiere und die 'Systematische Phylogenie' E. Haeckels." *Der Biologe* 8 (1939): 264–273.

Heinroth, Katharina. *Oskar Heinroth, Vater der Verhaltensforschung.* Stuttgart: Wissenschaftliche Verlagsgesellschaft, 1971.

Heinroth, Oskar. "Beiträge zur Biologie: namentlich Ethologie und Psychologie der Anatiden." In *Verhandlungen des 5. Internationalen Ornithologen-Kongresses in Berlin, 30. Mai bis 4. Juni 1910,* ed. Herman Schalow, pp. 589–702. Berlin: Deutsche Ornithologische Gesellschaft, 1911.

Heinroth, Oskar, and Magdalena Heinroth. *Die Vögel Mitteleuropas in allen Lebens- und Entwicklungsstufen photographisch aufgenommen und in ihrem Seelenleben bei der Aufzucht vom Ei ab beobachtet.* 4 vols. Berlin: H. Bermühler, 1924–1934.

Heinroth, Oskar, and Konrad Lorenz. *Wozu aber hat das Vieh diesen Schnabel? Briefe aus der frühen Verhaltensforschung, 1930–1940.* Edited by Otto Koenig. Munich: Piper, 1988.

Henning, Eckart, and Marion Kazemi. "Chronik der Max-Planck-Gesellschaft zur Förderung der Wissenschaften unter der Präsidentschaft Otto Hans (1946–1960)." *Veröffentlichungen aus dem Archiv zur Geschichte der Max-Planck-Gesellschaft* 4 (1992): 1–160.

Herrick, Francis. "The individual vs. the species in behavior studies." *Auk* 56 (1939): 244–249.

Hertz, Mathilde. "Beobachtungen an gefangenen Rabenvögeln." *Psychologische Forschung* 8 (1926): 336–397.

Hess, Eckhard H. "Ethology: an approach toward the complete analysis of behavior." In *New Directions in Psychology I,* by Roger Brown, Eugen Galanter, Eckhard H. Hess, and George Mandler, pp. 157–266. New York: Holt Rinehart and Winston, 1961.

Hicks, Lawrence E. "The fifty-sixth stated meeting of the American Ornithologists' Union." *Auk* 56 (1939): 112–123.

Hinde, R. A. "The behaviour of the great tit (*Parus major*) and some other related species." *Behaviour,* supp. 2 (1952): v–x, 1–201.

———. "Factors governing the changes in strength of a partially inborn response, as shown by the mobbing behaviour of the chaffinch." Pt. 1, "The nature of the response, and an examination of its course." *Proceedings of the Royal Society of London, Series B* 142 (1954): 306–331.

———. "Factors governing the changes in strength of a partially inborn response, as shown by the mobbing behaviour of the chaffinch." Pt. 2, "The waning of the response." *Proceedings of the Royal Society of London, Series B* 142 (1954): 331–358.

———. "Changes in responsiveness to a constant stimulus." *British Journal of Animal Behaviour* 2 (1954): 41–55.

———. "Ethological models and the concept of 'drive.'" *British Journal for the Philosophy of Science* 6 (1956): 321–331.

———. "The nature of aggression." *New Society* 9 (1967): 302–304.

———. *Ethology: Its Nature and Relations with Other Sciences.* New York: Oxford University Press, 1982.

———. "William Homan Thorpe." *Biographical Memoirs of Fellows of the Royal Society* 33 (1987): 621–639.

———. "Ethology in relation to other disciplines." In *Studying Animal Behavior: Autobiographies of the Founders,* ed. Donald A. Dewsbury, pp. 193–203. Chicago: University of Chicago Press, 1989.

———. "Nikolaas Tinbergen." *Biographical Memoirs of Fellows of the Royal Society* 36 (1990): 547–565.

Hinde, R. A., and W. H. Thorpe. "Nobel recognition for ethology." *Nature* 245 (1973): 346.

Hinde, R. A., and N. Tinbergen. "The comparative study of species-specific behavior." In *Behavior and Evolution,* ed. A. Roe and G. G. Simpson, pp. 251–268. New Haven, CT: Yale University Press, 1958.

Hippius, Rudolf, I. G. Feldman, K. Jellinek, and K. Leider. *Volkstum, Gesinnung und Charakter: Bericht über psychologische Untersuchungen an Posener deutsch-polnischen Mischlingen und Polen, Sommer 1942.* Stuttgart: W. Kohlhammer, 1943.

Holldobler, Bert, and Martin Lindauer, eds. *Experimental Behavioral Ecology and Sociobiology: In Memoriam Karl von Frisch, 1886–1982.* Sunderland, MA: Sinauer Associates, 1985.

Holst, Erich von. "Versuche zur Theorie der relativen Koordination." *Pflügers Archiv* 237 (1936): 93–121.

———. Review of "Das Tier als Individualität in der Großhirnforschung" (1936), by Emil Diebschlag. *Berichte über de wissenschaftliche Biologie* 41 (1937): 366.

———. "Quantitative Messung von Stimmungen im Verhalten der Fische." *Symposia of the Society for Experimental Biology* 4 (1950): 143–172.

———. "Albrecht Bethe." *Die Naturwissenschaften* 42 (1955): 99–101.

Holst, Erich von, and H. Mittelstaedt. "Das Reafferenzprincip." *Die Naturwissenschaften* 37 (1950): 464–476.

Hoogland, R., D. Morris, and N. Tinbergen. "The spines of sticklebacks (*Gasterosteus* and *Pygosteus*) as means of defence against predators (*Perca* and *Esox*)." *Behaviour* 10 (1957): 205–236.

Hoßfeld, Uwe. *Gerhard Heberer (1901–1973): Sein Beitrag zur Biologie im 20. Jahrhundert.* Berlin: Verlag für Wissenschaft und Bildung, 1997.

Howard, H. E. "The grasshopper-warbler (*Locustella naevia*) in north Worcestershire." *Zoologist* 5 (1901): 60–63.

———. "On Mr. Selous' theory of the origin of nests." *Zoologist* 6 (1902): 145–148.

———. "On sexual selection and the aesthetic sense in birds." *Zoologist* 7 (1903): 407–417.

———. *The British Warblers: A History with Problems of Their Lives.* London: R. H. Porter, 1907–1914.

———. *Territory in Bird Life.* London: John Murray, 1920.

———. *The Nature of a Bird's World.* Cambridge: Cambridge University Press, 1935.

———. *A Waterhen's Worlds.* Cambridge: Cambridge University Press, 1940.

Huntingford, Felicity. *The Study of Animal Behaviour.* London: Chapman and Hall, 1984.

Huxley, Julian S. *The Individual in the Animal Kingdom.* Cambridge: Cambridge University Press, 1912.

———. "A 'disharmony' in the reproductive habits of the wild duck (*Anas bochas*, L.)." *Biologisches Zentralblatt* 32 (1912): 621–623.

———. "A first account of the courtship of the redshank (*Totanus calidris* L.)." *Proceedings of the Zoological Society of London* (1912): 647–655.

———. "The great crested grebe and the idea of secondary sexual characteristics." *Science*, n.s., 36 (1912): 601–602.

———. "The courtship-habits of the great crested grebe (*Podiceps cristatus*); with an addition to the theory of sexual selection." *Proceedings of the Zoological Society of London* (1914): 491–562.

———. "Bird-watching and biological science: some observations on the study of courtship in birds." *Auk* 33 (1916): 143–161, 256–270.

———. "Some points in the sexual habits of the little grebe, with a note on the occurrences of vocal duets in birds." *British Birds* 13 (1919): 155–158.

———. "Obituary: W. Warde-Fowler." *British Birds* 15 (1921): 143–144.

———. "The accessory nature of many structures and habits associated with courtship." *Nature* 108 (1921): 565–566.

———. Review of *Territory in Bird Life*, by H. Eliot Howard. *Discovery* 2 (1921): 135–136.

———. "Preferential mating in birds with similar coloration in both sexes." *British Birds* 15 (1922): 99–101.

———. "Courtship activities in the red-throated diver (*Colymbus stellatus* Pontopp.); together with a discussion of the evolution of courtship in birds. *Journal of the Linnean Society of London, Zoology*, 35 (1923): 253–292.

———. "Ils n'ont que de l'âme: an essay on bird mind." *Cornhill Magazine* 54 (1923): 415–427.

———. "The outlook for biology." *Rice Institute Pamphlet* 11 (1924): 241–338.

———. "Some further observations on the courtship behaviour of the great crested grebe." *British Birds* 18 (1924): 129–134.

———. "Some points in the breeding behaviour of the common heron." *British Birds* 18 (1924): 155–163.

———. "The absence of 'courtship' in the avocet." *British Birds* 19 (1925): 88–94.

———. *Bird-Watching and Bird Behaviour.* London: Chatto and Windus, 1930.

———. "Memorable incidents with birds." *Listener* 3 (1930): 841–842.

———. "The courtship of birds." *Listener* 3 (1930), p. 935.

———. "Biology of bird courtship." In *Proceedings of the VIIth International Ornithological Congress at Amsterdam, 1930*, pp. 107–108. Amsterdam: Prof. Dr. L. F. de Beaufort, 1931.

———. "Threat and warning coloration in birds, with a general discussion of the biological functions of colour." In *Proceedings of the Eighth International Ornithological Congress, Oxford, July 1934*, pp. 430–455. Oxford: Oxford University Press, 1938.

———. "Making and using nature films." *Listener* 13 (1935): 595–597, 629.

———. "Natural selection and evolutionary progress." In *British Association for the Advancement of Science: Report of the Annual Meeting, 1936*, pp. 81–100. London: Office of the British Association, 1936.

———. "Darwin's theory of sexual selection and the data subsumed by it, in the light of recent research." *American Naturalist* 72 (1938): 416–433.

———. "The present standing of the theory of sexual selection." In *Evolution: Essays on Aspects of Evolutionary Biology Presented to Professor E. S. Goodrich on His Seventieth Birthday*, ed. G. R. de Beer, pp. 11–42. Oxford: Clarendon Press, 1938.

———. *Memories*. 2 vols. London: George Allen and Unwin, 1970, 1973.

Huxley, J. S., assisted by F. A. Montague. "Studies on the courtship and sexual life of birds." Pt. 5, "The oyster-catcher." *Ibis*, 12th ser., 1 (1925): 868–897.

Huxley, J. S., and G. J. van Oordt. "Some observations on the habits of the red-throated diver in Spitsbergen." *British Birds* 16 (1922): 34–46.

Huxley, Juliette. *Leaves of the Tulip Tree*. London: John Murray, 1986.

Iersel, J. J. A. Van. "An analysis of the parental behaviour of the male three-spined stickleback (*Gasterosteus aculeatus* L.). *Behaviour*, supp. 3 (1953): 1–159.

Jacob, François. *Of Flies, Mice, and Men*. Cambridge, MA: Harvard University Press, 1998.

Jaensch, E. R. "Einladung zur XVI. Tagung der Deutschen Gesellschaft für Psychologie 2.–4. Juli 1938 in Bayreuth." *Zeitschrift für Tierpsychologie* 2 (1938–1939): 99.

———. "Der Hühnerhof als Forschungs- und Aufklärungsmittel in menschlichen Rassenfragen." *Zeitschrift für Tierpsychologie* 2 (1938–1939): 223–258.

Jaynes, Julian. "The historical origins of 'ethology' and 'comparative psychology.'" *Animal Behaviour* 17 (1969): 601–606.

Joncich, Geraldine. *The Sane Positivist: A Biography of Edward L. Thorndike*. Middletown, CT: Wesleyan University Press, 1968.

Kalikow, Theodora J. "History of Konrad Lorenz's ethological theory, 1927–1939: the role of meta-theory, theory, anomaly and new discoveries in a scientific 'evolution.'" *Studies in History and Philosophy of Science* 6 (1975): 331–341.

———. "Konrad Lorenz's ethological theory, 1939–1943: 'explanations' of human thinking, feeling and behaviour." *Philosophy of the Social Sciences* 6 (1976): 15–34.

———. "Die ethologische Theorie von Konrad Lorenz: Erklärung und Ideologie, 1938 bis 1943." In *Naturwissenschaft, Technik und NS-Ideologie*, ed. S. Richter and H. Mehrtens, pp. 189–214. Frankfurt: Suhrkamp, 1980.

———. "Konrad Lorenz's ethological theory: explanation and ideology, 1938–1943." *Journal of the History of Biology* 16 (1983): 39–73.

Kalikow, Theodora J., and John A. Mills. "Wallace Craig (1876–1954), ethologist and animal psychologist." *Journal of Comparative Psychology* 103 (1989): 282–288.

Kammerer, Paul. "Breeding experiments on the inheritance of acquired characters." *Nature* 111 (1923): 637–640.

Kennedy, J. S. "Is modern ethology objective?" *British Journal of Animal Behaviour* 2 (1954): 12-19.

Kirkman, F. B., ed. *The British Bird Book: An Account of All the Birds, Nests and Eggs Found in the British Isles.* 4 vols. London: T. C. and E. C. Jack, 1910–1913.

———. "The study of animal mind and natural history." *Wild Life* 3 (1914): 146–157.

———. *Bird Behaviour: A Contribution Based Chiefly on a Study of the Black-Headed Gull.* London: T. Nelson and Sons and T. C. and E. C. Jack, 1937.

Klopfer, Peter H. "Konrad Lorenz and the National Socialists: on the politics of ethology." *International Journal of Comparative Psychology* 7 (1994): 202–208.

———. *Politics and People in Ethology.* Lewisburg, PA: Bucknell University Press; London: Associated University Presses, 1999.

Koehler, Otto. "Das Ganzheits Problem in der Biologie." *Schriften der Königsberger Gelehrten Gesellschaft* 9 (1932): 139–204.

———. "'Instinkt oder Verstand?' Wertung zweier neuer Bücher und ihrer Stellung in der heutigen Tierpsychologie." *Naturwissenschaften* 27 (1939): 179–184.

———. Review of *Die Tierseele auf der Grundlage der grundwissenschaftlichen Philosophie von Johannes Rehmke,* by Bernhard Hecke. *Zeitschrift für Tierpsychologie* 3 (1939): 374–384.

———. Review of *Tierseele und Schöpfungsgeheimnis,* by Herbert Fritsche. *Der Biologe* 9 (1940): 48–50.

———. Review of *Die tierischen Instinkte und ihr Umbau durch Erfahrung,* by J. A. Bierens de Haan. *Zeitschrift für Tierpsychologie* 4 (1940): 280–286.

———. "Die Aufgabe der Tierpsychologie." *Schriften der Königsberger Gelehrten Gesellschaft* 18 (1943): 79–113.

———. Review of "A critique of Konrad Lorenz's theory of instinctive behaviour," by Daniel S. Lehrman. *Zeitschrift für Tierpsychologie* 11 (1954): 330–334.

Koestler, Arthur. *The Case of the Midwife Toad.* London: Hutchinson, 1971.

"König Ferdinand von Bulgarien zum 75. Geburtstage am 26. Februar 1936." *Journal für Ornithologie* 84 (1936): 1–2.

Kortlandt, A. "Wechselwirkung zwischen Instinkten." *Archives néerlandaises de zoologie* 4 (1940): 443–520.

———. "Patterns of pair-formation and nest-building in the European cormorant *Phalacrocorax carbo sinensis.*" *Ardea* 83 (1995): 11–25.

Kottler, Malcolm. "Darwin, Wallace, and the origin of sexual dimorphism." *Proceedings of the American Philosophical Society* 124 (1980): 203–226.

Kramer, Gustav. "50. Jahresversammlung der Deutschen Ornithologischen Gesellschaft in Wien vom 1.–4. Oktober 1932." *Journal für Ornithologie* 81 (1933): 360–375.

Krebs, John R. "Animal communication: ideas derived from Tinbergen's activities." In *The Tinbergen Legacy,* ed. M. S. Dawkins, T. R. Halliday, and R. Dawkins, pp. 60–74. London: Chapman and Hall, 1991.

Krebs, John R., and N. B. Davies, eds. *Behavioural Ecology: An Evolutionary Approach.* Oxford: Blackwell Scientific Publications, 1978.

Kronacher, C. "Erklärung zur Frage der 'zahlensprechenden' Tiere." *Zeitschrift für Tierpsychologie* 1 (1937): 91.

Kruuk, Hans. *Niko's Nature: A Life of Niko Tinbergen and His Science of Animal Behaviour.* Oxford: Oxford University Press, 2003.

Kühl, Stefan. *The Nazi Connection: Eugenics, American Racism, and German National Socialism.* New York: Oxford University Press, 1994.

Kuo, Z. Y. "Ontogeny of embryonic behavior in Aves." *Journal of Experimental Zoology* 61 (1932): 395–430; 62 (1932): 453–489; *Journal of Comparative Psychology* 13 (1932): 245–272; 14 (1932): 109–122.

Lack, David. "The releaser concept in bird behaviour." *Nature* 145 (1940): 107–108.

———. "Some aspects of instinctive behavior and display in birds." *Ibis* 5 (1941): 407–441.

———. *The Life of the Robin.* London: Witherby, 1943.

———. "Some British pioneers in ornithological research, 1859–1939." *Ibis* 101 (1959): 71–81.

———. "Preface by the president." In *Proceedings of the XIV International Ornithological Congress: Oxford, 24–30 July 1966,* ed. D. W. Snow, p. vii. Oxford: Blackwell Scientific Publications, 1967.

———. "Presidential address: interrelationships in breeding adaptations as shown by marine birds." In *Proceedings of the XIV International Ornithological Congress: Oxford, 24–30 July 1966,* ed. D. W. Snow, pp. 3–42. Oxford: Blackwell Scientific Publications, 1967.

Lagerlöf, Selma. *Nils Holgerssons underbara resa genom Sverige.* 2 vols. Stockholm: Albert Bonniers Förlag, 1907–1910.

———. *The Wonderful Adventures of Nils.* New York: Grosset and Dunlap, 1907.

Latour, Bruno. *Science in Action.* Cambridge, MA: Harvard University Press, 1987.

Laubichler, Manfred D. "Biology and the state: the biopolitics of Julius Tandler and Jakob von Uexküll." Unpublished paper.

Lehrman, Daniel S. "A critique of Konrad Lorenz's theory of instinctive behavior." *Quarterly Review of Biology* 298 (1953): 337–363.

———. "Semantic and conceptual issues in the nature-nurture problem." In *Development and Evolution of Behavior: Essays in Memory of T. C. Schneirla,* ed. Lester Aronson et al., pp. 17–52. San Francisco: W. H. Freeman, 1970.

Leyhausen, P. "The cat who walks by himself." In *Studying Animal Behavior: Autobiographies of the Founders,* ed. Donald A. Dewsbury, pp. 225–256. Chicago: University of Chicago Press, 1989.

———. Review of *Konrad Lorenz,* by F. M. Wuketits. *Ethology* 89 (1991): 344–346.

Lillie, Frank R. "Charles Otis Whitman." *Journal of Morphology* 22 (1911): xv–lxxvii.

Lincoln, Frederick C. "The individual versus the species in migration studies." *Auk* 56 (1939): 250–254.

Lissmann, Hans-Werner. "Die Umwelt des Kampffisches (*Betta splendens* Regan)." *Zeitschrift für vergleichende Physiologie* 18 (1932–1933): 65–111.

Lorenz, Adolf. *My Life and Work: The Search for a Missing Glove.* New York: Charles Scribner's Sons, 1936.

Lorenz, Konrad Z. "Beobachtungen an Dohlen." *Journal für Ornithologie* 75 (1927): 511–519.

———. "Beiträge zur Ethologie sozialer Corviden." *Journal für Ornithologie* 79 (1931): 67–127.

———. "Betrachtungen über das Erkennen der arteigenen Triebhandlungen der Vögel." *Journal für Ornithologie* 80 (1932): 50–98.

———. "Beobachtetes über das Fliegen der Vögel und über die Beziehungen der Flügel- und Steuerform zur Art des Fluges." *Journal für Ornithologie* 81 (1933): 107–236.

———. "Fliegen mit dem Wind und gegen den Wind." *Journal für Ornithologie* 81 (1933): 596–607.

———. "Beobachtungen an freifliegenden zahmgehaltenen Nachtreihern." *Journal für Ornithologie* 82 (1934): 160–161.

———. "Der Kumpan in der Umwelt des Vogels: der Artgenosse als auslösendes Moment sozialer Verhaltungsweisen." *Journal für Ornithologie* 83 (1935): 137–215, 289–413.

———. "Über eine eigentümliche Verbindung branchialer Hirnnerven bei *Cypselus apus.*" *Gegenbaurs morphologisches Jahrbuch* 77 (1936): 305–325.

———. "Über die Bildung des Instinktbegriffes." *Die Naturwissenschaften* 25 (1937): 289–300, 307–318, 324–331.

———. "Über den Begriff der Instinkthandlung." *Folia Biotheoretica, Series B* 2 (1937): 17–50.

———. "Biologische Fragestellung in der Tierpsychologie." *Zeitschrift fur Tierpsychologie* 1 (1937): 24–32.

———. "The companion in the bird's world." *Auk* 54 (1937): 245–273.

———. "A contribution to the comparative sociology of colonial-nesting birds." In *Proceedings of the Eighth International Ornithological Congress, Oxford, July 1934,* pp. 207–218. London: Oxford University Press, 1938.

———. "Über Ausfallserscheinungen im Instinktverhalten von Haustieren und ihre sozialpsychologische Bedeutung." In *Charakter und Erziehung: Bericht über den 16. Kongress der Deutschen Gesellschaft für Psychologie in Bayreuth,* ed. Otto Klemm, pp. 139–147. Leipzig: Johann Ambrosius Barth, 1939.

———. "Vergleichende Verhaltensforschung." *Verhandlungen der Deutschen Zoologischen Gesellschaft, Zoologischer Anzeiger,* supp. 12 (1939): 69–102.

———. "Vergleichendes über die Balz der Schwimmenten." *Journal für Ornithologie* 87 (1939): 172–174.

———. "Die Paarbildung beim Kolkraben." *Zeitschrift für Tierpsychologie* 3 (1940): 278–292.

———. "Nochmals: Systematik und Entwicklungsgedanke im Unterricht." *Der Biologe* 9 (1940): 24–36.

———. "Durch Domestikation verursachte Störungen arteigenen Verhaltens." *Zeitschrift für angewandte Psychologie und Charakterkunde* 59 (1940): 2–81.

———. "Kants Lehre vom Apriorischen im Lichte gegenwärtiger Biologie." *Blätter für deutsche Philosophie* 15 (1941): 94–125. Translated as "Kant's Doctrine of the *a priori* in the light of contemporary biology," in *Konrad Lorenz: The Man and His Ideas*, by Richard I. Evans (New York: Harcourt Brace Jovanovich, 1975), pp. 181–217.

———. "Vergleichende Bewegungsstudien an Anatiden." In "Festschrift O. Heinroth." Ergänzungsband 3, *Journal für Ornithologie* 89 (1941): 194–293.

———. "Oskar Heinroth 70 Jahre!" *Der Biologe*, Heft $^2/3$ (1941): 45–47.

———. "Induktive und teleologische Psychologie." *Die Naturwissenschaften* 30 (1942): 133–143.

———. "Psychologie und Stammesgeschichte." In *Die Evolution der Organismen*, ed. Gerhard Heberer, pp. 105–127. Jena: G. Fischer, 1943.

———. "Die angeborenen Formen möglicher Erfahrung." *Zeitschrift für Tierpsychologie* 5 (1943): 235–409.

———. "The comparative method in studying innate behaviour patterns." *Symposia of the Society for Experimental Biology* 4 (1950): 221–268.

———. "Ganzheit und Teil in der tierischen und menschlichen Gemeinschaft." *Studium Generale* 3, Heft 9 (1950): 455–499.

———. *King Solomon's Ring*. London: Methuen, 1952.

———. "Über das Töten von Artgenossen." *Jahrbuch der Max-Planck-Gesellschaft zur Förderung der Wissenschaften* (1955): 105–140.

———. "Morphology and behavior patterns in closely related species." In *Group Processes: Transactions of the First Conference, September 26, 27, 28, 29, and 30, 1954, Ithaca, New York*, ed. Bertram Schaffner, pp. 168–220. New York: Josiah Macy, Jr. Foundation, 1955.

———. "The objectivistic theory of instinct." In *L'Instinct dans le comportement des animaux et de l'homme*, by M. Autuori et al., pp. 51–76. Paris: Masson, 1956.

———. "VII. Internationaler Ethologenkongreß." *Zeitschrift für Tierpsychologie* 18 (1961): 495–497.

———. "Phylogenetische Anpassung und adaptive Modifikation des Verhaltens." *Zeitschrift für Tierpsychologie* 18 (1961): 139–187.

———. *Evolution and Modification of Behavior*. Chicago: University of Chicago Press, 1965.

———. "Preface." In *The Expression of the Emotions in Man and Animals*, by Charles Darwin, pp. ix–xiii. Chicago: University of Chicago Press, 1965.

———. *On Aggression*. New York: Harcourt Brace and World, 1966.

———. *Studies in Animal and Human Behaviour*. Translated by Robert Martin. 2 vols. Cambridge, MA: Harvard University Press, 1970–1971.

———. "The fashionable fallacy of dispensing with description." *Naturwissenchaften* 60 (1973): 1–9.

———. *Civilized Man's Eight Deadly Sins*. New York: Harcourt Brace Jovanovich, 1974.

———. "Konrad Lorenz." In *Les Prix Nobel en 1973*, pp. 177–184. Stockholm: Imprimerie Royale P. A. Norstedt & Söner, 1974.

———. *Behind the Mirror: A Search for a Natural History of Human Knowledge*. London: Methuen, 1977.

———. *Vergleichende Verhaltensforschung: Grundlagen der Ethologie*. Vienna: Springer-

Verlag, 1978. Translated as *The Foundations of Ethology* (New York: Springer-Verlag, 1981).

———. "My family and other animals." In *Studying Animal Behavior: Autobiographies of the Founders,* ed. Donald A. Dewsbury, pp. 259–287. Chicago: University of Chicago Press, 1989.

———. *Here Am I—Where Are You? The Behavior of the Greylag Goose.* New York: Harcourt Brace Jovanovich, 1991.

———. *Die Naturwissenschaft vom Menschen: Eine Einführung in die vergleichende Verhaltensforschung: Das "Russische Manuskript" (1944–1948).* Munich: Piper, 1992. Translated as *The Natural Science of the Human Species: An Introduction to Comparative Behavioral Research: The "Russian Manuscript" (1944–1948)* (Cambridge, MA: MIT Press, 1996).

Lorenz, Konrad Z., and Niko Tinbergen. "Taxis und Instinkthandlung in der Eirollbewegung der Graugans." Pt. 1. *Zeitschrift für Tierpsychologie* 2 (1938): 1–29.

Lowe, Percy R. "Henry Eliot Howard: an appreciation." *British Birds* 34 (1941): 195–197.

Luza, Radomír. *Austro-German Relations in the Anschluss Era.* Princeton, NJ: Princeton University Press, 1975.

Macrakis, Kristie. *Surviving the Swastika: Scientific Research in Nazi Germany.* New York: Oxford University Press, 1993.

Maienschein, Jane. "Whitman at Chicago: establishing a Chicago style of biology?" In *The American Development of Biology,* ed. Ronald Rainger, Keith R. Benson, and Jane Maienschein, pp. 151–182. Philadelphia: University of Pennsylvania Press, 1988.

———, ed. *Defining Biology: Lectures from the 1890s.* Cambridge, MA: Harvard University Press, 1986.

Maienschein, Jane, Ronald Rainger, and Keith R. Benson. "Introduction: were American morphologists in revolt?" *Journal of the History of Biology* 14 (1981): 83–87.

Maier, N. R. F., and T. C. Schneirla. *Principles of Animal Psychology.* New York: McGraw-Hill, 1935.

Makkink, G. F. "An attempt at an ethogram of the European avocet (*Recurvirostra avosetta* L.), with ethological and psychological remarks." *Ardea* 25 (1936): 1–62.

Malthus, Thomas Robert. *An Essay on the Principle of Population, as It Affects the Future Improvement of Society.* London: J. Johnson, 1798.

Marinelli, Wilhelm. "Die Geschichte der Zoologie in Wien." *Verhandlungen der Deutschen Zoologischen Gesellschaft* 56 (1962): 50–55.

Marler, Peter. "The characteristics of some animal calls." *Nature* 176 (1955): 6–8.

———. "Studies of fighting in chaffinches." Pt. 4, "Appetitive and consummatory behaviour." *British Journal of Animal Behaviour* 5 (1957): 29–37.

———. "Hark ye to the birds: autobiographical marginalia." In *Studying Animal Behavior: Autobiographies of the Founders,* ed. Donald A. Dewsbury, pp. 315–345. Chicago: University of Chicago Press, 1989.

Matheson, Percy Ewing. "William Warde Fowler (1847–1921)." In *Dictionary of National Biography,* 1912–1921, pp. 194–195. London: Oxford University Press, 1927.

Mayer, Alfred Goldsborough. "Marine Biological Laboratory at Tortugas, Florida." *Carnegie Institution of Washington Yearbook* 3 (1904): 50–54.

Mayr, Ernst. "Bernard Altum and the territory theory." *Proceedings of the Linnaean Society of New York,* nos. 45–46 (1935): 24–38.

———. "Cause and effect in biology." *Science* 134 (1961): 1501–1506.

———. "Whitman, Charles Otis." *Dictionary of Scientific Biography* 14 (1976): 313–315.

———. *Evolution and the Diversity of Life: Selected Essays.* Cambridge, MA: Harvard University Press, 1976.

———. "Erwin Stresemann." *Dictionary of Scientific Biography* 18, supp. 2 (1990): 888–890.

McDougall, William. *An Outline of Psychology.* 6th ed. London: Methuen, 1933.

Medawar, Peter. "Is the scientific paper fraudulent?" *Saturday Review,* 1 August 1964, pp. 42–43.

Mitman, Gregg. "Evolution as gospel: William Patten, the language of democracy, and the Great War." *Isis* 81 (1990): 446–463.

———. "Dominance, leadership, and aggression: animal behavior studies during the Second World War." *Journal of the History of the Behavioral Sciences* 26 (1990): 3–16.

———. *The State of Nature: Ecology, Community and American Social Thought, 1900–1950.* Chicago: University of Chicago Press, 1992.

———. "Cinematic nature: Hollywood technology, popular culture, and the American Museum of Natural History." *Isis* 84 (1993): 637–661.

———. *Reel Nature: America's Romance with Wildlife on Film.* Cambridge, MA: Harvard University Press, 1999.

Mitman, Gregg, and Richard W. Burkhardt, Jr. "Struggling for identity: the study of behavior in America, 1930–1945." In *The Expansion of American Biology,* ed. Keith R. Benson, Jane Maienschein, and Ronald Rainger, pp. 164–194. New Brunswick, NJ: Rutgers University Press, 1991.

Montagu, Ashley, ed. *Man and Aggression.* New York: Oxford University Press, 1968.

Moore, James R. "Darwin of Down: the evolutionist as squarson-naturalist." In *The Darwinian Heritage,* ed. David Kohn, pp. 435–481. Princeton, NJ: Princeton University Press, 1985.

Morrell, Jack. *Science at Oxford, 1914–1939: Transforming an Arts University.* Oxford: Clarendon Press, 1997.

Morris, Desmond. "Homosexuality in the ten-spined stickleback (*Pygosteus pungitius* L.)." *Behaviour* 4 (1952): 233–261.

———. "The reproductive behaviour of the ten-spined stickleback (*Pygosteus pungitius* L.)." *Behaviour,* supp. 6 (1958): 1–154.

———. *Animal Days.* New York: William Morrow, 1980.

Morse, Edward S. "Charles Otis Whitman." *Biographical Memoirs, National Academy of Sciences* 70 (1912): 269–288.

Mosse, George L. *The Crisis of German Ideology.* New York: Grossett and Dunlap, 1964.

Muller, H. J. Review of *Human Heredity,* by Erwin Baur, Eugen Fischer, and Fritz Lenz. *Birth Control Review* 17 (January 1933): 19–21.

"New fields for an old hunter." *Saturday Review* 91 (1901): 82–83.

Nice, Margaret Morse. "Zur Naturgeschichte des Singammers." *Journal für Ornithologie* 81 (1933): 552–595; 82 (1934): 1–96.

———. "Edmund Selous: an appreciation." *Bird-Banding* 6 (1935): 90–96.

———. Review of "The *Kumpan* in the bird's world: the fellow-member of the species as releasing factor of social behavior," by Konrad Lorenz. *Bird-Banding* 6 (1935): 113–114, 146–147.

———. "Studies in the life history of the song sparrow." Pt. 1 "A Population study of the song sparrow." *Transactions of the Linnaean Society of New York* 4 (1937): 1–247.

———. "The social Kumpan and the song sparrow." *Auk* 56 (1939): 255–262.

———. "The role of territory in bird life." *American Midland Naturalist* 26 (1941): 441–487.

———. "Studies in the life history of the song sparrow." Pt. 2, "The behavior of the song sparrow and other passerines." *Transactions of the Linnaean Society of New York* 6 (1943): 1–328.

———. *Research Is a Passion with Me*. Toronto: Consolidated Amethyst Communications, 1979.

Nicholson, E. M. *How Birds Live*. London: Williams and Norgate, 1927.

Nicolai, Jurgen. *Bird Life*. New York: G. P. Putnam's Sons; London: Thames and Hudson, 1974.

Nisbett, Alec. *Konrad Lorenz*. New York: Harcourt Brace Jovanovich, 1976.

Noble, G. K. "The role of dominance in the social life of birds." *Auk* 56 (1939): 263–273.

Noble, G. K., and D. S. Lehrman. "Egg recognition by the laughing gull." *Auk* 57 (1940): 22–43.

Nyhart, Lynn. *Biology Takes Form: Animal Morphology and the German Universities, 1800–1900*. Chicago: University of Chicago Press, 1995.

———. "Natural history and the 'new' biology." In *Cultures of Natural History*, ed. N. Jardine, J. A. Secord, and E. C. Spary, pp. 426–443. Cambridge: Cambridge University Press, 1996.

O'Donnell, John M. *The Origins of Behaviorism: American Psychology, 1870–1920*. New York: New York University Press, 1985.

Oordt, G. J. van. "The ornithological reservations of the Netherlands in 1930." *Ardea* 19 (1930): 1–12.

Pauly, Philip J. "The Loeb-Jennings debate and the science of animal behavior." *Journal of the History of the Behavioral Sciences* 17 (1981): 504–515.

———. *Controlling Life: Jacques Loeb and the Engineering Ideal in Biology*. New York: Oxford University Press, 1987.

———. "From adventism to biology: the development of Charles Otis Whitman." *Perspectives in Biology and Medicine* 37 (1994): 395–408.

Peckham, George W., and Elizabeth G. Peckham. "Observations on sexual selection in spiders of the family Attidae." *Wisconsin Natural History Society Occasional Papers* 1 (1889–1990): 3–60.

———. "Additional observations on sexual selection in spiders of the family Attidae, with some remarks on Mr. Wallace's theory of sexual ornamentation." *Wisconsin Natural History Society Occasional Papers* 1 (1889–1990): 117–151.

Pelkwijk, J. J. ter, and N. Tinbergen. "Roodkaakjes." *De Levende Natuur* 41 (1936): 129–137.

———. "Eine reizbiologische Analyse einiger Verhaltensweisen von *Gasterosteus aculeatus* L." *Zeitschrift für Tierpsychologie* 1 (1937): 193–200.

Pfungst, Oskar. *Clever Hans (the Horse of Mr. Von Osten): A Contribution to Experimental Animal and Human Psychology.* New York: Henry Holt, 1911.

Pickering, Andrew. *The Mangle of Practice: Time, Agency, and Science.* Chicago: University of Chicago Press, 1995.

Polish Ministry of Information. *The German New Order in Poland.* London: Hutchinson and Co., n.d. [1942?].

Poll, Heinrich. "Über Vogelmischlinge." In *Verhandlungen des 5. Internationalen Ornithologen-Kongresses in Berlin, 30. Mai bis 4. Juni 1910,* ed. Herman Schalow. Berlin: Deutsche Ornithologische Gesellschaft, 1911.

Portielje, A. F. J. "Zur Ethologie bzw. Psychologie der *Rhea americana* L." *Ardea* 14 (1925): 1–14.

———. "Zur Ethologie bezw. Psychologie von *Botaurus stellaris* (L.)." *Ardea* 15 (1926): 1–15.

———. "Zur Ethologie bezw. Psychologie von *Phalacrocorax carbo subcormoranus* (Brehm)." *Ardea* 16 (1927): 107–123.

———. "Zur Ethologie bezw. Psychologie der Silbermöwe, *Larus argentatus argentatus* Pont." *Ardea* 17 (1928): 112–149.

Poulton, E. B. *The Colours of Animals, Their Meaning and Use, Especially Considered in the Case of Insects.* New York: Appleton, 1890.

———. *Essays on Evolution.* Oxford: Clarendon Press, 1908.

———. *Charles Darwin and the Origin of Species.* London: Longmans, Green, and Co., 1909.

Proctor, Robert N. *The Nazi War on Cancer.* Princeton, NJ: Princeton University Press, 1999.

Pycraft, W. P. *A History of Birds.* London: Methuen and Co., 1910.

———. *The Infancy of Animals.* London: Hutchinson, 1912.

———. *The Courtship of Animals.* London: Hutchinson, 1913.

Rand, A. L. "Lorenz's objective method of interpreting bird behavior." *Auk* 58 (1941): 289–291.

———. "Nest sanitation and an alleged releaser." *Auk* 59 (1942): 404–409.

Raven, Charles E. *In Praise of Birds: Pictures of Bird Life.* London: Martin Hopkinson and Co., 1925.

Ravetz, Jerome R. *Scientific Knowledge and Its Social Problems.* New York: Oxford University Press, 1971.

Ray, John. *The Wisdom of God Manifested in the Works of Creation.* London: Samuel Smith, 1691.

Renneberg, Monika, and Mark Walker, eds. *Science, Technology and National Socialism.* Cambridge: Cambridge University Press, 1994.

Richards, Robert J. "The innate and the learned: the evolution of Konrad Lorenz's theory of instinct." *Philosophy of the Social Sciences* 4 (1974): 111–133.

———. *Darwin and the Emergence of Evolutionary Theories of Mind and Behavior.* Chicago: University of Chicago Press, 1987.

Riedl, Rupert. "Die Wiener Schule." *Verhandlungen der Deutschen Zoologischen Gesellschaft* 78 (1985): 5–9.

Ringer, Fritz K. *The Decline of the German Mandarins: The German Academic Community, 1890–1933.* Cambridge, MA: Harvard University Press, 1969.

Roeder, K. D. "VII. Internationaler Ethologenkongreß." *Zeitschrift für Tierpsychologie* 18 (1961): 491–494.

Röell, D. R. *De wereld van instinct: Niko Tinbergen en het ontstaan van de ethologie in Nederland (1920–1950).* Rotterdam: Erasmus Publishing, 1996. Translated as *The World of Instinct: Niko Tinbergen and the Rise of Ethology in the Netherlands (1920–1950)* (Assen: Van Gorcum: 2002).

Rosenberg, Charles. "Toward an ecology of knowledge: on discipline, context, and history." In *The Organization of Knowledge in Modern America, 1860–1920,* ed. Alexander Oleson and John Voss, pp. 440–455. Baltimore: Johns Hopkins University Press, 1979.

Rosenblatt, Jay S. "Daniel Sanford Lehrman, June 1, 1919–August 27, 1972." *Biographical Memoirs of the National Academy of Sciences* 66 (1995): 227–246.

Rossner, Ferdinand. "Lebenskunde ist Tatsachenforschung!" *Der Biologe* 8 (1939): 73.

———. "Systematik und Entwicklungsgedanke im Unterricht." *Der Biologe* 8 (1939): 366–372.

Ruiter, L. de. "Some experiments on the camouflage of stick caterpillars." *Behaviour* 4 (1952): 222–232.

———. "Countershading in caterpillars: an analysis of its adaptive significance." *Archives néerlandaises de zoologie* 11 (1956): 285–342.

Russell, Edward Stuart. *The Behaviour of Animals.* London: E. Arnold and Co., 1934.

———. "Instinctive behaviour and bodily development." *Folia Biotheoretica, Series B* 2 (1937): 67–76.

Schaffner, Bertram, ed. *Group Processes: Transactions of the First Conference, September 26, 27, 28, 29, and 30, 1954, Ithaca, New York.* New York: Josiah Macy, Jr. Foundation, 1955.

Schiller, Claire H., ed. *Instinctive Behavior: The Development of a Modern Concept.* New York: International Universities Press, 1957.

Schleidt, Wolfgang M. "Reaktionen von Truthühnern auf fliegende Raubvögel und Versuche zur Analyse ihres AAM's." *Zeitschrift für Tierpsychologie* 18 (1961): 534–560.

———. "Politik gegen und mit Konrad Lorenz." In *Konrad Lorenz und seine verhaltensbiologischen Konzepte aus heutiger Sicht,* ed. K. Kotrschal, G. Müller, and H. Winkler, pp. 73–92. Fürth: Filander Verlag, 2001.

———. "Lorenz, Konrad (1903–1989)." In *International Encyclopedia of the Social and Behavioral Sciences,* 13:9083–9089. 2001.

Schneirla, T. C. "Contemporary American animal psychology in perspective." In *Twentieth Century Psychology,* ed. P. L. Harriman, pp. 306–316. New York: Philosophical Library, 1946.

———. "Interrelationships of the 'innate' and the 'acquired' in instinctive behavior." In *L'Instinct dans le comportement des animaux et de l'homme,* by M. Autuori et al., pp. 387–452. Paris: Masson, 1956.

———. "The study of animal behavior: its history and relation to the museum." Pts. 1 and 2. *Curator,* no. 4 (1958): 17–35; no. 1 (1959): 27–48.

————. *Selected Writings of T. C. Schneirla.* Edited by Lester R. Aronson et al. San Francisco: W. H. Freeman, 1972.

Schuyl, G., L. Tinbergen, and N. Tinbergen. "Ethologische Beobachtungen am Baumfalken (*Falco s. subbuteo* L.)." *Journal für Ornithologie* 84 (1936): 387–433.

Schwidetzky, Georg. *Sprechen Sie Schimpansisch? Einführung in die Tier- und Ursprachenlehre.* Leipzig: Deutsche Gesellschaft für Tier- und Ursprachenforschung, 1931.

————. *Unsere Ausstellung: Rasse und Sprache: Eine Führung für die Abwesenden.* Leipzig: Deutsche Gesellschaft für Tier- und Ursprachenforschung, 1938.

"Scientific affairs in Europe." *Nature* 156 (1945): 576–579.

Segerstråle, Ullica. *Defenders of the Truth: The Sociobiology Debate.* New York: Oxford University Press, 2000.

Seitz, Alfred. "Die Paarbildung bei einigen Cichliden." Pt. 1, "Die Paarbildung bei Astatotilapia strigigena Pfeffer." *Zeitschrift für Tierpsychologie* 4 (1940): 40–84.

————. "Die Paarbildung bei einigen Cichliden." Pt. 2, "Die Paarbildung bei Hemichromis bimaculatus Gill." *Zeitschrift für Tierpsychologie* 5 (1942): 74–101.

Selous, Edmund. "An observational diary of the habits of night jars (*Caprimulgus europaeus*), mostly of a sitting pair: notes taken at time and on spot." *Zoologist* 3 (1899): 388–402, 486–505.

————. "An observational diary of the habits of the great plover (*Oedicnemus crepitans*) during September and October." *Zoologist* 4 (1900): 173.

————. *Beautiful Birds.* London: J. M. Dent and Co., 1901.

————. *Bird Watching.* London: J. M. Dent and Co., 1901.

————. "The old zoo and the new." *Saturday Review* 91 (1901): 330–332, 365–366, 397–398, 433–434.

————. "Rabbits and hares." *Saturday Review* 92 (1901): 618–619, 677–679.

————. "An observational diary of the habits—mostly domestic—of the great crested grebe (*Podicipes cristatus*)." *Zoologist* 5 (1901): 161–183.

————. "An observational diary of the habits—mostly domestic—of the great crested grebe (*Podicipes cristatus*) and of the peewit (*Vanellus vulgaris*), with some general remarks." *Zoologist* 5 (1901): 339–350, 454–462; 6 (1902): 133–144.

————. "Variations in colouring of *Stercorarius crepidatus.*" *Zoologist* 6 (1902): 368–373.

————. *Bird Life Glimpses.* London: Allen, 1905.

————. *The Bird Watcher in the Shetlands: With some Notes on Seals—and Digressions.* London: J. M. Dent and Co.; New York: E. P. Dutton and Co., 1905.

————. "Observations tending to throw light on the question of sexual selection in birds, including a day-to-day diary on the breeding habits of the ruff (*Machetes pugnax*)." *Zoologist* 10 (1906): 201.

————. "An observational diary of the nuptial habits of the blackcock (*Tetrao tetrix*) in Scandinavia and England." *Zoologist* 13 (1909): 401–413; 14 (1910): 23–29, 51–56, 176–182, 248–265.

————. "The buntings." In *The British Bird Book,* ed. F. B. Kirkman, 1:169–198. London: T. C. and E. C. Jack, 1910.

————. "The finches." In *The British Bird Book,* ed. F. B. Kirkman, 1:83–156. London: T. C. and E. C. Jack, 1910.

———. "The pipits." In *The British Bird Book*, ed. F. B. Kirkman, 1:260–277. London: T. C. and E. C. Jack, 1910.

———. "The wagtails." In *The British Bird Book*, ed. F. B. Kirkman, 1:239–259. London: T. C. and E. C. Jack, 1910.

———. "Origin of the social antics and courting displays of birds." *Zoologist* 16 (1912): 197–199.

———. "The nuptial habits of the blackcock." *Naturalist* (1913): 96–98.

———. *Realities of Bird Life: Being Extracts from the Diaries of a Life-Loving Naturalist.* London: Constable, 1927.

———. *Thought-Transference (or What?) in Birds.* London: Constable, 1931.

———. *Evolution of Habit in Birds.* London: Constable, 1933.

Selous, Frederick Courteney. *A Hunter's Wanderings in Africa.* London: Richard Bentley and Son, 1881.

"The shrine of the naturalist." *Saturday Review* 88 (1899): 825–826.

Silver, Rae, and Jay S. Rosenblatt. "The development of a developmentalist: Daniel S. Lehrman." *Developmental Psychobiology* 20 (1987): 563–570.

Simmons, K. E. L. "Edmund Selous (1857–1934): fragments for a biography." *Ibis* 126 (1984): 595–596.

"South African fauna." *Saturday Review* 90 (1900): 397–398.

Spurway, Helen. "The causes of domestication: an attempt to integrate some ideas of Konrad Lorenz with evolution theory." *Journal of Genetics* 53 (1955): 325–362.

Stang, V., O. Koehler, and J. Effertz. "Prof. emer. Dr. Dr. h. C. Dr. agr. h. c. Carl Kronacher." *Zeitschrift für Tierpsychologie* 2 (1938–1939): i–iv.

Star, Susan Leigh, and James R. Griesemer. "Institutional ecology, 'translations' and boundary objects: amateurs and professionals in Berkeley's Museum of Vertebrate Zoology, 1907–1939." *Social Studies of Science* 19 (1989): 387–420.

Stauffer, Robert C.. "Ecology in the long manuscript version of Darwin's *Origin of Species* and Linnaeus' *Oeconomy of Nature*." *Proceedings of the American Philosophical Society* 104 (1960): 235–241.

Sullivan, Walter. "Questions raised on Lorenz's prize." *New York Times*, 15 December 1973, p. 9.

Taschwer, Klaus. "'Rendezvous mit Tier und Mensch': Wissenschaftssoziologische Anmerkungen zur Geschichte der Verhaltensforschung in Österreich." In "Soziologische und historische Analysen der Sozialwissenschaften," ed. Christian Fleck. *Österreichische Zeitschrift für Soziologie*, Sonderband 5 (2000): 309–341.

Taschwer, Klaus, and Benedikt Föger. *Konrad Lorenz: Biographie.* Vienna: Paul Zolnay Verlag, 2003.

Taylor, Stephen. *The Mighty Nimrod: A Life of Frederick Courteney Selous, African Hunter and Adventurer, 1851–1917.* London: Collins, 1989.

Thorndike, Edward Lee. "Edward Lee Thorndike." In *A History of Psychology in Autobiography*, vol. 3, ed. Carl Murchison, pp. 263–270. London: Oxford University Press, 1936.

Thorpe, W. H. "Biological races in insects and allied groups." *Biological Reviews of the Cambridge Philosophical Society* 5 (1930): 177–212.

———. "Further experiments on olfactory condition in a parasitic insect: the nature of the conditioning process." *Proceedings of the Royal Society of London, Series B* 126 (1938): 370–397.

———. "Ecology and the future of systematics." In *The New Systematics,* ed. Julian Huxley, pp. 341–364. Oxford, 1940.

———. "Types of learning in insects and other arthropods." *British Journal of Psychology* 33 (1943): 220–234; 34 (1943): 20–31; 34 (1944): 66–76.

———. "The evolutionary significance of habitat selection." *Journal of Animal Ecology* 14 (1945): 67–70.

———. "Animal learning and evolution." *Nature* 156 (1945): 46–47.

———. "The modern concept of instinctive behaviour." *Bulletin of Animal Behaviour,* no. 7 (1948): 2–12.

———. "The definition of some terms used in animal behaviour studies." *Bulletin of Animal Behaviour,* no. 9 (1951): 34–40.

———. "The learning of song patterns by birds, with especial reference to the song of the chaffinch (*Fringilla coelebs*)." *Ibis* 100 (1958): 535–570.

———. *Bird Song: The Biology of Vocal Communication and Expression in Birds.* Cambridge: Cambridge University Press, 1961.

———. *The Origins and Rise of Ethology.* London: Heinemann, 1979.

Thorpe. W. H., and F. G. W. Jones. "Olfactory condition in a parasitic insect and its relation to the problem of host selection." *Proceedings of the Royal Society of London, Series B* 124 (1937): 56–81.

Tinbergen, Niko. "Over het voedsel van de Blauwe Reiger (*Ardea cinerea cinerea* L.)." *Ardea* 19 (1930): 89–93.

———. "Zur Paarungsbiologie der Flussseeschwalbe (*Sterna hirundo hirundo* L.)." *Ardea* 20 (1931): 1–18.

———. "Beobachtungen am Baumfalken (*Falco subbuteo subbuteo* L.)." *Journal für Ornithologie* 80 (1932): 40–50.

———. "Über die Ernährung einer Waldohreulenbrut (*Asio otus otus* L.)." *Beiträge zur Fortpflanzungsbiologie der Vögel* 8 (1932): 54–55.

———. "Over het voedsel van de Sperwer (*Accipiter nisus nisus* L.) in de Nederlandsche duinstreek." *Ardea* 21 (1932): 77–89.

———. "Über die Orientierung des Bienenwolfes (*Philanthus triangulum* Fabr.)." *Zeitschrift für vergleichende Physiologie* 16 (1932): 304–334.

———. "Field observations of east Greenland birds." Pt. 1, "The behaviour of the red-necked phalarope (*Phalaropus lobatus* L.) in spring." *Ardea* 24 (1935): 1–42.

———. "Waarnemingen en proeven over de sociologie van een zilvermeeuwenkolonie." *De Levende Natuur* 40 (1935): 263–280, 304–308.

———. "Über die Orientierung des Beinenwolfes (*Philanthus triangulum* Fabr.)." Pt. 2, "Die Bienenjagd." *Zeitschrift für vergleichende Physiologie* 21 (1935): 699–716.

———. "Zur Soziologie der Silbermöwe, *Larus a. argentatus* Pont." *Beiträge zur Fortpflanzungsbiologie der Vögel* 12 (1936): 89–96.

———. "Why do birds behave as they do?" *Bird-Lore* 40 (1938): 389–395; 41 (1939): 23–30.

———. "In the life of a herring gull." *Natural History* 43 (1939): 222–229.

———. "The behavior of the snow bunting in spring." *Transactions of the Linnaean Society of New York* 5 (1939): 1–94.

———. "On the analysis of social organization among vertebrates, with special reference to birds." *American Midland Naturalist* 21 (1939): 210–234.

———. "Die Ethologie als Hilfswissenschaft der Oekologie." *Journal für Ornithologie* 88 (1940): 171–177.

———. "Die Übersprungbewegung." *Zeitschrift für Tierpsychologie* 4 (1940): 1–40.

———. "An objectivistic study of the innate behaviour of animals." *Bibliotheca Biotheoretica* 1 (1942): 39–98.

———. *De Natuur is sterker dan de leer, of de lof van het veldwerk.* Leiden: Universiteit te Leiden, 1947.

———. "Physiologische Instinktforschung." *Experientia* 4 (1948): 121–133.

———. "Social releasers and the experimental method required for their study." *Wilson Bulletin* 60 (1948): 6–51.

———. "The hierarchical organization of nervous mechanisms underlying instinctive behaviour." *Symposia of the Society of Experimental Biology* 4 (1950): 305–312.

———. "Recent advances in the study of bird behaviour." In *Proceedings of the Xth International Ornithological Congress, Uppsala, 1950*, ed. Sven Hörstadius, pp. 360–374. Uppsala: Almquist & Wiksell, 1951.

———. *The Study of Instinct.* Oxford: Clarendon Press, 1951.

———. "A note on the origin and evolution of threat display." *Ibis* 94 (1952): 160–162.

———. "'Derived' activities: their causation, biological significance, origin, and emancipation during evolution." *Quarterly Review of Biology* 27 (1952): 1–32.

———. *The Herring Gull's World.* London: Collins, 1953.

———. "Psychology and ethology as supplementary parts of a science of behavior." In *Group Processes: Transactions of the First Conference, September 26, 27, 28, 29, and 30, 1954, Ithaca, New York*, ed. Bertram Schaffner, pp. 75–167. New York: Josiah Macy, Jr. Foundation, 1955.

———. "On the functions of territory in gulls." *Ibis* 98 (1956): 400–411.

———. "5. Internationaler Ethologenkongreß." *Zeitschrift für Tierpsychologie* 14 (1957): 377–380.

———. *Curious Naturalists.* London: Country Life, 1958.

———. "Behaviour, systematics, and natural selection." *Ibis* 101 (1959): 318–330. Reproduced in *Evolution after Darwin*, ed. Sol Tax, vol. 1, *The Evolution of Life*, pp. 595–613. Chicago: University of Chicago Press, 1960.

———. "Comparative studies of the behaviour of gulls (Laridae): a progress report." *Behaviour* 15 (1959): 1–70.

———. "The work of the Animal Behaviour Research Group in the Department of Zoology, University of Oxford." *Animal Behaviour* 11 (1963): 206–209.

———. "On aims and methods of ethology." *Zeitschrift für Tierpsychologie* 20 (1963): 410–433.

———. "On adaptive radiation in gulls (tribe *Larini*)." *Zoologische Mededeelingen* 39 (1964): 209–223.

———. "Behavior and natural selection." In *Ideas in Modern Biology*, ed. John A. Moore, pp. 519–542. Garden City, NY: Natural History Press, 1965.

———. "Adaptive features of the black-headed gull *Larus ridibundus* L." In *Proceedings of the XIV International Ornithological Congress: Oxford, 24–30 July 1966*, ed. D. W. Snow, pp. 43–59. Oxford: Blackwell Scientific Publications, 1967.

———. "On war and peace in animals and man: an ethologist's approach to the biology of aggression." *Science* 160 (1968): 1411–1418.

———. "Ethology." In *Scientific Thought, 1900–1960*, ed. R. Harré, pp. 238–268. Oxford: Clarendon Press, 1969.

———. *The Study of Instinct: With a New Introduction*. Oxford: Clarendon Press, 1959.

———. "Functional ethology and the human sciences." *Proceedings of the Royal Society of London, Series B* 182 (1972): 385–410.

———. *The Animal in Its World: Explorations of an Ethologist, 1932–1972*. 2 vols. London: George Allen and Unwin, 1972, 1973.

———. "Nikolaas Tinbergen." In *Les Prix Nobel en 1973*, pp. 197–200. Stockholm: Imprimerie Royale P. A. Norstedt & Söner, 1974.

———. Review of *Sociobiology: The New Synthesis*, by Edward O. Wilson. *New Humanist*, October 1975, pp. 147–148.

———. "Ethology in a changing world." In *Growing Points in Ethology*, ed. P. P. G. Bateson and R. A. Hinde, pp. 507–527. Cambridge: Cambridge University Press, 1976.

———. "Watching and Wondering." In *Studying Animal Behavior: Autobiographies of the Founders*, ed. Donald A. Dewsbury, pp. 431–463. Chicago: University of Chicago Press, 1989. Originally published as *Leaders in the Study of Animal Behavior* (London: Associated University Presses, 1985).

———. "Aus der Kinderstube der Ethologie." In *Wozu aber hat das Vieh diesen Schnabel?* by Oskar Heinroth and Konrad Lorenz, ed. Otto Koenig, pp. 309–314. Munich: Piper, 1988.

Tinbergen, N., G. J. Broekuysen, F. Feekes, J. C. W. Houghton, H. Kruuk, and E. Szulc. "Egg shell removal by the black-headed gull, *Larus ridibundus* L.: a behaviour component of camouflage." *Behaviour* 19 (1962): 74–117.

Tinbergen, N., and W. Kruyt. "Über die Orientierung des Beinenwolfes (*Philanthus triangulum* Fabr.)." Pt. 3, "Die Bevorzugung bestimmter Wegmarken." *Zeitschrift für vergleichende Physiologie* 25 (1938): 292–334.

Tinbergen, N., and D. J. Kuenen. "Über die auslösenden und die richtunggebenden Reizsituationen der Sperrbewegung von jungen Drosseln (*Turdus m. merula* L. und *T. e. ericetorum* Turton)." *Zeitschrift für Tierpsychologie* 3 (1939): 37–60.

Tinbergen, N., B. J. D. Meeuse, L. K. Boerema, and W. Varossieau. "Die Balz des Samtfalters, *Eumenis semele* (L.)." *Zeitschrift für Tierpsychologie* 5 (1942): 182–226.

Tinbergen, N., and M. Moynihan. "Head flagging in the black-headed gull: its function and origin." *British Birds* 45 (1952): 19–22.

Tinbergen, N., and A. C. Perdeck. "On the stimulus situation releasing the begging response in the newly hatched herring gull chick (*Larus argentatus argentatus* Pont.)." *Behaviour* 3 (1950): 1–39.

Tinbergen, N., and R. J. van der Linde. "Über die Orientierung des Beinenwolfes (*Philanthus triangulum* Fabr.)." Pt. 4, "Heimflug aus unbekanntem Gebiet." *Biologisches Zentralblatt* 58 (1938): 425–435.

Tinbergen, N., and J. J. A. Van Iersel. "'Displacement reactions' in the three-spined stickleback." *Behaviour* 1 (1947): 56–63.

Tobach, Ethel, and Lester R. Aronson. "T. C. Schneirla: a biographical note." In *Development and Evolution of Behavior: Essays in Memory of T. C. Schneirla*, ed. Lester Aronson et al., pp. xi–xviii. San Francisco: W. H. Freeman, 1970.

Tolman, Edward Chace. *Purposive Behavior in Animals and Men.* New York: Century Co., 1932.

Tuchman, Arleen Marcia. "Institutions and disciplines: recent work in the history of German science." *Journal of Modern History* 69 (1997): 298–319.

Turner, Emma L. *Broadland Birds.* London: Country Life, 1924.

Uexküll, Jakob von. *Umwelt und Innenwelt der Thiere.* Berlin: J. Springer, 1909; 2nd ed., 1921.

———. "Umweltforschung." *Zeitschrift für Tierpsychologie* 1 (1937) pp. 33–34.

Uexküll, Jakob von, and Georg Kriszat. *Streifzüge durch die Umwelten von Tieren und Menschen.* Berlin: J. Springer, 1934.

"The unknown bird world." *Saturday Review* 91 (1901): 638.

Van Iersel, J. J. A. "An analysis of the parental behaviour of the male three-spined stickleback (*Gasterosteus aculeatus* L.)." *Behaviour,* supp. 3 (1953).

Verlaine, L. "Qu'est-ce que l'instinct?" *Folia Biotheoretica, Series B* 2 (1937): 51–64.

Verwey, Jan. "Die Paarungsbiologie des Fischreihers (*Ardea cinerea* L.)." In *Verhandlungen des VI. Internationalen Ornithologen-Kongresses in Kopenhagen 1926,* ed. F. Steinbacher, pp. 390–413. Berlin, 1929.

———. "Die Paarungsbiologie des Fischreihers." *Zoologische Jahrbücher* 48 (1930): 1–120.

Virchow, Rudolph. "Die Freiheit der Wissenschaft im modernen Staat." *Amtlicher Bericht der 50. Versammlung Deutscher Naturforscher und Ärzte in München,* pp. 65–77. Munich: F. Straub, 1877.

Vogt, William. "Preliminary notes on the behavior and ecology of the eastern willet." *Proceedings of the Linnaean Society of New York* 49 (1938): 8–42.

Voous, Karel H. "Report of the secretary-general." In *Proceedings of the XVth International Ornithological Congress, The Hague, The Netherlands, 30 August–5 September 1970,* ed. K. H. Voous, pp. 1–12. Leiden: E. J. Brill, 1972.

Warmbrunn, Werner. *The Dutch under German Occupation, 1940–1945.* Stanford, CA: Stanford University Press, 1963.

Waters, C. Kenneth, and Albert van Helden, eds. *Julian Huxley: Biologist and Statesman of Science.* Houston: Rice University Press, 1992.

Watson, John B. "The behavior of noddy and sooty terns." *Papers from the Tortugas Laboratory of the Carnegie Institution of Washington* 2 (1908): 187–255.

———. "The new science of animal behavior." *Harper's* 120 (1910): 346–353.

———. "Psychology as the behaviorist views it." *Psychological Review* 20 (1913): 158–177.

Watson, John B., and K. S. Lashley. "Homing and related activities of birds." *Papers from the Tortugas Laboratory of the Carnegie Institution of Washington* 8 (1915): 1–104.

Weart, Spencer. *Nuclear Fear: A History of Images.* Cambridge, MA: Harvard University Press, 1988.

Weindling, Paul. *Health, Race and German Politics between National Unification and Nazism, 1870–1945.* Cambridge: Cambridge University Press, 1989.

Weiss, Paul. "Self-differentiation of the basic patterns of coordination." *Comparative Psychology Monographs* 17 (1941): 1–96.

Weiss, Sheila Faith. *Race Hygiene and National Efficiency: The Eugenics of Wilhelm Schallmayer.* Berkeley: University of California Press, 1987.

West-Eberhard, Mary Jane. "Sexual selection, social competition, and speciation." *Quarterly Review of Biology* 58 (1983): 155–183.

Wheeler, William Morton. "'Natural history,' 'oecology' or 'ethology.'" *Science,* n.s., 15 (1902): 971–976.

White, Frederick R., ed. *Famous Utopias of the Renaissance.* Chicago: Packard and Company, 1946.

White, Gilbert. *The Natural History and Antiquities of Selborne.* London: Bensely, 1789.

Whitman, C. O. "The embryology of Clepsine." *Quarterly Journal of Microscopical Science* 18 (1878): 215–315.

———. "Specialization and organization, companion principles of all progress—the most important need of American biology." In *Biological Lectures Delivered at the Marine Biological Laboratory of Wood's Holl in the Summer Session of 1890,* pp. 1–26. Boston: Ginn and Company, 1891.

———. "Some of the functions and features of a biological station." In *Biological Lectures Delivered at the Marine Biological Laboratory of Wood's Holl, 1896–1897,* pp. 231–242. Boston: Ginn and Company, 1898.

———. "Animal behavior." In *Biological Lectures from the Marine Biological Laboratory Wood's Holl, Mass., 1898,* pp. 285–338. Boston: Ginn and Company, 1899.

———. "Myths in animal psychology." *Monist* 9 (1899): 524–537.

———. "A biological farm: for the experimental investigation of heredity, variation and evolution and for the study of life-histories, habits, instincts and intelligence." *Biological Bulletin* 3 (1902): 214–224.

———. "The problem of the origin of species." In *Congress of Arts and Science, Universal Exposition, St. Louis, 1904,* ed. Howard J. Rogers, 5:41–58. Boston: Houghton, Mifflin and Company, 1906.

———. *Posthumous Works of Charles Otis Whitman.* Vol. 1, *Orthogenetic Evolution in Pigeons,* ed. Oscar Riddle. Vol. 2, *Inheritance, Fertility, and the Dominance of Sex and Color in Hybrids of Wild Species of Pigeons,* ed. Oscar Riddle. Vol. 3, *The Behavior of Pigeons,* ed. Harvey A. Carr. Washington, DC: Carnegie Institution of Washington, 1919.

———. "The origin of species." *Bulletin of the Wisconsin Natural History Society* 5 (1907): 6–14.

Wilson, Edward O. *Sociobiology: The New Synthesis.* Cambridge, MA: Harvard University Press, 1975.

————. *Naturalist*. New York: Warner Books, 1995.

Witkowski, J. A. "Julian Huxley and the laboratory: embracing inquisitiveness and widespread curiosity." In *Julian Huxley: Biologist and Statesman of Science*, ed. C. Kenneth Waters and Albert Van Helden, pp. 79–103. Houston: Rice University Press, 1992.

Wuketits, Franz M. *Konrad Lorenz: Leben und Werk eines großen Naturforschers*. Munich: Piper, 1990.

————. *Die Entdeckung des Verhaltens: Eine Geschichte der Verhaltensforschung*. Darmstadt: Wissenschaftliche Buchgesellschaft, 1995.

Yerkes, Robert M. "Psychology in its relations to biology." *Journal of Philosophy* 7 (1910): 113–124.

Ziegler, Heinrich Ernst. *Die Naturwissenschaft und die socialdemokratische Theorie, ihr Verhältniss dargelegt auf Grund der Werke von Darwin und Bebel; zugleich ein Beitrag zur wissenschaftlichen Kritik der Theorien der derzeitigen Socialdemokratie*. Stuttgart: F. Enke, 1894.

————. *Der Begriff des Instinktes einst und jetzt*. 2nd ed. Jena: Gustav Fischer, 1910.

Zippelius, Hanna-Maria. *Die vermessene Theorie: Eine kritische Auseinandersetzung mit der Instinkttheorie von Konrad Lorenz und verhaltenskundlicher Forschungspraxis*. Braunschweig: Friedr. Vieweg & Sohn, 1992.